READING THE LANDSCAPE

Writing a World

The **cover stock** is 10# C1S recycled. It contains 50% waste papers, of which at least 10% is post-consumer waste and 50% is virgin fibers. The obvious environmental benefit is the reduction of waste that would ultimately end up in landfills. Another important benefit is the reduction of the use of forests not only as a result of using post-consumer waste but also from the highly efficient use of pre-consumer waste.

The **protective coating** used on the cover is an aqueous coating. Aqueous coating is environmentally friendly for the following reasons:

1. It is water based and will decompose much more readily than UV or laminations. UV is acrylic based and laminations are oil based.

2. It does not have to be reprocessed in order to be recycled. UV must go through expensive reprocessing before it can be recycled. Laminations go through a still more expensive reprocessing before they can be recycled. The processes that UV and laminations must go through are so expensive they are economically unfeasible and so, in all but a few cases, they end up un-recycled and in landfills.

The **insert paper** is 70# C2S recycled. It contains 50% waste papers, of which at least 10% is post-consumer waste and 50% is virgin fibers. The obvious environmental benefit is the reduction of waste that would ultimately end up in landfills. Another important benefit is the reduction of the use of forests not only as a result of using post-consumer waste but also from the highly efficient use of pre-consumer waste.

All parts of the book—text, insert, and cover—are printed with soy-based inks. Traditional inks are oil-based.

READING THE LANDSCAPE
Writing a World

Peter Valenti
Fayetteville State University

Harcourt Brace College Publishers

Fort Worth   Philadelphia   San Diego   New York   Orlando   Austin   San Antonio
Toronto   Montreal   London   Sydney   Tokyo

Publisher	TED BUCHHOLZ
Editor in Chief	CHRISTOPHER P. KLEIN
Acquisitions Editor	STEPHEN T. JORDAN/JOHN P. MEYERS
Developmental Editor	MICHELL PHIFER
Project Editor	BARBARA MORELAND
Production Manager	DEBRA A. JENKIN
Senior Art Director	DAVID A. DAY
Electronic Publishing Coordinator	JILL A. PETERSON

Cover Image:
Thomas Moran, Mist in Kanab Canyon, Utah, 1892. ©National Museum of American Art, Washington DC/Art Resource, NY.

ISBN: 0-15-501432-3

Library of Congress Catalog Card Number: 95-60357

Copyright © 1996 by Harcourt Brace & Company

Address for Editorial Correspondence: Harcourt Brace College Publishers, 301 Commerce Street, Suite 3700, Fort Worth, TX 76102

Address for Orders: Harcourt Brace & Company, 6277 Sea Harbor Drive, Orlando, FL 32887-6777, 1-800-782-4479, or 1-800-433-0001 (in Florida)

(Copyright acknowledgments begin on page 581, which constitutes a continuation of this copyright page.)

Printed in the United States of America

5 6 7 8 9 0 1 2 3 4   039   10 9 8 7 6 5 4 3 2 1

PREFACE

A landscape is both the world we know, the source of information for all our writing, as well as a specific site in the natural world, singled out for its beauty or significance. Understanding the relationship between the landscapes of personal experience and those of a global community is part of the process of learning to respond to the land as human environment.

Reading the Landscape serves as both a composition textbook and a reader: each chapter introduces the student to stages of the writing process, beginning with techniques for invention and moving toward editing of the final draft. Students read and practice with various rhetorical modes, moving from relatively simple narration and description to more complex argumentation and research.

The textbook as a whole is organized into the Writing Process, Chapters 1–4; Rhetorical Strategies, Chapters 5–10; Argumentation and Research, Chapters 11 and 12; Documentation Guide; and the Visual Texts with color plates, discussion questions, and writing assignments.

The writing process covered in the first four chapters is divided into the following main parts:

- *Prewriting*, which includes journals and freewriting
- *Planning*, which covers collaboration, brainstorming, and mapping
- *Organizing*, which discusses methods of arranging an essay
- *Revising*, which presents stages of drafting through final editing

The reading selections in the first four chapters not only move from the backyard, home, or protected valley out into the larger world, but they also provide formats for exploring and developing the writing process. Journals provide a means of amassing detail and recording experiences that will be used in formal writing assignments. Memoirs emphasize the function of memory in recalling detail; prewriting techniques show how the writer can use various means to tap memory and unlock creative responses. Oral history involves student writers with the larger lives of family and community and helps them to experiment with sifting through information to glean important details. The process of writing demonstrated in these selections parallels writers' struggles to locate a voice and a place in the world.

The authors included in *Reading the Landscape* represent the widest diversity of gender and ethnic identities of any text focusing on the relations between people and their physical environment. Not only does it present traditional nature writers such as Rachel Carson, John Muir, Henry David Thoreau, Mary Austin, Aldo Leopold, Annie Dillard, and Barry Lopez, but it also includes such new voices as Clarissa Pinkola Estes, Terry Tempest Williams, David Quammen, Gretel Ehrlich, and Gary Nabhan. Other authors such as W. E. B. Dubois, Richard Wright, Mary Wilkins Freeman, Alice Walker, and Leslie Marmon Silko are not traditionally considered "nature writers," but they have produced some effective writing about the relationship between our human and natural environments. And finally, writers such as Gloria Anzaldúa, William Least Heat Moon, Luci Tapahonso, Marie DeSantis, Dayton Duncan, and Kathleen Stocking show how particular regions and cultures affect the way we see and write about our worlds.

The book offers many means of reading and writing about landscape:

- *Exploring Journals and Responding Journals* to engage students with the readings
- *Collaborative Activities* to provide peer interaction and different perspectives
- *Sequencing* to enable students from any region to explore their links to the land
- *Rhetorical strategies* grounded in the reading selections on landscape
- *Argumentation and research* procedures for polemical readings of landscape
- *Instructor's manual* to provide ideas for course outlines, writing sequences, portfolios, and additional writing activities and essay topics

Studying essays written by others will help student writers generate ideas for their own arguments, so they can relate their responses to the responses of other writers. In this way, students will move from reading about the landscape to writing a landscape of their own, from observing the world to interacting with it and affecting the outcome. Their final writing assignments should not only demonstrate this movement but will also show how student writers' voices can derive their power from the land.

ACKNOWLEDGMENTS

Many people have had a hand in the genesis of this book; and without their help it could not have happened. The ideas first took shape as a book in 1991 at West Point, where Cols. Paul Christopher and Joe Cox encouraged the first form of *Reading the Landscape*. Our joint expeditions provided outdoor parallels, usually unexpected, to the reading and writing activities. Joe Trimmer generously showed me how a writing text works and guided me through the publication process. At Harcourt Brace, Senior Acquisitions Editor Stephen Jordan demonstrated great faith and patience as he helped me navigate through various versions of the text to reach the final form. Likewise, Developmental Editor Michell Phifer provided support and direction as I tried her good will and forbearance in ways too numerous to catalog. Project Editor Barbara Moreland prepared the final text with extreme care and thoroughness.

At Fayetteville State University, many people have helped me not only with this project but with all of my work. Reference Librarian Mrs. Eloise Cave was in the first class I taught at FSU, and since then she has cheerfully helped me in more ways than I can count. My chair, Booker T. Anthony, has responded to every request I have made, from material assistance to release time. Elaine Newsome continues her support into retirement. Bob Eddy helped me to bring something approaching order from a state of confusion; the late Izola Young provided many suggestions that only now am I beginning to fully appreciate.

More than anyone else, my students over the past twenty years have indicated in their writing the crucial importance of a sense of place. Ellen Lively and Arjay Hinek showed me once again how teachers learn more from their students than they could possible have taught them. My academic homeplace has been extremely fertile, and I alone am responsible for any shortcomings.

My most pervasive debt is to the people at the NTE readings who looked at various parts of the manuscript, sent me materials, and generally showed me how valuable friends in professional settings really are. Ed Uehling, Tom Bonner, Tom Colonnese, Kathy Bell, Vara Neverow, Ed Borowiec, Joy Passanante, Laura McGowan, Bill King, Bob Litchfield, John Ruff, Alison Wheatley, and Louise Wheatley all provided materials that helped me complete the book.

Thanks to my family—Patricia, Christine, and Marco—for helping me to get this work done. Patricia was always ready with constructive criticism, Christine helped get material in the mail on time, and Marco handled the first round of permissions while I was in the hospital. This book is, again, for the three of you.

I was extremely fortunate in the reviewers who responded so helpfully to the manuscript. In addition to Paul Christopher and Joe Trimmer, Mike Hogan and Peter White gave valuable direction at the beginning of the project. I wish also to thank the formal reviewers of the manuscript: Joe Trimmer, Ball State; Cheryll Glotfelty, University of Nevada; Michael Branch, Florida International; Alison Wallace, Unity College of Maine; Barbara Lounsberry, University of Northern Iowa; Sally Reagan, University of Missouri at St. Louis; David Teague, University of Houston; Zita Ingham, State University Arkansas; Kathleen Bell, University of Central Florida; Carl Herndl, New Mexico State; Barry Greer, Linn-Benton Community College; Fred Waage, East Tennessee State; Jim Killingsworth, Texas A&M; Thomas Scanlon, University of Minnesota.

*for my mother
and in memory of
my father*

BRIEF TABLE OF CONTENTS

TABLE OF CONTENTS

INTRODUCTION:
READING THE LANDSCAPE

If I were now to visit another country, I would ask my local companion, before I saw any museum or library, any factory or fabled town, to walk me in the country of his or her youth, to tell me the names of things and how, traditionally, they have been fitted together in a community. I would ask for the stories, the voice of memory over the land. I would ask to taste the wild nuts and fruits, to see their fishing lures, their bouquets, their fences. I would ask about the history of storms there, the age of the trees, the winter color of the hills. Only then would I ask to see the museum. I would want first the sense of a real place, to know that I was not inhabiting an idea. I would want to know the lay of the land first, the real geography, and take some measure of the love of it in my companion before I stood before the paintings or read works of scholarship. I would want to have something real and remembered against which I might hope to measure their truth.

—BARRY LOPEZ,
"MAPPING THE REAL GEOGRAPHY"

Thinking about how we respond to the landscape—the rural hills and rivers, the suburban development, the block of apartments, the distant mountains— enables us to begin to understand our relationship to the physical space we occupy. Just as importantly, understanding the real geography allows us to understand how we construct human relationships. The ecology of the natural world is also the ecology of the human world.

In *36 Children* (1967), Herbert Kohl explains how he began the process of freeing up his students to write about their lives—and to begin the process of experiencing the excitement of learning—by writing about their neighborhoods. Since his students lived in Harlem, they envisioned their immediate world in terms of blocks. By giving expression to the frustration they felt about the places where they lived (the only place most of them had ever been), they began to understand themselves and their relationships to the larger world around them. Not only for the students in Kohl's class, but for virtually all of us the experience of writing about our neighborhood and our fit within it ushers us into a new realm of possibilities.

In the epilogue to *The Way to Rainy Mountain* (1969), N. Scott Momaday talks about the way a given landscape insinuates itself into a person's being: "None of us lives apart from the land entirely; such an isolation is unimaginable. We have sooner or later to come to terms with the world around us—and I mean especially the physical world, not only as it is revealed to us immediately through our senses, but also as it is perceived more truly in the long turn of seasons and of years." In our searches to understand our places in this physical world, we must all attempt to find a logical starting point, a point from which we can begin to see how we have grown. Communication through writing is one way to help us understand just what we have become by revealing where we have been and what the individual details of the experience tell us about ourselves. We write to learn not only who we are, but where we have come from; in this way, we will be better able to understand where we are headed.

In the sections that follow, numerous writers explore the relationships between themselves and the lands they encounter. They often marvel at what they see, and the experience of seeing often reveals much to them about their connections with other human beings. In much of this writing the authors discover meaning that differs from their original understanding. Such serendipity occurs so frequently, in fact, that writing about personal experiences with land must have some marvelous power to unlock memories and trigger responses. The classical myth of Antaeus (a Libyan giant born of Poseidon, god of the sea, and Gaia, personification of the earth and universal symbol of motherhood) tells of a man who regains tremendous strength every time he makes contact with his mother the earth; Hercules is able to defeat Antaeus only when he holds the giant up in the air, thus disconnecting him from the restorative powers of the land. To some extent, perhaps we all feel this benevolent and nurturing quality of the landscape when we stop to think about how we associate powerfully positive feelings with specific places.

What this textbook attempts to do, then, is to suggest how you might move through the writing process. We begin by reading selections in which writers trace their experiences of making a connection with that landscape. This remembered place is not necessarily a spot of natural beauty such as a park; it could as easily be a block, a building, or a backyard, any place that holds special meaning. Next, the book presents several other activities that help in thinking about how to explore responses to the world more systematically and fully. The activities below will follow each reading selection in a fairly consistent manner because such a structure will help you to develop patterns of response and writing that will facilitate your task.

The writing process itself falls into four main parts:

Prewriting

Drafting, Revising, and Editing

Rhetorical Strategies

Argument and Research

These categories reflect not only the organization of this book, but they also indicate the order of the writing stages. The first two categories are self-explanatory in terms of the activities involved. The third category refers to the different strategic choices available to you as a writer; you will see how different methods of presentation can strengthen your work. Finally, the argument and research section will help you learn how to construct your presentations with different forms of evidence as support.

The categories of the apparatus offered along with the selections reflect crucial ways of reacting to the text and working through the steps of the composing process. Questions, suggestions, and directions accompanying the selections fall into these four categories:

Journal Entry Suggestions: Exploring and Responding

Questions for Critical Thinking

Collaborative Writing Activities

Suggestions for Formal Writing Assignments

JOURNAL

Your instructor may ask you to keep separate journals or to subdivide a single journal into different sections. Separating a journal, or keeping separate volumes, offers you the chance to reserve some entries as personal—to be seen only by you—and use others as material you will share and work on with classmates. Whatever your class does, you will want to be aware of at least two sorts of journal entries, which appear here as *Exploring* and *Responding* to differentiate between those composed before and after reading a selection.

In both categories, you will frequently see suggestions for *freewriting* on various topics. Freewriting simply means recording your thoughts on paper as they occur to you, jotting down ideas as quickly as possible, keeping your pen or pencil moving across the page. The idea is to use the allotted time in actually writing, getting the feeling of producing words as quickly as thoughts come to you. At first, this process may seem strange, but the prompts should direct you so that you will have enough ideas to move you through the freewriting assignment.

The Exploring assignment is a little like a jump-start for a car battery: an aid to get you thinking about a writing situation, to get you moving along. These exploratory journal entries prime you to begin productive thinking about the upcoming reading selection and also to begin recording your thoughts in some form of writing. This writing may or may not find its way directly into more formal writing assignments, but it will definitely help you to think about the ideas inherent in the following selection or selections.

The Responding assignment provides you the opportunity to record your ideas and reactions in virtually any form, from lists to carefully composed prose, at any level of detail. As the reader not only of the selections in the text, but of those texts that you and your fellow students produce, your reactions to what you read become the focus of your journal. You will soon see that there is no "correct" or "right" method to read any of the texts in this book, and that many of your classmates may respond to the texts in ways that differ markedly from your own. Since each of us has had different and unique experiences, we are all likely to come to the reading of an essay with a different set of expectations, body of knowledge, and values. Because many of us do experience similar responses to texts, some literary critics have used the term *interpretive communities* to designate those responses that a group of readers share; such interpretive communities, for example, will respond to an issue such as animal rights in different ways if they were raised on a farm or if they experienced animals only in zoos. Thus, one interpretive community sees animals as a resource grown to feed the family, while another may see animals as cruelly caged simply to satisfy human curiosity. Your own responses to a text, then, are crucial because they reveal not only specific reactions to a given text but also because they provide an index to many of your assumptions and values. Keeping such responses in a journal allows you to look back at a certain point and see how much your views have changed over a period of time.

QUESTIONS FOR CRITICAL THINKING

This category is divided into two groups: *Strategy* and *Issues*. Strategy questions direct your attention to the means by which the authors of the selections achieve their purposes. Issues questions relate directly to the ideas presented in the essays and the applications that may come out of them.

By initiating and focusing discussion, both kinds of questions encourage you to consider what the selection says explicitly as well as what it implies. For example, when reading one of the selections in this textbook, you might begin your critical thinking about what you are reading by breaking your written responses to the essay into two groups: those dealing with the *strategy* the author employs to make the essay effective, and those dealing with the *issues* that the essay raises for you as a concerned reader.

The first category, the strategy, comprises those responses that indicate something about the way an essay works: the way it is structured, the ways the author offers examples and details, the use of figurative language, the syntax and other hallmarks of style, and the extent to which you feel you understand the author's points.

The second category, the issues, relates to any ideas that the writer seems to want to get across in the essay. These include the main point, the surface meaning, and also the secondary messages that the essay raises for you. Often, questions involving issues seem open to more extensive discussion than those questions dealing with the structure of a work. These two categories are not meant to be mutually exclusive; in fact, they are much more likely to represent relative positions where some overlapping may occur. Their main function is to encourage you to consider how the essay works on you and makes you think.

Your written responses include all the markings you make with your pencil or pen on the text, the margins, or another sheet of paper. When you underline, you make a value judgment by highlighting certain parts of the essay. When you draw a question mark in the margin, you indicate your uncertainty with an idea the text has raised: perhaps you do not literally understand what the author is saying, perhaps there is a specific word you do not know or a phrase you do not understand. Perhaps you simply cannot believe that anyone would think, let alone write, such a foolish thing! Other markings such as the comments you make in the margins are more obviously records of your reading of the text.

COLLABORATIVE WRITING

Writing as part of a group offers you numerous advantages; the exercises in this text are designed to help you see how you can be part of a writing support group whose members encourage one another, thus making everyone more productive. This group situation will help you in two ways. Not only will you have the benefit of numerous responses to your work, but you will also gain from the experience of responding to the writing of others and discussing your own work and that of others in a positive, non-threatening situation. When you think of how your writing will be judged beyond the composition classroom, this consideration looms important indeed: in other classes whatever you turn in—particularly your written work—is *you*. Finally, writing collaboratively will furnish you with immediate models to apply to some of the interpersonal situations that occur in the readings you will be studying.

None of this collaboration, however, absolves you from bearing the primary responsibility for your writing. You ultimately decide what to use, what subject you will pursue, and how to treat that subject. After working through reader-response activities and collaborative undertakings, you will have a good many options at your fingertips.

Collaborative writing can take many forms. Your instructor might have you form a new group almost every class period, or you can form groups at the beginning of the semester that will remain intact throughout the course. Early in the semester, you can brainstorm ways of helping one another become better writers. People chosen as recorders of each group can present the list of ideas that each individual group has generated and then help to compile a master list of all the ideas produced in the class.

SUGGESTIONS FOR FORMAL WRITING ASSIGNMENTS

The final category of material at the end of each chapter is a series of suggestions for writing assignments. These assignments range, generally, from more readily accessible forms such as description and narration through potentially more complex forms until the final sections deal with the construction of complex arguments that require application of various combinations of the techniques and strategies practiced earlier. In plainer terms, your assignments will move from relatively simple to relatively complex. While each section will have some variation in levels of complexity, the readings and the activities in the early chapters will focus primarily on description and narration, while later chapters will feature more prominently argumentative and polemical writing.

In terms of forms, likewise, the emphasis increases on more complex structures. Chapter 1 begins with some short, intensely personal narratives that describe a certain landscape at a certain time. These selections give readers the opportunity to think about how the authors of these pieces understood the new worlds they were experiencing. Other narratives afford opportunities for writers to explore their homes, backyards, or neighborhoods; we might use these as a means of examining similar images for ourselves. Moving gradually away from the sense of a place that is specifically to be recorded through memory, we come to more diverse territory and see how, as writers, we assimilate new impressions and new material. Thus we move from those personal descriptions in journals, memoirs, and oral histories, to other forms of writing that offer the raw material for more consciously developed pieces in various genres.

Not only personal experience finds its way into written responses to the landscape and the people who inhabit it. Various other means of exploring human relationship to the physical world, including advertising, paintings and photographs of scenes of natural beauty, male-female differences in response to landscapes, waterscapes, and even revisited landscapes all offer much grist for our reader-response and collaborative mills. Analyzing such stimuli should result in the ability to articulate positions supported by mature and reasoned arguments.

In all of these writing assignments, we will be thinking about the three key concepts of your thesis, your audience, and your role as writer. As author, you take on certain responsibilities. You have to present your writing in a way that will produce the desired effect upon your reader. If you have thought carefully about what precisely you want to say, you will be able to articulate a thesis—the point you are making in the piece of writing—so that a reader understands it fully. This thesis usually takes the form of a single sentence that explains and justifies your main idea. Here is one example: "The state of Florida should prevent towns on the Keys from disposing of incompletely treated sewage into the Atlantic because raw sewage is killing the reefs and coral colonies." In making your presentation, you would need to consider not only your own beliefs and feelings on the subject, but you would also need to consider your audience; how should you approach those people who will read your essay? Thus, you consider your *thesis* in terms of your *audience* and the *role* you take as the author in this equation.

Audience can be a complex issue because you want to encourage as many readers as possible to believe that your thesis is not only acceptable but preferable to other possible means of looking at the subject you are treating. For some writing, audience concerns may be simple. When you are writing a summary, you need only to make sure that you have presented your facts accurately and in a clearly organized fashion. When you want to present your position on a debatable point, however, you need to take your audience into careful consideration. In this matter, your work with collaborative activities will be a great help because you can see the immediate reaction your work has on your classmates. Your instructor is another important member of your audience, and her or his responses will guide your developing sense of how to be sure you are reaching your targeted audience.

READING VISUAL TEXTS

The paintings and photographs in this textbook parallel the written selections in various ways. We can begin with the same triad of writing characteristics that we covered earlier: the thesis, or the point the author of the text wishes to make; the role, or the stance the author takes in relation to the work created; and the audience, or the people at whom the text is aimed. The same ideas apply when we read visual texts. We assume that the painter or photographer in each instance feels that something worthwhile is happening within the frame, and so will offer that enclosed space to us as a text to be read. The choice of precisely where to set the boundaries of the frame is the choice of the creator. We also have to take in mind the question of audience: who will see the image, and how will they be affected? Thus, we can examine our own feelings and responses to come to terms with the idea of audience.

We are always put in the position of the "implied spectator" and we have to determine why the artist or photographer chose to put us in this precise spot. Thinking about how the artist decided to position us in relation to the visual image helps us to understand the nature of representation in art. A painted landscape of an actual place is a landscape twice removed from us. The original landscape exists (or existed once), and the artist has presented his or her version of that landscape. Our vision of the landscape is thus the third stage: we see what the artist has presented, but we might interpret that landscape in a different way than the artist intended. With visual texts as with written texts, our response to the work determines much of the work's meaning. Our meaning may not coincide with the artist's meaning. Often in verbal texts, the author presents us directly with a thesis statement. Unless a visual text's title or accompanying description furnishes us a thesis, we have to infer the thesis for ourselves. Likewise, we must infer the artist's purpose in determining the particular composition and point of view.

The most successful visual texts are those that invite us into the frame, those that create in us a desire to enter the world of the frame and participate imaginatively in what transpires within that frame. In order to accomplish this, the visual artist must organize the elements within the frame in such a way that we can comprehend whatever narrative elements might exist in that frame. Sensitizing ourselves to the intricacies of visual composition is the best way to begin the process of becoming close readers of visual texts.

The first painting presented here, *Robert Andrews and His Wife* was painted by the esteemed British artist Thomas Gainsborough in 1760. Not only does it demonstrate the "high classical" painting artists in America tried to imitate, but it also shows how landscape was presented in formal painting. As we would in a photo, we see foreground, middle distance, and horizon or background, and here we also see human figures presented in the landscape in a manner very close to portrait-painting. Mr. and Mrs. Andrews, through their posture and expressions, seem to take a certain attitude toward the land they own. Look at the placement of Mr. Andrews' hands and the arrangement of his

Thomas Gainsborough, *Robert Andrews and His Wife,* 1760

Marion Post Wolcott, *Pursglove, West Virginia,* 1938

limbs; what does his body language convey? What do you see in his dress and accessories or in Mrs. Andrews' carefully arranged dress? How might they regard their lands or the harvest scene depicted on their fields?

Paintings such as this one were underwritten by wealthy patrons, people who would support artists financially and often commission paintings. The painting then took on a very different life from one intended for public display. What sorts of differences might you expect to see between a painting for a patron and one done for the sake of satisfying the artist's vision? How hard would it be to combine the two? Most landscape paintings tend to work toward the horizon, that is, the effect of the canvas is most often to take the viewer's eye toward a point in the painting that suggests the farthest reach of the painting, often the point of final detail before the scene melts into indistinctness. This point is often referred to as the *vanishing point;* painters manage perspective in their work by organizing the detail in a painting around this imaginary line.

Landscape painting tends to emphasize horizontal lines, with the vertical shapes often occurring in the foreground, sometimes in the middle distances, with human figures distributed throughout the painting depending on their function. Frequently, as in the paintings by Colman and Whittredge, the human figures in the middle distances confer a sense of scale on the grand natural forms in the painting. Other times, as is the case in the painting by Durand, the human figures are larger in the middle distance because they represent key human presence in the natural scene: Durand is presenting two of his best and most honored friends. In Gainsborough's painting, we have the two human figures at the extreme left, taking up nearly half of the canvas. They dominate the landscape, perhaps because they own it. Other verticals reinforce this image: the dog plays at its master's feet, and the tree behind the couple suggests long-term stability. Perhaps you will want to discuss the implications of this painting in terms of its owners' sense of the natural world.

Photography operates on similar principles. For example, the first photograph in this section—*Pursglove, West Virginia*—was taken by a young aspiring professional photographer named Marion Post Wolcott in 1938.

In her photo Wolcott places her subject, the young girl, at the lower center of the frame. You might ask why she does this; after all, if she had wanted to show us what the coal miner's daughter looks like, she could have taken a shot from the front and shown the girl's face. She might also have moved in closer for a tighter shot: that is, she might have made the figure of the girl larger, made it take up more of the frame, by eliminating the New York Central coal cars on the left or the coal miner's house on the right. We have to ask ourselves, then, exactly what Wolcott gains by setting up her photo as she does.

Often, some verbal text will help us in reading a visual text. The full caption of the photo reads, "Coal miner's child taking home kerosene for lamps, used in company houses—coal tipple [a swinging device to load coal] in the background. Pursglove, on Scotts Run, West Virginia, 1938." Think about the narrative detail provided for you in this description. What does it do to suggest why Wolcott would have included so much of the house on the right in the photo, and why the child is close to the center of the photo?

The overall composition of the photo—as is true of any visual text—can be broken up into three general areas. Though this schematization is oversimplified, it will help orient us to the practice of reading pictures. Think of a visual text like a photograph or painting as being roughly divided into three horizontal areas or planes. This structure consists of a foreground (the third closest to us), a middle distance (the next closest area), and a background or horizon (the far area). Most visual representations of landscape work in this way. Often as viewers of paintings or photos, we see the foreground close up, as in the photograph of Childress County, and when we look at such a photo we feel we are on the edge of that field in front of the house. This orientation can have the effect of making us feel as though we are a part of the scene. In the Wolcott photo, we can see the foreground clearly in the lower left corner. Think about how this small detail orients us to the picture overall. Doesn't it align us with the path the little girl has taken, in effect making us as viewers walk the same route the little girl has traveled? The coal tipple hovers above the girl. It is a symbol of the employment in Pursglove; is it not also a symbol of another force in the world of the photo? Look at the shape of the girl. Could it be the can of kerosene she is carrying that bends her body, her head slanted to the left, while the road bends back toward the right? What do these opposing forces suggest? If possible, take some time to discuss the narrative possibilities of this image with your classmates.

Moving beyond the child to the overall picture indicates the importance of other elements in the frame. The coal tipple, just above the center of the frame, towers above the landscape. We can make out the hills beyond the town and railroad cars, but can the child in the photo see them? Look at her vantage point, or the level of her sight as we imagine it; she can see nothing but the coal cars and the dismal company houses. As a matter of fact, the imposition of forms on this landscape shows what has happened not only to this child but to this region of West Virginia: the rolling hills have been transformed. The mining of the coal resembles the boring in of the camera into the life of the mines. Wolcott's photo gets at the heart of the life in this village both literally and figuratively. The photo shows a compromised landscape under a leaden or late afternoon sky; foreground, middle distance, and horizon are blended in a manner that suggests why Wolcott took this picture at precisely the moment she did. You might want to discuss with your classmates as well why she chose this moment; what would other shots have gained or lost, if for example she had taken the shot when the girl was about to turn the corner to the right? If she had taken the photo as soon as she could get the kerosene can in the frame?

Other visual texts reproduced in this book demostrate similar properties. You might draw a parallel between the house situated in plowed fields appearing in the photograph by Dorothea Lange and the cemetery-backyard image captured by Walker Evans. You can think about how the vanishing point is used in the Lange photo of the road west to California, the "migrant trail" of the Great Depression. Think as well about how the details that make each painting unique are organized. For example, similar shapes link natural and human forms in Blumenschein's painting. Treatments of color—though with very different effects—dominate O'Keefe's red sky at Lake George and the striking combination of clouds and sky offered by Dixon. Inness' Lackawanna painting parallels Delano's photo of western Pennsylvania in their central depiction of railroad tracks; how do the two media differ in the feeling they create for the viewer?

Other examples to consider together in order to see how different artists manage style are the winter paintings. Bodmer and Granda Moses use human activity on snowscapes as the central narrative element in each painting. What do you think the story to be told in each might be? How do style and manner of presentation affect each painting? How do the human figures help to establish the mood of each painting? How would you contrast the human narrative elements in these two paintings with the human dimensions of Bywater's image of the oilfield girls? Considering possible answers to such questions provides the focus necessary to see how all visual texts operate on their viewers.

PART ONE

Prewriting Techniques

Chapter 1

JOURNAL, MEMOIR, AND ORAL HISTORY: THE LARGER WORLD AND THE COMMUNITY

Dorothea Lange, *Westward to the Pacific Coast on U.S. 80,* 1938

1. How does the composition of the scene affect you as a viewer? How effectively do the different elements of the image work together?

2. How does the lack of human presence affect your experience of the photograph?

3. Explain how this photograph might suggest both the historical movement toward the West and the psychological or emotional ideas of "moving on."

Chapter 1

JOURNAL, MEMOIR, AND ORAL HISTORY: THE LARGER WORLD AND THE COMMUNITY

The first three forms of writing you encounter—journal, memoir, and oral history—offer various opportunities to get you started with your writing. One of the purposes of this first chapter is to suggest to you the wide variety of possibilities for writing. You will see how early explorers and travelers looked at unknown lands, and you will also see how minute inspection of your block, town, and backyard will reveal new material as long as you explore them. We first cast a wide net to open the world of possibility, and then we look more directly in the following chapters at how our sense of our place in the world will make us more effective writers.

We begin our exploration of the writing process in the early experiences of people wherever these experiences are located: in exotic mountain ranges, the tenement backyard, or the suburban garden. In this way, we will have a truly ecological perspective because we have located our best-loved writers of the natural world—our Thoreaus, Dillards, Momadays, Walkers, Lopezes—in the larger contexts of American life. It goes without saying, then, that the focus of this book is not only strongly multicultural, but it also betrays a keen attention (a "sharp lookout," as John Burroughs described his sense of observation of natural phenomena) to traditions of Native American responses to the land and those organisms which exist together on it. Their direct relationships with natural cycles parallel our instinctive reactions to the natural world that we experienced early in our lives.

This book provides an opportunity to write about the natural world—our environment, the landscapes and waterscapes we inhabit. Thus, we have two concerns: as writers you will write both about what you now know and about what you are learning. Your journals will give you opportunities to explore your ideas and your experiences of environments. Since your home and your neighborhood—your personal environments—are what you know best, you can start looking there for journal entry material. As you go through the following selections, you will find suggestions for journal entries; these are intended as prompts, as means of getting you started. In your journal, you are in control of what you record; outside of topics assigned by your instructor, you provide the raw material for continuing thought.

JOURNALS

In order to chronicle experiences of fantastic places, writers have traditionally relied on journals or, as they are sometimes called, travels. Journals offer an opportunity to set down all sorts of experience. We will encounter wild rapids, unknown animals, people with unaccustomed habits and modes of thought, and close consideration of interesting objects for study found right around home. Some journal writers find significance in snow and surrounding hills; others find deep silence to be as productive of ideas as the most exotic places.

You might ask, "How can I learn about home, about writing, by going all over the place like an explorer?" You may well find that far-flung exploits give you helpful perspective on the most familiar landscapes; just look at Thoreau or Marcia Bonta and their entries. Journals offer the most intensely *expressive* writing situations. They allow us to "write to learn," to discover what (and how) we think, what we feel, and what we actually know. By chronicling our feelings at one time and by recording information at times when we may be slow in coming up with our own ideas, we are forming a hybrid, creating a form of writing that expresses our individuality.

Journals serve a variety of purposes beyond what is required in the syllabus for this course. Since they are for your own use and most often never see publication unless you wish to make them available to other people, journals can take whatever form you decide. Your instructor will help you to determine how you will be using your journal in relation to the categories outlined in the introduction, but the most likely case will be a range of purposes for your journals. The activities that you will carry out in your journals will be various: you may list several ideas at one time, or you may record a single impression, or you may keep a running account that may span several days, even weeks, in your journals.

Think about the journals as a sort of workout location for your mind. You will go through a number of activities that will exercise different capabilities in your array of writing talents, from description through narration through different kinds of comparison and analysis. You will observe, you will record, and you will make judgments.

To begin thinking about the exercises in your workout journal, don't worry that you have nothing to write about, or that you cannot immediately think of something worth immortalizing in deathless prose. Instead, try a technique that will serve as a warm-up, something that will stretch your thinking in a way that will ease the further exercising of your mind. Try making lists. For example, think of the most dominant characteristics of your home; or think of a few places you would like to visit. From your first list, choose the two characteristics that have done the most to determine your personality. From the second list, pick two that you feel inclined to explain at some length by examining why you would like to visit there. What do you expect to find? What would be the best route to take to get there? What will you need to bring with you, and who should accompany you on the shopping trip to get what you need?

You can also respond to events taking place in the world around you. Look at a morning paper to see events related to the environment, to land use, to the relationship of people to their land. For example, you might find that an industry accused of some form of pollution has its case come up in court, you might see a zoning issue for a variance requested by a developer considered, or you might see residents of an area getting together to protest a rise in local crime and subsequent changes in their daily routines. Have gangs and their turf affected the ways in which people live their daily lives in your neighborhoods? After perhaps quoting a headline and briefly summarizing what the paper has to say, you could then go on to offer your own suggestions and solutions. What position—mayor, head of city council, chief of police, state director of environmental protection—would you have to occupy in order to effect a logical resolution? Put yourself in that position and explain in your journal how you would proceed.

Other possible journal topics include the following:

- Explain the phrase "Think globally, act locally." Have you heard this phrase used before? What does it mean to you? Do you think that its originators had other ideas in mind? How should we interpret the phrase today?
- Ask yourself, "What do I need to do to help my world?" "What can I do to help my world, right now, right where I am?" Explain your response.
- Think about how your presentation can change emphasis when you consider carefully your audience. Look at this list of words and choose a few as the basis for some comments:

tree-hugger	fresh-air fiend
eco-warrior	nature freak
eco-feminist	nature enthusiast
environmental activist	flower sniffer
naturalist	animal rights activist

- Name six things that should never be allowed to happen to the environment. Choose one that you are afraid might happen and explain what your community (however large) should do to prevent its happening.

In the journals reprinted in this text, you have a wide variety of authors and purposes. Some are intended only to chronicle the authors' responses to nature, while others were originally written with the hope of publication, and still others were written as part of the authors' official duties in service of their governments. The purpose of these entries is not to show you precisely how to set up your journal; rather it is to suggest how other writers have found journals useful in their own writing. These journals should also serve as suggestions of the ways you might explore ideas that are important to you. Your journal writing may begin a process you will find valuable and continue far beyond this class.

MEMOIRS

Memoirs (literally, "memories") are the recording of one's life or experiences; therefore, they are very closely allied to such forms of history as biography and autobiography. However, memoirs are different from biography because they are selective. They may represent only brief times in a person's life, or they may represent a recurrent theme in a person's life or career. A memoir can focus on one incident or issue because the term itself suggests a single memory, a single recollection, or in the plural, a series of memories.

The memoirs presented here indicate the importance of specific places. Laura Green describes how valuable she found a regular place to visit with neighbors. Zitkala-Sa examines her decision to leave her native reservation for a Quaker school and compares what she has learned about the two. Old Lady Horse describes, in a combination of actual experience and folktale, a key period in Kiowa history. All describe effectively the effects of a particular site on the people they have become. A memoir, it seems, allows you to explore a chapter in your life, or an aspect of your life that you find compelling.

ORAL HISTORY

Oral history probably offers the most efficient method of learning about the traditions and background of the place you inhabit or want to explore. Like the ideas that Barry Lopez offers in the quotation that opens this book's introduction, oral history gives the strong native flavors of people and places. It tells you how people actually lived, how the land affected them, how the land makes them think.

Most simply defined, oral history involves the documenting of words spoken (or, less likely, written) to an interviewer. The interviewer bears the responsibility of asking useful questions that will help the interviewee provide, as far as possible, the information that will constitute the details of the oral history. The usual method is to begin with a set of questions and a tape recorder; as the interviewer asks the questions, the interviewee responds, the tape rolls, and the oral history then exists on that tape. Depending on the purpose and length of the interview, the tape may or may not be transcribed into hard copy. You may have family histories that are based on oral history interviews; after working with the examples here, you may well want to begin compiling oral histories in your family and community. Since many oral traditions are handed down through generations and never recorded, your older relatives may provide you with startling new material.

JOURNALS

The journals included here demonstrate several attitudes toward landscape as the journal writers explore their worlds. Mrs. Colt kept her journal to record her activities and reactions as her family traveled from New York to Kansas to establish a new community, while Thoreau recorded his thoughts about how he would keep an accurate record of his responses to the world about him. Muir and Stanwell-Fletcher describe unusual, even exhilirating, situations they encounter in wild landscapes. Even if you never write about anything as exciting as Muir's dangerous peering down into the Yosemite, your journal will provide a place for you to record important thoughts and describe reactions to the subjects you encounter. In a journal entry, the drama is often the surprise of a new insight that has come as a result of writing about a response to a particular, sometimes even familiar, stimulus. The act of gaining knowledge through your writing constantly provides fresh opportunity.

EXPLORING JOURNAL

Recall what you can remember from other classes, your reading, movies and television, or accounts told by older people about moving across the landscape, whether pioneers heading toward the horizon or families searching for a better life. Do all of your ideas come from history books? What are your mental pictures of pioneers like? Try a ten-minute freewrite on this subject. If you have any difficulty with it, try thinking back to your earliest experiences and move through your school career. Did you have strong images of explorers or pilgrims? How about Columbus and the Native Americans who helped him and other explorers, often to their own detriment? Perhaps thinking about certain holidays will help you with this assignment; once you get started, though, you may find that ten minutes is not enough. In that case, keep on going.

from *Went to Kansas*
Miriam Davis Colt

Born in 1817, Miriam Davis married William Colt in 1845. Davis put herself through school by working at several jobs, and after her marriage she taught with her husband in Montreal. Though she was happy with her life and two children in Montreal, she acquiesced to her husband's desires to join a vegetarian community just beginning in Kansas. They sold all they had, invested in the new colony, and began the difficult overland journey by leaving her relatives in upstate New York in mid-April of 1856.

Disaster is the only word to describe Miriam Davis Colt's experiences on the frontier. Poor administration doomed the planned vegetarian colony, and sickness quickly drained her family of its strength. By October, husband William and son Willie had died; she had to sell all of her possessions to buy headstones and return to New York with her daughter. Only when finances again became difficult for Colt in 1862 did she publish the journals of her trip as Went to Kansas. *No precise figures are available, but it is likely that more prospective pioneers in pre-Civil war days met a fate like Colt's than managed to survive and ultimately succeed.*

January 5th, 1856.   We are going to Kansas. The Vegetarian Company that has been forming for many months, has finally organized, formed its constitution, elected its directors, and is making all necessary preparations for the spring settlement. . . . We can have, I think, good faith to believe, that our directors will fulfill on their part; and we, as settlers of a new country, by going in a company will escape the hardships attendant on families going in singly, and at once find ourselves surrounded by improving society in a young and flourishing city. It will be better for ourselves pecuniarily, and better in the future for our children.

My husband has long been a practical vegetarian, and we expect much from living in such a genial clime, where fruit is so quickly grown, and with people whose tastes and habits will coincide with our own.

January 15th.   We are making every necessary preparation for our journey, and our home in Kansas. My husband has sold his farm, purchased shares in the company, sent his money as directed by H.S. Clubb. . . . I am very busy in repairing all of our clothing, looking over bags of pieces, tearing off and reducing down, bringing everything into as small a compass as possible, so that we shall have no unnecessary baggage.

April 15th.   Have been here in West Stockholm, at my brother's, since Friday last. Have visited Mother very hard, for, in all probability, it is the last visit we shall have until we meet where parting never comes—believe we have said everything we can think of to say.

April 16th.   Antwerp, N.Y. Bade our friends good bye, in Potsdam, this morning, at the early hour of two o'clock. . . .

May 12th.   Full of hope, as we leave the smoking embers of our camp-fire this morning. Expect tonight to arrive at our new home.

It begins to rain, rain, rain, like a shower; we move slowly on, from high prairie, around the deep ravine—are in sight of the timber that skirts the

Neosho river. Have sent three men in advance to announce our coming; are looking for our Secretary, (Henry S. Clubb) with an escort to welcome us into the embryo city. If the booming of cannon is not heard at our approach, shall expect a salute from the firing of Sharp's rifles, certainly.

No escort is seen! no salute is heard! We move slowly and drippingly into town just at nightfall—feeling not a little nonplussed on learning that our worthy, or unworthy Secretary was out walking in the rain with his *dear* wife. We leave our wagons and make our way to the large camp-fire. It is surrounded by men and women cooking their suppers—while others are busy close by, grinding their hominy in hand mills.

Look around, and see the grounds all around the camp-fire are covered with tents, in which the families are staying. Not a house is to be seen. In the large tent here is a cook stove—they have supper prepared for us; it consists of hominy, soft Johnny cake (or corn bread, as it is called here), stewed apple, and tea. We eat what is set before us, "asking no questions for conscience' sake."

The ladies tell us they are sorry to see us come to this place; which shows us that all is not right. Are too weary to question, but with hope depressed go to our lodgings, which we find around in the tents, and in our wagons.

May 13th.   Can any one imagine our disappointment this morning, on learning from this and that member, that no mills have been built; that the directors, after receiving our money to build mills, have not fulfilled the trust reposed in them, and that in consequence, some families have already left the settlement...?

As it is, we find the families, some living in tents of cloth, some of cloth and green bark just peeled from the trees, and some wholly of green barn, stuck up on the damp ground, without floors or fires. Only two stoves in the company. . . .

We see that the city grounds, which have been surveyed ... contain only one log cabin, 16 by 16, mudded between the logs on the inside, instead of on the outside; neither door nor window; the roof covered with "shakes" (western shingles), split out of oak I should think, 3 1/2 feet in length, and about as wide as a sheet of fools cap paper.

May 14th.   Some improvements are being made in the "centre octagon" to-day. *[The octagon plan designated sixteen farms around a communal eight-sided building.]* My husband has put up some shelves on one side, by boring holes into the logs, putting in long and strong wooden pins, and laying on some of the "shakes" for shelves.

May 15th.   A cold, drizzling rain. The prairie winds come whizzing in. Have hung up an Indian blanket at the door, but by putting trunks and even stones on to the end that drags, can hardly make it answer the purpose of a door. It is dark, gloomy, cheerless, uncomfortable and cold inside.

Have a fire out of doors to cook by; two crotches driven into the ground, with a round pole laid thereon, on which to hang our kettles and camp pails, stones laid up at the ends and back to make it as much as it can be in the form of a fireplace, so as to keep our fire, ashes and all, from blowing high and dry, when these fierce prairie winds blow. It is not very agreeable work, cooking out of doors in this windy, rainy weather, or when the scorching sun shines.

The bottoms of our dresses are burnt full of holes now, and they will soon be burnt off. If we stay here we must needs don the Bloomer costume. Our bill of fare is limited—hommy, Johnny cake, Graham pudding, some white bread, now and then stewed apple, a little rice, and tea occasionally for the old people. . . .

Father has got a broom stick, and is peeling a broom. He says, "I intend you shall keep this stone floor swept up clean." ...

May 20th.   I went with my husband two miles, to his claim, to plant corn. A bright and lovely day came in with the rising sun, not a cloud in the heavens above. . . . Not a stump, fence, stone or log, to mar the beautiful picture. . . . We sat in the wagon, while my hopeful husband planted corn and garden seeds. After the ploughing, the planting is done by just cutting through the sod with an axe, and dropping in the seeds—no hoeing the first year; nothing more is to be done until the full yellow ears are gathered in the autumn time. . . .

After we had eaten our dinner in the wagon, we went and selected a site for our log cabin, a little way from the clear, stony-bottomed creek that flows through our claim. . . . My husband says we shall have an elegant building spot, and that he will build a neat little log cabin; that he will get the large flat stones from the creek that will cleave apart, for walks to the creek and around our cabin. . . . I do not like to hear the voice which whispers, "This never will be;" but still it will whisper. . . .

May 30th.   Am wearing the Bloomer dresses now; find they are well suited to a wild life like mine. Can bound over the prairies like an antelope, and am not in so much danger of setting my clothes on fire while cooking when these prairie winds blow. Have had Mrs. Herriman's baby here for a few days, she is so very sick. Mr. H. wanted to plant his corn and garden seeds; he could leave his sick wife with the little two-year-old boy, but could not leave the little one to cry when its mother could not stir to take care of it.

Have been over to see Mrs. H; she is some better. Picked another bouquet of very rich flowers on my way, and placed them for her to look at; there were Japan lilies, large beautiful snake's head, larkspurs of many colors, and much larger than those we cultivate at home ... wild peas and beans are scattered broadcast over these green fields; their blossoms are very pretty; they are eaten by the Indians—are said to make good coffee, and when green they are some-times pickled. Beds and beds of onions are growing here and there, with the little onions all clustered in on the top, not larger than kernels of wheat.

June 3d.   A most terrific thunder-storm came up last night; the thunder tumbled from the sky, crash upon crash ... the rain came in torrents, and the wind blew almost tornadoes. . . . When we heard the storm approaching, we dressed ourselves, wrapping Indian blankets about us, and made ready to protect our children from the rain that was then dripping through the roof. We put all our bedding around them and all we could see to get by the glare of the lightning, (could not keep a candle lit), spread our umbrellas, (five in number), placed about, and held over them. We all got wet, and were obliged to lie in our wet beds till morning. This morning all was calm; the bright sun ascended up into a cloudless sky, as majestically as though there had been no war in the elements through the night. But the rain had dissolved our mud chinking, and the wind had strewed it all over and in our beds, on our clothes, over our dishes, and into every corner of the house. Have had all our sheets to wash, beds and blankets to dry in the sun and rub up, our log walls to sweep down, our shelves and dishes to clean, and our *own selves* to brush up. "Such is prairie life," so they say. . . .

June 16th.   What are we to look for, and what fear next? The mosquitoes have come upon us all of a sudden. They troubled us very much at the creek to-day while washing. . . . Our bed being short, in the night they have a good chance to nibble away at our protruding extremities. I lie awake. . . . I try to keep my children covered, so they wont eat them all up before morning. As for myself, I get so infuriated that I get up, descend the ladder, make my way out into the wet grass upon the run, not minding what reptiles may be under my bare feet; I then return from my dewy bath, lie down and try to sleep, but it is almost in vain.

June 17th.   The soil of rich layers of vegetable mould is throwing up the rows of dark green blades of corn. Our cornfield of six acres looks promising, as do all cornfields around. Pumpkins, squashes, melons, cucumbers, beans, peas, potatoes, and tomatoes are thriving finely. The next work our settlers will find to do, will be fencing cornfields, splitting rails, cutting poles, and drawing them from the bottom lands to do it with. . . .

June 26th.   Several members of our company have suddenly been taken with the chills and [*malarial*] fever; and here in our own cabin it has fallen upon Mr. V. and wife, mother, sister L. and Mema. It is sorrowful to see what a change comes over them in one day. Mr. V. thinks this is too much for him to stand—will leave cornfield, and all the prospects of this beautiful country, and hasten to the North again. . . .

June 29th.   A lovely Sunday. I, too, have fallen victim to the dreaded disease. Mr. & Mrs. V., Mema and myself, have occupied the loft today. Dishes of water have been set near our heads, so that we could help ourselves to drink when it seemed as though we should burn up with fever. My head has ached dreadfully; am glad to crawl down the ladder, with weakened limbs get out door here, sit down on a stone, lean my dizzy head against the logs of the cabin, breathe a little fresh air, see the sun go down, and ask, "Can this be the same sun that shines [on] our Northern friends, who are enjoying the blessings and comforts they know not how to appreciate?"

June 30th.   Mr. V. has spoken for a passage in an ox-wagon, for himself and wife, to Kansas City; my husband thinks we had better make preparations to leave the Territory with him, and not wait till he and father get sick; so he has done the washing to-day himself, and packed our trunks. And now, since our paroxysms of chills and fever went off for the day, sister L. has packed their trunks, and I have been trying to help my husband cook a little to take on our journey Northward. I have mixed up some bread, using baking powder for risings, for I could not tend to rising yeast. I find myself very weak—was obliged to sit down twice while mixing my bread. When I began to feel faint and dizzy, I would sit down on a stone, and when the dizziness passed off, go on with my mixing.

My husband is now baking the bread in the Dutch oven. This is the first lesson he has taken in baking in my big kitchen, and it troubles him to keep the coals on the oven. I believe we shall be ready to start any time now, when the command is given. . . .

July 1st.   The water is fast drying up; the spring that was cleaned out and dug deeper, in the gulch below our cabin, is almost dry; the water is not fit to drink.

from *Journals*
Henry David Thoreau

Henry David Thoreau (1817–1862) has been widely revered as America's premier naturalist, though he himself would probably scorn such a title. Never traveling for long from his native Concord, he was one of the circle of liberal thinkers—along with Ralph Waldo Emerson, William Ellery Channing, Margaret Fuller, Bronson Alcott, and William Lloyd Garrison—who came to be associated with transcendentalism. Loosely described, this belief saw mankind as essentially good and capable of knowing truth intuitively, without prior sensory experience. A form of idealism, transcendentalism was influenced by Eastern thought and German and British romanticism. It was first made popular in America by Thoreau's patron Emerson.

During his lifetime, Thoreau's only published books were A Week on the Concord and Merrimack Rivers *(1849) and* Walden *(1854), though he did publish some poetry and nonfiction essays in periodicals. His books were little noticed during his lifetime; in fact, when more than 700 unsold copies of* A Week *were returned to him, he claimed that he had a library of 1,000 books, and he had written over 700 of them himself!*

Thoreau's preparation for writing was his journalkeeping, which he began at Emerson's suggestion. Gradually, he found that his characteristic mode of introspective nature writing was best developed by the sort of meditations on people, places, things, and ideas that he recorded in his journal. Excerpted here are Thoreau's meditations on the nature of journalkeeping.

Feb. 8, 1841

My Journal is that of me which would else spill over and run to waste, gleanings from the field which in action I reap. I must not live for it, but in it for the gods. They are my correspondent, to whom daily I send off this sheet postpaid. I am clerk in their counting-room, and at evening transfer the account from day-book to ledger. It is as a leaf which hangs over my head in the path. I bend the twig and write my prayers on it; then letting it go, the bough springs up and shows the scrawl to heaven. As if it were not kept shut in my desk, but were as public a leaf as any in nature. It is papyrus by the riverside; it is vellum in the pastures; it is parchment on the hills. I find it everywhere as free as the leaves which troop along the lanes in autumn. The crow, the goose, the eagle carry my quill, and the wind blows the leaves as far as I go. Or, if my imagination does not soar, but gropes in slime and mud, then I write with a reed.

Nov. 16, 1850

In literature it is only the wild that attracts us. Dullness is only another name for tameness. It is the untamed, uncivilized, free, and wild thinking in Hamlet, in thc Iliad, and in all the scriptures and mythologies that delights us—not learned in the schools, not refined and polished by art. A truly good book is something as wildly natural and primitive, mysterious and marvellous, ambrosial and fertile, as a fungus or a lichen.

My Journal should be the record of my love. I would write in it only of the things I love, my affection for any aspect of the world, what I love to think of. I have no more distinctness or pointedness in my yearnings than an expanding bud, which does indeed point to flower and fruit, to summer and autumn, but is aware of the warm sun and spring influence only. I feel ripe for something, yet do nothing, can't discover what that thing is. I feel fertile merely. It is seed-time with me. I have lain fallow long enough.

Notwithstanding a sense of unworthiness which possesses me, not without reason, notwithstanding that I regard myself as a good deal of a scamp, yet for the most part the spirit of the universe is unaccountably kind to me, and I enjoy perhaps an unusual share of happiness.

Nov. 12, 1851

Write often, write upon a thousand themes, rather than long at a time, not trying to turn too many feeble somersets in the air—and so come down upon your head at last. Antaeus-like, be not long absent from the ground. Those sentences are good and well discharged which are like so many little resiliencies from the spring floor of our life—a distinct fruit and kernel itself, springing from *terra firma.* Let there be as many distinct plants as the soil and the light can sustain. Take as many bounds in a day as possible. Sentences uttered with your back to the wall.

"July 15, 1869—The Yosemite"
John Muir

John Muir (1838–1914) is one of the first and most important nature writers. Born in Scotland, he spent much of his youth laboring on a Wisconsin farm before he ventured west to the Sierras, the range to which he developed lifetime ties. As a result of his intense commitment to saving the wild places of the world, he is regarded as the "father" of the national parks system. Muir founded the Sierra Club in 1892 in order to preserve the scenic and geographic resources of America for future generations; he has become a heroic symbol of the individual fighting bureaucratic systems on behalf of the natural world.

In 1869, Muir took his first trip to the California Sierra mountain country. There he worked as a sheepherder, taking the sheep from the Central Valley to their summer grazing lands in the Sierras and keeping the journal that became My First Summer in the Sierra *when he published it in 1874. He tied the small journal around his waist so that it would not be in his way while he worked, yet he could still have it handy when he wished to record his experiences. The entry printed here records his experience looking down into the Yosemite from the break of the Yosemite Creek falls at the top of the canyon.*

July 15   Followed the Mono Trail up the eastern rim of the basin nearly to its summit, then turned off southward to a small shallow valley that extends to the edge of the Yosemite, which we reached about noon, and encamped. After luncheon I made haste to high ground, and from the top of the ridge on the west side of Indian Cañon gained the noblest view of the summit peaks I have ever yet enjoyed. Nearly all the upper basin of the Merced was displayed, with its sublime domes and cañons, dark upsweeping forests, and glorious array of white peaks deep in the sky, every feature glowing, radiating beauty that pours into our flesh and bones like heat rays from fire. Sunshine over all; no breath of wind to stir the brooding calm. Never before had I seen so glorious a landscape, so boundless an affluence of sublime mountain beauty.

The most extravagant description I might give of this view to any one who has not seen similar landscapes with his own eyes would not so much as hint its grandeur and the spiritual glow that covered it. I shouted and gesticulated in a wild burst of ecstasy, much to the astonishment of St. Bernard Carlo, who came running up to me, manifesting in his intelligent eyes a puzzled concern that was very ludicrous, which had the effect of bringing me to my senses. A brown bear, too, it would seem, had been a spectator of the show I had made of myself, for I had gone but a few yards when I started one from a thicket of brush. He evidently considered me dangerous, for he ran away very fast, tumbling over the tops of the tangled manzanita bushes in his haste. Carlo drew back, with his ears depressed as if afraid, and kept looking me in the face, as if expecting me to pursue and shoot, for he had seen many a bear battle in his day.

Following the ridge, which made a gradual descent to the south, I came at length to the brow of that massive cliff that stands between Indian Cañon and Yosemite Falls, and here the far-famed valley came suddenly into view throughout almost its whole extent. The noble walls—sculptured into endless variety of domes and gables, spires and battlements and plain mural precipices—all a-tremble with the thunder tones of the falling water. The level bottom seemed to be dressed like a garden—sunny meadows here and there, and groves of pine and oak; the river of Mercy sweeping in majesty through the midst of them and flashing back the sunbeams. The great Tissiack, or Half-Dome, rising at the upper end of the valley to a height of nearly a mile, is nobly proportioned and life-like, the most impressive of all the rocks, holding the eye in devout admiration, calling it back again and again from falls or meadows, or even the mountains beyond—marvelous cliffs, marvelous in sheer dizzy depth and sculpture, types of endurance. Thousands of years have they stood in the sky exposed to rain, snow, frost, earthquake and avalanche, yet they still wear the bloom of youth.

I rambled along the valley rim to the westward; most of it is rounded off on the very brink, so that it is not easy to find places where one may look clear down the face of the wall to the bottom. When such places were found, and I had cautiously set my feet and drawn my body erect, I could not help fearing a little that the rock might split off and let me down, and what a down!—more than three thousand feet. Still my limbs did not tremble, nor did I feel the least uncertainty as to the reliance to be placed on them. My only fear was that a flake of the granite, which in some places showed joints more or less open and running parallel with the face of the cliff, might give away. After withdrawing from such places, excited with the view I had got, I would say to myself, "Now don't go out on the verge again." But in the face of Yosemite scenery cautious remonstrance is vain; under its spell one's body seems to go where it likes with a will over which we seem to have scarce any control.

After a mile or so of this memorable cliff work I approached Yosemite Creek, admiring its easy, graceful, confident gestures as it comes bravely forward in its narrow channel, singing the last of its mountain songs on its way to its fate—a few rods more over the shining granite, then down half a mile in showy foam to another world, to be lost in the Merced, where climate, vegetation, inhabitants, all are different. Emerging from its last gorge, it glides in wide lace-like rapids down a smooth incline into a pool where it seems to rest and compose its gray, agitated waters before taking the grand plunge, then slowly slipping over the lip of the pool basin, it descends another glossy slope with rapidly accelerated speed to the brink of the tremendous cliff, and with sublime, fateful confidence springs out free in the air.

I took off my shoes and stockings and worked my way cautiously down alongside the rushing flood, keeping my feet and hands pressed firmly on the polished rock. The booming, roaring water, rushing past close to my head, was very exciting. I had expected that the sloping apron would terminate with the perpendicular wall of the valley, and that from the foot of it, where it is less steeply inclined, I should be able to lean far enough out to see the forms and behavior of the fall all the way down to the bottom. But I found that there was yet another small brow over which I could not see, and which appeared to be too steep for mortal feet. Scanning it keenly, I discovered a narrow shelf about three inches wide on the very brink, just wide enough for a rest for one's heels. But there seemed to be no way of reaching it over so steep a brow. At length, after careful scrutiny of the surface, I found an irregular edge of a flake of the rock some distance back from the margin of the torrent. If I was to get down to the brink at all that rough edge, which might offer slight finger-holds, was the only way. But the slope beside it looked dangerously smooth and steep, and the swift roaring flood beneath, overhead, and beside me was very nerve-trying. I therefore concluded not to venture farther, but did nevertheless. Tufts of artemisia were growing in clefts of the rock near by, and I filled my mouth with the bitter leaves, hoping they might help to prevent giddiness. Then, with a caution not known in ordinary circumstances, I crept down safely to the little ledge, got my heels well planted on it, then shuffled in a horizontal direction twenty or thirty feet until close to the outplunging current, which, by the time it had descended thus far, was already white. Here I obtained a perfectly free view down into the heart of the snowy, chanting throng of comet-like streamers, into which the body of the fall soon separates.

While perched on that narrow niche I was not distinctly conscious of danger. The tremendous grandeur of the fall in form and sound and motion, acting at close range, smothered the sense of fear, and in such places one's body takes keen care for safety on its own account. How long I remained down there, or how I returned, I can hardly tell. Anyhow I had a glorious time, and got back to camp about dark, enjoying triumphant exhilaration soon followed by dull weariness. Hereafter I'll try to keep from such extravagant nerve-straining places. Yet such a day is well worth venturing for. My first view of the High Sierra, first view looking down into Yosemite, the death song of Yosemite Creek, and its flight over the vast cliff, each one of these is of itself enough for a great life-long landscape fortune—a most memorable day of days—enjoyment enough to kill if that were possible.

from *Driftwood Valley*
Theodora C. Stanwell-Fletcher

In August 1937, Theodora C. Stanwell-Fletcher and her husband Jack began their trip to Driftwood Valley, in central British Columbia. There they would build a cabin and live with wilderness, with the animals and forces of nature they would encounter far from other human beings. Despite the smashed kidney Jack had suffered in Pennsylvania the previous winter, the two were adamant in their desire to experience true solitude in a building of their own construction.

After being educated in his native England, Jack experienced outdoor life in various parts of the world as an explorer, hunter, and trapper. Mrs. Stanwell-Fletcher's fascination with little-known parts of the earth ranging as far afield as the South Pacific and the sub-arctic tundra paralleled her husband's travels. Trained as a naturalist and a writer at Mt. Holyoke and Cornell, she enjoyed considerable success with her books The Tundra World *and* Clear Lands and Icy Seas. *She was awarded the John Burroughs Medal for her writing on natural history; her work here as she describes Driftwood Valley shows how well she accepted the challenges of the wild. Theodora Stanwell-Fletcher edited the diaries she kept at Lake Tetana to produce her book; here follow two important days in the journals of the stay by the uninhabited frozen lake.*

Christmas Eve *December 24*

We've had a long spell of warm, melting weather. Until today we almost despaired of seeing the sun again. The temperature has been 32 or 34 degrees above, and masses of snow plus warm air created a blanket-like fog and a dampness that seemed to stifle our very breathing. We could have been living on a waterless, treeless, hill-less plain for all the sign we ever saw of lake, forest, or mountain. Walking on snowshoes, which were heavy and sodden with melting snow, was exhausting; getting firewood an ordeal of strength and endurance.

There have, however, been a few partially clear nights when forests and hills and mountains, glittering with white crust, have appeared again for brief intervals. It has been full moon and, for a few hours at a time, the world was almost too lovely for mortal eyes. Shafts of moonlight, coming between clouds, turned the high edges of the Driftwoods to lines of white gold; lower slopes were deep and black. And luminous beams, like a searchlight, shot across the lake and forest. Sometimes one peak alone was all lit up, while others, soft, hazy dreams, were smudged with violet shadow.

Today we are greeted by a cold snap—just in time for Christmas. It is brilliantly clear and the mountains stand out once more—so distinct that they seem almost on top of the cabin. This afternoon the sun went down behind the

Driftwoods at two-fifteen and the mercury began to drop and drop. It went from 36 above to 39 below, a drop of 75 degrees in a few hours. I keep wondering how on earth our bodies can possibly adapt themselves to such extreme changes, but we appear to be in perfect health.

And the cabin, piled round with snow, now keeps warm with only a moderate fire in the stove, no matter what the temperatures. Tonight with the thermometer outside at 40 below, the thermometer inside is 59 above. During the few daylight hours when there is bright sunlight, we scarcely need even a small fire, because the sun, shining in all our windows, creates so much heat.

After supper, when stars were flashing above the snow piled to within a foot of the top windowpanes, I went outside to view the world.

"You won't feel cold," remarked J., "*at first*, but watch your lungs."

As I opened the door I wondered what he was talking about. With the first breath, I knew. I choked and gasped and sputtered. In this temperature one's breath freezes as one inhales and less oxygen than usual is taken into the lungs. Except for this, I simply was unconscious of the cold. By taking little short breaths I found that I could breathe sufficiently well. The snow underfoot was so hard that it didn't seem like snow at all. As I stepped on it, it tinkled musically like pieces of metal striking together.

In these very low temperatures, the air is crystal clear. Over the absolute stillness of the icy night, the stars looked as though they had come alive. These were not the serene, peaceful, far-off stars of summer skies; these were flashing and sparkling and burning, fanned by invisible fires to dazzling life. These were more brilliant than I had ever seen them anywhere, in the tropics or on high mountain tops; the light they shed across the earth was as revealing as clear moonlight. The white lake, the white mountains, the white forests, were glittering in their radiance. At the back of the cabin I saw the Great and Little Bears, the Big and Little Dippers, etched brilliantly and enormously on the sky. The Milky Way was not a narrow band of white light, but a broad twinkling path of individual shining stars stretched across the whole zenith. And to the south, most marvelous of all, was the giant Orion, followed by the Dog Star, Sirius, marching above the Driftwood Valley. The blue-white, moving, living fire of sirius seemed to light the whole of Lake Tetana. Other stars, flashing darts of red and blue and yellow, danced on the highest peaks and white knife-like edges of the mountains. I could *hear* the stars as they pulsed and moved above me.

When I realized suddenly that I was almost too stiff to move, I went in, and J., who has seen before the sky of an arctic night, smiled in understanding at the expression on my face. I thought of the verse from Paulus Gerhardt's old hymn:

> Now all the heav'nly splendor
> Breaks forth in starlight tender
> From myriad worlds unknown;
> And man, the marvel seeing,
> Forgets his selfish being,
> For joy of beauty not his own.

Christmas Day

Last night when we went to bed the windows on the inside were covered with frost an inch thick; the logs in the walls, and the shakes in the roof, cracked like gunshots, as they were split by the cold; and out on the lake the ice kept up an almost steady booming, interspersed with the horrid ripping and tearing that always makes my spine tingle. During the night I was waked repeatedly by such terrific cracks in the logs that I thought the cabin was coming down on our heads. When the temperature is falling, we expect a drop of 15 or 25 degrees during the night, beginning at sunset, but last night it broke all records.

This morning I was the first one out of bed. These days I can hardly wait to get up. Whether this is because I'm always hungry, or because the night is so long, I don't know, but it is refreshing to be able to jeer at J. as he lies lazily in bed with his morning smoke. Although when we went to bed last night we piled up the stove with slow-burning green wood, this morning the fire was practically out. It was still dark outside; dawn had not yet begun although it was long past eight. The windows were so densely frosted that it seemed as if daylight, even if it were there, could never penetrate the cabin. I lit the lamp, then carefully laid the pile of shavings, as always unfailingly prepared by J. the evening before, in the front of the stove and applied a match. After which I delicately laid on more shavings and then larger and larger sticks. Everything was ice cold and I was careful not to touch any metal with bare hands, having learned from bitter experience that skin, especially moist skin, freezes fast and is sometimes peeled right off at the slightest contact with very cold metal. Still clad in bathrobe and slippers, I went to scrape away the frost and read the thermometer outside. I realized then that, although I didn't feel chilled, I could hardly move my arms.

"I'll just see what the temperature is *inside* first," I thought, and went to peer at the thermometer hanging above our dining table. It read 25 below. Gosh! That couldn't be right! How could we sleep like that, how could I be wandering around with only a wrapper on? I *must* see what it was outside!

I couldn't even find the mercury. It dawned on me, after a time, that it had gone its limit and jammed at 50 below. My exclamations roused J. and we were so busy arguing over the thermometer that we forgot to say "Merry Christmas." I was convinced that if the mercury could have gone beyond 50 it would have read at least 60 below. J. said it was not much colder than 50; that he could tell from past experience in the Arctic just how many degrees the temperature drops when it gets beyond 45 below. For one thing, if it is 55 or more below, when hot water is poured out of a window, it freezes solid before it reaches the snow level. I could hardly wait until we had hot water to try this experiment. Sure enough—when, later, I poured a stream from the teakettle onto the snow outside, the water steamed and twisted into threads, but did not turn actually solid till it reached the ground. We must put out our unused thermometer whose scale goes to 60 below.

By the time dawn was coming we had scraped two peepholes in the frost on the panes; and we stood quiet to watch the winter sunrise. The radiant peaks of the Driftwoods, cut like white icing into pinnacles and rims against the apple-green sky, were brushed with pink, that, even as we watched, spread down and down and turned to gold. Rays of the rising sun, coming between the pointed firs of the east shore, stretched straight across the white lake, and as they touched it huge crystals, formed by the intense cold, burst into sparkling, scintillating light. The snow-bowed trees of the south and west shores were hung with diamonds; and finally the willows, around our cabin, were decked with jewels as large as robins' eggs that flashed red and green and blue. No Christmas trees decorated by human hands were ever so exquisite as the frosted trees of this northern forest. The sky turned to deep, deep blue, and the white world burst into dazzling, dancing colors as the sun topped the forest. The dippers, undismayed by a cold that froze dumb all other living things, broke into their joyous tinkling melody by the open water patch below the bank. And our first Christmas Day in the wilderness was upon us.

After a breakfast of canned grapefruit which we had been saving especially, and pancakes with the last of our syrup, also preserved scrupulously for Christmas, we did our usual chores. I cleaned the cabin and began a round of baking. In addition to bannock, which I bake daily either in the drum oven or in an open frypan on top of the stove, I made tarts of strawberry jam and a chocolate cake. As these favorite articles of diet make inroads on a meager supply of crisco, jam, and sugar we have them only for very special celebrations.

When I went outside to scatter crumbs for our furred and feathered friends, the jays were almost too stiff to move. Instead of flying down to snatch the food before it left my hand, they sat on the spruce branches, their feathers so fluffed up that I could hardly distinguish head from tail. Sometimes they moved near the smoke from our stovepipe, which was giving forth some warmth. The chickadees, tiny as they are, though also tremendously puffed out, were slightly more active than the jays.

Why are some animals in this country so much better equipped for cold than others? There are the dippers, for example, whose nerves and organs are, seemingly, completely unaffected by an almost arctic temperature. Twice, during this month, we've watched a pair performing the act of coition on a snowbank.

Toward noon the temperature moderated enough for us to enjoy a tramp. That is, it had gone up from 50, or whatever below it was, to 36 below. Our snowshoes tossed up clouds of crystals. Young trees which, in autumn, had reached above our heads had been completely covered with fresh snow, so that they were transformed into great mounds and small hills. Wherever we looked our eyes were dimmed by the twinkling brilliants scattered before us. The azure of the sky above, the unsullied whiteness below, the mountains and the woods, the intense pureness of the air, were exhilarating beyond imagining. And there was not a sound or a motion, anywhere, to distract our senses of sight and feeling.

Soon after noon the temperature began dropping again, fast. Our faces, which we rubbed constantly with wool mitts, began to show a tendency toward frostbite, and J.'s right big toe, once badly frozen in arctic tundras, was starting to pain severely. So we turned homeward.

As daylight faded, the rays of the sinking sun tinted the snow with red and lavender. The mountains grew purple and then came that period which, if I could make a choice of the wonders of all the twenty-four hours of a winter's day, seems the most wonderful of all. It is that moment of white twilight which comes on a particularly clear afternoon, after the last colors of sunset fade and just before the first stars shine out. I don't suppose its like can be seen anywhere except in the snowbound, ice-cold arctic places. Everything in the universe becomes a luminous white. Even the dark trees of the forest, and the sky overhead, are completely colorless. It is the ultimate perfection of purity and peace. But even as one looks and wonders, the white sky takes on a faint pale green, there are the stars, and then the great winter's night is upon one.

We had our Christmas dinner at five: dehydrated potatoes and onions and a bit of moose steak, especially saved and tendered, baked in a pan with stuffing. For dessert there were the jam tarts and chocolate cake. With these vanished the last vestiges of Christmas, the things which made it a little different from our other days.

Have we greatly missed the things that make Christmas Day in civilization? Other loved human beings, Christmas carols, wonderful food? I suppose so, but I think that this lack is more than made up for by the deep contentment of our healthy minds and bodies, by our closeness to and awareness of the earth, and of each other.

RESPONDING JOURNAL

Which of these journals seems most like one that you would keep yourself? Why? Explain your reasons for selecting one over other possible choices. Write your response in the form of a freewriting entry in your journal.

QUESTIONS FOR CRITICAL THINKING

STRATEGY

1. What seem to you to be the identifying characteristics of this group of journal entries? On what topics do the writers focus? How polished do their entries appear to be? Jot down a list of these qualities, and then rank them in order of importance.
2. Jot down a brief statement about who you judge to be the audience for the journal entries that you have read. What clues do you find to suggest who the author has in mind as he or she writes? Does the writer seem to feel a need to write, to communicate with the journal itself?

1. Think about the attitudes toward the land that you find here. What seem to be the important concerns? Do you find any consistent patterns of response? As a nation, to what extent have we followed or not followed the spirit of these journalkeepers?
2. How do you feel about issues beyond the landscape being treated in the journals? For example, consider the difficulties of home life for Davis or the exposition of bird activities in Stanwell-Fletcher's journal. Does the landscape seem to call forth such thinking, or are the journalkeepers simply writing about what matters to them at the moment?

COLLABORATIVE WRITING ACTIVITY

Discuss the interest factor in the early journal selections that you have read. What qualities do they possess that seem to make them good writing? In turn, allow each member of the group to suggest what he or she feels to be the most compelling part of the journal writing, and then open the passage for discussion by the group. Try to see why a particular passage would be of interest to some people, and why it might not be so interesting for others.

MEMOIRS

Memoirs provide opportunity to look back over a part of your life and explore not only the experience but your reaction to it. You may begin with scraps of information from journals, diaries, or even your memory. You may flesh out your memoir with research at levels ranging from scanning newspapers to full-scale archival explorations. In order to suggest for your reader the importance of the narrative you are providing, you will comment on the significance of the events recorded and explain what you learned from the process. The "confession" is a form of writing parallel to the memoir, though confessions often carry the connotation of something racy. A memoir is a more neutral and therefore a more flexible form of writing.

EXPLORING JOURNAL

Try to imagine a landscape in your experience that suggests to you a powerful emotion. Is there a place that you associate with the love of a parent, the short-comings of a parent or relative, or the experience of being separated from a family member you love? How about a general quality that you associate with a particular place—say the depression that may come from growing up in poverty, or the feeling that you are associated by some people with a place that brands you as "inferior" to others? Explore this idea in a journal entry; give yourself at least ten minutes to freewrite after you have thought enough to come up with a likely topic.

⇒·◦·⇐

"Stoops"
Laura Green

In 1991, Laura Green and five collaborators published Reinventing Home: Six Working Women Look at Their Home Lives. *In this fascinating collection of memoirs written by the six women, they present different versions of what* home *means to women who have to work outside the home and still find time to create a space that reflects their concept of* home *for themselves and their families. Green and her co-editor, Carroll Stoner, tell their readers in the preface that "This book is an invitation, if not a plea, to reconsider domestic life and reinvent the home as a place to nourish the scenes we've so carefully created" (p. 4). Home means our first scene, our first landscape; in this selection Laura Green offers a memoir that suggests how merely sitting on front-porch stoops with neighbors can create for us a sense of community, of belonging to a group of people who provide a sense of home within larger social units.*

When I moved to Chicago, I lived in a neighborhood that was as densely packed as Paris and just as charming in its own solid, redbrick way. When I lived there, the rents were low; everyone was young and starting out, or old and in the last place they would live before the nursing home. I was drawn by the architecture, a confident hodgepodge of turrets, carvings, gargoyles, and bays. I loved the quirky front yards with English ivy and hosta spilling onto the brick walks and the sense of the past from the gas jets and wood paneling that remained in some of the converted parlors. The apartments were cramped, but they rarely stayed on the market more than a week because atmosphere, not space, was what my friends and I coveted. Instead of long living rooms we didn't have furniture for, we got original stained-glass windows, big cottonwoods that arched over the sidewalks, and front stoops.

If baseball is the American pastime, then "stoop-sitting," as they call it in Chicago, is the neighborhood pastime, a pleasure as simple as music. If there is one event, one tradition that turns a block into a community, sitting on the stoop is it.

On our block, stoop etiquette was as rigid as a tea ceremony. We met on one stoop only, the steep wooden steps of a woman who grew up in one of those tight xenophobic Chicago neighborhoods that gave birth to stoop-sitting as a balm for life in the factories and stockyards. Her house was on the east side of the street, where the stoop caught the late afternoon sun. Because she was the owner and not a tenant, it was okay for the rest of us to ease our tired hams on her wooden steps. We met there and nowhere else. We congregated only at certain times of the day and only if she was already there.

It would have been unthinkable to gather on Carol's stoop without Carol. We rarely went to Inge's, and we never met on Judy's porch or my porch, which lay in deep, cold, five o'clock shade. We weren't comfortable there. For one thing, my landlord sometimes sat at the grand piano in the first-floor bay window above our heads. Wearing only his boxer shorts, he banged out show tunes, as oblivious of us as we were aware of him. Even if he had been out in the garage puttering in his workshop, my landlady, a transplant from the suburbs, would have been uncomfortable with us. She was a nice person, but she observed what I have come to think of as suburban proprieties. We were expected to keep our distance. It would have been an invasion of her privacy for anyone, tenant or not, to sit on her front steps. She never would have called it a stoop; the idea of a stoop was alien to her.

Not to me. I took to stoop-sitting the way a small boy takes to dirt. I would have sat all day if I could, but that was out of the question. We never began before about 5 P.M., when everyone was returning to the block. The mothers who didn't work were finishing the rounds of kids' activities. The ones like me, who did work, were trudging home from the office. I collected the baby from the babysitter and often, without dropping my stuff off at my place (we lived on the third floor), plunked down on Carol's stoop to unwind and regroup.

Stoop-sitting is one of the few ways to bring the unhurried intimacy of domestic life outdoors. There's no hurry, no pressure to talk, no need to dress up. You just sit and drink your coffee, or read your mail, or balance your checkbook. You can be your unembellished self on someone's front stoop in the same way that you can be in your own living room.

Early evenings were for the women because we were the first ones home. Carol and I sat in the late sun and had long theoretical discussions about the psychology of raising children and the Freudian ebbing and flowing of our friends' lives. From time to time we would be interrupted by our neighbor down the street, who was in her mid-twenties and terribly earnest, a woman who boned up for dinner parties by reading the *New York Times.* "What do you think of intermediate-range missiles?" she would ask brightly, one or two toddlers clinging to her shorts. We thought she was a dolt, having mastered the art of dinner party discussions many years earlier.

Eventually we took the children home for dinner (something ethnic with lots of grains in it), then came back again with our coffee and our families. Conversationally, one of the great things about living in Chicago is its inescapable, Byzantine politics that reach right down to the street corner. Chicago politics is the one area where trickle-down theories produce floods of events. Its intricacy, its corruption, and its endless, awful local ramifications gave us plenty to talk about. As it grew darker and the fireflies came out, we nursed our coffee while the kids rode up and down the block on their tricycles, fell, bled, fought, giggled, got bitten by mosquitoes, and had a fine time.

Twelve years ago we had a second baby and moved to a bigger place on a busy street. There were no stoops worthy of the name, but the apartment complex had something almost as good—a communal yard with a redbrick patio, several picnic tables, and a few crusty grills. We must have looked crazy grilling hamburgers a few steps from the traffic under the shadow of the curved high-rise across the street that looked like it belonged on a Caribbean beach. We gossiped, drank our beer, and ate our dinners in a fog of car exhaust as the kids wobbled by on their new two-wheelers.

The patio wasn't much different in spirit from Carol's stoop. We lived elbow to elbow and got to know one another in an extended-family, neighbor-hood way. When my son was a toddler, he called the place "my village." He got it right, the way kids do. The complex held four generations, including a hand-ful of obdurate misanthropes and one or two nasty eccentrics who drew us together by being people nearly all of us could hate. All the place really need-ed to be like a village was a well, a few steep, cobbled streets, and three or four skulking dogs.

It has been years since I walked across the street to drink the last cup of coffee I would have on Carol's front stoop. The toddlers are now in high school. The crowd has scattered, to Florida, to Washington, to New York, split and re-formed through divorce and remarriage. Carol lives in another city house with steep front steps, but now she sits on her back porch, where she can be alone and enjoy her privacy. She is still my good friend, but it is just not the same when you have to meet in restaurants—as we did for breakfast for a full year after I moved away.

The sad truth is that none of us have much time for sitting on stoops, which takes an hour here and a half hour there if you're going to do it right. An indoor domestic life is all I can manage. Stoop life had time built into it for talk and for the quiet that lets bonds knit. When you invite people into your house, you have an obligation to entertain them, but sitting on a stoop is like pot luck. It just happens. It recharges the batteries in a more natural way than an evening dinner party and sends us home feeling connected to the human village, wher-ever it is, that our deepest memories come from.

I miss it.

"Memoirs of an Indian Girlhood"
Zitkala-Sa

Zitkala-Sa, or Red Bird, was actually born Gertrude Simmons in 1876 and raised on the Yankton Sioux reservation in South Dakota. As she explains in the following memoirs, she left her Sioux home to board at a Quaker school when she was eight. Her strong sense of connection to her land and people, however, brought her back to reservation life after she completed her formal education. Zitkala-Sa devoted much of her life to improving life for Indians, working both on reservations and later with the Society of American Indians and the National Council of American Indians, which she founded in 1926 in Washington, D. C.

She published two books, Old Indian Legends *in 1901 and* American Indian Stories *in 1921, in addition to the autobiographical material reprinted here from the 1900* Atlantic Monthly. *She would certainly have written more if she had been less of a political activist, but her substantial contributions to more enlightened treatment for Indians often prevented her from writing. During her lifetime, much of her work was published under her married name, Gertrude Simmons Bonnin.*

IMPRESSIONS OF AN INDIAN CHILDHOOD

I. MY MOTHER

A wigwam of weather-stained canvas stood at the base of some irregularly ascending hills. A footpath wound its way gently down the sloping land till it reached the broad river bottom; creeping through the long swamp grasses that bent over it on either side, it came out on the edge of the Missouri.

Here, morning, noon, and evening, my mother came to draw water from the muddy stream for our household use. Always, when my mother started for the river, I stopped my play to run along with her. She was only of medium height. Often she was sad and silent, at which times her full arched lips were compressed into hard and bitter lines, and shadows fell under her black eyes. Then I clung to her hand and begged to know what made the tears fall.

"Hush; my little daughter must never talk about my tears;" and smiling through them, she patted my head and said, "Now let me see how fast you can run to-day." Whereupon I tore away at my highest possible speed with my long black hair blowing in the breeze.

I was a wild little girl of seven. Loosely clad in a slip of brown buckskin, and light-footed with a pair of soft moccasins on my feet, I was as free as the wind that blew my hair, and no less spirited than a bounding deer. These were my mother's pride—my wild freedom and overflowing spirits. She taught me no fear save that of intruding myself upon others.

Having gone many paces ahead I stopped, panting for breath, and laughing with glee as my mother watched my every movement. I was not wholly conscious of myself, but was more keenly alive to the fire within. It was as if I were the activity, and my hands and feet were only experiments for my spirit to work upon.

Returning from the river, I tugged beside my mother, with my hand upon the bucket I believed I was carrying. One time, on such a return, I remember a bit of conversation we had. My grown-up cousin, Warca-Ziwin (Sunflower), who was then seventeen, always went to the river alone for water for her mother. Their wigwam was not far from ours; and I saw her daily going to and from the river. I admired my cousin greatly. So I said: "Mother, when I am tall as my cousin Warca-Ziwin, you shall not have to come for water. I will do it for you."

With a strange tremor in her voice which I could not understand, she answered, "If the paleface does not take away from us the river we drink."

"Mother, who is this bad paleface?" I asked.

"My little daughter, he is a sham—a sickly sham! The bronzed Dakota is the only real man."

I looked up into my mother's face while she spoke; and seeing her bite her lips, I knew she was unhappy. This aroused revenge in my small soul. Stamping my foot on the earth, I cried aloud, "I hate the paleface that makes my mother cry!"

Setting the pail of water on the ground, my mother stooped, and stretching her left hand out on the level with my eyes, she placed her other arm about me; she pointed to the hill where my uncle and my only sister lay buried.

"There is what the paleface has done! Since then your father too has been buried in a hill nearer the rising sun. We were once very happy. But the paleface has stolen our lands and driven us hither. Having defrauded us of our land, the paleface forced us away.

"Well, it happened on the day we moved camp that your sister and uncle were both very sick. Many others were ailing, but there seemed to be no help. We traveled many days and nights; not in the grand happy way that we moved camp when I was a little girl, but we were driven, my child, driven like a herd of buffalo. With every step, your sister, who was not as large as you are now, shrieked with the painful jar until she was hoarse with crying. She grew more and more feverish. Her little hands and cheeks were burning hot. Her little lips were parched and dry, but she would not drink the water I gave her. Then I discovered that her throat was swollen and red. My poor child, how I cried with her because the Great Spirit had forgotten us!

"At last, when we reached this western country, on the first weary night your sister died. And soon your uncle died also, leaving a widow and an orphan daughter, your cousin Warca-Ziwin. Both your sister and uncle might have been happy with us to-day, had it not been for the heartless paleface."

My mother was silent the rest of the way to our wigwam. Though I saw no tears in her eyes, I knew that was because I was with her. She seldom wept before me. . . .

VI. THE GROUND SQUIRREL

In the busy autumn days, my cousin Warca-Ziwin's mother came to our wigwam to help my mother preserve foods for our winter use. I was very fond of my aunt, because she was not so quiet as my mother. Though she was older, she was more jovial and less reserved. She was slender and remarkably erect. While my mother's hair was heavy and black, my aunt had unusually thin locks.

Ever since I knew her, she wore a string of large blue beads around her neck—beads that were precious because my uncle had given them to her when she was a younger woman. She had a peculiar swing in her gait, caused by a long stride rarely natural to so slight a figure. It was during my aunt's visit with us that my mother forgot her accustomed quietness, often laughing heartily at some of my aunt's witty remarks.

I loved my aunt threefold: for her hearty laughter, for the cheerfulness she caused my mother, and most of all for the times she dried my tears and held me in her lap, when my mother had reproved me.

Early in the cool mornings, just as the yellow rim of the sun rose above the hills, we were up and eating our breakfast. We awoke so early that we saw the sacred hour when a misty smoke hung over a pit surrounded by an impassable sinking mire. This strange smoke appeared every morning, both winter and summer; but most visibly in midwinter it rose immediately above the marshy spot. By the time the full face of the sun appeared above the eastern horizon, the smoke vanished. Even very old men, who had known this country the longest, said that the smoke from this pit had never failed a single day to rise heavenward.

As I frolicked about our dwelling, I used to stop suddenly, and with a fearful awe watch the smoking of the unknown fires. While the vapor was visible, I was afraid to go very far from our wigwam unless I went with my mother.

From a field in the fertile river bottom my mother and aunt gathered an abundant supply of corn. Near our tepee, they spread a large canvas upon the grass and dried their sweet corn in it. I was left to watch the corn, that nothing should disturb it. I played around it with dolls made of ears of corn. I braided their soft fine silk for hair, and gave them blankets as various as the scraps I found in my mother's workbag.

There was a little stranger with a black-and-yellow-striped coat that used to come to the drying corn. It was a little ground squirrel, who was so fearless of me that he came to one corner of the canvas and carried away as much of the sweet corn as he could hold. I wanted very much to catch him, and rub his pretty fur back, but my mother said he would be so frightened if I caught him that he would bite my fingers. So I was as content as he to keep the corn between us. Every morning he came for more corn. Some evenings I have seen him creeping about our grounds; and when I gave a sudden whoop of recognition, he ran quickly out of sight.

When mother had dried all the corn she wished, then she sliced great pumpkins into thin rings; and these she doubled and linked together into long chains. She hung them on a pole that stretched between two forked posts. The wind and sun soon thoroughly dried the chains of pumpkin. Then she packed them away in a case of thick and stiff buckskin.

In the sun and wind she also dried many wild fruits—cherries, berries, and plums. But chiefest among my early recollections of autumn is that one of the corn drying and the ground squirrel.

I have few memories of winter days, at this period of my life, though many of the summer. There is one only which I can recall.

Some missionaries gave me a little bag of marbles. They were all sizes and colors. Among them were some of colored glass. Walking with my mother to the river, on a late winter day, we found great chunks of ice piled all along the bank. The ice on the river was floating in huge pieces. As I stood beside one large block, I noticed for the first time the colors of the rainbow in the crystal ice. Immediately I thought of my glass marbles at home. With my bare fingers I tried to pick out some of the colors, for they seemed so near the surface. But my fingers began to sting with the intense cold, and I had to bite them hard to keep from crying.

From that day on, for many a moon, I believed that glass marbles had river ice inside of them.

VII. THE BIG RED APPLES

The first turning away from the easy, natural flow of my life occurred in an early spring. It was in my eighth year; in the month of March, I afterward learned. At this age I knew but one language and that was my mother's native tongue.

From some of my playmates I heard that two paleface missionaries were in our village. They were from that class of white men who wore big hats and carried large hearts, they said. Running direct to my mother, I began to question her why these two strangers were among us. She told me, after I had teased much, that they had come to take away Indian boys and girls to the East. My mother did not seem to want me to talk about them. But in a day or two, I gleaned many wonderful stories from my playfellows concerning the strangers.

"Mother, my friend Judéwin is going home with the missionaries. She is going to a more beautiful country than ours; the palefaces told her so!" I said wistfully, wishing in my heart that I too might go.

Mother sat in a chair, and I was hanging on her knee. Within the last two seasons my big brother Dawée had returned from a three years' education in the East, and his coming back influenced my mother to take a farther step from her native way of living. First it was a change from the buffalo skin to the white man's canvas that covered our wigwam. Now she had given up her wigwam of slender poles, to live, a foreigner, in a home of clumsy logs.

"Yes, my child, several others besides Judéwin are going away with the palefaces. Your brother said the missionaries had inquired about his little sister," she said, watching my face very closely.

My heart thumped so hard against my breast, I wondered if she could hear it.

"Did he tell them to take me, mother?" I asked, fearing lest Dawée had forbidden the palefaces to see me, and that my hope of going to the Wonderland would be entirely blighted.

With a sad, slow smile, she answered: "There! I knew you were wishing to go, because Judéwin has filled your ears with the white men's lies. Don't believe a word they say! Their words are sweet, but, my child, their deeds are bitter. You will cry for me, but they will not even soothe you. Stay with me, my little one! Your brother Dawée says that going East, away from your mother, is too hard an experience for his baby sister."

Thus my mother discouraged my curiosity about the lands beyond our eastern horizon; for it was not yet an ambition for Letters that was stirring me. But on the following day the missionaries did come to our very house. I spied them coming up the footpath leading to our cottage. A third man was with them, but he was not my brother Dawée. It was another, a young interpreter, a paleface who had a smattering of the Indian language. I was ready to run out to meet them, but I did not dare to displease my mother. With great glee, I jumped up and down on our ground floor. I begged my mother to open the door, that they would be sure to come to us. Alas! They came, they saw, and they conquered!

Judéwin had told me of the great tree where grew red, red apples; and how we could reach out our hands and pick all the red apples we could eat. I had never seen apple trees. I had never tasted more than a dozen red apples in my life; and when I heard of the orchards of the East, I was eager to roam among them. The missionaries smiled into my eyes, and patted my head. I wondered how mother could say such hard words against them.

"Mother, ask them if little girls may have all the red apples they want, when they go East," I whispered aloud, in my excitement.

The interpreter heard me and answered: "Yes, little girl, the nice red apples are for those who pick them; and you will have a ride on the iron horse if you go with these good people."

I had never seen a train, and he knew it.

"Mother, I'm going East! I like big red apples, and I want to ride on the iron horse! Mother, say yes!" I pleaded.

My mother said nothing. The missionaries waited in silence; and my eyes began to blur with tears, though I struggled to choke them back. The corners of my mouth twitched, and my mother saw me.

"I am not ready to give you any word," she said to them. "To-morrow I shall send you my answer by my son."

With this they left us. Alone with my mother, I yielded to my tears, and cried aloud, shaking my head so as not to hear what she was saying to me. This was the first time I had ever been so unwilling to give up my own desire that I refused to hearken to my mother's voice.

There was a solemn silence in our home that night. Before I went to bed I begged the Great Spirit to make my mother willing I should go with the missionaries.

The next morning came, and my mother called me to her side. "My daughter, do you still persist in wishing to leave your mother?" she asked.

"Oh, mother, it is not that I wish to leave you, but I want to see the wonderful Eastern land," I answered.

My dear old aunt came to our house that morning, and I heard her say "Let her try it."

I hoped that, as usual, my aunt was pleading on my side. My brother Dawée came for mother's decision. I dropped my play, and crept close to my aunt.

"Yes, Dawée, my daughter, though she does not understand what it all means, is anxious to go. She will need an education when she is grown, for then there will be fewer real Dakotas, and many more palefaces. This tearing her away, so young, from her mother is necessary, if I would have her an educated woman. The palefaces, who owe us a large debt for stolen lands, have begun to pay a tardy justice in offering some education to our children. But I know my daughter must suffer keenly in this experiment. For her sake, I dread to tell you my reply to the missionaries. Go, tell them that they may take my little daughter, and that the Great Spirit shall not fail to reward them according to their hearts."

Wrapped in my heavy blanket, I walked with my mother to the carriage that was soon to take us to the iron horse. I was happy. I met my playmates, who were also wearing their best thick blankets. We showed one another our new beaded moccasins, and the width of the belts that girdled our new dresses. Soon we were being drawn rapidly away by the white man's horses. When I saw the lonely figure of my mother vanish in the distance, a sense of regret settled heavily upon me. I felt suddenly weak, as if I might fall limp to the ground. I was in the hands of strangers whom my mother did not fully trust. I no longer felt free to be myself, or to voice my own feelings. The tears trickled down my cheeks, and I buried my face in the folds of my blanket. Now the first step, parting me from my mother, was taken, and all my belated tears availed nothing.

Having driven thirty miles to the ferryboat, we crossed the Missouri in the evening. Then riding again a few miles eastward, we stopped before a massive brick building. I looked at it in amazement, and with a vague misgiving, for in our village I had never seen so large a house. Trembling with fear and distrust of the palefaces, my teeth chattering from the chilly ride, I crept noiselessly in my soft moccasins along the narrow hall, keeping very close to the bare wall. I was as frightened and bewildered as the captured young of a wild creature.

THE SCHOOL DAYS OF AN INDIAN GIRL

I. THE LAND OF RED APPLES

There were eight in our party of bronzed children who were going East with the missionaries. Among us were three young braves, two tall girls, and we three little ones, Judéwin, Thowin, and I.

We had been very impatient to start on our journey to the Red Apple Country, which, we were told, lay a little beyond the great circular horizon of the Western prairie. Under a sky of rosy apples we dreamt of roaming as freely and happily as we had chased the cloud shadows on the Dakota plains. We had anticipated much pleasure from a ride on the iron horse, but the throngs of staring palefaces disturbed and troubled us.

On the train, fair women, with tottering babies on each arm, stopped their haste and scrutinized the children of absent mothers. Large men, with heavy bundles in their hands, halted near by, and riveted their glassy blue eyes upon us.

I sank deep into the corner of my seat, for I resented being watched. Directly in front of me, children who were no larger than I hung themselves upon the backs of their seats, with their bold white faces toward me. Sometimes they took their forefingers out of their mouths and pointed at my moccasined feet. Their mothers, instead of reproving such rude curiosity, looked closely at me, and attracted their children's further notice to my blanket. This embarrassed me, and kept me constantly on the verge of tears.

I sat perfectly still, with my eyes downcast, daring only now and then to shoot long glances around me. Chancing to turn to the window at my side, I was quite breathless upon seeing one familiar object. It was the telegraph pole which strode by at short paces. Very near my mother's dwelling, along the edge of a road thickly bordered with wild sunflowers, some poles like these had been planted by white men. Often I had stopped, on my way down the road, to hold my ear against the pole, and, hearing its low moaning, I used to wonder what the paleface had done to hurt it. Now I sat watching for each pole that glided by to be the last one.

In this way I had forgotten my uncomfortable surroundings, when I heard one of my comrades call out my name. I saw the missionary standing very near, tossing candies and gums into our midst. This amused us all, and we tried to see who could catch the most of the sweetmeats. The missionary's generous distribution of candies was impressed upon my memory by a disastrous result which followed. I had caught more than my share of candies and gums, and soon after our arrival at the school I had a chance to disgrace myself, which, I am ashamed to say, I did.

Though we rode several days inside of the iron horse, I do not recall a single thing about our luncheons.

It was night when we reached the school grounds. The lights from the windows of the large buildings fell upon some of the icicled trees that stood beneath them. We were led toward an open door, where the brightness of the lights within flooded out over the heads of the excited palefaces who blocked the way. My body trembled more from fear than from the snow I trod upon.

Entering the house, I stood close against the wall. The strong glaring light in the large whitewashed room dazzled my eyes. The noisy hurrying of hard shoes upon a bare wooden floor increased the whirring in my ears. My only safety seemed to be in keeping next to the wall. As I was wondering in which direction to escape from all this confusion, two warm hands grasped me firmly, and in the same moment I was tossed high in midair. A rosy-cheeked paleface woman caught me in her arms. I was both frightened and insulted by such trifling. I stared into her eyes, wishing her to let me stand on my own feet, but she jumped me up and down with increasing enthusiasm. My mother had never made a plaything of her wee daughter. Remembering this I began to cry aloud.

They misunderstood the cause of my tears, and placed me at a white table loaded with food. There our party were united again. As I did not hush my crying, one of the older ones whispered to me, "Wait until you are alone in the night."

It was very little I could swallow besides my sobs that evening.

"Oh, I want my mother and my brother Dawée! I want to go to my aunt!" I pleaded; but the ears of the palefaces could not hear me.

From the table we were taken along an upward incline of wooden boxes, which I learned afterward to call a stairway. At the top was a quiet hall, dimly lighted. Many narrow beds were in one straight line down the entire length of the wall. In them lay sleeping brown faces, which peeped just out of the coverings. I was tucked into bed with one of the tall girls, because she talked to me in my mother tongue and seemed to soothe me.

I had arrived in the wonderful land of rosy skies, but I was not happy, as I had thought I should be. My long travel and the bewildering sights had exhausted me. I fell asleep, heaving deep, tired sobs. My tears were left to dry themselves in streaks, because neither my aunt nor my mother was near to wipe them away.

"Everything the Kiowa Had Came from the Buffalo"
Old Lady Horse

By the 1870s, widespread slaughter of the buffalo on the Great Plains put tremendous strains on the traditional lifestyles of the Plains Indians. Not only were white settlers edging forever closer to Native American villages and centuries-old sacred grounds, but these new arrivals were also eliminating the main source of food and clothing for Plains Indians: the great buffalo herds that once roamed the region. Unimaginable to the Indians, the white man wasted not only buffalo meat and hides, but often took only the tongue and even killed the buffalo purely for sport.

Old Lady Horse was a girl when the great buffalo hunts were going on, and she recalls how horribly these "adventures" affected her people. She provides in the second half of her memoir an account of a recurrent folktale motif—the paradise inside the mountain. Notice how in her memoir Old Lady Horse moves so easily from describing human behavior to describing buffalo behavior as dignified and resigned.

Everything the Kiowas had came from the buffalo. Their tipis were made of buffalo hides, so were their clothes and moccasins. They ate buffalo meat.

Their containers were made of hide, or of bladders or stomachs. The buffalo were the life of the Kiowas.

Most of all, the buffalo was part of the Kiowa religion. A white buffalo calf must be sacrificed in the Sun Dance. The priests used parts of the buffalo to make their prayers when they healed people or when they sang to the powers above.

So, when the white men wanted to build railroads, or when they wanted to farm or raise cattle, the buffalo still protected the Kiowas. They tore up the railroad tracks and the gardens. They chased the cattle off the ranges. The buffalo loved their people as much as the Kiowas loved them.

There was war between the buffalo and the white men. The white men built forts in the Kiowa country, and the woolly-headed buffalo soldiers shot the buffalo as fast as they could, but the buffalo kept coming on, coming on, even into the post cemetery at Fort Sill. Soldiers were not enough to hold them back.

Then the white men hired hunters to do nothing but kill the buffalo. Up and down the plains those men ranged, shooting sometimes as many as a hundred buffalo a day. Behind them came the skinners with their wagons. They piled the hides and bones into the wagons until they were full, and then took their loads to the new railroad stations that were being built, to be shipped east to the market. Sometimes there would be a pile of bones as high as a man, stretching a mile along the railroad track.

The buffalo saw that their day was over. They could protect their people no longer. Sadly, the last remnant of the great herd gathered in council, and decided what they would do.

The Kiowas were camped on the north side of Mount Scott, those of them who were still free to camp. One young woman got up very early in the morning. The dawn mist was still rising from Medicine Creek, and as she looked across the water, peering through the haze, she saw the last buffalo herd appear like a spirit dream.

Straight to Mount Scott the leader of the herd walked. Behind him came the cows and their calves, and the few young males who had survived. As the woman watched, the face of the mountain opened.

Inside Mount Scott the world was green and fresh, as it had been when she was a small girl. The rivers ran clear, not red. The wild plums were in blossom, chasing the red buds up the inside slopes. Into this world of beauty the buffalo walked, never to be seen again.

RESPONDING JOURNAL

As you look back over these memoirs, think about the different reasons people have for writing memoirs. Describe the specific qualities of each memoir you have read. In a journal entry of approximately two pages, first list the hallmarks of memoirs and then explain which of these qualities you would feel most comfortable adapting as part of your writing techniques. Try to think at least as much about the *method* of writing as you do about the *subject* of the writing.

QUESTIONS FOR CRITICAL THINKING

STRATEGY

1. What constitutes a memoir? What do these selections have in common? What would you want to include in your own memoir that these pieces suggest?
2. Describe the differences in structure and writing style that you find among these memoirs. What seems to be the main distinguishing feature?

ISSUES

1. Describe the sense of relationship among home, family, and writer that you see in these memoirs. Do all of the writers value both home and family equally?
2. Laura Green describes a situation that may be changing. Does stoop-sitting still exist in your neighborhood? The question assumes that people sat on stoops in every neighborhood at some point; is this assumption correct? Do you live in a neighborhood where stoop-sitting is a thing of the past? What accounts for this change?

COLLABORATIVE WRITING ACTIVITY

Sketch a history of stoop-sitting in your neighborhood, or of some other neighborhood or community custom. What research will you have to do to find this history? Who would be some good sources? For every change that you know (or can guess) has occurred, provide an explanation. Read your history to your group. After you have heard all the histories, decide whether you can detect persistent patterns of development.

ORAL HISTORY

Oral history—the recollections of people on certain points in the past—serves its users in geography, history, anthropology, sociology, ecology, and other disciplines. The subjects who provide the data in oral history interviews may not have directly experienced all of the situations they describe because they may retell what other people have told them. Such information reveals what ideas and traditions are passed down, and in turn these ideas demonstrate what some people find important in a community or family.

The following guidelines list the sorts of questions that will lead to the basic information necessary to preserve recollections of the details of community and family history.

1. *Events* What events of family, local, or world importance do your subjects recall? What was their relation to these events? Can they suggest other local people who might be able to provide additional information?
2. *Stories* What are the important stories in your family or community? Are there stories of fortunes won or lost? Of courtships? Of colorful events in the lives of colorful people? Of family members or local people who achieved some sort of fame that went beyond your immediate area? Of places that have changed over time? Can you verify those parts of stories relating directly to world events, such as World War I or II, the Great Depression of the 1930s, or extending women the right to vote?
3. *People* Who else in your family or area might have the sort of knowledge you seek? How about your extended family, your great-aunts and uncles, for example, or your second and third cousins? Their families? Present and former neighbors?
4. *Names* Tell what you know about the names in your family. Have any first names or middle names been passed down in your family? Do you carry the name of anyone in your family? Are there any stories about the first or middle names in your family?

EXPLORING JOURNAL

What could you tell an interviewer living fifty years from now regarding what was important about where you live right now? What descriptions of where you live would help people fifty years in the future to understand what life is like now for you? Give this imaginary interviewer all the necessary information to understand you as a person living at this particular time in this particular place.

Interview with Mr. Bryce McAlister on Savoy Heights, 11/17/93

Savoy Heights, a residential area of Fayetteville, North Carolina, is located in the southwestern corner of the city. As the following interview indicates, most of the inhabitants of Savoy Heights are black; many residents are related to people who once worked in the mills that provided employment in the area from the turn of the century through 1938.

During early 1993, several drug-related incidents involving shootings and beatings were reported in local newspapers. Some reports indicated that murders and drug busts had occurred in Savoy Heights, when in fact the incidents occurred in adjoining neighborhoods. Just north of Savoy Heights is an area called Branson street, a decaying area that extends to the downtown area, also called Hay Street for the main street in the old city. Residents of Savoy Heights complained that their neighborhood, which they maintain carefully and where the sort of Branson Street violence reported in the papers does not occur, was being maligned. Perhaps reporters were not being sufficiently careful to determine boundaries of neighborhoods before they filed their stories.

In an effort to understand better how Savoy Heights residents felt about their neighborhood, I interviewed several people. My long-range objective was first to understand and then to communicate a sense of how the neighborhood is a living, growing, nurturing resource in the community. The following interview shows a life-long resident of the area who believes strongly in maintaining a sense of what has happened in the neighborhood and how people have lived their lives on this site.

VALENTI: What are some of the changes that you've seen here?

McALISTER: Well, this area is known as Silk Mill Hill. The mill was the Central Weaving Company; it was started about 1890 and it had all black workers. My mother worked there thirty years; she was thirteen when she started. They worked ten hours a day, six days a week, for $15 every two weeks.

Back then, the man who owned the mill was Dingle Ashley. Ashley School was named for him. His partner, Weiss, built fifty houses on Old Plank Road—what's now Turnpike Road—and Weiss Avenue was named for him. The rent on the houses was $1.50 a week. These two men owned all of where Savoy Heights is now, and it was called Ashley Heights. There were maybe three houses there then. In 1942, they built Ashley School across Robeson St. from where Savoy Heights is now; later they moved the school to where it is now.

In 1939, they closed the mill down. By then, Mack Lefnovich and Lou Anatola owned it. Mr. Ashley and Mr. Weiss had sold it.

VALENTI: When did they sell it?

McALISTER: About 1930. In 1939, they had two shifts working. But when they had to start paying standard wages, Lefnovich closed down the mill rather than pay standard wages. Things were bad for some people then because the mill was the only place you had to work and earn any money. Everything else was pretty poor.

VALENTI: What did the people do who lost their jobs when the mill closed?

McALISTER: Well, they were pretty lucky because right after the mill closed, Fort Bragg started to build up. A lot of people who worked in the mill got jobs out there. Some worked at the laundry out there, or other things. One man who had bought one of the houses from Mr. Anatola over on Weiss Avenue went to work for the railroad. People had to go on some way. Of course, you could get by with a little money back then. You could get a lot of groceries for your money, and there wasn't much else you could spend it on after the rent. You'd need some kerosene for your lamps, because of course there was no electricity, and you had to split wood for your heating in the winter, but that didn't cost you very much. There was no fancy-dancy stuff, like the medicines that they make people take now. If you were sick, you'd make some tea from some sassafras, or horehound, or even burdock [laughs]. We kept a few hogs and chickens, and a lot of people did. We had a turnip patch right over there, and we had corn and beans. When I told my daddy that I was sick of ham and eggs for breakfast and wanted something else, he told me that when I was grown up and hungry I'd wish I had some of those ham and eggs. It was what we had, and I didn't know how good it was.

VALENTI: Were there a lot of other activities, like community activities, centered around the mill?

McALISTER: There was Arthur McNatt's swimming pond behind the mill. You'd pay to swim there, and we used to fish there and in the pond on the other side of the mill where they used to get the water for the boiler room. There was a commissary at the store, and another one downtown, and a clothing store. You could get a card, for $5 or $10, they would let you get your food, and they would punch out whatever you got, fifteen cents or two dollars or whatever, on the card. And they would take the money out of your next pay.

VALENTI: How were the prices at the commissary? Were they fair?

McALISTER: Oh, yes, back then for two dollars at the grocery store, you could get more bags of groceries than you could carry. Not like today.

VALENTI: Were there any other mill activities?

McALISTER: Oh, yes. There were Fourth of July parties, where everyone would come, and there was a baseball team, the Silk Mill Giants, who would play all over the place, as far away as places like Norfolk.

VALENTI: Was it a good team? Did they ever have any players who went on to play at higher levels?

McALISTER: Oh, yes, they were good. There was a center fielder who could have gone on, but of course in those days you couldn't. You couldn't go on up, because they wouldn't let you. Not back then; nobody could.

VALENTI: You have a great memory. You seem to remember everything that happened at the mill and what went on at every spot around here.

McALISTER: I used to listen to my daddy and my grandfather. And I used to pay attention.

VALENTI: Well, I hope young people will take the time to find out what happened in their neighborhoods and will make sure that all that people like you know isn't forgotten. From what you've said, it seems like the life people had around here was satisfying.

Mr. McAlister and his friend, Mr. Cleveland, discussed where they had gone to school and the places where they were able to pursue various activities. After several minutes of discussing these places, Mr. Cleveland finished his reminiscence with the following statement.

"Well, Bryce, Fayetteville's been good to you and me, hasn't it?"
Mr. McAlister smiled and said nothing.

"The Sun Dagger on Fajada Butte"
Anna Sofaer

Jack Loeffler's introduction to his oral history interview with Anna Sofaer explains both her training as an artist and her interest in Fajada Butte as an outstanding example of Native American culture. The sun dagger suggests for Sofaer the importance of seasonal ritual in Chacoan culture. Learning about such careful reactions to the natural world dramatically suggests how we can find significance in our home landscapes by paying attention to these places during different seasons.

That the western hemisphere has been inhabited for many millennia by fellow humans is not contested. However, many of us who belong to the monoculture of mainstream America have an absurd tendency to overlook, or worse, to deny that the inhabitants of the New World prior to the advent of the Spanish conquest were capable of major intellectual achievements. Only recently, for example, have scholars begun to comprehend the scope of ancient Mayan culture whose denizens independently invented the concept of zero!

In an arid region in what is currently considered northwestern New Mexico is the canyon called Chaco where the ancient Anasazi people architected spectacular buildings whose ruins have piqued the interest of everyone who has ever stood in their shadow. In 1977, Anna Sofaer, an artist from Washington, D.C., visited Chaco Canyon. She was interested in petroglyphs and carried in her mind the recollection of a recent seminar concerning archaeoastronomy. She climbed 440-foot-high Fajada Butte to look at a spiral petroglyph that had been pecked into the southern face of the top of the butte. It was lunchtime on a day in late spring and Anna Sofaer and her companion sat in the shade near the petroglyph to relax and eat. As she watched, a dagger of sunlight passed near the center of the spiral. In a flash of intuition, Anna recognized the site as an Anasazi solstice marker.

It is an astonishing intellectual feat that a group of the Anasazi, ancestors of today's Puebloans, were able to calculate the time of the solstices using the noonday sun and a petroglyph enshadowed by great slabs of rock. It was also a moment of inspired intelligence when, a thousand years later, Anna Sofaer was able to extrapolate the significance of the ephemeral dagger of light.

After a decade of intense research, Anna has studied both the characteristics of the land around Chaco Canyon and the accompanying Anasazi ruins. She has discovered that the structures were almost perfectly oriented to the passage of the sun and the moon. Her tentative hypothesis is "that the Chacoans oriented, proportioned and located their buildings in a pattern focused on the sun and the moon into the ceremonial center of their regional culture, Chaco Canyon."

Since her discovery, Anna Sofaer has produced a documentary film entitled *The Sundagger* and has written numerous articles which have appeared in scientific journals.

LOEFFLER: Anna, I'd like to ask you to describe your first encounter with the Sun Dagger on Fajada Butte.

SOFAER: In 1977 I went into Chaco Canyon. I had already become involved in Mayan astronomy and had a strong interest in prehistoric astronomy and rock art. I finally decided after years of work in my studio that it was time to get out and actually see some of the rock art I had only looked at in books. I subscribed to a newsletter called "La Pintora," which was the American Rock Art Research Association's newsletter. It told about a group of volunteers who were going to Chaco for a couple of weeks in June of '77, and that they needed people. I just signed up. I had no idea what Chaco Canyon was. I had never been in the desert, or really to the Southwest. It was on that trip that the director of the group, Jim Bain, a retired colonel from Albuquerque who had organized the whole thing, said, "Well, nobody's ever recorded the rock art on Fajada Butte."

A fellow from California who was an excellent climber and seemed to be well-equipped and ready to go, needed somebody to accompany him. Everybody else was a little scared of the snakes and the height, and were savvy about what to be afraid of. I was so naive I just went along. He seemed to know what he was doing.

The first day we went up and recorded about twenty sites. We decided to go to the top out of curiosity, and see what else might be there for the next day. We got there kind of late in the day and saw a spiral, a large spiral, about fifteen inches across, behind the rock slabs. But it was late, so we decided to come back the next day and record it. When we came back, we happened to get there at about noon, about a week from summer solstice. And the dagger of light was right through the center of the spiral—very, very close to the center. It was so vivid that at that moment I said, "It's recording the highest point of the sun in the day and the year, being close to noon and summer solstice. My partner said, "Mmmmm. . . . whatever you say!" But I felt certain from the imagery of it.

Three weeks earlier, in my craze for rock art, I had gone to a conference of the American Rock Art Research Association, where they showed endless slides, which I loved, of rock art. One fellow showed that remarkable site in Baja, California, where a shaman figure with horns is painted on the wall of the cave. As he found out, at winter solstice sunrise, light comes through the cave and a horizontal dart crosses the eyes of the shaman just at that time of year, marking winter solstice sunrise. That image came to mind as I saw the Sun Dagger on Fajada, and helped me understand what it was. Plus the fact that three months earlier I'd had the opportunity to be in Yucatan, and to see the great pyramid of Chichen Itza. At equinox as the sun sets, the shadow form gives shape to the great serpent that's carved at the base—so you get the head of it carved in stone, and going up the carved steps to the top of the pyramid is the light shape of the rest of the serpent. It's absolutely marvelous, as the sun sets at equinox, casting the shadow from the edge of the pyramid to make that rippling shape of the snake. So those two experiences plus a little bit of study of Mayan astronomy, and having been at other sites which use shadow and light, gave me some assurance.

I didn't have any previous knowledge of the Chaco culture, of Southwest Indian culture. But there was a linkage; the shadow and light spirals had some connection to earth-sky imagery. I'd just finished a work of art of my own, which dealt with the same imagery. It was called "Stone Serpent in the Sky." It developed a lot out of the imagery of the Mayan astronomy I had studied in Washington. I had tried to bring together an almost ambiguous quality of earth and stone and sky, so that the shape was sculptural and also galaxy-like. When you think about it, the spiral with the dagger of light and the connection of the sky and earth there was so similar.

LOEFFLER: Can you describe the site on top of Fajada Butte?

SOFAER: Yes. It's three large rock slabs that are six to nine feet in height; they lean against the cliff face in a nearly vertical position so that the openings between them, which are oriented to the south sky, channel light in on the cliff face behind the rocks. That light comes in as vertical forms. There are two spirals carved on the cliff face behind the rocks. About midday, every day, light will come in in vertical forms on the spirals, but in a very particular form at summer solstice, equinox, and winter solstice. At summer solstice, a light dagger bisects the larger spiral right through the center, coming down in a vertical shape. Then at winter solstice, that light dagger has moved far to the right and a second light dagger has joined it, so that they bracket the large spiral, holding the center empty of light at the lowest time of the sun—winter solstice. At equinox that second light dagger pierces the center of the little spiral. So you have the three points in the solar cycle marked with the same kind of imagery each time.

LOEFFLER: What's your theory with regard to how this all came about?

SOFAER: Somewhere between 900 and 1300 A.D.—and possibly we can pin it down to around 1000 to 1150, when the Chaco culture was in its fluorescence and building solar-aligned, elaborate architecture—the Pueblo people carved the spirals. That is, they pecked them out on the cliff face. They shaped the surfaces of the rocks that cast the shadows that formed the light patterns, and may also have adjusted the rocks to some extent to make the particular forms they needed. The site is really quite complex in the number of light patterns.

We've found in the last three or four years, in addition to the light daggers at midday, that as the sun or the moon rises, there is a separate set of shadows on the same spirals. Not only on the same spirals, but in the same key positions. So the center, where the light dagger comes at summer solstice noon, is crossed by a shadow at the minor standstill position of the moon. And the left edge of the same spiral is marked with a shadow for the major standstill position of the moon. Each of these shadows is aligned with a diagonal groove pecked by the Anasazi, probably to emphasize them as markings. We've written in detail about the solar and lunar markings in science articles and presented them at conferences of archaeoastronomy.

So, the site marks five separate solar and lunar positions in a set of six markings, that repeatedly use three key points—the center and two outer edges of the large spiral—which really define the spiral's size and shape. There's no way to know with certainty whether the rock slabs were moved and shaped, but it seems likely that it would have been easier for the Anasazi to do it with some adjustment to the slabs' shapes and their positions.

LOEFFLER: Have you extrapolated any reason for all of this from the point of view of the ancient Chacoans?

SOFAER: Well, yes, now that we have noted more markings and we know more about the orientations of the pueblos. We've read a lot of the ethnography and talked to Indian people. First of all, there are seven other markings on Fajada Butte, using five other petroglyphs, just the way the Sun Dagger does. These are spiral forms, often crossed with dagger shapes of light, and a rattlesnake, aligned with the shadow edge. These markings give the time of solar noon—that's when the sun passes across the meridian— in distinctive configurations for each season. So you have a separate marking for summer solstice noon, winter solstice noon and equinox noon. And not just one marking—but multiple markings for each of these times. For instance, at equinox noon, at a site on the east side of the butte, the snake and a spiral are simultaneously crossed by shadow, while at a west side site, a spiral and rectangular form are marked with a dagger of light going through the center of a spiral. This pattern of multiple markings, so often in pairs, clearly seems intentional. We have now recorded a total of thirteen markings on the butte, several of which involve meridian passage, which is solar noon. Now we are looking in the canyon and analyzing, as others have already, Pueblo Bonito, probably the most central and important of all the structures of the Chaco system, which extends through the San Juan Basin, 30,000 square miles. Many of the archaeologists feel that

Pueblo Bonito may be the very center of that system. Certainly the canyon is, and Pueblo Bonito is the most important pueblo in the canyon. Perhaps the roads are coming right to Bonito and joining at its great kiva. When you survey the orientations of major walls in that pueblo, you find that the east-west wall is within a quarter of a degree of east-west; and the north-south wall that divides the pueblo is within a quarter of a degree north-south. Now the reason that ties into the Fajada markings, and begins to show a kind of pattern of cosmology and the use of the same astronomical concepts, is that to know noon to within a minute, you need to know north-south within a quarter of a degree. There's an equivalent accuracy; and the same is true for east-west and equinox. If you know the day of equinox, which seems to be true from the markings on Fajada Butte, you know east-west within a quarter of a degree, as is shown in Pueblo Bonito. But it isn't only the equivalent accuracy. These alignments in Bonito are, in a sense, a spatial expression that correspond to the temporal expression of the light markings. For instance, the mid-wall could be seen as the axis of the day the way the solar markings of noon are axes of the day. The east-west wall could be seen as the divider of the year, just the way equinox is the mid-point of the year. The joining of those cardinal alignments at the most sacred, central place in the culture is, I think, kind of a complementary structuring of space and time on the canyon floor, equivalent to the markings on Fajada Butte, which is that vertical axis drawing in the sky markings from above.

LOEFFLER: So these are people who were totally oriented to the sun?

SOFAER: The sun and the moon and the cardinal directions, and finding themselves in time and space by marking the center points and the extremes of the astronomical cycles. The solstices are the extremes; the equinoxes are the mid-point in the solar cycle; and the moon's two standstill positions are its extremes. All of these points have been marked on Fajada. There's yet another shadow as the sun and moon rise that comes at zero declination, the midpoint of their cycle, to the far right edge of the large spiral. And again, the seasonal markings join together the sun's annual daily cycles. It's a constant bringing together of these complementary positions. It's so integrated.

LOEFFLER: How would they have used this information?

SOFAER: Well, I suppose in the beginning my colleagues thought of it as a device or an instrument. I resisted that a bit, but I think not enough. More and more, it seems to me, and the Pueblo people with whom I've spoken, including the Hopi people, have expressed this feeling—that it's a sacred site—the Sun Dagger and the other ones as well. The butte itself is sacred. What was happening there was in a ceremonial context, of commemoration of these important times. There's such a redundancy in the markings. They're so remote. They have such vivid imagery—degrees of light through spirals are not just instruments or devices or calendars. They're really, I suppose, cosmic symbols for finding one's place in the center of time and space.

We've had the chance now to look throughout the San Juan Basin. With Mike Marshall, an archaeologist very knowledgeable about the Basin, I have now climbed or done an aerial survey of all of the prominent geological forms—anything like Fajada Butte. We've done about twenty of these surveys, and found only low-walled structures that appear to be shrine-like sites. But we've found almost no rock art, and nothing of explicit astronomy. So it's as though the butte were the unique center in the Chaco culture for that kind of ritual astronomy. Recently the director of the Chaco Research Center, the archaeology study center of the Chaco culture, has come out with his conclusion, after many years of study, that Chaco was not really lived in consistently. It was visited periodically, in what he called "ritual pilgrimages" to the canyon. One wonders if the ritual activity that centered in Chaco wasn't timed or linked in some way with those markings on Fajada. Hopi people have told me that one or two people might watch something like the dagger of light at that site, and that those people would have the responsibility to then tell the people in the canyon when to begin the ceremonies.

The complexity of the astronomical information is so striking—and so important to grapple with, to understand. And the accuracy, for naked-eye astronomy, is astounding too. But I think more than just the complexity and accuracy is the aesthetic context—the beautiful quality of putting all that on what in some cultures might be like a pyramid or massive temple. It is a beautiful natural structure, Fajada Butte.

RESPONDING JOURNAL

Think about what you would like to ask one of the people interviewed in the section above. Write down a few questions, and write a brief explanation with each question in which you tell the interviewee why you feel this information is important.

QUESTIONS FOR CRITICAL THINKING

STRATEGY

1. Much of the strategic planning in an oral history interview will be reflected in both the wording of the questions and the order in which the questions are asked. Generate a set of guidelines—perhaps five ideas—that an interviewer should consult when planning to conduct an interview.

2. Do you think the authors of these oral history interviews conducted their work effectively? What else would you have asked the subjects of their interviews? What else do you want to know about these subjects?

Issues

1. Here we have presented two very different sorts of interviews, yet both show us how we can learn about our landscapes by looking at them carefully at various times, at different periods. What categories should we apply to landscapes as we examine them during different seasons?
2. Look at your daily newspaper to find situations you feel could use some of the discussion generated in oral history interviews. Try to find a wide range of people and parts of the newspaper's circulation area, if possible.

COLLABORATIVE WRITING ACTIVITY

Generate a list of three people either in your family or in your community whom you feel should be interviewed because their experiences should be recorded. For each person, write a brief justification to suggest why this individual should be considered. When you get to your group, read your list with the accompanying justification. Explain any additional details that you think would help the members of your group understand why your choices should be interviewed formally. After you have heard all the members, compile a list of characteristics that you believe represent the range of people suggested in your group. Add any comments to this list that you think other potential interviewers might find helpful.

SUGGESTIONS FOR FORMAL WRITING ASSIGNMENTS

1. If you wanted to leave as part of your legacy some record of what was important to you in terms of home and family, what would it be? Look back over the forms presented in this chapter; would you want to leave a series of journal entries, a memoir, or an oral history interview? If you choose the last one, who would you select to interview you? What would that person be able to bring out in you? As part of a paper proposal or prospectus, explain why the format you have chosen would work well for your purposes.

2. Discuss your proposal with your collaborative learning group. Make sure that you have met your instructor's requirements for planning and prewriting. Write the journal entries or memoir or complete the oral history interview you have proposed in #1.

3. Think about some person you know or some place you value that you think should be remembered. Spend some time thinking about the sort of detail that will convey to a stranger your sense of the importance of the person or place you have chosen before you begin writing the substantial essay in memoir form.

Chapter 2

PLANNING YOUR WORK: DIFFERENT PERSPECTIVES

Solomon D. Butcher, *Harvey Andrews family at the grave of their child Willie*, 1887

1. How do the members of Willie's family seem to regard the photographer? What differences do you detect?

2. Explain the effect of the composition here. For example, how do the horizontals in the picture draw us spectators into the frame?

3. Discuss how people relate death to the earth, to the land. What do changing customs with regard to death tell us about our relationship to the earth?

Chapter 2

PLANNING YOUR WORK:
DIFFERENT PERSPECTIVES

There is no one trick to sitting down in front of a piece of paper and writing. Even professional writers need some assistance, need to use several special devices, to get started. Finding a subject and then developing it are the two tasks that face a writer; prewriting techniques will get a writer started.

First of all, you need to get moving physically. When the prospect of writing daunts you, it is hard to begin what you feel will be the ultimate essay. Instead of letting such needless worries slow you down, take the process one step at a time. Write down just the few words that come to you, or type the title page with the assignment, your name, class section, and date; sometimes the mere mechanical motion of beginning the process of typing and seeing words appear on the computer screen or blank sheet of paper will be enough to get you off and running.

The most important part of writing is writing. You will probably want to try a number of the following tips for beginning your writing. After some experimentation, you will see what works best for you, or perhaps you will modify the suggestions here to generate your own process. But regardless of the steps you follow, you will want to be aware of the different stages through which your writing will move. Since writing is a process, the steps you follow will determine how your essay will develop.

Beyond the journal, there are several strategies you can try to help you move along with your writing. These include brainstorming, clustering or mapping, and freewriting.

Brainstorming is another way of making a list, at least initially. Once you think of a topic, you begin to write down whatever pops into your mind, in the order that the ideas come to you, so that you can accommodate the furious pace of your thoughts. When brainstorming, you include details, examples, associated words, or other ideas that come to you as a result of thinking about the topic. Remember that when you brainstorm an idea, you leave your censoring equipment behind; do not eliminate any items that pop into your head, no matter how ridiculous they appear, because sometimes the most ridiculous idea or association can in turn trigger another association that will work well, that may indeed be just the key to open up a dynamite angle on a subject for you.

Particularly at first, you may want to get together in a group with your class-mates. Your ideas will encourage one another in unpredictable ways, and you are certain to come up with new angles after a group session. Use either a tape recorder or (second choice) be sure to take careful notes of what is being said. After you have given everyone a good chance to discuss the subject, look back over the transcription of the session. If you can, duplicate the transcription for the group so that you can then discuss how you could derive topics and issues for essays from the session.

Such an activity will show you the importance both of brainstorming in the early stages of the writing process and the helpful nature of collaboration through group work. The practice of brainstorming originated in creativity-enhancing situations held in the world of business to encourage new ideas or different twists to existing concepts in order to explore new fields. Since writing is a form of exploration of new ideas or of familiar ideas seen in new ways, brainstorming is a logical way to begin. Once you begin brainstorming, continue to do it regularly. You will be surprised to see the topics you will generate; this happens because you are accessing your unconscious through the process.

At this point, you might want to try a technique that has been used for generations, particularly by newspaper reporters: the five W's. This strategy consists of writers asking questions of themselves in order to be certain they have supplied in their stories all the information required to answer a reader's usual questions, which are:

Who

What

When

Where

Why

Added to these five is an H, the how, that often enters into the reporters' thinking. By making sure they have answered each of these five (or six) categories, reporters can be reasonably sure they have covered the main points in any given story.

Related to brainstorming is another method that allows you to get a visual sense of how your material is shaping up. Clustering or mapping may be the most useful of the prewriting techniques because it allows you to assign a visual value to what you are doing; it allows you to chart your lists and get a sense of what they will look like. Through clustering, you can find new ways to sequence your materials. You can spread out your ideas in a way that you cannot when you are listing in a vertical fashion.

The method works like this. You begin with a general concept, the first sense of the idea or topic you want to develop, and you place that word or brief phrase in the middle of a plain sheet of paper. For example, if you want to write about how you remember your childhood backyard, as Mary Mebane does in her description in this chapter, you might start with the big rock in the center of the yard that serves as her vantage point. You then branch off that main circle with other circles and other ideas as they occur to you. Let's try again with the example of Mary's rock in her backyard.

This second stage is most important; don't worry about exactly how another circle, with another idea, will relate to the first. Just concentrate on getting all of these ideas down, in some fashion, on that blank sheet of paper. Of course, if you can begin preliminary groupings by clustering related details, go ahead and do so; this step will save you time later.

After you have jotted down the related ideas all around the main concept bubble, go back and see what you have; draw lines and arrows among the different circles. At this point, you may start jotting down on another sheet of paper some possible relationships among the bubbles that you want to explore.

Thinkers from John Locke, the seventeenth-century political scientist and empiricist philosopher, through Gabrielle Rico, the modern writer who has

FIGURE 1
Clustering: Mary's Rock

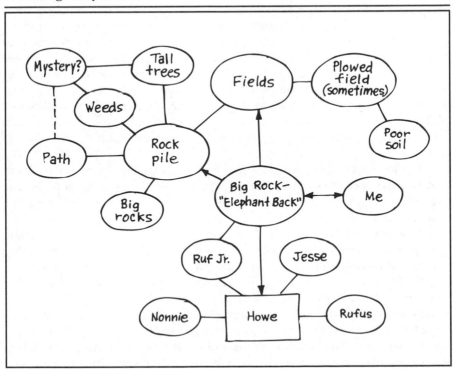

explored most fully the concept that she calls "clustering," have emphasized the tendency of our thinking system to group ideas together. Locke's "association of ideas" explained how ideas could be linked together in the mind. For example, when one idea is considered, others come along with that idea because they are *associated* with that idea, much as pulling one coat hanger from a box of hangers will bring up a bunch of hangers, or hearing one Beatles song will make you think of others. The next task, of course, is to sort out the hangers or songs and make order from the random arrangement. Clustering recognizes the tendency of the mind to work in this fashion and takes advantage of the mind's habitual desire to connect or associate.

Freewriting can be done either on separate sheets of paper or in your journal; the main idea to remember with freewriting is not to stop until the time limit is up. You need to keep writing, to record even the fact that you cannot think of anything to say: "I can't think of anything to say, why can't I think of anything to say, I don't want to keep just writing that I don't have anything to say, why can't I just stop doing this because nothing will come of it. I wish that I could find some topics because this is beginning to get real boring and besides my hand is starting to ache, and how can I write anything in here for ten minutes if my hand is all messed up. . . ." You get the idea; keep writing until your time is up. If you are able to write—either a planned entry or a piece of freewriting—for ten minutes a day in your journal, you will be surprised to see how much this practice will help your writing after a month. For one thing, you will not be afraid of the blank page or computer screen facing you because you will have some effective strategies to deal with it. Like stretching your muscles before you work out, freewriting warms you up. Freewriting also puts you in the habit of writing, of committing words to paper (or to the computer screen) so that it becomes almost a reflexive activity. If you do your work with a word processor, try this method. Turn off your monitor while you freewrite, and then turn it back on after your freewriting period is over so that you can look at what you have done. Your text will seem almost like the work of someone else.

The writings in this section suggest how other writers have used their formative influences of context to help them with the composing process. The area in which you grew up, or any setting that evokes strong feelings for you can serve as the basis not only for the content of your writing, but also for the way in which writing is organized. In a chronological organization, you might arrange your materials on a day-by-day basis, as Gretel Ehrlich has done in her account of three days of sheepherding, or you might focus more intensively on the activities of a single day, or even a few hours, as Richard Wright has done in his short story "Silt." The straightforward technique of the five W's that Steinbeck uses in the chapter from *The Harvest Gypsies,* which originally appeared as a newspaper article, might work well for you. As you look at these selections, think about how the authors have ordered their materials in accordance with the ideas and impressions you take away with you from their writing.

"From a Sheepherder's Notebook: Three Days"
Gretel Ehrlich

Gretel Ehrlich is a Californian educated at Bennington, the University of California at Los Angeles, and the New School in New York. When she found herself transplanted to Wyoming in 1976 to make a film, the stark western landscape changed her orientation toward writing. Ehrlich describes her philosophy of writing: "The truest art I would strive for in any work would be to give the page the same qualities as earth: weather would land on it harshly; light would elucidate the most difficult truths; wind would sweep away obtuse padding. Finally, the lessons of impermanence taught me this: loss constitutes an odd kind of fullness; despair empties out into an unquenchable appetite for life."

Of her considerable writing about the west, Ehrlich is best known for her essay "The Solace of Open Spaces," which also serves as the title for a collection of her essays about life in Wyoming. In that essay she explains what the immensity of the Wyoming landscape has taught her. She has also co-published with Edward Hoagland a collection of stories entitled City Tales, Wyoming Stories. Here, in the account of three days spent sheepherding, she gives us some clues to the sort of experience she undergoes on these lonely western lands. People may think the days of feuds between cattleranchers and sheepherders were finished early in the century, but Ehrlich writes of men who still spend their lives freely on the range in ways that have not changed in a hundred years.

EXPLORING JOURNAL

Read again the headnote quotation above in which Gretel Ehrlich describes her philosophy relating writing and the earth. In a substantial journal entry, freewrite what you think she might have in mind for the ways that writing mimics natural processes. Have you ever felt any of the emotions she describes? How difficult do you think it would be to write in such an honest way? If you disagree with her statements, use your journal entry to explain your position.

When the phone rang, it was John: "Maurice just upped and quit and there ain't nobody else around, so you better get packed. I'm taking you out to herd sheep." I walked to his trailerhouse. He smoked impatiently while I gathered my belongings. "Do you know *anything* about herding sheep after all this time?" he asked playfully. "No, not really." I was serious. "Well, it's too late now. You'll just have to figure it out. And there ain't no phones up there either!"

He left me off on a ridge at five in the morning with a mare and a border collie. "Last I saw the sheep, they was headed for them hills," he said, pointing up toward a dry ruffle of badlands. "I'll pull your wagon up ahead about two miles. You'll see it. Just go up that ridge, turn left at the pink rock, then keep agoing. And don't forget to bring the damned sheep."

Morning. Sagesmell, sunsquint, birdsong, cool wind. I have no idea where I am, how to get to the nearest paved road, or how to find the sheep. There are tracks going everywhere so I follow what appear to be the most definite ones. The horse picks a path through sagebrush. I watch the dog. We walk for several miles. Nothing. Then both sets of ears prick up. The dog looks at me imploringly. The sheep are in the draw ahead.

Move them slow or fast? Which crossing at the river? Which pink rock? It's like being a first-time mother, but mother now to two thousand sheep who give me the kind of disdainful look a teenager would his parent and, with my back turned, can get into as much trouble. I control the urge to keep them neatly arranged, bunched up by the dog, and, instead, let them spread out and fill up. Grass being scarce on spring range, they scatter.

Up the valley, I encounter a slalom course of oil rigs and fenced spills I hadn't been warned about. The lambs, predictably mischievous, emerge dripping black. Freed from those obstacles, I ride ahead to find the wagon which, I admit, I'm afraid I'll never see, leaving the sheep on the good faith that they'll stay on their uphill drift toward me.

"Where are my boundaries?" I'd asked John.

"Boundaries?" He looked puzzled for a minute. "Hell, Gretel, it's all the outfit's land, thirty or forty miles in any direction. Take them anywhere they want to go."

On the next ridge I find my wagon. It's a traditional sheepherder's wagon, rounded top, tiny wood cookstove, bed across the back, built-in benches and drawers. The rubber wheels and long tongue make it portable. The camp tender pulls it (now with a pickup, earlier with teams) from camp to camp as the feed is consumed, every two weeks or so. Sheep begin appearing and graze toward me. I picket my horse. The dog runs for shade to lick his sore feet. The view from the dutch doors of the wagon is to the southeast, down the long slit of a valley. If I rode north, I'd be in Montana within the day, and next week I'll begin the fifty-mile trail east to the Big Horns.

Three days before summer solstice; except to cook and sleep I spend every waking hour outside. Tides of weather bring the days and take them away. Every night a bobcat visits, perched at a discreet distance on a rock, facing me. A full moon, helium-filled, cruises through clouds and is lost behind rimrock. No paper cutout, this moon, but ripe and splendid. Then Venus, then the North Star. Time for bed. Are the sheep bedded down? Should I ride back to check them?

Morning. Blue air comes ringed with coyotes. The ewes wake clearing their communal throats like old men. Lambs shake their flop-eared heads at leaves of grass, negotiating the blade. People have asked in the past, "What do you do out there? Don't you get bored?" The problem seems to be something else. There's too much of everything here. I can't pace myself to it.

Down the valley the sheep move in a frontline phalanx, then turn suddenly in a card-stacked sequential falling, as though they had turned themselves inside out, and resume feeding again in whimsical processions. I think of town, of John's trailerhouse, the clean-bitten lawn, his fanatical obsession with neatness and work, his small talk with hired hands, my eyesore stacks of books and notes covering an empty bed, John smoking in the dark of early morning, drinking coffee, waiting for daylight to stream in.

After eating I return to the sheep, full of queasy fears that they will have vanished and I'll be pulled off the range to face those firing-squad looks of John's as he says, "I knew you'd screw up. Just like you screw up everything." But the sheep are there. I can't stop looking at them. They're there, paralyzing the hillside with thousands of mincing feet, their bodies pressed together as they move, saucerlike, scanning the earth for a landing.

Thunderstorm. Sheep feed far up a ridge I don't want them to go over, so the dog, horse, and I hotfoot it to the top and ambush them, yelling and hooting them back down. Cleverly, the horse uses me as a windbreak when the front moves in. Lightning fades and blooms. As we descend quickly, my rein-holding arm looks to me like a blank stick. I feel numb. Numb in all this vividness. I don't seem to occupy my life fully.

Down in the valley again I send the dog "way around" to turn the sheep, but he takes the law into his own hands and chases a lamb off a cliff. She's wedged upside down in a draw on the other side of the creek. It will take twenty minutes to reach her, and the rest of the sheep have already trailed ahead. This numbness is a wrist twisting inside my throat. A lone pine tree whistles, its needles are novocaine. "In nature there are neither rewards nor punishments; there are only consequences." I can't remember who said that. I ride on.

One dead. Will she be reborn? And as what? The dog that nips lambs' heels into butchering chutes? I look back. The "dead" lamb convulses into action and scrambles up the ledge to find his mother.

Twin terrors: to be awake; to be asleep.

All day clouds hang over the Beartooth Mountains. Looking for a place out of the wind, I follow a dry streambed to a sheltered inlet. In front of me, there's something sticking straight up. It's the shell of a dead frog propped up against a rock with its legs crossed at the ankles. A cartoonist's idea of a frog relaxing, but this one's skin is paper-thin, mouth opened as if to scream. I lean close. "It's too late, you're already dead!"

Because I forgot to bring hand cream or a hat, sun targets in on me like frostbite. The dog, horse, and I move through sagebrush in unison, a fortress against wind. Sheep ticks ride my peeling skin. The dog pees, then baptizes himself at the water hole—full immersion—lapping at spitting rain. Afterward, he rolls in dust and reappears with sage twigs and rabbit brush strung up in his coat, as though in disguise—a Shakespearian dog. Above me, oil wells are ridge-top jewelry adorning the skyline with ludicrous sexual pumps. Hump, hump go the wells. Hump, hump go the drones who gather that black soup, insatiable.

We walk the fuselage of the valley. A rattlesnake passes going the other way; plenty of warning but so close to my feet I hop the rest of the day. I come upon the tin-bright litter of a former sheep camp: Spam cans flattened to the ground, their keys sticking up as if ready to open my grave.

Sun is in and out after the storm. In a long gully, the lambs gambol, charging in small brigades up one side, then the other. Ewes look on bored. When the lamb-fun peters out, the whole band comes apart in a generous spread the way sheep ranchers like them. Here and there lambs, almost as big as their mothers, kneel with a contagiously enthusiastic wiggle, bumping the bag with a goatlike butt to take a long draw of milk.

Night. Nighthawks whir. Meadowlarks throw their heads back in one ecstatic song after another. In the wagon I find a piece of broken mirror big enough to see my face: blood drizzles from cracked lips, gnats have eaten away at my ears.

To herd sheep is to discover a new human gear somewhere between second and reverse—a slow, steady trot of keenness with no speed. There is no flab in these days. But the constant movement of sheep from water hole to water hole, from camp to camp, becomes a form of longing. But for what?

The ten other herders who work for this ranch begin to trail their sheep toward summer range in the Big Horns. They're ahead of me, though I can't see them for the curve of the earth. One-armed Red, Grady, and Ed; Bob, who always bakes a pie when he sees me riding toward his camp; Fred, wearer of rags; "Amorous Albert"; Rudy, Bertha, and Ed; and, finally, Doug, who travels circuslike with a menagerie of goats, roosters, colts, and dogs and keeps warm in the winter by sleeping with one of the nannies. A peaceful army, of which I am the tail end, moving in ragtag unison across the prairie.

A day goes by. Every shiver of grass counts. The shallows and dapples in air that give grass life are like water. The bobcat returns nightly. During easy jags of sleep the dog's dreampaws chase coyotes. I ride to the sheep. Empty sky, an absolute blue. Empty heart. Sunburned face blotches brown. Another layer of skin to peel, to meet myself again in the mirror. A plane passes overhead—probably the government trapper. I'm waving hello, but he speeds away.

Now it's tomorrow. I can hear John's truck, the stock racks speak before I can actually see him, and it's a long time shortening the distance between us.

"Hello."

"Hello."

He turns away because something tender he doesn't want me to see registers in his face.

"I'm moving you up on the bench. Take the sheep right out the tail end of this valley, then take them to water. It's where the tree is. I'll set your wagon by that road."

"What road?" I ask timidly.

Then he does look at me. He's trying to suppress a smile but speaks impatiently.

"You can see to hell and back up there, Gretel."

I ride to the sheep, but the heat of the day has already come on sizzling. It's too late to get them moving; they shade up defiantly, their heads knitted together into a wool umbrella. From the ridge there's whooping and yelling and rocks being thrown. It's John trying to get the sheep moving again. In a dust blizzard we squeeze them up the road, over a sharp lip onto the bench.

Here, there's wide-open country. A view. Sheep string out excitedly. I can see a hundred miles in every direction. When I catch up with John I get off my horse. We stand facing each other, then embrace quickly. He holds me close, then pulls away briskly and scuffles the sandy dirt with his boot.

"I've got to get back to town. Need anything?"

"Naw . . . I'm fine. Maybe a hat. . ."

He turns and walks his long-legged walk across the benchland. In the distance, at the pickup, an empty beer can falls on the ground when he gets in. I can hear his radio as he bumps toward town. Dust rises like an evening gown behind his truck. It flies free for a moment, then returns, leisurely, to the habitual road—that bruised string which leads to and from my heart.

RESPONDING JOURNAL

Imagine yourself in Ehrlich's place. Could you see yourself as a sheepherder, even for three days? What would be the decisive factors in your decision to try it? What does your decision say about your sense of the land and those who gain their livelihood from it?

QUESTIONS FOR CRITICAL THINKING

STRATEGY

1. What benefits does Ehrlich derive from the idea of three days' work as sheepherder? How does this relatively brief period allow her to develop certain ideas more effectively than if she had not set up such a clearly defined boundary? Does this structure limit her in any way?
2. Recall Ehrlich's earlier comments about her writing and the land. Explain any parallels you see between her three-day journal and the ideas that run through her head as she does the work.

ISSUES

1. Characterize the relationship between John and Gretel. Does their feeling for one another in any way suggest how they will look after

the sheep? What aspects of this romance are conventional, and which are not?

2. Ehrlich offers several observations, such as, "I don't seem to occupy my life fully" and "Twin terrors: to be awake; to be asleep." What effect does such commentary have on the tone of the piece? What effect does it have on you as a reader?

COLLABORATIVE WRITING ACTIVITY

Think of a time when you were left alone to do a job you were afraid you might not be able to handle. How did you feel when you realized what was happening? What did you think about at the moment you realized that you could/could not get the job done? Share your response with the group and discuss any common themes or situations you find.

"My Backyard" from *Mary: An Autobiography*
Mary Mebane

Durham, North Carolina, was essentially a large rural town when Mary Mebane was born there in 1933. She attended North Carolina Central University in Durham for her undergraduate work and moved ten miles west to Chapel Hill to attend the University of North Carolina, where she completed the Ph.D. in English. She taught in the English Department at the University of South Carolina until shortly before her death in 1992.

In 1981 Mebane published her first autobiography, Mary: An Autobiography, *from which "My Backyard" is excerpted. She followed that book with a sequel,* Mary: Wayfarer, *in 1983. In the text she tells us how she felt in her backyard and in the wider world verging on that backyard plot.*

EXPLORING JOURNAL

Think about a favorite backyard you recall from some period in your childhood. Think of a particular spot in that yard from which you might view the important events which happened there, and see the people who were important to you. Freewrite for ten minutes in order to recall that spot as precisely as you can.

My name is Mary.

When I first opened my eyes to the world, on June 26, 1933, in the Wildwood community in Durham County, North Carolina, the world was a green Eden—and it was magic. My favorite place in the whole world was a big rock in the backyard that looked like the back of a buried elephant. I spent a lot of time squatting on that rock. I realize now that I probably selected it because it was in the *center* of our yard, and from it, by shifting ever so slightly, this way and that, I could see *everything*. I liked to look. Mama must have told me several thousand times that I was going to die with my eyes open, looking.

When I sat on the rock with my back to the house, the fields were in front of me. On the left was another lot that we called the Rock Pile, and to the right was an untended strip of land, strewn with rocks but cleared enough to be plowed sometimes. The Rock Pile, full of weeds and tall trees, was a place of mystery. It had so many rocks and some of them were so large that it was left uncleared with just a path through it. Behind me I could overhear voices coming from the back porch and kitchen. I could see who was chopping or picking something in the garden and I could see who was coming through the Rock Pile.

The road in front of our house was a dirty strip swirling with the thick red dust of state trucks going to and from the rock quarry. I saw the quarry once. To me it was one of the Seven Wonders of the World—a very wide hole, dug deep in the ground. The trucks around it looked like little toys. It was a mountain going in the wrong direction, with me standing at the top. Good-looking Edmund, the one with the limp from polio, died at the rock quarry. Explosives. They didn't want anyone to see his body. Someone said that they found only pieces. He was a grown man and I was a child at the time, but I remember him. He came down from Virginia to stay with his sister and find work. Like his sisters, he was very light-skinned and had thick, curly brown hair. I used to wonder about his hair. Did they find a lock of curly brown hair after the explosion?

Then there was the rich white contractor who lived on the highway. His oldest son had died as a young man, in another part of the state, backing a truck too near such an opening. The truck had started to slide and he couldn't stop it. I saw his picture once, when I went as a teenager to baby-sit at their house. His brother, then a grown man and my employer for the evening, told me about the accident. In the photograph the dead heir was still a child, about nine years old, sitting on a horse, smiling at the camera.

The world consisted of me at first; then, when Ruf Junior, the baby, was big enough to walk, he joined me on the rock. My older brother, Jesse, was a big boy who came dashing by to or from some adventure, and he might say something or he might not. My mother, Nonnie, and my father, Rufus, were the grown people who called me to dinner, to bed, or to do chores.

One day a car came up to the house and a lady got out, while a man put suitcases on the porch. The lady was Aunt Jo, and she stayed with us for several years. Sometimes other people came, but they were visitors, and unless they had someone my age for me to play with, they didn't affect me one way or the other.

I would squat on that rock, my stick legs poking through the openings of my dirt-stained bloomers, my birdlike head turning from side to side, my gaze, unblinking, focusing up, down, in front of me, in back of me, now zooming in on the lower yard, then penetrating deeper into the garden, then rising up ever so slightly to where the corn was planted on the hill. I was in the center of life and I didn't miss a thing; nothing slipped by unobserved or unnoted. My problems started when I began to comment on what I saw. I insisted on being accurate. But the world I was born into didn't want that. Indeed, its very survival depended on not knowing, not seeing—and, certainly, not saying anything at all about what it was really like.

The whole backyard slanted down. It started at the well and sloped down to the lettuce patch. When it rained, water ran in gullies clear down to a ditch Daddy had dug parallel to the flow to make the water run off. Later he decided to stop the flow up higher, and Ruf Junior and Jesse and I toted rocks and formed a little dam between the big sloping rock and the two less-big rocks that lay on either side of it.

The rock on the left was big, but it looked like a rock, not like the back of a buried elephant. I could see all around it, but it was big. My brothers and I couldn't move it, not even when we all pushed together. The one on the right of the big sloping rock looked like its brother. You couldn't see where it started or stopped. Its back was like the back of a smaller gray buried elephant—the younger brother of the large gray sloping elephant that was buried in the middle.

The little dam, built up with sticks and rocks, held fast; later, when it rained, the running dirt stopped there and backed up and covered some of the hundreds of rocks that studded the backyard and on which we stubbed innumerable bare toes in the May-to-September, school-is-out summer.

The well was in the upper part of the backyard, before the slope started. The wooden box that was the superstructure of the well was partly rotted; there were wide spaces between the boards, and my brothers and I had been warned not to lean too hard on them when we pulled the bucket up or we would fall in. The bucket was beaten up from banging on the rocks that lined the narrow well, and when the bucket came out, water sloshed over the side and spurted out of the little holes in it. The rocks in the well had wet green moss on them as far down as the eye could see. The well was about the width of a giant inner tube, but rough where the rocks stuck out all the way down. When you looked over the boards, the water below looked like quicksilver in the sun. I would look into the well and think deep thoughts and smell the wet moss and rotting wood.

Hanging on a nail on the well was a gourd to drink out of. This was full and round at one end and tubular at the other, and had tiny ridges in its mud-brown interior. A drink of cool well water from a sweet-tasting gourd when you're thirsty is the best drink in the world.

Sometimes I liked to lean over the edge of the well and look past the box where the wood had rotted and splintered from the water, past the moss lining the bottom of the wood and the rocks paving the well. I thought that under-neath the moving water was China. I read a story in the second grade that said that if you dug a deep, deep hole down from where you were and put your eye down, clear through the hole on the other side you would see China.

I believed stories like that. Just as once I read a story about Mexico and was struck by the bold designs on the pottery. That night I dreamed that I was in heaven and God looked like the father of a Mexican family, dark, with black hair and a long, colorful robe, with his wife and child by his side. They were standing in front of a clay house, and nearby were enormous pottery jars with bright designs on them.

Our well wasn't as deep as it should have been, and a couple of summers it went dry, so we would go to one of the springs. The nearest, at the bottom of a little slope, bubbled clear water up out of clean sand. The shallow encir-cling wall was about three hands high and a child could curl up and hold his feet with his hands around its circumference. The water there bubbled up end-lessly, clear and sweet, shaded by tall North Carolina pines. The other spring was farther away. It was larger, more conventional, and not nearly so romantic.

In the morning the sun glistened on the long grass in the vacant lot that was never mowed. The cow ate the grass and kept it low, all except the spot where I emptied the peepots. She wouldn't eat there unless Mama sprinkled a lot of salt on it to fool her. Then sometimes she would eat it and sometimes she wouldn't. This grass was on the lot next to the main yard. Daddy sometimes plowed it all up and harrowed it and sowed it with grain and got mad when Mr. Jake's chickens came over from next door and ate it all up. But it was dry, hard, cracked land and never grew much of anything.

On the left side of the main yard was the Rock Pile where I picked black-berries. Aunt Jo made purple dye from the pokeberries that also grew there, and sometimes she cooked poke salad, which I hated; it tasted like cooked leaves. But I liked the dye. When Aunt Jo dipped a white gunnysack in the dark water and it came up a beautiful purple, I was filled with wonder.

Sometimes my brothers and I played jumping from rock to rock. If you stepped off a rock you were "out." Sometimes snakes slid out of the Rock Pile. My brothers and I couldn't run through it; the briars tore too bad. But we had a path and as long as I stayed in the path I felt safe. One step either to the right or the left of the path and I felt scared. Not only of snakes and other natural dangers but of something else. I didn't know what.

At the bottom of the yard there was a vegetable garden with green growing things: lettuce and cabbage and cucumbers and squash. Little sticks held up the vines of tomatoes and string beans and butter beans. Beyond it, the field ran up a little hill from the garden a long way and then down to some pines. After the rows of pines, the Bottom started. It was low there and wet most of the time, but I liked it; my father worked hard there with Suki, the mule, plowing. I followed him there all the time, but I liked mainly to play near the creek in the Bottom and on the pine-straw-covered mounds, where I slid down into the gullies.

RESPONDING JOURNAL

What does Mary's description make you think of? Are you or is anyone in your family like Mary in the way she chronicles her childhood? In a substantial journal entry, tell who in your family would make the best family historian and explain why.

QUESTIONS FOR CRITICAL THINKING

STRATEGY

1. Look at the essay from the vantage point of the big rock; does all of the detail radiate out from that center? Explain what advantages you think Mebane derives from setting up her description in this manner.
2. Think about how Mebane relates her backyard to people. Explain how she introduces people into her writing, and think about how effective her manner of characterization seems to be. Does she make her characters more vivid because of her organization?

ISSUES

1. Consider the language and sentence structure Mebane uses here. Try to explain how her style of writing effectively presents her account of childhood remembered on the basis of a particular place.
2. Think about the ending of the section, where Mebane describes what happened on the Bottom. In what ways is this an effective means of ending her backyard reverie? Does she open up other possible leads for development here in this final paragraph?

COLLABORATIVE WRITING ACTIVITY

As a child, what playmate do you remember as having the backyard, or yard, or other special place that you most enjoyed visiting? Perhaps it was the people there, or the special geographical features, or some special objects or equipment someone kept in an apartment. Decide what this favorite playmate place might be, and describe it for your group in a few paragraphs. Read your description to the group; after everyone has read, try to generalize about what the most important categories for deciding on significant childhood places seem to be. Present these to the entire class, along with the detail that convinced your group of the importance of your points.

from *The Harvest Gypsies*
John Steinbeck

Winner of the Nobel Prize for Literature in 1962, John Steinbeck (1902–1968) is best known for his fiction set in California and dealing with people often overlooked: immigrants, the poor, the transient working class. In Of Mice and Men *(1937) and* The Grapes of Wrath *(1939) Steinbeck gave us two of the most accurate and affecting pictures of the lives of migrant farm workers and of the "Okies," people who had been forced by the Dust Bowl of the Great Depression in the early 1930s to leave their native Oklahoma in search of work in California.*

In 1936, Steinbeck accepted an assignment from the San Francisco News *to provide a series of articles on the flood of migrants pouring into California from other states. These six articles were later collected under the titles* The Harvest Gypsies *and* Their Blood Is Strong, *titles which emphasize the migrant workers' function and character. Steinbeck spent several weeks with the migrant workers in their shanty towns developing his acquaintance with them in order to describe them for his newspaper audiences. This writing was the first step in producing what was to become the novel* The Grapes of Wrath. *The selection appearing here is the second essay in the series; to some extent Steinbeck is creating composite characters, but his models are real.*

EXPLORING JOURNAL

Think of the term "migrant workers." What do you know about the condition of their lives? Have migrants worked in the area where you live? How much do you know about the work they do and the pay they receive for it? What opportunities for school do their children have? Do a focused freewrite for at least ten minutes on the subject of migrant workers.

The squatters' camps are located all over California. Let us see what a typical one is like. It is located on the banks of a river, near an irrigation ditch or on a side road where a spring of water is available. From a distance it looks like a city dump, and well it may, for the city dumps are the sources for the material of which it is built. You can see a litter of dirty rags and scrap iron, of houses built of weeds, of flattened cans or of paper. It is only on close approach that it can be seen that these are homes.

Here is a house built by a family who have tried to maintain a neatness. The house is about 10 feet by 10 feet, and it is built completely of corrugated paper. The roof is peaked, the walls are tacked to a wooden frame. The dirt floor is swept clean, and along the irrigation ditch or in the muddy river the wife of the family scrubs clothes without soap and tries to rinse out the mud in muddy water. The spirit of this family is not quite broken, for the children, three of them, still have clothes, and the family possesses three old quilts and a soggy, lumpy mattress. But the money so needed for food cannot be used for soap nor for clothes.

With the first rain the carefully built house will slop down into a brown, pulpy mush; in a few months the clothes will fray off the children's bodies while the lack of nourishing food will subject the whole family to pneumonia when the first cold comes.

Five years ago this family had fifty acres of land and a thousand dollars in the bank. The wife belonged to a sewing circle and the man was a member of the grange. They raised chickens, pigs, pigeons and vegetables and fruit for their own use; and their land produced the tall corn of the middle west. Now they have nothing.

If the husband hits every harvest without delay and works the maximum time, he may make four hundred dollars this year. But if anything happens, if his old car breaks down, if he is late and misses a harvest or two, he will have to feed his whole family on as little as one hundred and fifty.

But there is still pride in this family. Wherever they stop they try to put the children in school. It may be that the children will be in a school for as much as a month before they are moved to another locality.

Here, in the faces of the husband and his wife, you begin to see an expression you will notice on every face; not worry, but absolute terror of the starvation that crowds in against the borders of the camp. This man has tried to make a toilet by digging a hole in the ground near his paper house and surrounding it with an old piece of burlap. But he will only do things like that this year. He is a newcomer and his spirit and decency and his sense of his own dignity have not been quite wiped out. Next year he will be like his next door neighbor.

This is a family of six; a man, his wife and four children. They live in a tent the color of the ground. Rot has set in on the canvas so that the flaps and the sides hang in tatters and are held together with bits of rusty baling wire. There is one bed in the family and that is a big tick lying on the ground inside the tent.

They have one quilt and a piece of canvas for bedding. The sleeping arrangement is clever. Mother and father lie down together and two children lie between them. Then, heading the other way, the other two children lie, the littler ones. If the mother and father sleep with their legs spread wide, there is room for the legs of the children.

There is more filth here. The tent is full of flies clinging to the apple box that is the dinner table, buzzing about the foul clothes of the children, particularly the baby, who has not been bathed nor cleaned for several days. This family has been on the road longer than the builder of the paper house. There is no toilet here, but there is a clump of willows nearby where human feces lie exposed to the flies—the same flies that are in the tent.

Two weeks ago there was another child, a four-year-old boy. For a few weeks they had noticed that he was kind of lackadaisical, that his eyes had been feverish. They had given him the best place in the bed, between father and mother. But one night he went into convulsions and died, and the next morning the coroner's wagon took him away. It was one step down.

They know pretty well that it was a diet of fresh fruit, beans and little else that caused his death. He had no milk for months. With this death there came a change of mind in his family. The father and mother now feel that paralyzed dullness with which the mind protects itself against too much sorrow and too much pain.

And this father will not be able to make a maximum of four hundred dollars a year any more because he is no longer alert; he isn't quick at piece-work, and he is not able to fight clear of the dullness that has settled on him. His spirit is losing caste rapidly.

The dullness shows in the faces of this family, and in addition there is a sullenness that makes them taciturn. Sometimes they still start the older children off to school, but the ragged little things will not go; they hide in ditches or wander off by themselves until it is time to go back to the tent, because they are scorned in the school.

The better-dressed children shout and jeer, the teachers are quite often impatient with these additions to their duties, and the parents of the "nice" children do not want to have disease carriers in the schools.

The father of this family once had a little grocery store and his family lived in back of it so that even the children could wait on the counter. When the drought set in there was no trade for the store any more.

This is the middle class of the squatters' camp. In a few months this family will slip down to the lower class. Dignity is all gone, and spirit has turned to sullen anger before it dies.

The next door neighbor family of man, wife and three children of from three to nine years of age, have built a house by driving willow branches into the ground and wattling weeds, tin, old paper and strips of carpet against them. A few branches are placed over the top to keep out the noonday sun. It would not turn water at all. There is no bed. Somewhere the family has found a big piece of old carpet. It is on the ground. To go to bed the members of the family lie on the ground and fold the carpet up over them.

The three-year-old child has a gunny sack tied about his middle for clothing. He has the swollen belly caused by malnutrition.

He sits on the ground in the sun in front of the house, and the little black fruit flies buzz in circles and land on his closed eyes and crawl up his nose until he weakly brushes them away.

They try to get at the mucous in the eye-corners. This child seems to have the reactions of a baby much younger. The first year he had a little milk, but he has had none since.

He will die in a very short time. The older children may survive. Four nights ago the mother had a baby in the tent, on the dirty carpet. It was born dead, which was just as well because she could not have fed it at the breast; her own diet will not produce milk.

After it was born and she had seen that it was dead, the mother rolled over and lay still for two days. She is up today, tottering around. The last baby, born less than a year ago, lived a week. This woman's eyes have the glazed, far-away look of a sleep walker's eyes. She does not wash clothes any more. The drive that makes for cleanliness has been drained out of her and she hasn't the energy. The husband was a share-cropper once, but he couldn't make it go. Now he has lost even the desire to talk. He will not look directly at you for that requires will, and will needs strength. He is a bad field worker for the same reason. It takes him a long time to make up his mind, so he is always late in moving and late in arriving in the fields. His top wage, when he can find work now, which isn't often, is a dollar a day.

The children do not even go to the willow clump any more. They squat where they are and kick a little dirt. The father is vaguely aware that there is a culture of hookworm in the mud along the river bank. He knows the children will get it on their bare feet. But he hasn't the will nor the energy to resist. Too many things have happened to him. This is the lower class of the camp.

This is what the man in the tent will be in six months; what the man in the paper house with its peaked roof will be in a year, after his house has washed down and his children have sickened or died, after the loss of dignity and spirit have cut him down to a kind of subhumanity.

Helpful strangers are not well-received in this camp. The local sheriff makes a raid now and then for a wanted man, and if there is labor trouble the vigilantes may burn the poor houses. Social workers, survey workers have taken case histories. They are filed and open for inspection. These families have been questioned over and over about their origins, number of children living and dead. The information is taken down and filed. That is that. It has been done so often and so little has come of it.

And there is another way for them to get attention. Let an epidemic break out, say typhoid or scarlet fever, and the country doctor will come to the camp and hurry the infected cases to the pest house. But malnutrition is not infectious, nor is dysentery, which is almost the rule among the children.

The county hospital has no room for measles, mumps, whooping cough; and yet these are often deadly to hunger-weakened children. And although we hear much about the free clinics for the poor, these people do not know how to get the aid and they do not get it. Also, since most of their dealings with authority are painful to them, they prefer not to take the chance.

This is the squatters' camp. Some are a little better, some much worse. I have described three typical families. In some of the camps there are as many as three hundred families like these. Some are so far from water that it must be bought at five cents a bucket.

And if these men steal, if there is developing among them a suspicion and hatred of well-dressed, satisfied people, the reason is not to be sought in their origin nor in any tendency to weakness in their character.

RESPONDING JOURNAL

Compare this particular image of life on the road, without a clear sense of where a family will find the money for its next meal—or whether it will even have a next meal—with pictures of life on the road you have seen on television news or read about elsewhere. What qualities suggest desperation in each? What differences do you find? Explore these ideas in a focused freewriting entry.

QUESTIONS FOR CRITICAL THINKING

STRATEGY

1. Consider this essay as a feature appearing in a daily newspaper. How does Steinbeck achieve his purpose of informing a wide audience?
2. Look at the first two paragraphs of the article. What is the effect of Steinbeck's technique? What does he cause you as a reader to *see*? How much does the visual quality of the description contribute to the reader's ability to empathize with the plight of the people being depicted?

ISSUES

1. Look up the words *sociology* and *ecology*, preferably in an unabridged dictionary. Provide an illustration of as many definitions of the words as you can by referring to this excerpt from *The Harvest Gypsies*.
2. Look at chapters 3, 5, and 15 from *The Grapes of Wrath*. Generate a series of words you find Steinbeck using in each of his two texts to show his audiences the important characteristics of the people he is describing. Evaluate the relative strengths of the two presentations based on the vocabularies you have examined. If you were to take your study further, what other elements of the two pieces of writing would you have to take into account?

COLLABORATIVE WRITING ACTIVITY

Look at the last paragraph of Steinbeck's article; find a contemporary instance of a parallel or similar situation. Bring your parallel example to class (or a fairly detailed summary if you are bringing in an example from broadcast media) and read it to your group. Discuss commonalities in the selections the group has presented.

"Silt"
Richard Wright

Richard Wright (1908–1960) is best known for his powerful and popular novel Native Son, *first published in 1940. In this searing portrayal of the grim life of a young Black man driven to murder on Chicago's South Side, Wright explored the cityscape of a great metropolis where liberal white attitudes toward race come into conflict with the reality of life for the Black citizens who live desperate lives in poverty and hopelessness. Born in Mississippi, Wright experienced stark terror as his mother was forced to move him and his siblings to Tennessee and later to Arkansas in search of even minimal stability in their lives. After his uncle was murdered by a White mob while Wright was staying with his aunt's family in Arkansas, he moved back to Mississippi to live with his grandmother. Wright records much of the anguish of these years in his autobiographical memoir* Black Boy, *published in 1945.*

In 1927 Wright moved to Chicago, where most of his surviving family members later joined him. Here he began to see first-hand the nuances of race relations that he would present in his major work. Wright's roots in rural culture provide both contrasts and parallels to his experiences in the black ghettoes of Chicago, and his work offers many avenues for exploring rural and urban poverty and racism.

By 1933 Wright had joined the Communist Party, and he began submitting his writing for publication. In 1937 he moved to New York in order to serve as Harlem editor for the Daily Worker. *His commitment to communism continued to grow throughout the decade, as his work in this period attests. "Silt" first appeared in the communist publication* New Masses *in 1937. It was later reprinted as "The Man Who Saw the Flood."*

EXPLORING JOURNAL

What images do you recall most vividly from media coverage of recent floods? If you do not recall the details of the floods in the midwest during the summer of 1993, think about other accounts of flooding you have read. Put yourself in the place of victims of flooding. What do you think would be the most devastating aspects of the flood for you personally? Think about those possessions or things you value that could be harmed by flooding. How might you respond to such an experience? Compose a focused journal entry on this topic.

At last the flood waters had receded. A black father, a black mother, and a black child tramped through muddy fields, leading a tired cow by a thin bit of rope. They stopped on a hilltop and shifted the bundles on their shoulders. As far as they could see the ground was covered with flood-silt. The little girl lifted a skinny finger and pointed to a mud-caked cabin.

"Look, Pa! Ain' that our home?"

The man, round-shouldered, clad in blue, ragged overalls, looked with bewildered eyes. Without moving a muscle, scarcely moving his lips, he said: "Yeah."

For five minutes they did not speak or move. The flood waters had been more than eight feet high here. Every tree, blade of grass, and stray stick had its flood-mark: caky, yellow mud. It clung to the ground, cracking thinly here and there in spider-web fashion. Over the stark fields came a gusty spring wind. The sky was high, blue, full of white clouds and sunshine. Over all hung a first-day strangeness.

"The hen house is gone," sighed the woman.

"N the pig pen," sighed the man.

They spoke without bitterness.

"Ah reckon them chickens is all done drowned."

"Yeah."

"Miz Flora's house is gone, too," said the little girl.

They looked at a clump of trees where their neighbor's house had stood.

"Lawd!"

"Yuh reckon anybody knows where they is?"

"Hard t' tell."

The man walked down the slope and stood uncertainly.

"There wuz a road erlong here somewheres," he said.

But there was no road now. Just a wide sweep of yellow, scalloped silt.

"Look, Tom!" called the woman. "Here's a piece of our gate!"

The gate-post was half buried in the ground. A rusty hinge stood stiff, like a lonely finger. Tom pried it loose and caught it firmly in his hand. There was nothing in particular he wanted to do with it; he just stood holding it firmly. Finally he dropped it, looked up, and said:

"C'mon. Le's go down n see whut we kin do."

Because it sat in a slight depression, the ground about the cabin was soft and slimy.

"Gimme tha' bag o' lime, May," he said.

With his shoes sucking in mud, he went slowly around the cabin, spreading the white lime with thick fingers. When he reached the front again he had a little left; he shook the bag out on the porch. The fine grains of floating lime flickered in the sunlight.

"Tha' oughta hep some," he said.

"Now, yuh be careful, Sal!" said May. "Don' yuh go n fall down in all this mud, yuh hear?"

"Yessum."

The steps were gone. Tom lifted May and Sally to the porch. They stood a moment looking at the half-opened door. He had shut it when he left, but somehow it seemed natural that he should find it open. The planks in the porch floor were swollen and warped. The cabin had two colors: near the bottom it was a solid yellow; at the top it was the familiar grey. It looked weird, as though its ghost were standing beside it.

The cow lowed.

"Tie Pat t' the pos' on the en' of the porch, May."

May tied the rope slowly, listlessly. When they attempted to open the front door, it would not budge. It was not until Tom had placed his shoulder against it and gave it a stout shove that it scraped back jerkily. The front room was dark and silent. The damp smell of flood-silt came fresh and sharp to their nostrils. Only one-half of the upper window was clear, and through it fell a rectangle of dingy light. The floors swam in ooze. Like a mute warning, a wavering flood-mark went high around the walls of the room. A dresser sat cater-cornered, its drawers and sides bulging like a bloated corpse. The bed, with the mattress still on it, was like a casket forged of mud. Two smashed chairs lay in a corner, as though huddled together for protection.

"Le's see the kitchen," said Tom.

The stove-pipe was gone. But the stove stood in the same place.

"The stove's still good. We kin clean it."

"Yeah."

"But where's the table?"

"Lawd knows."

"It must've washed erway wid the rest of the stuff, Ah reckon."

They opened the back door and looked out. They missed the barn, the hen house, and the pig pen.

"Tom, yuh bettah try tha ol' pump 'n see ef any watah's there."

The pump was stiff. Tom threw his weight on the handle and carried it up and down. No water came. He pumped on. There was a dry, hollow cough. Then yellow water trickled. He caught his breath and kept pumping. The water flowed white.

"Thank Gawd! We's got some watah."

"Yuh bettah boil it fo yuh use it," he said.

"Yeah. Ah know."

"Look, Pa! Here's yo ax," called Sally.

Tom took the ax from her. "Yeah. Ah'll need this."

"N here's somethin else," called Sally, digging spoons out of the mud.

"Waal, Ahma git a bucket n start cleanin," said May. "Ain no use in waitin, cause we's gotta sleep on them floors tonight."

When she was filling the bucket from the pump, Tom called from around the cabin. "May, look! Ah done foun mah plow!" Proudly he dragged the silt-caked plow to the pump. "Ah'll wash it n it'll be awright."

"Ah'm hongry," said Sally.

"Now, yuh jus wait! Yuh et this mawnin," said May. She turned to Tom. "Now, whutcha gonna do, Tom?"

He stood looking at the mud-filled fields.

"Yuh goin back t Burgess?"

"Ah reckon Ah have to."

"Whut else kin yuh do?"

"Nothin," he said. "Lawd, but Ah sho hate t start all over wid tha white man. Ah'd leave here ef Ah could. Ah owes im nigh eight hundred dollahs. N we needs a hoss, grub, seed, n a lot mo other things. Ef we keeps on like this tha white man'll own us body n soul. . . ."

"But, Tom, there ain nothin else t do," she said.

"Ef we try t run erway they'll put us in jail."

"It coulda been worse," she said.

Sally came running from the kitchen. "Pa!"

"Hunh?"

"There's a shelf in the kitchen the flood didn't git!"

"Where?"

"Right up over the stove."

"But, chile, ain nothin up there," said May.

"But there's somethin on it," said Sally.

"C'mon. Le's see."

High and dry, untouched by the flood-water, was a box of matches. And beside it a half-full sack of Bull Durham tobacco. He took a match from the box and scratched it on his overalls. It burned to his fingers before he dropped it.

"May!"

"Hunh?"

"Look! Here's muh 'bacco n some matches!"

She stared unbelievingly. "Lawd!" she breathed.

Tom rolled a cigarette clumsily.

May washed the stove, gathered some sticks, and after some difficulty, made a fire. The kitchen stove smoked, and their eyes smarted. May put water on to heat and went into the front room. It was getting dark. From the bundles they took a kerosene lamp and lit it. Outside Pat lowed longingly into the thickening gloam and tinkled her cowbell.

"Tha old cow's hongry," said May.

"Ah reckon Ah'll have t be gitting erlong t Burgess."

They stood on the front porch.

"Yuh bettah git on, Tom, fo it gits too dark."

"Yeah."

The wind had stopped blowing. In the east a cluster of stars hung.

"Yuh goin, Tom?"

"Ah reckon Ah have t."

"Ma, Ah'm hongry," said Sally.

"Wait erwhile, honey. Ma knows yuh's hongry."

Tom threw his cigarette away and sighed.

"Look! Here comes somebody!"

"Tha's Mistah Burgess now!"

A mud-caked buggy rolled up. The shaggy horse was splattered all over. Burgess leaned his white face out of the buggy and spat.

"Well, I see you're back."

"Yessuh."

"How things look?"

"They don look so good, Mistah."

"What seems to be the trouble?"

"Waal, Ah ain got no hoss, no grub, nothing. . . . The only thing Ah is got is tha ol cow there. . . . "

"You owe eight hundred dollahs down at the store, Tom."

"Yessuh, Ah know. But, Mistah Burgess, can't yuh knock somethin off of tha, seein as how Ahm down n out now?"

"You ate that grub, and I got to pay for it, Tom."

"Yessuh, Ah know."

"It's going to be a little tough, Tom. But you got to go through with it. Two of the boys tried to run away this morning and dodge their debts, and I had to have the sheriff pick em up. I wasn't looking for no trouble out of you, Tom. . . . The rest of the families are going back."

Leaning out of the buggy, Burgess waited. In the surrounding stillness the cowbell tinkled again. Tom stood with his back against a post.

"Yuh got t go on, Tom. We ain't got nothing here," said May.

Tom looked at Burgess.

"Mistah Burgess, Ah don wanna make no trouble. But this is jus *too* hard. Ahm worse off now than befo. Ah got to start from scratch. . . ."

"Get in the buggy and come with me. I'll stake you with grub. We can talk over how you can pay it back." Tom said nothing. He rested his back against the post and looked at the mud-filled fields.

"Well," asked Burgess. "You coming?" Tom said nothing. He got slowly to the ground and pulled himself into the buggy. May watched them drive off.

"Hurry back, Tom!"

"Awright. "

"Ma, tell Pa, t bring me some 'lasses, " begged Sally.

"Oh, Tom!"

Tom's head came out of the side of the buggy.

"Hunh?"

"Bring some 'lasses!"

"Hunh?"

"Bring some 'lasses fer Sal!"

"Awright!"

She watched the buggy disappear over the crest of the muddy hill. Then she sighed, caught Sally's hand, and turned back into the cabin.

———◆◆———

RESPONDING JOURNAL

In a substantial journal entry, describe what life is like for the family a year after the story takes place. What has happened in their lives? How have things changed for them? You may find that once you begin, you will want to continue for several pages. Try to use as many details as Wright does; show how the yard and house have changed.

QUESTIONS FOR CRITICAL THINKING

STRATEGY

1. Wright has decided on a particular structure to use in introducing his readers to his subject. How would you describe his strategy to involve you as a sympathetic reader in the plight of Tom and May and their family? What other means might he have used? Think of a term to describe this means of introducing a subject. Where else have you seen similar structures?
2. Consider the effect of dialect in relation to landscape. What does Wright show you about the land and the people's relation to that land that he could not have done without the dialect? How effective do you find it?

ISSUES

1. Consider the racial relations presented in this story. Explain where you think the issue of the land itself enters the equation between Blacks and Whites. To what extent do you think this story might depict typical events?
2. If a similar situation were to take place today, what would the conditions be like? How could you imagine a parallel event happening? Would such a sequence of events be more likely to take place in a city than on a farm? Explain your reasoning.

COLLABORATIVE WRITING ACTIVITY

How would small farmers like the family in "Silt" have fared during the floods of 1993 or some other devastating natural disaster? Update the account of a struggling family in the face of a natural disaster such as a flood; depict the family returning and the problems of moving back into their former home. Read your account to your group; discuss among yourselves what you feel you have learned from hearing other members' accounts of the post-disaster situations. Report on your group findings to the assembled class.

"January Journal"
Marcia Bonta

Author of Outbound Journeys in Pennsylvania *and* Women in the Field, *editor of the University of Pittsburgh Press's Pitt Nature Series, Marcia Bonta has written extensively on the natural world. A native of southern New Jersey, she nevertheless identified strongly with her father's native Pennsylvania. After "coming home" to college in Pennsylvania, and after brief side trips to Washington, D.C. for three years and Maine for five, she and her family found the place in central Pennsylvania near State College where she has been able to develop into, as she says, "a writer of place."*

Bonta's "January Journal" covers a month on her farm, which covers several hundred acres on a Pennsylvania ridge. Her dealings with neighbors, both human and animal, and the land itself furnish her with material for musings similar in kind to other writers who feel strong attachments to place—writers like Thoreau at Concord, Burroughs in the Catskills, or Dillard at Tinker Creek. You may find it interesting to compare her accounts of similar events to those of the other keepers of journals; for example, Thoreau shared Bonta's delight in tracking the trails of foxes in winter and musing on what their intricate paths might tell us about their habits and activities.

EXPLORING JOURNAL

Think about how you might describe the history of a month in your journal. What month would you choose? What makes that month so interesting? Explore these ideas in an entry.

January 1, 1991

Seventeen degrees at dawn and crystal clear. The gloom of December has been cast off, at least momentarily. I can only hope that the new weather pattern of light and peace will symbolize the new year, which everyone is fearing with its looming threat of war in the Middle East. Now that we have celebrated the birth of a Man who symbolized light and peace with our usual mix of idealism and materialism, we have resumed business as usual in the real world.

I sat, at sunset, tucked among the Norway spruce trees while dark-eyed juncos zipped in over my head so closely that I felt the wind from their wings. They made their clicking, scolding sounds when they saw me. But eventually they settled in for the night in nearby spruces, one or two to a tree.

Suddenly a loud screaming rent the pre-dusk stillness somewhere below me in the vicinity of the Far Field Road. At first I stayed where I was, since usually my investigations of odd noises are fruitless. But the sounds continued unabated, and I finally stood up and walked to an area overlooking the road. Still I could see nothing, but the screaming went on and on. Then I sat down and scoped the area below with my binoculars. Almost at once I spotted a large bird, looking mostly dark in the dim light except for a white line above its eye, leaping around on the ground and screaming like an angry troll. As I started toward it, it took off fast and low to the ground and was gone in an instant. Fixing my sights on the place it had been, I climbed down to find one small, downy, gray feather. Had the bird been pummeling small prey, such as a junco, or did the feather belong to the bird? Although I searched the area, I found no other clues to the bird's identity or hints regarding the behavior I had observed.

I hurried home to check my field guides. The most likely bet, I decided after studying the pictures, was an immature northern goshawk, but the experts claimed that northern goshawks are silent birds except during the mating season. Next I listened to my bird call record and easily identified the screaming as that of a northern goshawk. No other bird of prey sounded even remotely like it. To cinch the matter, I called our local bird expert, and he told me he had seen a couple of northern goshawks in the valley below our property a couple of days earlier. Later, I talked to a graduate student who is studying northern goshawks and who has been a licensed falconer for many years. I had witnessed, he said, what biologists call "play behavior" in young northern goshawks.

What a wonderful beginning for the new year!

January 2, 1991

Twenty degrees and cloudless at dawn, warming up to forty-eight degrees in the sun by mid-afternoon with clouds rolling up from the south. I sat in the woods overlooking the Far Field thicket and listened to the earth sigh and mutter with every vagrant breeze. In the distance pileated woodpeckers called. Other birds cheeped occasionally, but mostly the woods were wrapped in winter silence. Walking back up the Far Field Trail, I heard a quiet tapping and looked up to see a female hairy woodpecker working over a red oak tree branch. She was the only creature I saw during my two-hour walk.

I wondered, as I moved along, why I feel compelled to be outside every day, and I finally decided that my job is to bear witness to the beauty of the earth. I am neither a scientist nor an environmental writer. To me, the outdoors means just that—being outdoors—not to hunt, fish, hike, canoe, bike, or bird, but to be, even in the winter—especially in the winter—when nature is stripped to its bare essentials and few people are abroad.

Sometimes I'm tempted to move somewhere else, to another country even, and to learn a whole new culture, a whole new concept of nature. But I cannot bring myself to say goodbye to these old hills forever, despite the fact that they have been degraded by humanity's misuse. So I remain, a stubborn naturalist of place, content to look for the unusual in what many see as a commonplace and unexciting area that lacks the glamour of the north woods, the seacoast, the Rocky Mountains, or the desert. But there is always more to discover here, as I learned yesterday.

Again I sat in the spruces after sunset, but instead of a northern goshawk I heard a pair of great horned owls calling back and forth while occasional dark-eyed juncos flicked into the grove to roost for the night.

January 16, 1991

Thirty-six degrees. I awoke to hear it raining this morning, and all day I have been encased in fog up to my doorstep. A song sparrow flew into the feeder. So did a single American goldfinch in his winter garb. Both are new arrivals to the feeder for the year. In mid-afternoon I watched a male downy woodpecker examining the dead, bent-over stalks of pokeweed below our back porch, but he was not tempted by the suet on the feeder.

Near dusk I took a walk in the fog and had about fifty feet of visibility. Nothing stirred except the juncos in the spruce grove at the top of First Field. I sat there, wrapped in muffled silence, the usual sounds from the valley muted by the fog. Raindrops hung in solitary splendor from the tips of the spruce needles. Juncos chirped at me from their sheltered perches deep within the spruce boughs. They seem to come into their night roosts one by one and from different directions, not in a concentrated flock as some researchers have reported. The same is true of the juncos who use the juniper bush beside the house.

A good day to start a war—gloomy with rain and fog—and so the Gulf War began early this evening with bombing raids on Baghdad. Yet here on our mountain it is so peaceful and beautiful, even on a rainy, foggy day. I am grateful to be living in such a place during these troubled, tumultuous times. Yet when, in the history of civilized humanity, have the times *not* been troubled and tumultuous? To take solace from nature instead of humanity is the true balm of Gilead for the human spirit, distraught as some of us are with the warrior mentality of so many human beings. Where is the peace of God that passeth all understanding? Why does the Prince of Peace's message fall mostly on deaf ears, even among Christians? I believe it is because religion, to most people, is helpful only in times of trouble. Otherwise, it is highly impractical, and its basic tenets are easy to ignore if you look at humanity with clear-eyed vision. In fact, religion has served as a rallying war cry more than once and is now doing so again. . . .

January 29, 1991

Twenty-four degrees at dawn with Sapsucker Ridge lit scarlet from the rising sun. Jet trails remained like long white fingers in the southeast while the rest of the sky was blue. The machines at the limestone quarry in the valley were so loud that their noise penetrated the walls of the house this morning, and later, when I sat above the Far Field Road, I could still hear them. Other valley sounds also funneled up distinctly, so the illusion of winter's peace was shattered.

Several days ago the war took on an even more horrifying aspect as Iraq began flooding the Persian Gulf with Kuwaiti oil, creating the largest oil slick in history. Experts on cleaning up oil spills were rushed to the Gulf, and a bombing raid was launched to destroy the pumps emitting the oil. But what madness! Destroying an entire ecosystem from which all the Gulf area residents prosper. Birds and fish are dead or dying, and the slick is moving inexorably toward Saudi Arabia, where desalination plants process saltwater into drinking water. And there lies the explanation for what seemed at first like unreasoning madness. Destroy a civilization's water source and you destroy a civilization, especially a desert people's. But such a deed demonstrates how sick humanity is—always putting the petty ambitions of people above the nurturing of the natural world. The planet seems too small to support so many rapacious humans bent on victory at all costs. Of course, we did the same thing in the Vietnam War with defoliants, and we tried to talk Peru into using them to wipe out the drug trade. Why must nature pay for humanity's deeds? Or, more to the point, when will we discover that everything *is* connected to everything else and that when we fiddle with it we imperil our own survival?

An incident from William Warner's *Beautiful Swimmers* says it all regarding humanity's relationship with nature. A Maryland biologist tried to explain new conservation measures to the Tangier Sound Watermen's Union on Smith Island and asked for discussion. According to Warner, one islander finally rose to his feet and said, in answer, "Mr. Manning, there is something you don't understand. These here communities on the Shore, our little towns here on the island and over to mainland, was all founded on the right of free plunder. If you follow the water, that's how it was and that's how it's got to be." The right of free plunder has beggared us and will continue to do so.

I sat quietly in the corner of First Field listening to the tapping of an unseen woodpecker. After a time I found it—a female pileated quietly tapping and then lifting her head as if she were listening for the sound of carpenter ants in the bark. She repeated that several times before flying off. Next, a pair of foraging golden-crowned kinglets flew into the area, followed by a white-breasted nuthatch, several black-capped chickadees, and a female downy woodpecker. The downy landed in a tree next to me and made more noise than the pileated. All ignored my presence as they went about their business, so I had the rare pleasure of feeling as if I were an integral part of the scene and not an unwanted interloper.

Continuing my exploration of Sapsucker Ridge, I discovered a new fox hole at the base of an old stump near the Sapsucker Ridge powerline right-of-way. I also disturbed two deer basking in the sunshine of the Sapsucker Ridge thicket. One bounded quickly up the hill, while the other leaped slowly along. The bottom half of its right front leg was missing, probably shot off during hunting season. Several years ago we watched a doe that had lost most of her left hind leg not only learn to run nearly as fast as the other deer she associated with, but give birth to twins and raise them to maturity.

In the warm sun of mid-afternoon the thermometer registered fifty degrees on the veranda. I heard a raven croaking through the walls of the house and went out to watch as it flew from its perch in the woods.

January 30, 1991

Forty-eight degrees by mid-morning and absolutely clear. I sunbathed against the Far Field Road bank in utter peace. Winter silence on top of a worn-down, ancient Pennsylvania mountaintop, with only the cries of occasional birds to keep me company, can be as purifying an experience as forty days in the desert. Life is stripped to its barest essentials to survive the winter, and the sun pours from the sky unimpeded by tree leaves. I can see far in any direction, so I tend to spot the shyer creatures, such as porcupines, foxes, even a bobcat once, in the winter. But my companions today were more commonplace—a pileated woodpecker flying silently past, a golden-crowned kinglet calling and fluttering in front of me, a tree sparrow flying up from the brush to shake its feathers at me, four deer filing up Sapsucker Ridge.

Later, I sat at the crest of First Field surveying the landscape and thinking, I hold all the joys of life in two hands—a loving, encouraging husband who has nurtured my growth and unselfishly lived here for my pleasure, three sons who were my close companions in their childhood and youth and who still share most of their lives with me, a home that is warm and simply furnished with minimal fancy furniture and maximum books, a handful of friends who share my concerns and interests, and a stable life-style dependent more on spiritual than material things.

January 31, 1991

Thirty-three degrees at dawn. An absolutely radiant day, not a cloud in the sky, and I was out at eight in the morning to walk down Laurel Ridge toward the Tyrone Gap. As I reached the end of the ridge, I had a graphic view of the limestone quarry. Layer after layer of gouged-out hillside with all the noisy, earth-eating machines sitting at rest in the bottom of the desolation. Without that sight and the sound of vehicles from the valley, the view of blue-misted hills and rolling farm fields would have been timeless. But people seem to adjust to the gradual degradation of their environment. Anything to further the technological success of our civilization. What? Question the folly of removing whole mountainsides as quickly as possible to fix up and build more roads? You can't hold back progress.

Winter is the time to see visions. I seem to hear prophetic voices, railing at humanity's failings, prophesying the end of the world. Light pours from the sky, bathing my thoughts with the heat of conviction. They are voices crying in the wilderness, preparing the way for doom that no one believes and no one heeds. The earth withers away as I watch, and I am stunned by the rapidity with which it destructs once it begins to unravel from too much use and too little love.

———◦•◦———

RESPONDING JOURNAL

Think about the references to the Gulf War in Bonta's journal. Is this a new means of exploring landscape in journals, or do you remember seeing journalkeepers' thoughts on issues of the day in other journals? What is the relationship between the land and the war?

QUESTIONS FOR CRITICAL THINKING

STRATEGY

1. How does Bonta's entry for January 1 establish a sense of her personality and the way she goes about her work as a writer? What sense of beginning, middle, and end do you see in her entry?

2. Bonta notes in her entry for January 16 that she is "grateful to be living in such a place during these troubled, tumultuous times." How does the journal serve the purpose of relieving her mind? Do you find other similar instances here?

ISSUES

1. Explain the main idea that you see Bonta developing in her January 2 entry. What seems to be the point of her musing about her role and her feeling of belonging to a certain place?

2. Consider Bonta's last comment: "The earth withers away as I watch, and I am stunned by the rapidity with which it destructs once it begins to unravel from too much use and too little love." To what extent do your experiences affirm or deny the truth of this statement?

COLLABORATIVE WRITING ACTIVITY

Write a journal entry or the draft of an essay on Bonta's comment in her January 31 entry that "people seem to adjust to the gradual degradation of their environment. Anything to further the technological success of our civilization." To what extent do you agree with her position? Bring your drafts back to your group and exchange them; discuss the differences you see and the ways in which your classmates have interpreted Bonta's ideas as she expresses them in her journal.

"Sawdust"
Jane Shore

Jane Shore's poetry has been collected in several volumes, most recently in Eye Level *and* The Minute Hand. *A keen observer of the natural world of New England, she focuses on human connections to the larger world.*

"Sawdust" may well echo other poems, essays, or fiction you have read. Do you recall any experiences of your own as you read about the felling of the sugar maple? What does the title do for the recording of the sawing, for

Shore's reactions, or for the concept of writing about human movement through the natural world?

Exploring Journal

What are some of the uses of sawdust? What do you know about what happens to it as a byproduct? What sorts of connotations does it have for you? Give your mind the chance to play over the idea of sawdust.

———◆◆———

They've cut down the old sugar maple
that stood for more than one hundred years
on our lawn in a line with her five other sisters
since at least my grandmother was born.
Late October, the leaves lay in rusting heaps
under a sky raw with storm.

The tree *looked* ill.
All summer it ailed, its trunk
filigreed with lichen.
Last year's drought, and years of acid rain—
the hundred gallons of fertilizer
we pumped into the ground around its roots
was like water down a drain.
We got a second opinion.
The next big storm, if a big branch fell,
and crushed a passing car,
we were responsible.

Late morning, the truck arrived
with heavy equipment,
and two men and a crane.
After filling their chain-saws with gasoline,
one man climbed into the elevator bucket
which levitated him high into the branches.
To the hard-hatted men, tan as surfers,
trees are a religion.

They switched on their saws
and began the chain-saw mating call,
the deafening whines
pushing higher and higher in tandem.
The man in the tree lashed himself, like Odysseus
to the mast, to his bucket in the branches,

while the other shouted,
from the ground, and supervised.

There is a strategy
for cutting down a tree.
You start from the bottom up,
making your selection,
then work your way up the trunk,
roping the main branches,
lopping off a branch, a larger branch.
The dead ones hit the ground with a hollow sound.

High in the second story bedroom,
my breath fogged the window.
Sawdust sprinkling the grass like snow.

Then the tree stood, a naked torso.
A deep V cut into the trunk, down low.

Then the tree fell.

No need to get sentimental.

After they cut the trunk into logs and raked the lawn
and stacked the kindling,
and fed the twigs and branches to the grinder
that chewed them into chips,

it was quiet,
quieter than before,
quieter than the world was, ever.

Old tree, old grandmother,
your stump staring straight up into the sky.

For the first time we had
a clear view of the road,

the rising moon.
A consolation.

And for the next two days,
our elderly neighbor, Morris,
split logs that would last him the winter;
his pounding ax
a heartbeat shaking our house.

RESPONDING JOURNAL

When the tree falls, Shore says, "No need to get sentimental." What does she mean? Is she thinking about herself or about us, her readers? Respond in a journal entry.

QUESTIONS FOR CRITICAL THINKING

STRATEGY

1. Explain what Shore accomplishes in her first stanza, lines 1–6. How else might she have started her poem? What does she gain from her opening lines?
2. What effect does the final image of the poem have on you? What sort of "poetic justice" does this last line have for the reader of the poem?

ISSUES

1. What is Shore's opinion of the men who cut down the tree? What shows you how she feels about them? Which images do you find most effective?
2. How could the time after the cutting down of the tree be "quieter than the world was, ever"? How accurate is Shore being here? What is the point of this statement?

COLLABORATIVE WRITING ACTIVITY

Take the idea "There is a strategy for cutting down a tree" and apply it to another activity that relates to the natural world. Describe this comparison in a journal entry; bring the entry to your group and allow each member to ask questions about the entry in order to understand it more fully. After that, use another journal entry to explain what you learned from the exchange of ideas.

SUGGESTIONS FOR FORMAL WRITING ASSIGNMENTS

1. Decide on an experience that you have had in the world of nature as you moved away from your "comfort zone," the immediate area of home. You'll want to choose a situation in which you felt threatened, scared, or severely challenged. Write down several possible means of presenting an account of this experience; use the essays in this chapter as models, modifying the organizational patterns as you see the need. Look over the short list of organizational patterns, and choose the one that you think will work the best to write your essay.

2. Think of your backyard when you were a child, or an area that would be like a backyard. (I recall vividly a patch of bare ground next to an alley between tenement buildings; I was five or six and had a chance to play there with an older cousin, whom I revered.) Take your reader through that backyard and explain the significance of the objects and memories the place holds for you. Think of a logical spatial plan to cover these important points, and think as well of a good order in which to describe these points. What would be the best vantage point for you to situate yourself while you conduct your description? What relative advantages would different positions hold? Write the descriptive essay that results.

PART TWO

Assembling Your Materials

Chapter 3

ORGANIZING AND SEQUENCING: CONCEPTUALIZING THE WORLD

Walker Evans, *Graveyard and Steel Mill, Bethlehem, Pennsylvania,* **1936**

1. What effect does the cross in the foreground have on the rest of the image? How does it cause you to respond to the photo?

2. If the photo were in color, how would its mood and meaning change? Does the photograph draw any strength from its black-and-white nature? If so, what?

3. Explain the idea of topography you get from looking at this photograph. How does this sense of the shape of the land relate to the lives of the people who must inhabit it? Explain how the repeated vertical and horizontal lines might reinforce some of these ideas.

Chapter 3

ORGANIZING AND SEQUENCING: CONCEPTUALIZING THE WORLD

Once you have generated some ideas and have a sense of the details and examples you will be using, you can decide how best to present this specific information. Next you will need to sharpen your *focus,* your sense of the key point you will be making in your essay. This focal point will be expressed in your thesis, the basic statement of your ideas.

Many experienced teachers of writing, such as Donald Graves and Donald Murray, promote what they call "rehearsal" as a stage in the writing process. Rehearsal is the process of setting up in your mind what it is you plan to write. You go through the steps of your essay or presentation in your head before you commit them to paper or computer. You prepare yourself mentally for your work, just as you would if you were getting ready to go on stage to act in a play. In this way, you have the chance to run through the ideas and organization of your essay several times so you can adjust and modify it before you actually begin setting words to paper.

This process of rehearsing to see exactly how you want to proceed is a means of determining your point of view as well. You have to decide the best position from which to present your essay, to "tell your story." In this chapter, Mary Austin imagines the ground-level view of a ground squirrel to make her points about the desert. You may not be quite so unusual in your decision of how to approach your subject, but you will see how different ways of presenting your ideas to a reader can help.

You will need to get a definite reader in mind. If you are writing in a letter to a friend an elaboration of your ideas on a subject of special interest to you, or if you are presenting an argument in writing to the CEO of a local corporation that you fear is not managing its waste properly, then you are addressing a single person. You can easily picture how this person might react, and you can plan your essay accordingly. Aiming an essay at a single reader is an effective technique even when you are writing for a more diverse group. When writing for class, for example, think about how a classmate you respect would respond to what you are writing. Create a worthy audience for your writing so that you will see the need to make your work as effective as possible. Since your sense of audience and your purpose in writing are so directly related, you will want, for example, to have your classmates and instructor clearly in mind as you focus your writing for this course.

In order to be sure you have covered your working ideas to the satisfaction of your audience, you can focus on two keys:

1. What questions will my reader ask?
2. What should I do to answer them?

You will probably notice right away that the questions your reader will ask will depend on your thesis. What is it that you want to say? How might you best express these ideas to satisfy your reader and to convince her of your points? Your classmates can be a tremendous help here. If you work in peer or collaborative groups, you are already accustomed to seeing what others can tell you about your work, and what you can tell your classmates about their ideas and writing. Using peer responses is a crucial early step in the process of organizing to make your writing as effective as it can be.

Probably the most time-honored method of determining effective order and sequence is outlining. Many writers, however, find the formal outline—with its strictly prescribed conventions—confining. They prefer to modify the outline concept to find a way of setting down their ideas or visualizing their work, without having to stop to worry about whether or not they have satisfied all the requirements of the outline (if you have an "A" you must have a "B," if you have a "1" you must have a "2"). Such modified, informal outlines still provide the benefits of the more formal ones in that they give you an overview of your projected writing and allow you to pick out potential problems or suggest where you need more work. As you work along, you see where you want to modify your original plan and then change the organization as necessary. Several software programs provide an outline-maker, a form that will take you through the steps of an outline in order to help you see how your proposed project will work in outline form.

Think of an outline as an educated guess. You will not have to face any penalties if you decide halfway through the writing of your essay that your outline is not doing the job for you and you have to dismiss it. Regard your outline as your potential guide, but one that you can readily cast aside when you find that it has stopped helping you. Lists can be as helpful as outlines, and they cause less stress because you do not have the mental baggage of obligation to the formal outline structure. Just jot down the ideas as they come to you and take a look at them; then, if you wish, you can arrange the items that you select from the list into an outline.

Some writers like to use notecards instead of outlines. They put each major heading on a separate card, and then they lay them out on a flat surface to see how they look together. When the writer begins to see how different arrangements might work and starts to shuffle the cards around, the benefits of this system jump out. With notecards, you can easily add or subtract headings, and of course the cards are right there if you decide a discarded heading should go back in at a different point in your organization. This is an especially helpful technique if you are having trouble seeing how a fairly detailed or complex presentation of material might work. The time spent making the cards will be well repaid by the quality of your final effort.

Finally, remind yourself of a few basic principles that will give you the confidence to feel you have a solid plan for writing. When composing, you are trying to get as much of your thinking on paper as possible. For this reason, many people talk about "writing as discovery" or "writing to learn." You are trying to see where your ideas on a particular topic will take you, and pushing your pencil across the paper or generating page after page on the word processor will help you to focus your thoughts. This is where journals come in: they serve as sounding boards and experimental sites where you can try out ideas.

As you can see, a writer can choose among several different ways to organize and order materials, and you are probably beginning to realize the complexity of overlaying different patterns of organization on your material so that you can discover how to present your ideas most effectively to your reader. After we generate sufficient material to have something to write about, we have to decide what will work effectively to present these ideas for other readers. The main means of development are:

Chronological

Sequential

Spatial

Chronological refers solely to time. Any sort of organization that follows the order in which events occur or have occurred is chronological.

Sequential organization is not tied to a single concept, like time. Rather, the sequence can be organized on several principles such as the following:

1. Ascending importance occurs when the sequence of events, examples, details, or illustrations moves from least important to most important.
2. Descending importance occurs when the sequence of events, examples, details, or illustrations moves from most important to least important.

Spatial organization depends on the distribution of objects or features in physical space. This scheme of organization is particularly important in discussing landscapes, of course, but it can also help to present a logical path through a neighborhood or community.

Other patterns of sequence can be based on other criteria. Much modern art, particularly literature and film, provides examples of sequencing that are difficult to understand if we cannot see, for example, the importance of the events to the consciousness of a central character. The details in Barry Lopez' *River Notes,* for example, depend on our knowing that they are perceived through the eyes of a raven. This knowledge makes the organizational pattern clear.

In addition to these main organizing principles, other useful patterns fall within the three main categories, or even share characteristics with more than one group. The most obvious pattern is the 5 W's (who-what-when-where-why plus how), which most news stories illustrate. You can also establish patterns in your own writing by considering recollections of people or places, relations among parts, process in the natural or mechanical world, or more linear patterns such as following old trails or moving outward from a central point. All of these strategies appear in this chapter or in the preceding one. Can you identify them?

The writers in this section have gone through the process both of rehearsing, or of thinking through the subjects they cover, and organizing their efforts in order to achieve the desired effects upon the reader. Annie Dillard begins her meditations on Tinker Creek by suggesting the elementally savage nature of her pet, bearing evidence of bloody deeds, as it snuggles up to her. W.E.B. Du Bois sets the problems of American racism against a background of universal suffering in a foreign war while establishing a backdrop of the emotionally stirring landscapes of Manhattan and the Grand Canyon. Steven Vincent Benét imagines what a pioneer would focus on as he looked back at his life from the vantage point of the grave. David Rains Wallace finds parallels between the features of a familiar hammock and his own mind. Mary Austin sees rodent trails in much the same way we regard our own highways, while John Burroughs finds in the young boys he meets on a river voyage much that reminds him of his own youth on that river. These different organizational patterns have in common the desire to present ideas in a form suitable to the point of view, thesis, and audience of the essay.

"Heaven and Earth in Jest"
from *Pilgrim at Tinker Creek*
Annie Dillard

Annie Dillard has become one of the most acclaimed contemporary writers of the natural world, and her 1974 book Pilgrim at Tinker Creek *has become an American classic. She has explored the twin processes of observing the natural world and writing in her other nonfiction works, including her memoir* An American Childhood *and the collection of essays* Teaching a Stone to Talk. *She is also a poet; many of her poems are collected in* Tickets for a Prayer Wheel.*

Tinker Creek actually exists, in western Virginia, near where Dillard attended Hollins College. There she began writing about her intense concentration on the natural world. Reprinted here is the first chapter from Pilgrim at Tinker Creek; *many other sections of the book have been frequently reprinted.*

EXPLORING JOURNAL

Exactly what do you think of when you read the word *pilgrim*? What kind of response to the world of nature would you expect from a pilgrim? Give yourself 12–15 minutes to do a structured freewrite on this topic.

———•◦•———

I used to have a cat, an old fighting tom, who would jump through the open window by my bed in the middle of the night and land on my chest. I'd half-awaken. He'd stick his skull under my nose and purr, stinking of urine and blood. Some nights he kneaded my bare chest with his front paws, powerfully, arching his back, as if sharpening his claws, or pummeling a mother for milk. And some mornings I'd wake in daylight to find my body covered with paw prints in blood; I looked as though I'd been painted with roses.

It was hot, so hot the mirror felt warm. I washed before the mirror in a daze, my twisted summer sleep still hung about me like sea kelp. What blood was this, and what roses? It could have been the rose of union, the blood of murder, or the rose of beauty bare and the blood of some unspeakable sacrifice or birth. The sign on my body could have been an emblem or a stain, the keys to the kingdom or the mark of Cain. I never knew. I never knew as I washed, and the blood streaked, faded, and finally disappeared, whether I'd purified myself or ruined the blood sign of the passover. We wake, if we ever wake at all, to mystery, rumors of death, beauty, violence. . . . "Seem like we're just set down here," a woman said to me recently, "and don't nobody know why."

These are morning matters, pictures you dream as the final wave heaves you up on the sand to the bright light and drying air. You remember pressure, and a curved sleep you rested against, soft, like a scallop in its shell. But the air hardens your skin; you stand; you leave the lighted shore to explore some dim headland, and soon you're lost in the leafy interior, intent, remembering nothing.

I still think of that old tomcat, mornings, when I wake. Things are tamer now; I sleep with the window shut. The cat and our rites are gone and my life is changed, but the memory remains of something powerful playing over me. I wake expectant, hoping to see a new thing. If I'm lucky I might be jogged awake by a strange birdcall. I dress in a hurry, imagining the yard flapping with auks, or flamingos. This morning it was a wood duck, down at the creek. It flew away.

I live by a creek, Tinker Creek, in a valley in Virginia's Blue Ridge. An anchorite's hermitage is called an anchor-hold; some anchor-holds were simple sheds clamped to the side of a church like a barnacle to a rock. I think of this house clamped to the side of Tinker Creek as an anchor-hold. It holds me at anchor to the rock bottom of the creek itself and it keeps me steadied in the current, as a sea anchor does, facing the stream of light pouring down. It's a good place to live; there's a lot to think about. The creeks—Tinker and

Carvin's—are an active mystery, fresh every minute. Theirs is the mystery of the continuous creation and all that providence implies: the uncertainty of vision, the horror of the fixed, the dissolution of the present, the intricacy of beauty, the pressure of fecundity, the elusiveness of the free, and the flawed nature of perfection. The mountains—Tinker and Brushy, McAfee's Knob and Dead Man—are a passive mystery, the oldest of all. Theirs is the one simple mystery of creation from nothing, of matter itself, anything at all, the given. Mountains are giant, restful, absorbent. You can heave your spirit into a mountain and the mountain will keep it, folded, and not throw it back as some creeks will. The creeks are the world with all its stimulus and beauty; I live there. But the mountains are home.

The wood duck flew away. I caught only a glimpse' of something like a bright torpedo that blasted the leaves where it flew. Back at the house I ate a bowl of oatmeal; much later in the day came the long slant of light that means good walking.

If the day is fine, any walk will do; it all looks good. Water in particular looks its best, reflecting blue sky in the flat, and chopping it into graveled shallows and white chute and foam in the riffles. On a dark day, or a hazy one, everything's washed-out and lackluster but the water. It carries its own lights. I set out for the railroad tracks, for the hill the flocks fly over, for the woods where the white mare lives. But I go to the water.

Today is one of those excellent January partly cloudies in which light chooses an unexpected part of the landscape to trick out in gilt, and then shadow sweeps it away. You know you're alive. You take huge steps, trying to feel the planet's roundness arc between your feet. Kazantzakis says that when he was young he had a canary and a globe. When he freed the canary, it would perch on the globe and sing. All his life, wandering the earth, he felt as though he had a canary on top of his mind, singing.

West of the house, Tinker Creek makes a sharp loop, so that the creek is both in back of the house, south of me, and also on the other side of the road, north of me. I like to go north. There the afternoon sun hits the creek just right, deepening the reflected blue and lighting the sides of trees on the banks. Steers from the pasture across the creek come down to drink; I always flush a rabbit or two there; I sit on a fallen trunk in the shade and watch the squirrels in the sun. There are two separated wooden fences suspended from cables that cross the creek just upstream from my tree-trunk bench. They keep the steers from escaping up or down the creek when they come to drink. Squirrels, the neighborhood children, and I use the downstream fence as a swaying bridge across the creek. But the steers are there today.

I sit on the downed tree and watch the black steers slip on the creek bottom. They arc all bred beef: beef heart, beef hide, beef hocks. They're a human product like rayon. They're like a field of shoes. They have cast-iron shanks and tongues like foam insoles. You can't see through to their brains as you can with other animals; they have beef fat behind their eyes, beef stew.

I cross the fence six feet above the water, walking my hands down the rusty cable and tight roping my feet along the narrow edge of the planks. When

I hit the other bank and terra firma, some steers are bunched in a knot between me and the barbed wire fence I want to cross. So I suddenly rush at them in an enthusiastic sprint, flailing my arms and hollering, "Lightning! Copperhead! Swedish meatballs!" They flee, still in a knot, stumbling across the flat pasture. I stand with the wind on my face.

When I slide under a barbed-wire fence, cross a field, and run over a sycamore trunk felled across the water, I'm on a little island shaped like a tear in the middle of Tinker Creek. On one side of the creek is a steep forested bank; the water is swift and deep on that side of the island. On the other side is the level field I walked through next to the steers' pasture; the water between the field and the island is shallow and sluggish. In summer's low water, flags and bulrushes grow along a series of shallow pools cooled by the lazy current. Water striders patrol the surface film, crayfish hump along the silt bottom eating filth, frogs shout and glare, and shiners and small bream hide among roots from the sulky green heron's eye. I come to this island every month of the year. I walk around it, stopping and staring, or I straddle the sycamore log over the creek, curling my legs out of the water in winter, trying to read. Today I sit on dry grass at the end of the island by the slower side of the creek. I'm drawn to this spot. I come to it as to an oracle; I return to it as a man years later will seek out the battlefield where he lost a leg or an arm.

A couple of summers ago I was walking along the edge of the island to see what I could see in the water, and mainly to scare frogs. Frogs have an inelegant way of taking off from invisible positions on the bank just ahead of your feet, in dire panic, emitting a froggy "Yike!" and splashing into the water. Incredibly, this amused me, and, incredibly, it amuses me still. As I walked along the grassy edge of the island, I got better and better at seeing frogs both in and out of the water. I learned to recognize, slowing down, the difference in texture of the light reflected from mudbank, water, grass, or frog. Frogs were flying all around me. At the end of the island I noticed a small green frog. He was exactly half in and half out of the water, looking like a schematic diagram of an amphibian, and he didn't jump.

He didn't jump; I crept closer. At last I knelt on the island's winterkilled grass, lost, dumbstruck, staring at the frog in the creek just four feet away. He was a very small frog with wide, dull eyes. And just as I looked at him, he slowly crumpled and began to sag. The spirit vanished from his eyes as if snuffed. His skin emptied and drooped; his very skull seemed to collapse and settle like a kicked tent. He was shrinking before my eyes like a deflating football. I watched the taut, glistening skin on his shoulders ruck, and rumple, and fall. Soon, part of his skin, formless as a pricked balloon, lay in floating folds like bright scum on top of the water: it was a monstrous and terrifying thing. I gaped bewildered, appalled. An oval shadow hung in the water behind the drained frog; then the shadow glided away. The frog skin bag started to sink.

I had read about the giant water bug, but never seen one. "Giant water bug" is really the name of the creature, which is an enormous, heavy-bodied brown bug. It eats insects, tadpoles, fish, and frogs. Its grasping forelegs are mighty and hooked inward. It seizes a victim with these legs, hugs it tight, and paralyzes it with enzymes injected during a vicious bite. That one bite is the only bite it ever takes. Through the puncture shoot the poisons that dissolve the victim's muscles and bones and organs—all but the skin—and through it the giant water bug sucks out the victim's body, reduced to a juice. This event is quite common in warm fresh water. The frog I saw was being sucked by a giant water bug. I had been kneeling on the island grass; when the unrecognizable flap of frog skin settled on the creek bottom, swaying, I stood up and brushed the knees of my pants. I couldn't catch my breath.

Of course, many carnivorous animals devour their prey alive. The usual method seems to be to subdue the victim by downing or grasping it so it can't flee, then eating it whole or in a series of bloody bites. Frogs eat everything whole, stuffing prey into their mouths with their thumbs. People have seen frogs with their wide jaws so full of live dragonflies they couldn't close them. Ants don't even have to catch their prey: in the spring they swarm over newly hatched, featherless birds in the nest and eat them tiny bite by bite.

That it's rough out there and chancy is no surprise. Every live thing is a survivor on a kind of extended emergency bivouac. But at the same time we are also created. In the Koran, Allah asks, "The heaven and the earth and all in between, thinkest thou I made them *in jest?*" It's a good question. What do we think of the created universe, spanning an unthinkable void with an unthinkable profusion of forms? Or what do we think of nothingness, those sickening reaches of time in either direction? If the giant water bug was not made in jest, was it then made in earnest ? Pascal uses a nice term to describe the notion of the creator's, once having called forth the universe, turning his back to it: *Deus Absconditus*. Is this what we think happened? Was the sense of it there, and God absconded with it, ate it, like a wolf who disappears round the edge of the house with the Thanksgiving turkey? "God is subtle," Einstein said, "but not malicious." Again, Einstein said that "nature conceals her mystery by means of her essential grandeur, not by her cunning." It could be that God has not absconded but spread, as our vision and understanding of the universe have spread, to a fabric of spirit and sense so grand and subtle, so powerful in a new way, that we can only feel blindly of its hem. In making the thick darkness a swaddling band for the sea, God "set bars and doors" and said, "Hitherto shalt thou come, but no further." But have we come even that far? Have we rowed out to the thick darkness, or are we all playing pinochle in the bottom of the boat?

Cruelty is a mystery, and the waste of pain. But if we describe a world to compass these things, a world that is a long, brute game, then we bump against another mystery: the inrush of power and light, the canary that sings on the skull. Unless all ages and races of men have been deluded by the same mass hypnotist (who?), there seems to be such a thing as beauty, a grace wholly gratuitous. About five years ago I saw a mockingbird make a straight vertical descent from the roof gutter of a four-story building. It was an act as careless and spontaneous as the curl of a stem or the kindling of a star.

The mockingbird took a single step into the air and dropped. His wings were still folded against his sides as though he were singing from a limb and not falling, accelerating thirty-two feet per second per second, through empty air. Just a breath before he would have been dashed to the ground, he unfurled his wings with exact, deliberate care, revealing the broad bars of white, spread his elegant, white-banded tail, and so floated onto the grass. I had just rounded a corner when his insouciant step caught my eye; there was no one else in sight. The fact of his free fall was like the old philosophical conundrum about the tree that falls in the forest. The answer must be, I think, that beauty and grace are performed whether or not we will or sense them. The least we can do is try to be there.

Another time I saw another wonder: sharks off the Atlantic coast of Florida. There is a way a wave rises above the ocean horizon, a triangular wedge against the sky. If you stand where the ocean breaks on a shallow beach, you see the raised water in a wave is translucent, shot with lights. One late afternoon at low tide a hundred big sharks passed the beach near the mouth of a tidal river in a feeding frenzy. As each green wave rose from the churning water, it illuminated within itself the six- or eight-foot-long bodies of twisting sharks. The sharks disappeared as each wave rolled toward me; then a new wave would swell above the horizon, containing in it, like scorpions in amber, sharks that roiled and heaved. The sight held awesome wonders: power and beauty, grace tangled in a rapture with violence.

We don't know what's going on here. If these tremendous events are random combinations of matter run amok, the yield of millions of monkeys at millions of typewriters, then what is it in us, hammered out of those same typewriters, that they ignite? We don't know. Our life is a faint tracing on the surface of mystery, like the idle, curved tunnels of leaf miners on the face of a leaf. We must somehow take a wider view, look at the whole landscape, really see it, and describe what's going on here. Then we can at least wail the right question into the swaddling band of darkness, or, if it comes to that, choir the proper praise.

At the time of Lewis and Clark, setting the prairies on fire was a well-known signal that meant, "Come down to the water." It was an extravagant gesture, but we can't do less. If the landscape reveals one certainty, it is that the extravagant gesture is the very stuff of creation. After the one extravagant gesture of creation in the first place, the universe has continued to deal exclusively in extravagances, flinging intricacies and colossi down aeons of emptiness, heaping profusions on profligacies with ever-fresh vigor. The whole show has been on fire from the word go. I come down to the water to cool my eyes. But everywhere I look I see fire; that which isn't flint is tinder, and the whole world sparks and flames.

I have come to the grassy island late in the day. The creek is up; icy water sweeps under the sycamore log bridge. The frog skin, of course, is utterly gone. I have stared at that one spot on the creek bottom for so long, focusing past the rush of water, that when I stand, the opposite bank seems to stretch before my eyes and flow grassily upstream. When the bank settles down I cross the sycamore log and enter again the big plowed field next to the steers' pasture.

The wind is terrific out of the west; the sun comes and goes. I can see the shadow on the field before me deepen uniformly and spread like a plague. Everything seems so dull I am amazed I can even distinguish objects. And suddenly the light runs across the land like a comber, and up the trees, and goes again in a wink: I think I've gone blind or died. When it comes again, the light, you hold your breath, and if it stays you forget about it until it goes again.

It's the most beautiful day of the year. At four o'clock the eastern sky is a dead stratus black flecked with low white clouds. The sun in the west illuminates the ground, the mountains, and especially the bare branches of trees, so that everywhere silver trees cut into the black sky like a photographer's negative of a landscape. The air and the ground are dry; the mountains are going on and off like neon signs. Clouds slide east as if pulled from the horizon, like a tablecloth whipped off a table. The hemlocks by the barbed-wire fence are flinging themselves east as though their backs would break. Purple shadows are racing east; the wind makes me face east, and again I feel the dizzying, drawn sensation I felt when the creek bank reeled.

At four-thirty the sky in the east is clear; how could that big blackness be blown? Fifteen minutes later another darkness is coming overhead from the northwest; and it's here. Everything is drained of its light as if sucked. Only at the horizon do inky black mountains give way to distant, lighted mountains— lighted not by direct illumination but rather paled by glowing sheets of mist hung before them. Now the blackness is in the east; everything is half in shadow, half in sun, every clod, tree, mountain, and hedge. I can't see Tinker Mountain through the line of hemlock, till it comes on like a streetlight, ping, *ex nihilo*. Its sandstone cliffs pink and swell. Suddenly the light goes; the cliffs recede as if pushed. The sun hits a clump of sycamores between me and the mountains; the sycamore arms light up, and *I can't see the cliffs*. They're gone. The pale network of sycamore arms, which a second ago was transparent as a screen, is suddenly opaque, glowing with light. Now the sycamore arms snuff out, the mountains come on, and there are the cliffs again.

I walk home. By five-thirty the show has pulled out. Nothing is left but an unreal blue and a few banked clouds low in the north. Some sort of carnival magician has been here, some fast-talking worker of wonders who has the act backwards. "Something in this hand," he says, "something in this hand, something up my sleeve, something behind my back . . ." and abracadabra, he snaps his fingers, and it's all gone. Only the bland, blank-faced magician remains, in his unruffled coat, barehanded, acknowledging a smattering of baffled applause. When you look again the whole show has pulled up stakes and moved on down the road. It never stops. New shows roll in from over the mountains and the magician reappears unannounced from a fold in the curtain you never dreamed was an opening. Scarves of clouds, rabbits in plain view, disappear into the black hat forever. Presto chango. The audience, if there is an audience at all, is dizzy from head-turning, dazed.

Like the bear who went over the mountain, I went out to see what I could see. And, I might as well warn you, like the bear, all that I could see was the other side of the mountain: more of same. On a good day I might catch a glimpse of another wooded ridge rolling under the sun like water, another bivouac. I propose to keep here what Thoreau called "a meteorological journal of the mind," telling some tales and describing some of the sights of this rather tamed valley, and exploring, in fear and trembling, some of the unmapped dim reaches and unholy fastnesses to which those tales and sights so dizzyingly lead.

I am no scientist. I explore the neighborhood. An infant who has just learned to hold his head up has a frank and forthright way of gazing about him in bewilderment. He hasn't the faintest clue where he is, and he aims to learn. In a couple of years, what he will have learned instead is how to fake it: he'll have the cocksure air of a squatter who has come to feel he owns the place. Some unwonted, taught pride diverts us from our original intent, which is to explore the neighborhood, view the landscape, to discover at least where it is that we have been so startlingly set down, if we can't learn why.

So I think about the valley. It is my leisure as well as my work, a game. It is a fierce game I have joined because it is being played anyway, a game of both skill and chance, played against an unseen adversary—the conditions of time—in which the payoffs, which may suddenly arrive in a blast of light at any moment, might as well come to me as anyone else. I stake the time I'm grateful to have, the energies I'm glad to direct. I risk getting stuck on the board, so to speak, unable to move in any direction, which happens enough, God knows; and I risk the searing, exhausting nightmares that plunder rest and force me face down all night long in some muddy ditch seething with hatching insects and crustaceans.

But if I can bear the nights, the days are a pleasure. I walk out; I see something, some event that would otherwise have been utterly missed and lost; or something sees me, some enormous power brushes me with its clean wing, and I resound like a beaten bell.

I am an explorer, then, and I am also a stalker, or the instrument of the hunt itself. Certain Indians used to carve long grooves along the wooden shafts of their arrows. They called the grooves "lightning marks," because they resembled the curved fissure lightning slices down the trunks of trees. The function of lightning marks is this: if the arrow fails to kill the game, blood from a deep wound will channel along the lightning mark, streak down the arrow shaft, and spatter to the ground, laying a trail dripped on broadleaves, on stones, that the barefoot and trembling archer can follow into whatever deep or rare wilderness it leads. I am the arrow shaft, carved along my length by unexpected lights and gashes from the very sky, and this book is the straying trail of blood.

Something pummels us, something barely sheathed. Power broods and lights. We're played on like a pipe; our breath is not our own. James Houston describes two young Eskimo girls sitting cross-legged on the ground, mouth on mouth, blowing by turns each other's throat cords, making a low, unearthly music. When I cross again the bridge that is really the steers' fence, the wind has thinned to the delicate air of twilight; it crumples the water's skin. I watch the running sheets of light raised on the creek's surface. The sight has the appeal of the purely passive, like the racing of light under clouds on a field, the beautiful dream at the moment of being dreamed. The breeze is the merest puff, but you yourself sail headlong and breathless under the gale force of the spirit.

RESPONDING JOURNAL

What impresses you most about Dillard's writing? She uses many techniques and brings in ideas from many subjects as different as magicians and members of religious orders. Do some of these ideas work better for you than others? Explain your response as fully as you can in your journal entry.

QUESTIONS FOR CRITICAL THINKING

STRATEGY

1. Why does Dillard open a book on the beauties of nature with a smelly, bruised cat? What could she have in mind? How well does this strategy seem to you to work?
2. Waking up is the central idea for Dillard's opening pages. Why would this be an effective way to begin her book?

ISSUES

1. How accessible is the sort of experience with nature that Dillard describes? How much do you think this quality adds to the appeal of her writing?
2. "Cruelty is a mystery," Dillard says. How well does she succeed in convincing you that seeming cruelty is part of some great universal pattern?

COLLABORATIVE WRITING ACTIVITY

What do you think of Dillard's image of the universe "in jest"? Does this idea cause you to think about revising your sense of how people look at the world of nature? Describe an example from your own experience that shows how the patterns of life in the universe are deadly serious, even when they seem most whimsical. Share your work with your group and then collectively draft a letter to Dillard that either reinforces or challenges her idea of "heaven and earth in jest."

from "Of Beauty and Death"
W.E.B. Du Bois

William Edward Burghardt Du Bois (1868-1963) was an author, scholar, teacher, sociologist, and political leader. A key fighter for the rights of blacks in America, he helped in 1905 to found the Niagara Movement, which three years later became the National Association for the Advancement of Colored People (NAACP). From 1910 to 1932 he edited Crisis, *the NAACP journal. His best-known writing is* The Souls of Black Folk, *a collection of essays on the plight of Blacks in the early twentieth century. In this collection appeared "Of Mr. Booker T. Washington and Others," a crucial essay in the history of the civil rights movement. Here Du Bois argues against accommodation and a narrow trade school education for blacks; he presses for full equality and access to education without regard to race.*

"Of Beauty and Death" appeared in Du Bois' essay collection Darkwater: Voices from Within the Veil *(1920)—his reflections on race relations, World War I, and other social issues of the first two decades of the century. In this excerpt, Du Bois relates his recollections of the great landforms of the Grand Canyon to his experiences in World War I in France. Recurring ideas of beauty and death preoccupy him even as he visits great landmarks after the war; regardless of his surroundings, he cannot forget the issues delaying true equality for Blacks.*

EXPLORING JOURNAL

What does the Grand Canyon mean to you? If you have not seen it, what do you know about it? Explain in a ten-minute freewrite why you think it has become such a favored spot to visit.

———◦◦◦———

Once upon a time I took a great journey in this land to three of the ends of our world and over seven thousand mighty miles. I saw the grim desert and the high ramparts of the Rocky Mountains. Three days I flew from the silver beauty of Seattle to the somber whirl of Kansas City. Three days I flew from the brute might of Chicago to the air of the Angels in California, scented with golden flowers, where the homes of men crouch low and loving on the good, broad earth, as though they were kissing her blossoms. Three days I flew through the empire of Texas, but all these shall be tales untold, for in all this journey I saw but one thing that lived and will live eternal in my soul—the Grand Cañon.

It is a sudden void in the bosom of earth, down to its entrails—a wound where the dull titanic knife has turned and twisted in the hole, leaving its edges livid, scarred, jagged, and pulsing over the white, and red, and purple of its mighty flesh, while down below—down, down below, in black and severed vein, boils the dull and sullen flood of the Colorado.

It is awful. There can be nothing like it. It is the earth and sky gone stark and raving mad. The mountains up-twirled, disbodied and inverted, stand on their peaks and throw their bowels to the sky. Their earth is air; their ether blood-red rock engreened. You stand upon their roots and fall into their pinnacles, a mighty mile.

Behold this mauve and purple mocking of time and space! See yonder peak! No human foot has trod it. Into that blue shadow only the eye of God has looked. Listen to the accents of that gorge which mutters: "Before Abraham was, I am." Is yonder wall a hedge of black or is it the rampart between heaven and hell? I see greens—is it moss or giant pines? I see specks that may be boulders. Ever the winds sigh and drop into those sun-swept silences. Ever the gorge lies motionless, unmoved, until I fear. It is a grim thing, unholy, terrible! It is human—some mighty drama unseen, unheard, is playing there its tragedies or mocking comedy, and the laugh of endless years is shrieking onward from peak to peak, unheard, unechoed, and unknown.

One throws a rock into the abyss. It gives back no sound. It falls on silence—the voice of its thunders cannot reach so far. It is not—it cannot be a mere inert, unfeeling, brute fact—its grandeur is too serene—its beauty too divine! It is not red, and blue, and green, but, ah! the shadows and the shades of all the world, glad colorings touched with a hesitant spiritual delicacy. What does it mean—what does it mean? Tell me, black and boiling water!

It is not real. It is but shadows. The shading of eternity. Last night yonder tesselated palace was gloom—dark, brooding thought and sin, while hither rose the mountains of the sun, golden, blazing, ensanguined. It was a dream. This blue and brilliant morning shows all those burning peaks alight, while here, shapeless, mistful, brood the shadowed towers.

I have seen down into the entrails of earth—down, down by straight and staring cliffs—down by sounding waters and sun-strewn meadows; down by great, steep chasms—down by the gnarled and twisted fists of God to the deep, sad moan of the yellow river that did this thing of wonder—a little winding river with death in its depth and a crown of glory in its flying hair.

I have seen what eye of man was never meant to see. I have profaned the sanctuary. I have looked upon the dread disrobing of the Night, and yet I live. Ere I hid my head she was standing in her cavern halls, glowing coldly westward—her feet were blackness: her robes, empurpled, flowed mistily from shoulder down in formless folds of folds; her head, pine-crowned, was set with jeweled stars. I turned away and dreamed—the cañon—the awful, its depths called; its heights shuddered. Then suddenly I arose and looked. Her robes were falling. At dim-dawn they hung purplish-green and black. Slowly she stripped them from her gaunt and shapely limbs—her cold, gray garments shot with shadows stood revealed. Down dropped the black-blue robes, gray-pearled, and slipped, leaving a filmy, silken, misty thing, and underneath I glimpsed her limbs of utter light.

My God! For what am I thankful this night? For nothing. For nothing but the most commonplace of commonplaces; a table of gentlewomen and gentlemen—soft-spoken, sweet-tempered, full of human sympathy, who made me, a stranger, one of them. Ours was a fellowship of common books, common knowledge, mighty aims. We could laugh and joke and think as friends and the Thing—the hateful, murderous, dirty Thing which in America we call "Nigger-hatred" was not only not there—it could not even be understood. It was a curious monstrosity at which civilized folk laughed or looked puzzled. There was no elegant and elaborate condescension of—"We once had a colored servant"—"My father was an Abolitionist"—"I've always been interested in *your people*"—there was only the community of kindred souls, the delicate reverence for the Thought that led, the quick deference to the guest. You left in quiet regret, knowing that they were not discussing you behind your back with lies and license. God! It was simply human decency and I had to be thankful for it because I am an American Negro and white America, with saving exceptions, is cruel to everything that has black blood—and this was Paris, in the years of salvation, 1919. Fellow blacks, we must join the democracy of Europe.

Toul! Dim through the deepening dark of early afternoon, I saw its towers gloom dusky toward the murk of heaven. We wound in misty roads and dropped upon the city through the great throats of its walled bastions. There lay France—a strange, unknown, unfamiliar France. The city was dispossessed. Through its streets—its narrow, winding streets, old and low and dark, carven and quaint—poured thousands upon thousands of strange feet of khaki-clad foreigners, and the echoes threw back awkward syllables that were never French. Here was France beaten to its knees yet fighting as never nation fought before, calling in her death agony across the seas till her help came and with all its strut and careless braggadocio saved the worthiest nation of the world from the wickedest fate ever plotted by Fools.

Tim Brimm was playing by the town-pump. Tim Brimm and the bugles of Harlem blared in the little streets of Maron in far Lorraine. The tiny streets were seas of mud. Dank mist and rain sifted through the cold air above the blue Moselle. Soldiers—soldiers everywhere—black soldiers, boys of Washington, Alabama, Philadelphia, Mississippi. Wild and sweet and wooing leapt the strains upon the air. French children gazed in wonder—women left their washing. Up in the window stood a black Major, a Captain, a Teacher, and I—with tears behind our smiling eyes. Tim Brimm was playing by the town-pump.

The audience was framed in smoke. It rose ghost-like out of memories—bitter memories of the officer near dead of pneumonia whose pain was lighted up by the nurses waiting to know whether he must be "Jim-Crowed" with privates or not. Memories of that great last morning when the thunders of hell called the Ninety-second to its last drive. Memories of bitter humiliations, determined triumphs, great victories, and bugle-calls that sounded from earth to heaven. Like memories framed in the breath of God, my audience peered in upon me—good, brown faces with great, kind, beautiful eyes—black soldiers of America rescuing beloved France—and the words came in praise and bene-diction there in the "Y," with its little stock of cigarettes and candies and its rusty wood stove.

"*Alors,*" said Madame, "*quatre sont morts*"—four dead—four tall, strong sons dead for France—sons like the sweet and blue-eyed daughter who was hid-ing her brave smile in the dusk. It was a tiny stone house whose front window lipped the passing sidewalk where ever tramped the feet of black soldiers marching home. There was a cavernous wardrobe, a great fireplace invaded by a new and jaunty iron stove. Vast, thick piles of bedding rose in yonder corner. Without was the crowded kitchen and up a half-stair was our bedroom that gave upon a tiny court with arched stone staircase and one green tree. We were a touching family party held together by a great sorrow and a great joy. How we laughed over the salad that got brandy instead of vinegar—how we ate the golden pile of fried potatoes and how we pored over the post-card from the Lieutenant of the Senegalese—dear little vale of crushed and risen France, in the day when Negroes went "over the top" at Pont-à-Mousson.

Paris, Paris by purple façade of the opera, the crowd on the Boulevard des Italiens and the great swing of the Champs Elysées. But not the Paris the world knows. Paris with its soul cut to the core—feverish, crowded, nervous, hurried; full of uniforms and mourning bands, with cafés closed at 9:30—no sugar, scarce bread, and tears so intertwined with joy that there is scant difference. Paris has been dreaming a nightmare, and though she awakes, the grim terror is upon her—it lies on the sand-closed art treasures of the Louvre. Only the flowers are there, always the flowers, the Roses of England and the Lilies of France.

New York! Behind the Liberty that faces free France rise the white cliffs of Manhattan, tier on tier, with a curving pinnacle, towers square and twin, a giant inkwell daintily stoppered, an ancient pyramid enthroned; beneath, low ram-parts wide and mighty; while above, faint-limned against the turbulent sky, looms the vast grace of that Cathedral of the Purchased and Purchasing Poor, topping the world and pointing higher.

Yonder the gray cobwebs of the Brooklyn bridges leap the sea, and here creep the argosies from all earth's ends. We move to this swift home on dun and swelling waters and hear as we come the heartbeats of the new world.

New York and night from the Brooklyn Bridge: The bees and fireflies flit and twinkle in their vast hives; curved clouds like the breath of gods hover between the towers and the moon. One hears the hiss of lightnings, the deep thunder of human things, and a fevered breathing as of some attendant and invincible Powers. The glow of burning millions melts outward into dim and fairy outlines until afar the liquid music born of rushing crowds drips like a benediction on the sea.

New York and morning; the sun is kissing the timid dew in Central Park, and from the Mountain of Plenty one looks along that world street, Fifth-Avenue, and walks toward town. The earth lifts and curves graciously down from the older mansions of princes to the newer shops of luxury. Egypt and Abyssinia, Paris and Damascus, London and India caress you by the way; churches stand aloof while the shops swell to emporiums. But all this is nothing. Everything is mankind. Humanity stands and flies and walks and rolls about—the poor, the priceless, the world-known and the forgotten; child and grandfather, king and leman—the pageant of the world goes by, set in a frame of stone and jewels, clothed in scarlet and rags. Princes Street and the Elysian Fields, the Strand and the Ringstrasse—these are the Ways of the World today.

New York and twilight, there where the Sixth Avenue "L" rises and leaps above the tenements into the free air at 110th Street. It circles like a bird with heaven and St. John's above and earth and the sweet green and gold of the Park beneath. Beyond lie all the blue mists and mysteries of distance; beneath, the city rushes and crawls. Behind echo all the roar and war and care and maze of the wide city set in its sullen darkening walls, flashing weird and crimson farewells. Out at the sides the stars twinkle.

Again New York and Night and Harlem. A dark city of fifty thousand rises like magic from the earth. Gone is the white world, the pale lips, the lank hair; gone is the West and North—the East and South is here triumphant. The street is crowd and leisure and laughter. Everywhere black eyes, black and brown, and frizzled hair curled and sleek, and skins that riot with luscious color and deep, burning blood. Humanity is packed dense in high piles of close-knit homes that lie in layers above gray shops of food and clothes and drink, with here and there a moving-picture show. Orators declaim on the corners, lovers lark in the streets, gamblers glide by the saloons, workers lounge wearily home. Children scream and run and frolic, and all is good and human and beautiful and ugly and evil, even as Life is elsewhere.

And then—the Veil. It drops as drops the night on southern seas—vast, sudden, unanswering. There is Hate behind it, and Cruelty and Tears. As one peers through its intricate, unfathomable pattern of ancient, old, old, design, one sees blood and guilt and misunderstanding. And yet it hangs there, this Veil, between Then and Now, between Pale and Colored and Black and White—between You and Me. Surely it is a thought-thing, tenuous, intangible; yet just as surely is it true and terrible and not in our little day may you and I lift it. We may feverishly unravel its edges and even climb slow with giant shears to where its ringed and gilded top nestles close to the throne of God. But as we work and climb we shall see through streaming eyes and hear with aching ears, lynching and murder, cheating and despising, degrading and lying, so flashed and fleshed through this vast hanging darkness that the Doer never sees the Deed and the Victim knows not the Victor and Each hates All in wild and bitter ignorance. Listen, O Isles, to these Voices from within the Veil, for they portray the most human hurt of the Twentieth Cycle of that poor Jesus who was called the Christ!

There is something in the nature of Beauty that demands an end. Ugliness may be indefinite. It may trail off into gray endlessness. But Beauty must be complete—whether it be a field of poppies or a great life—it must end, and the End is part and triumph of the Beauty. I know there are those who envisage a beauty eternal. But I cannot. I can dream of great and never-ending processions of beautiful things and visions and acts. But each must be complete or it cannot for me exist.

On the other hand, Ugliness to me is eternal, not in the essence but in its incompleteness; but its eternity does not daunt me, for its eternal unfulfilment is a cause of joy. There is in it nothing new or unexpected; it is the old evil stretching out and ever seeking the end it cannot find; it may coil and writhe and recur in endless battle to days without end, but it is the same human ill and bitter hurt. But Beauty is fulfilment. It satisfies. It is always new and strange. It is the reasonable thing. Its end is Death—the sweet silence of perfection, the calm and balance of utter music. Therein is the triumph of Beauty.

So strong is the spell of beauty that there are those who, contradicting their own knowledge and experience, try to say that all is beauty: They are called optimists, and they lie. All is not beauty. Ugliness and hate and ill are here with all their contradiction and illogic; they will always be here—perhaps, God send, with lessened volume and force, but here and eternal, while beauty triumphs in its great completion—Death. We cannot conjure the end of all ugliness in eternal beauty, for beauty by its very being and definition has in each definition its ends and limits; but while beauty lies implicit and revealed in its end, ugliness writhes on in darkness forever. So the ugliness of continual birth fulfils itself and conquers gloriously only in the beautiful end, Death.

At last to us all comes happiness, there in the Court of Peace, where the dead lie so still and calm and good. If we were not dead we would lie and listen to the flowers grow. We would hear the birds sing and see how the rain rises and blushes and burns and pales and dies in beauty. We would see spring, summer, and the red riot of autumn, and then in winter, beneath the soft white snow, sleep and dream of dreams. But we know that being dead, our Happiness is a fine and finished thing and that ten, a hundred, and a thousand years, we shall lie at rest, unhurt in the Court of Peace.

RESPONDING JOURNAL

Have you remembered at odd times landscapes or scenes in the natural world that made strong impressions on you? If you can recall one of these instances, record in your journal why you think this particular scene came up at this time.

If you cannot recall an instance like this, explain why an experience you had with a natural scene might one day recur to you in similar fashion.

QUESTIONS FOR CRITICAL THINKING

STRATEGY

1. Du Bois often shifts his subject radically. What reasons might he have for such shifts? How effective do you find them as a means of development?
2. Look at Du Bois' sentences announcing the subjects of his paragraphs: "Touls," "Paris," "New York." He follows these descriptions with another short opening, "And then—the Veil." Explain the effect of these short statements on the paragraphs they begin, and explain also their effect on you as a reader. What potential ideas for your own writing do you see here?

ISSUES

1. Explain in your own words Du Bois' idea of beauty. How does he relate his concept of beauty to the places that he describes?
2. Carry out a little research in the library. Check to learn what you can about Black soldiers in World War I. Did any see combat ? How were they housed in the army? What surprises you as you do this research?

COLLABORATIVE WRITING ACTIVITY

Imagine that Du Bois has returned to the United States to comment on race relations today. What landscapes would he find that would serve as starting points for his comments? What would have changed for him? Would he take issue with anyone the way he took issue with Booker T. Washington? Bring your responses in the forms of sketches or lists with explanations of the individual items listed so that when you get into your groups, you can pair up and discuss how the two of you have envisioned Du Bois' response. Working together from your raw material and the ideas generated in your discussion, compose a statement of how you think Du Bois would respond today.

"The Ballad of William Sycamore"
Stephen Vincent Benét

Stephen Vincent Benét (1898-1943) was an American poet who used historical material in much of his work. Born in Pennsylvania to an army family,

he lived in various parts of the country and graduated from Yale in 1919.
By then he had already published two volumes of poetry and quickly gained
an enthusiastic following, particularly among younger readers.

He often takes for his subjects various aspects of western frontier life;
many people feel he helped to establish our national sense of the glory of the
western pioneers. He demonstrates this power perhaps most successfully in
his long narrative poem John Brown's Body, *for which he was awarded the*
Pulitzer Prize in 1928. His collection of poems entitled The Ballad of William
Sycamore *appeared in 1923; the title poem was widely taught in schools for*
decades after it appeared, and is still often included in anthologies of
American poetry.

EXPLORING JOURNAL

What do you see as the proper role of governmental policies in your life? Should
you as a taxpayer, for example, be partially responsible for paying for reclamation
of polluted areas or waging a military action? Explore your feeling in an entry.

My father, he was a mountaineer,
His fist was a knotty hammer;
He was quick on his feet as a running deer,
And he spoke with a Yankee stammer.

My mother, she was merry and brave,
And so she came to her labor,
With a tall green fir for her doctor grave
And a stream for her comforting neighbor.

And some are wrapped in the linen fine,
And some like a godling's scion;
But I was cradled on twigs of pine
And the skin of a mountain lion.

And some remember a white, starched lap
And a ewer with silver handles;
But I remember a coonskin cap
And the smell of bayberry candles.

The cabin logs, with the bark still rough,
And my mother who laughed at trifles,
And the tall, lank visitors, brown as snuff,
With their long, straight squirrel-rifles.

I can hear them dance, like a foggy song,
Through the deepest one of my slumbers,

The fiddle squeaking the boots along
And my father calling the numbers.

The quick feet shaking the puncheon-floor
And the fiddle squealing and squealing,
Till the dried herbs rattled above the door
And the dust went up to the ceiling.

There are children lucky from dawn till dusk,
But never a child so lucky!
For I cut my teeth on "Money Musk"
In the Bloody Ground of Kentucky!

When I grew tall as the Indian corn,
My father had little to lend me,
But he gave me his great, old powder-horn
And his woodsman's skill to befriend me.

With a leather shirt to cover my back,
And a redskin nose to unravel
Each forest sign, I carried my pack
As far as a scout could travel.

Till I lost my boyhood and found my wife,
A girl like a Salem clipper!
A woman straight as a hunting-knife
With eyes as bright as the Dipper!

We cleared our camp where the buffalo feed,
Unheard-of streams were our flagons;
And I sowed my sons like the apple-seed
On the trail of the Western wagons.

They were right, tight boys, never sulky or slow,
A fruitful, a goodly muster.
The eldest died at the Alamo.
The youngest fell with Custer.

The letter that told it burned my hand.
Yet we smiled and said, "So be it!"
But I could not live when they fenced the land,
For it broke my heart to see it.

I saddled a red, unbroken colt
And rode him into the day there;
And he threw me down like a thunderbolt
And rolled on me as I lay there.

The hunter's whistle hummed in my ear
As the city-men tried to move me,
And I died in my boots like a pioneer
With the whole wide sky above me.

Now I lie in the heart of the fat, black soil,
Like the seed of a prairie-thistle;
It has washed my bones with honey and oil
And picked them clean as a whistle.

And my youth returns, like the rains of Spring,
And my sons, like the wild-geese flying;
And I lie and hear the meadow-lark sing
And have much content in my dying.

Go play with the towns you have built of blocks.
The towns where you would have bound me!
I sleep in my earth like a tired fox,
And my buffalo have found me.

RESPONDING JOURNAL

How well do you think Benét's poem holds up today? Does its sentiments of an American heritage strike you as valid, or has the world changed so much that the thinking and values of Sycamore and his family seem out of date?

QUESTIONS FOR CRITICAL THINKING

STRATEGY

1. What use of proper names does Benét seem to make? Do the names have any particular qualities that contribute to the poem's effect? How about the title itself?
2. Explain why you feel Benét has Sycamore speak to us from the grave. What does the poet gain by presenting his main character as someone long dead?

ISSUES

1. How does the reference to "city-men" affect the poem's meaning? Why might Benét mention them?

2. To whom does the poem seem to be addressed? What tells you about how the speaker seems to regard his audience?

COLLABORATIVE WRITING ACTIVITY

Locate a color plate of Thomas Hart Benton's painting "After Many a Spring" and compare it to Benét's poem. What similarities in mood, tone, or subject do you find? What details in the poem seem to parallel visual details in the painting? Bring your ideas to your group in the form of notes, and together write an essay explaining the connections your group has discerned between poem and painting. If you wish, refer back to "Reading Visual Texts" on page 8.

from *Bulow Hammock*
David Rains Wallace

A highly esteemed contemporary nature writer, David Rains Wallace has written on numerous subjects for the Sierra Club as well as other commercial publishers. In his writing, he is wary of allowing his speculations to become as philosophical as those of many other nature writers; he prefers the direct and the straightforward stimuli that come from actual experience.

In Bulow Hammock, *Wallace presents a series of views of a Florida landscape that he first visited as a child and continues to visit. Like many landscapes you may remember from your own experience, Bulow Hammock has changed and suffered over the years because of nearby housing developments and, further away, the growth of Daytona Beach as a resort. But enough remains in the hammock to continue to challenge Wallace's powers of observation. As his series of sketches of the hammock at various points in his life indicates, the area still has more than enough to stimulate Wallace's curiosity and provide material for writing. Presented here are the chapters entitled "The Green Tunnel" from the book's opening and "The Otter," a later chapter which incorporates Wallace's revised ideas of the hammock.*

EXPLORING JOURNAL

What sort of landscape experience do you think you would enjoy in a swamp? Do you feel that some experiences, on certain kinds of terrain, are more desirable than others? In a journal entry, explain your reasons.

THE GREEN TUNNEL

The thing that first struck me about Bulow Hammock is the hardest to describe: the smell. Hammocks are woodlands (the name refers to hardwood groves that punctuate the more open marshes and pine woods of Florida, and may derive from Indian words for "shady place," "garden place," or "floating plants"), but Bulow Hammock didn't smell like any woodlands I knew. I was used to the brisk, humus-and-chlorophyll tang of New England woods with their associations of uplifting weekend hikes. The hammock was different.

I must have been about nine years old when I encountered the hammock, so I didn't articulate any of this. Yet I clearly remember my sensations on stepping out of my parents' car into the shade of the magnolias and cabbage palmettoes. I was fascinated but daunted. The Connecticut woods I'd played in had been inviting, welcoming. The hammock was . . . seductive. It smelled sweet, a perfumery sweetness that reminded me of the hotel lobbies and cocktail lounges I'd occasionally been in with my parents.

Smells are hard to describe because we can't really remember them as we do sights and sounds, we can only recognize them. Smells lie deeper than our remembering, thinking neocortex, in the olfactory lobe we inherited from the early vertebrates. Yet smells are related to thought in profound ways because our nocturnal ancestors, the early mammals, lived by smell. The human ability to relate present to past and future may stem from their scent-tracking of food, an activity which takes place in time as well as space, unlike a hawk's immediate striking on sight, and thus implies planning. The curious resonance the olfactory senses have in memory, as when Proust tasted an epoch in a teacake, suggests that we have a great deal to learn from them.

Complex smells are the hardest to describe. Bulow Hammock smelled stranger than liquor and perfume. It smelled intricately spicy, with a sweetness not so much of flowers as of romantic bark and leaves. There also was an air of decay in the sweetness; not the rich, sleepy, somewhat bitter decay of New England woods, but more of a nervous, sour atmosphere. When I scraped my foot over fallen leaves on the ground, I didn't uncover the soft brown dirt I was used to, but white sand and a network of fine, blackish roots like the hair of a buried animal. The sand was part of the smell too, a dusty, siliceous undertone to the spice and decay.

There was something dangerous about the smell, something inhibiting to my nine-year-old mind. I didn't want to rush into the hammock as I'd have wanted to rush into an unfamiliar Connecticut woodland. It wasn't that the hammock seemed ugly or repellent—on the contrary. The seductiveness was part of the inhibition. Perhaps it was just that the hammock was *so* unfamiliar. It's easy to read things into childhood memories. But the smell was powerful.

Society is suspicious of wild places because it fears a turning away from human solidarity toward a spurious, sentimental freedom. It is interesting, in this regard, to recall how *little* of freedom there was in my first perception of Bulow Hammock, how little of the unfettered feeling I got in sand dunes, hill meadows, pine woods or other open places that promised release from streets and classrooms. I wonder if the hammock inhibited me because there was more of humanity about it than a dune, meadow, or pine forest has; not of humanity in the sense of society and civilization, which (however irrationally, given the history of civilization) we associate with safety, but of animal humanity, of the walking primate that has spent most of its evolution in warm places like Florida: spicy, moldy, sandy places. Perhaps it wasn't the strangeness of the hammock that made it seem dangerously seductive, but a certain familiarity. It is, after all, dangerous to be human.

We'd come to Florida to visit my father's mother, who had a retirement cottage in Ormond-by-the-Sea, an early geriatric enclave complete with shuffleboard court (which, three decades later, has become somebody's driveway). On the drive south, we passed another stretch of coastal hammock that was being burned and bulldozed during some kind of road construction involving sweaty convicts in gray twill. There'd been something very malignant looking about that stretch of charred palmetto. Blackened fronds had thrust at the sky like fire-sharpened spears. As though to heighten the effect, someone had erected a doll's head, also charred, on a crooked stick.

I couldn't have looked at this scene for more than a few seconds, but it made a big impression. At nine, I had no very firm grasp of its rational implications, of the likelihood that the head had been stuck up there by some whimsically ghoulish convict who'd found it while grubbing in the brush. I must have been aware of that likelihood, but other things seemed possible: that it was a real head, a baby's or a monkey's; that it manifested an unknown savage world in the uncut hammock farther from the road, of which there was a lot more in Florida then. The southern landscape threw the human and wild together more than the northern. I remember a great loneliness in it, brown fields of broomsedge reaching almost to the horizon and unpainted shacks against ragged woods over which circled vultures in numbers out of proportion to the vacancy beneath them. The blackwater swamps that the road periodically passed over seemed cheerful in comparison, albeit dangerous.

Of course, my response to the road construction—fire, sticks, head, uncut green wall in the distance—was an educated one, as was my response to Bulow Hammock's smell. It would be banal to assert that the smell awakened atavistic race memories of life in the jungle. We'd been getting our first taste of human evolution in my fourth grade class, and I'd found *that* pretty spicy, all those skeletons and hairy people: Piltdown Man (we must have been the last class to learn about Piltdown Man, since the hoax was discovered around that year), Java Man, Peking Man. A normally bloodthirsty fourth-grader, I'd thrilled to learn that Peking Man had scooped out and probably eaten the brains of other Peking men. I'd seen the "green hell" jungle movies of the early 1950s: Charlton Heston in *The Naked Jungle,* Jeff Chandler in *Green Fire.* I had a whole set of cultural preconceptions ready for Bulow Hammock.

Yet banality is a kind of fossilized reality, the bones of insights buried in the silt of intellectual fashion. I wouldn't dismiss my nine-year-old perceptions just because they were culturally conditioned. Classrooms and movie theaters teach little about smell, for one thing, and, sophisticated as they are, they still share with nine-year-olds a descent from spicy, moldy, sandy places. We don't know enough about that descent to dismiss anything. Fire, sticks, head, and green wall have been at the center of things for most of human experience, and they still are, in a sense, although the green wall may have receded.

A green wall is what Bulow Hammock seemed as my father drove down the low sand road leading into it, or rather a green arch, a tunnel. Its surfaces seemed much more solid than the crumbly coquina of the nineteenth-century sugar plantation ruins we had come to the hammock to see. The mill was roof-less while the hammock enclosed us completely, from its ground-hugging coontie, dog hobble, and saw palmetto to its undergrowth of bayberry, horn-beam, and dahoon to its canopy of live oak, red bay, magnolia and cabbage pal-metto. Glimpses of the hammock interior lacked perspective: they had the wavery, spotty aspect of underwater things. The plant forms were too eccen-tric for geometry—palm, spike, spray, corkscrew, club, plume, lace, spiral. It was beautiful, but the intricacy was like the complexity of smell. It inhibited. Its seductiveness was also a warning because it hinted at passionate entangle-ment more than freedom or tranquillity.

I followed my parents around the sugar mill ruins like a good little boy. The Seminoles had burned the plantation in 1835: that was interesting. There were displays of implements found in the ruins, and a brochure about the planta-tion's history. There wasn't any explanation of the hammock. There may have been signs identifying birds or plants, but if there were, they did little to eluci-date the fearful seductiveness of the place, a seductiveness to which the adult world seemed curiously immune. But then, children are used to being sur-rounded by powerful, unexplained seductions.

I never did venture into the hammock as a child, although I was wandering miles through the Connecticut woods. I don't recall going more than a few yards even into the barrier island scrub that grew behind my grandmother's cot-tage in the fifties, before the Ormond Mall was built. The mailman had put his hand into a pile of leaves (trusting children, we didn't ask why) and withdrawn it with a coral snake attached to the skin between his fingers. Coral snakes, Grandmother told my sister and me, had to hold and chew their victims to inject their almost invariably fatal poison.

Grandmother wasn't a snake-hater: her deepest antipathies were for the British royal family (her father was Irish) and J. Edgar Hoover (her former employer). She was more passionate in her opinions than most grandmothers, always applauding when Harry Truman appeared in movie newsreels whether or not anybody else did. Perhaps because of this, her dictums had considerable authority, and we weren't about to put our hands in any dead leaves, or our feet. There were poisonous copperheads in the Connecticut woods, of course, but they didn't chew on you. We contented ourselves with watching big toads eat little toads in her backyard.

THE OTTER

My attempt to repeat the Bulow Hammock experiment seemed modestly suc-cessful. I'd not only found alligators, but I'd found turkeys and tortoises, the whole program the ranger had announced, not to mention softshell turtles in swamps and toads in stumps.

I wasn't sure what use it all was, aside from providing me some perhaps temporary sense that by finding these things I'd found parts of myself, not just the cultural parts, but little-known, innate ones. At least, repeating the experiment had been useful in demonstrating that it could be repeated. Evolution and life can be seen as experiments, with success rewarded by the opportunity for further experiment.

I visited the hammock again in early January of 1987. There had been forest fires on the coast in 1985, and I was afraid of finding a blackened ruin. Hammocks aren't fire-adapted ecosystems as southern pinelands and prairies are. I found, however, that the fire had stopped at the pine woods just to the north. The hammock's sponginess seemed to make it as impervious to fire as it was to flood. A scattering of charred pines and palmettoes on the marsh horizon was the only vestige of the destruction.

The hammock was drier than I'd ever seen it, despite persistent rain the day before. A depression near the trailhead that had always been a deep puddle wasn't even muddy, and the swamp rivulets were very low. The dryness hadn't stopped the hammock's fungus populations from responding exuberantly to the rain. There'd never been so many mushrooms: big, pearly amanitas, some with yellow warts; red-topped russulas; clumps of smallish golden armillarias. Velvety ochre bracket fungi grew out of saw palmetto roots. Chestnut brown brackets covered dead cabbage palmetto roots.

Here, it seemed, was yet another case of organs relating independently to environment while the organism took a back seat, but this was an even more flagrant case than brains or flowers. Mushrooms are merely the reproductive organs of fungi, but they are so much more specialized, colorful, and elaborate than the amorphous, thread like hyphae that comprise the main organism as to appear the fungus's raison d'être rather than vice versa. Mushrooms busily attract and repel animals like flies and humans with smells, shapes, and chemistry, while their hyphae just sit there in the ground. That such "wayward" organs could have evolved at the "primitive" fungal level, as well as at the seed, plant, and hominid levels, again suggested some kind of basic tendency for such evolution.

Many of the fungi that brandished extravagant mushrooms about the hammock had stopped being organisms in the true sense. Their specialized hyphal cells, mycorrhizae, had become organs of the trees, in effect. They had replaced the root cells of most of the hammock's tree species in the function of drawing water and nutrients from the soil, taking food from the trees' sap in exchange. Thus, their mushrooms had become part of the flowers in the canopy, each attracting spore-eating flies or beetles, or pollen-eating bees or flies to perpetuate a superorganism, a forest. If mushrooms and flowers were linked in this subtle way, it made me wonder where brains might fit in. I could think of at least one way: the acorn-hiding brains of birds and squirrels also perpetuate the forest.

Aside from the mushrooms, the hammock seemed a little shrunken and dull in its dryness. Dessicated epiphytes littered the ground. A pileated woodpecker squawked peevishly somewhere, and as I approached, an armadillo sat up and sniffed nearsightedly, pink snout twitching. The only plants that had grown noticeably were the canes in the glades, which now stood over my head.

The sinkhole where I'd found the stinkpots copulating was almost dry. The only sign of turtles was an abandoned-looking gopher tortoise hole in the bank. The sweetwater swamp where I'd found the softshell was even drier, and willows had grown and spread to obscure the vista that had impressed me three years before. Once mysteriously luminescent with its sheet of obsidian water, the dried swamp seemed disheveled and deserted, hardly the same place, and I wasn't altogether sure it was, although the location seemed right.

I still had hopes for alligators on Bulow Creek. It was the right time of year, right weather. I pushed through the brush on the north side of the drainage ditch until I reached the red cedar and baccharis along the creek, but the view wasn't any better than it had been in April. I heard some birds flying away, and glimpsed a lone green heron, but I was too far upstream to see the mudbank where the alligators might be sunning.

I decided to wade the ditch. It was still full of water, but the sand bottom seemed firm. The cold water made my legs ache, but I got across without getting wetter than that. I followed the ditch east to the mound from which I'd watched the alligators five years before. My caution in peeking over the mound was wasted. The tide was high. Not only did I see no alligators, but I saw no mudbank, just water and saw grass. Even the green heron had disappeared.

I returned to my ford, waded back across, and lay down in the sun to dry off. Vultures circled overhead: I tried to see if they were blacks or turkeys. Then I heard something from the ditch a few yards away and looked up. An otter had surfaced in the water near where I'd climbed out on the other bank. It was growling at me, a small, irascible sound. Then it turned and began swimming upstream.

I'd noticed an underwater burrow entrance in the bank there. It seemed I'd disturbed the otter's afternoon nap in its den by clambering over the roof. I could understand its annoyance, especially when it emerged to find a stranger sprawled across its doorstep.

I watched the V of its wake receding up the ditch, supposing that would be the last I'd see of it. Then the otter did an unexpected thing. When it was a couple of hundred feet away, it suddenly turned and swam back. A few dozen feet upstream of me, it dived, but I could see its continued downstream course by a trail of bubbles. The bubbles came abreast of the burrow entrance and stopped. The otter surfaced, looked me straight in the face, growled again, then swam toward the burrow entrance, dived, and disappeared, evidently back to its interrupted nap.

I felt I understood perfectly what had gone on in the otter's head. It was what would have gone on in mine if I'd been disturbed in my residence by a noisy clod, had started to flee to quieter surroundings, then had gotten righteously indignant at the imposition and gone back to reclaim my rightful place.

I couldn't recall a more convincing display of conscious volition in a wild animal. When the sunbathing alligators had fled into the water after I appeared on the mound above Bulow Creek, they had certainly shown consciousness, but there'd been an element of automatism in their prompt reappearance on the bank when I'd kept still. The reappearance seemed motivated as much by short attention spans as by a determination to keep sunbathing. The turkey hens that had returned to get their chicks had certainly shown volition, but there'd been a strong instinctual element in their concern for their young. In contrast, the otter's refusal to be driven from its den seemed entirely personal, willful. Its awareness of my presence had never wavered, yet it had come back anyway. Its irritable attachment to its own space and comfort suggested a sense of self like my own.

It was strange to think of such a consciousness inhabiting the hammock full-time. We tend to think of personality and selfhood as cultural rather than natural, but here was a creature living in a hole in the bank who seemed as individuated as my neighbors in Ormond Beach. It implied that the individual integrity we feel as organisms is not something that emerges from ecosystems, but something that fits *in* them.

Yet how can something that seems as separate as individual consciousness fit into something that seems as indifferent as an ecosystem? If the brain is the organism's survival computer, perhaps it can't fit for very long. If the brain is a semi-independent mediator between organism and ecosystem, however, the idea of a fit between the two makes more sense. If the brain is a mediator, then consciousness is not really separate.

Such ideas may seem mystical, but we know so little about the relationship of consciousness and life that they can't justly be discarded. We live daily with unanswered questions which make the "abominable mystery" of flowering play origins seem fit for elementary school. I passed one on the way out of the hammock. I turned over a log and found a dark brown anole under it.

I wondered how the anole had managed to match the color of the leaf mold under the log when, presumably, there had been no light under the log and thus no color. But that wasn't the main question, which remained as to how the lizard's brain was able to assimilate information about the color of the environment and program it back into the color of its skin. We know the mechanism for the skin's color change—pigment cells—but not for the basic transferral of information.

The hammock was very still as I walked out. This should have allowed me to remember that I'd wanted to see if I'd smell the old seductive, dangerous smell again after being away for three years. I didn't think of it until I was bumping away down the sand road. I realized that I probably hadn't smelled it if I hadn't even remembered to see if I smelled it, but I stopped the car and opened the window anyway. I sniffed. The January air smelled only faintly of sand and spice and not at all of 1950s cocktail lounges. The green tunnel looked as seductive and impenetrable as ever, though. I felt little more certainty about what lay inside it than I had in 1953.

RESPONDING JOURNAL

Think of a place you visited as a child. If possible, think of a place you anticipated visiting, or one that surprised you in some way when you reached it.
Freewrite for ten minutes on the impressions you can recall from that place or,
better yet, the experience of visiting it.

QUESTIONS FOR CRITICAL THINKING

STRATEGY

1. Structuring the opening of his book on the basis of our sense of smell,
   Wallace invites us to think of parallel experiences we have had. Think
   about his sentence on page 126, "There was something dangerous
   about the smell, something inhibiting to my nine-year-old mind. " Try
   writing about this idea by modifying the sentence; fill in the blanks:
   "There was something _____ about the smell, something _____ to
   my ____-year-old mind."
2. Think about alternative strategies Wallace might have used to describe
   how he felt as a child first experiencing the hammock. Explain how he
   might have pursued another of these possible strategies; would he have
   done better to choose another way of describing the nature of the
   hammock?

ISSUES

1. Wallace certainly seems to find much of interest in the often unnoticed
   parts of the hammock. Think of some parts of landscapes with which
   you are familiar that are worth concentrating on, but that other people
   might not notice. What makes these details important to you?
2. Think about Wallace's statement that "the individual integrity we feel as
   organisms is not something that emerges from ecosystems, but something that fits *in* them." How far are you willing to accept this pro-
   nouncement? Explain what Wallace means in his sentence, "Fire, sticks,
   head, and green wall have been at the center of things for most of
   human experience, and they still are, in a sense, although the green wall
   may have receded." Think about the connotations of the first four
   nouns, and see how you would restate the sentence in your own words.
   Would you want to use another sentence to explain the last clause in
   the sentence? If so, what would that sentence say?

COLLABORATIVE WRITING ACTIVITY

In your group, appoint a recorder to write down an outline for your combined
suggestions about what Wallace will do in the rest of his book. After deciding

what he might be attempting in the two chapters presented here, build on that discussion to project what your group thinks will occupy Wallace further along in his account of the hammock.

"Water Trails of the Ceriso"
from *The Land of Little Rain*
Mary Austin

Mary Austin (1868-1934) was a devoted feminist and fierce opponent of commercial development of her adoptive home in the California desert country. Born Mary Hunter in Illinois to a family of moderate means, she moved in 1888 to California to join a brother who had begun a homestead near the San Joaquin Valley. She had attended Blackburn College before leaving Illinois, where she studied natural history and wrote poetry.

She fell in love with the California desert of the Owens Valley, where she lived with her husband Wallace Austin in straitened financial circumstances. Frustrated by an unhappy marriage and the grind of poverty, she began writing and in 1903 published her first book, The Land of Little Rain. *The book became immediately successful, and she followed it up with numerous other works of essays, poetry, fiction, and autobiography. Her first book remains her most widely read; the minute descriptions of the desert's inhabitants, human and nonhuman, have impressed many readers.*

EXPLORING JOURNAL

Freewrite about your sense of how animals in the desert get to water. What problems do you imagine they might experience? What predators would be able to stake out likely places and prey on animals who need water?

———◆◆◆———

By the end of the dry season the water trails of the Ceriso are worn to a white ribbon in the leaning grass, spread out faint and fanwise toward the homes of gopher and ground rat and squirrel. But however faint to man-sight, they are sufficiently plain to the furred and feathered folk who travel them. Getting down to the eye level of rat and squirrel kind, one perceives what might easily be wide and winding roads to us if they occurred in thick plantations of trees three times the height of a man. It needs but a slender thread of barrenness to make a mouse trail in the forest of the sod. To the little people the water trails are as country roads, with scents as signboards.

It seems that man-height is the least fortunate of all heights from which to study trails. It is better to go up the front of some tall hill, say the spur of Black Mountain, looking back and down across the hollow of the Ceriso. Strange how long the soil keeps the impression of any continuous treading, even after grass has overgrown it. Twenty years since, a brief heyday of mining at Black Mountain made a stage road across the Ceriso, yet the parallel lines that are the wheel traces show from the height dark and well defined. Afoot in the Ceriso one looks in vain for any sign of it. So all the paths that wild creatures use going down to the Lone Tree Spring are mapped out whitely from this level, which is also the level of the hawks.

There is little water in the Ceriso at the best of times, and that little brackish and smelling vilely, but by a lone juniper where the rim of the Ceriso breaks away to the lower country, there is a perpetual rill of fresh sweet drink in the midst of lush grass and watercress. In the dry season there is no water else for a man's long journey of a day. East to the foot of the Black Mountain, and north and south without counting, are the burrows of small rodents, rat and squirrel kind. Under the sage are the shallow forms of the jackrabbits, and in the dry banks of washes, and among the strewn fragments of black rock, lairs of bobcat, fox, and coyote.

The coyote is your true water-witch, one who snuffs and paws, snuffs and paws again at the smallest spot of moisture-scented earth until he has freed the blind water from the soil. Many water-holes are no more than this detected by the lean hobo of the hills in localities where not even an Indian would look for it.

It is the opinion of many wise and busy people that the hillfolk pass the ten-month interval between the end and renewal of winter rains, with no drink; but your true idler, with days and nights to spend beside the water trails, will not subscribe to it. The trails begin, as I said, very far back in the Ceriso, faintly, and converge in one span broad, white, hard-trodden way in the gully of the spring. And why trails if there are no travelers in that direction?

I have yet to find the land not scarred by the thin, far roadways of rabbits and what not of furry folks that run in them. Venture to look for some seldom-touched water-hole, and so long as the trails run with your general direction make sure you are right, but if they begin to cross yours at never so slight an angle, to converge toward a point left or right of your objective, no matter what the maps say, or your memory, trust them; they *know*.

It is very still in the Ceriso by day, so that were it not for the evidence of those white beaten ways, it might be the desert it looks. The sun is hot in the dry season, and the days are filled with the glare of it. Now and again some unseen coyote signals his pack in a long-drawn, dolorous whine that comes from no determinate point, but nothing stirs much before mid-afternoon. It is a sign when there begin to be hawks skimming above the sage that the little people are going about their business.

We have fallen on a very careless usage, speaking of wild creatures as if they were bound by some such limitation as hampers clockwork. When we say of one and another, they are night prowlers, it is perhaps true only as the things they feed upon are more easily come by in the dark, and they know well how to adjust themselves to conditions wherein food is more plentiful by day. And their accustomed performance is very much a matter of keen eye, keener scent, quick ear, and a better memory of sights and sounds than man dares boast. Watch a coyote come out of his lair and cast about in his mind where he will go for his daily killing. You cannot very well tell what decides him, but very easily that he has decided. He trots or breaks into short gallops, with very perceptible pauses to look up and about at landmarks, alters his tack a little, looking forward and back to steer his proper course. I am persuaded that the coyotes in my valley, which is narrow and beset with steep, sharp hills, in long passages steer by the pinnacles of the sky-line, going with head cocked to one side to keep to the left or right of such and such a promontory.

I have trailed a coyote often, going across country, perhaps to where some slant-winged scavenger hanging in the air signaled prospect of a dinner, and found his track such as a man, a very intelligent man accustomed to a hill country, and a little cautious, would make to the same point. Here a detour to avoid a stretch of too little cover, there a pause on the rim of gully to pick the better way—and it is usually the best way—and making his point with the greatest economy of effort. Since the time of Seyavi the deer have shifted their feeding ground across the valley at the beginning of deep snows, by way of the Black Rock, fording the river at Charley's Butte, and making straight for the mouth of the cañon that is the easiest going to the winter pastures on Waban. So they still cross, though whatever trail they had has been long broken by ploughed ground; but from the mouth of Tinpah Creek, where the deer come out of the Sierras, it is easily seen that the creek, the point of Black Rock, and Charley's Butte are in line with the wide bulk of shade that is the foot of Waban Pass. And along with this the deer have learned that Charley's Butte is almost the only possible ford, and all the shortest crossing of the valley. It seems that the wild creatures have learned all that is important to their way of life except the changes of the moon. I have seen some prowling fox or coyote, surprised by its sudden rising from behind the mountain wall, slink in its increasing glow, watch it furtively from the cover of near-by brush, unprepared and half uncertain of its identity until it rode clear of the peaks, and finally make off with all the air of one caught napping by an ancient joke. The moon in its wanderings must be a sort of exasperation to cunning beasts, likely to spoil by untimely risings some fore-planned mischief.

But to take the trail again; the coyotes that are astir in the Ceriso of late afternoons, harrying the rabbits from their shallow forms, and the hawks that sweep and swing above them, are not there from any mechanical promptings of instinct, but because they know of old experience that the small fry are about to take to seed gathering and the water trails. The rabbits begin it, taking the trail with long, light leaps, one eye and ear cocked to the hills from whence a coyote might descend upon them at any moment. Rabbits are a foolish people. They do not fight except with their own kind, nor use their paws except for feet, and appear to have no reason for existence but to furnish meals for meat-eaters. In flight they seem to rebound from the earth of their own elasticity, but keep a sober pace going to the spring. It is the young watercress that tempts them and the pleasures of society, for they seldom drink. Even in localities where there are flowing streams they seem to prefer the moisture that collects on herbage, and after rains may be seen rising on their haunches to drink delicately the clear drops caught in the tops of the young sage. But drink they must, as I have often seen them mornings and evenings at the rill that goes by my door. Wait long enough at the Lone Tree Spring and sooner or later they will all come in. But here their matings are accomplished, and though they are fearful of so little as a cloud shadow or blown leaf, they contrive to have some playful hours. At the spring the bobcat drops down upon them from the black rock, and the red fox picks them up returning in the dark. By day the hawk and eagle overshadow them, and the coyote has all times and seasons for his own.

Cattle, when there are any in the Ceriso, drink morning and evening, spending the night on the warm last lighted slopes of neighboring hills, stirring with the peep o' day. In these half wild spotted steers the habits of an earlier lineage persist. It must be long since they have made beds for themselves, but before lying down they turn themselves round and round as dogs do. They choose bare and stony ground, exposed fronts of westward facing hills, and lie down in companies. Usually by the end of the summer the cattle have been driven or gone of their own choosing to the mountain meadows. One year a maverick yearling, strayed or overlooked by the vaqueros, kept on until the season's end, and so betrayed another visitor to the spring that else I might have missed. On a certain morning the half-eaten carcass lay at the foot of the black rock, and in moist earth by the rill of the spring, the foot-pads of a cougar, puma, mountain lion, or whatever the beast is rightly called. The kill must have been made early in the evening, for it appeared that the cougar had been twice to the spring; and since the meat-eater drinks little until he has eaten, he must have fed and drunk, and after an interval of lying up in the black rock, had eaten and drunk again. There was no knowing how far he had come, but if he came again the second night he found that the coyotes had left him very little of his kill.

Nobody ventures to say how infrequently and at what hour the small fry visit the spring. There are such numbers of them that if each came once between the last of spring and the first of winter rains, there would still be water trails. I have seen badgers drinking about the hour when the light takes on the yellow tinge it has from coming slantwise through the hills. They find out shallow places, and are loath to wet their feet. Rats and chipmunks have been observed visiting the spring as late as nine o'clock mornings. The larger spermophiles that live near the spring and keep awake to work all day, come and go at no particular hour, drinking sparingly. At long intervals on half-lighted days, meadow and field mice steal delicately along the trail. These visitors are all too small to be watched carefully at night, but for evidence of their frequent coming there are the trails that may be traced miles out among the crisping grasses. On rare nights, in the places where no grass grows between the shrubs, and the sand silvers whitely to the moon, one sees them whisking to

and fro on innumerable errands of seed gathering, but the chief witnesses of
their presence near the spring are the elf owls. Those burrow-haunting, speck-
led fluffs of greediness begin a twilight flitting toward the spring, feeding as
they go on grasshoppers, lizards, and small, swift creatures, diving into burrows
to catch field mice asleep, battling with chipmunks at their own doors, and getting down in great numbers toward the lone juniper. Now owls do not love
water greatly on its own account. Not to my knowledge have I caught one
drinking or bathing, though on night wanderings across the mesa they flit up
from under the horse's feet along stream borders. Their presence near the
spring in great numbers would indicate the presence of the things they feed
upon. All night the rustle and soft hooting keeps on in the neighborhood of the
spring, with seldom small shrieks of mortal agony. It is clear day before they
have all gotten back to their particular hummocks, and if one follows cautiously,
not to frighten them into some near-by burrow, it is possible to trail them far
up the slope.

The crested quail that troop in the Ceriso are the happiest frequenters of
the water trails. There is no furtiveness about their morning drink. About the
time the burrowers and all that feed upon them are addressing themselves to
sleep, great flocks pour down the trails with that peculiar melting motion of
moving quail, twittering, shoving, and shouldering. They splatter into the shallows, drink daintily, shake out small showers over their perfect coats, and melt
away again into the scrub, preening and pranking, with soft contented noises.

After the quail, sparrows and ground-inhabiting birds bathe with the
utmost frankness and a great deal of splutter; and here in the heart of noon
hawks resort, sitting panting, with wings aslant, and a truce to all hostilities
because of the heat. One summer there came a road-runner up from the lower
valley, peeking and prying, and he had never any patience with the water baths
of the sparrows. His own ablutions were performed in the clean, hopeful dust
of the chaparral; and whenever he happened on their morning splatterings, he
would depress his glossy crest, slant his shining tail to the level of his body,
until he looked most like some bright venomous snake, daunting them with
shrill abuse and feint of battle. Then suddenly he would go tilting and balancing down the gully in fine disdain, only to return in a day or two to make sure
the foolish bodies were still at it.

Out on the Ceriso about five miles, and wholly out of sight of it, near where
the immemorial foot trail goes up from Saline Flat toward Black Mountain, is a
water sign worth turning out of the trail to see. It is a laid circle of stones large
enough not to be disturbed by any ordinary hap, with an opening flanked by
two parallel rows of similar stones, between which were an arrow placed,
touching the opposite rim of the circle, it would point as the crow flies to the
spring. It is the old, indubitable water mark of the Shoshones. One still finds it
in the desert ranges in Salt Wells and Mesquite valleys, and along the slopes of
Waban. On the other side of Ceriso, where the black rock begins, about a mile
from the spring, is the work of an older, forgotten people. The rock hereabout
is all volcanic, fracturing with a crystalline whitish surface, but weathered outside to furnace blackness. Around the spring, where must have been a gathering

place of the tribes, it is scored over with strange pictures and symbols that have no meaning to the Indians of the present day; but out where the rock begins, there is carved into the white heart of it a pointing arrow over the symbol for distance and a circle full of wavy lines reading thus: "In this direction three [units of measurement unknown] is a spring of sweet water; look for it."

RESPONDING JOURNAL

Austin gives us a sense of animal roads or streets that would correspond to those that a human would know and use. Brainstorm other parallels between what an animal requires and what the human parallel might be. Decide on one to explore at greater length in a journal entry.

QUESTIONS FOR CRITICAL THINKING

STRATEGY

1. Explore the strategy Austin uses here. Does the entire essay rest on comparison and contrast? Or does Austin begin with a clear comparison and contrast and then shift to a strategy in which the comparisons and contrasts are more implied than they are explicit? How well does her strategy work here? Does she use comparison and contrast effectively?
2. Explain how effective you find Austin's use of animal points of view as a way of elaborating the life in the desert. What other devices or strategies might she have used?

ISSUES

1. Consider why Austin writes that, when following the trails, if the maps or your memory tell you something about another direction, be sure to obey the trails: "no matter what the maps say, or your memory, trust them; they *know*." What point is she making here?
2. What does Austin's focus on the animal life of the desert tell you about the land itself? Explain the feeling you have for the desert after having thought about it from the point of view of the animal life present there.

COLLABORATIVE WRITING ACTIVITY

Think of a structure other than water trails that could be used to chart imaginatively the comings and goings of creatures in the natural world. Give yourself some time to brainstorm your idea so you can come up with a list or matrix of what the different animals will be doing. The group should, after hearing all reports, decide on the best idea to present to the assembled class.

"Pepacton"
John Burroughs

When he died in 1921, John Burroughs was an immensely popular American writer—perhaps the most widely read author of that day. At his home, Slabsides, in the Hudson River Valley, he was visited by thousands of people, on virtual pilgrimages to a writer who espoused a gentle, unobtrusive relationship with the natural world. Theodore Roosevelt and Henry Ford paid their respects at Slabsides. Burroughs was one of ten "Most Prominent Americans," along with Washington, Lincoln, and Edison, chosen as the subjects of a ten-volume biography series intended to illustrate the development of the American character.

Burroughs' diminishing popularity began in the late 1930s, the days of the Great Depression before World War II. Some have speculated that Burroughs, who wrote hundreds of essays, never had a single long work like Walden *that would capture a long-term reading public. Others have suggested that his essays demonstrate a gentle familiarity and reverence that leans too far toward sentimentality and that his writing thus lacks the hard edge that would win it sustained popularity in an age more cynical than that of Burroughs' lifetime. You may decide that you find some of these qualities in Burroughs' work, or that you see some qualities in his writing that should resurrect it for new generations of readers.*

Born in the Catskills in Roxbury, New York, in 1837, Burroughs was first a rural schoolteacher and then spent a decade as a government clerk in Washington, D.C. He began writing in earnest when he returned to New York after ending his government service; his first major collection, Wake-Robin, *was published in 1871. "Pepacton," which is also the title of one of his many volumes of essays, was issued to elementary school children in the first decade of the twentieth century and was used through the 1940s to introduce nature study to the classroom. The essay is his backward glance at the area in which he was born. That area, along with much of the rest of the course of the East or Pepacton branch of the Delaware down which Burroughs travels, is now under Pepacton Reservoir, though the project was never dreamed of during Burroughs' lifetime.*

EXPLORING JOURNAL

If you reflect back on your own childhood, what companions do you remember best? What people were most important to you? In a freewriting entry, explain why these particular people meant so much to you.

PEPACTON: A SUMMER VOYAGE

When one summer day I bethought me of a voyage down the east or Pepacton branch of the Delaware, I seemed to want some excuse for the start, some send-off, some preparation, to give the enterprise genesis and head. This I found in building my own boat. It was a happy thought. How else should I have got under way, how else should I have raised the breeze? The boat-building warmed the blood; it made the germ take; it whetted my appetite for the voyage. There is nothing like serving an apprenticeship to fortune, like earning the right to your tools. In most enterprises the temptation is always to begin too far along; we want to start where somebody else leaves off. Go back to the stump, and see what an impetus you get. Those fishermen who wind their own flies before they go a-fishing—how they bring in the trout; and those hunters who run their own bullets or make their own cartridges—the game is already mortgaged to them.

When my boat was finished—and it was a very simple affair—I was as eager as a boy to be off; I feared the river would all run by before I could wet her bottom in it. This enthusiasm begat great expectations of the trip. I should surely surprise Nature and win some new secrets from her. I should glide down noiselessly upon her and see what all those willow screens and baffling curves concealed. As a fisherman and pedestrian I had been able to come at the stream only at certain points: now the most private and secluded retreats of the nymph would be opened to me; every bend and eddy, every cove hedged in by swamps or passage walled in by high alders, would be at the beck of my paddle.

Whom shall one take with him when he goes a-courting Nature? This is always a vital question. There are persons who will stand between you and that which you seek: they obtrude themselves; they monopolize your attention; they blunt your sense of the shy, half-revealed intelligences about you. I want for companion a dog or a boy, or a person who has the virtues of dogs and boys—transparency, good-nature, curiosity, open sense, and a nameless quality that is akin to trees and growths and the inarticulate forces of nature. With him you are alone, and yet have company; you are free; you feel no disturbing element; the influences of nature stream through him and around him; he is a good conductor of the subtle fluid. The quality or qualification I refer to belongs to most persons who spend their lives in the open air—to soldiers, hunters, fishers, laborers, and to artists and poets of the right sort. How full of it, to choose an illustrious example, was such a man as Walter Scott!

But no such person came in answer to my prayer, so I set out alone.

It was fit that I put my boat into the water at Arkville, but it may seem a little incongruous that I should launch her into Dry Brook; yet Dry Brook is here a fine large trout stream, and I soon found its waters were wet enough for all practical purposes. The Delaware is only one mile distant, and I chose this as the easiest road from the station to it. A young farmer helped me carry the boat to the water, but did not stay to see me off; only some calves feeding alongshore witnessed my embarkation. It would have been a godsend to boys, but there were no boys about. I stuck on a rift before I had gone ten yards, and saw with misgiving the paint transferred from the bottom of my little scow to the tops of the stones thus early in the journey. But I was soon making fair headway, and taking trout for my dinner as I floated along. My first mishap was when I broke the second joint of my rod on a bass, and the first serious impediment to my progress was when I encountered the trunk of a prostrate elm bridging the stream within a few inches of the surface. My rod mended and the elm cleared, I anticipated better sailing when I should reach the Delaware itself; but I found on this day and on subsequent days that the Delaware has a way of dividing up that is very embarrassing to the navigator. It is a stream of many minds: its waters cannot long agree to go all in the same channel, and whichever branch I took I was pretty sure to wish I had taken one of the others. I was constantly sticking on rifts, where I would have to dismount, or running full tilt into willow banks where I would lose my hat or endanger my fishing-tackle. On the whole, the result of my first day's voyaging was not encouraging. I made barely eight miles, and my ardor was a good deal dampened, to say nothing about my clothing. In mid-afternoon I went to a well-to-do-looking farmhouse and got some milk, which I am certain the thrifty housewife skimmed, for its blueness infected my spirits, and I went into camp that night more than half persuaded to abandon the enterprise in the morning. The loneliness of the river, too, unlike that of the fields and woods, to which I was more accustomed, oppressed me. In the woods, things are close to you, and you touch them and seem to interchange something with them; but upon the river, even though it be a narrow and shallow one like this, you are more isolated, farther removed from the soil and its attractions, and an easier prey to the unsocial demons. The long, unpeopled vistas ahead; the still, dark eddies; the endless monotone and soliloquy of the stream; the unheeding rocks basking like monsters along the shore, half out of the water, half in; a solitary heron starting up here and there, as you rounded some point, and flapping disconsolately ahead till lost to view, or standing like a gaunt spectre on the umbrageous side of the mountain, his motionless form revealed against the dark green as you passed; the trees and willows and alders that hemmed you in on either side, and hid the fields and the farmhouses and the road that ran near by—these things and others aided the skimmed milk to cast a gloom over my spirits that argued ill for the success of my undertaking. Those rubber boots, too, that parboiled my feet and were clogs of lead about them—whose spirits are elastic enough to endure them? A malediction upon the head of him who invented them! Take your old shoes, that will let the water in and let it out again, rather than stand knee-deep all day in these extinguishers.

I escaped from the river, that first night, and took to the woods, and profited by the change. In the woods I was at home again, and the bed of hemlock boughs salved my spirits. A cold spring run came down off the mountain, and beside it, underneath birches and hemlocks, I improvised my hearthstone. In sleeping on the ground it is a great advantage to have a back-log; it braces and supports you, and it is a bedfellow that will not crumble when, in the middle of the night, you crowd sharply up against it. It serves to keep in the warmth, also. A heavy stone or other *point de résistance* at your feet is also a help. Or, better still, scoop out a little place in the earth, a few inches deep, so as to admit your body from your hips to your shoulders; you thus get an equal bearing the whole length of you. I am told the Western hunters and guides do this. On the same principle the sand makes a good bed, and the snow. You make a mound in which you fit nicely. My berth that night was between two logs that the barkpeelers had stripped ten or more years before. As they had left the bark there, and as hemlock bark makes excellent fuel, I had more reasons than one to be grateful to them.

In the morning, I felt much refreshed, and as if the night had tided me over the bar that threatened to stay my progress. If I can steer clear of skimmed milk, I said, I shall now finish the voyage of fifty miles to Hancock with increasing pleasure.

When one breaks camp in the morning, he turns back again and again to see what he has left. Surely he feels, he has forgotten something; what is it? But it is only his own sad thoughts and musings he has left, the fragment of his life he has lived there. Where he hung his coat on the tree, where he slept on the boughs, where he made his coffee or broiled his trout over the coals, where he drank again and again at the little brown pool in the spring run, where he looked long and long up into the whispering branches overhead, he has left what he cannot bring away with him—the flame and the ashes of himself.

Of certain game-birds it is thought that at times they have the power of withholding their scent; no hint or particle of themselves goes out upon the air. I think there are persons whose spiritual pores are always sealed up, and I presume they have the best time of it. Their hearts never radiate into the void; they do not yearn and sympathize without return; they do not leave themselves by the wayside as the sheep leaves her wool upon the brambles and thorns.

This branch of the Delaware, so far as I could learn, had never before been descended by a white man in a boat. Rafts of pine and hemlock timber are run down on the spring and fall freshets, but of pleasure-seekers in boats I appeared to be the first. Hence my advent was a surprise to most creatures in the water and out. I surprised the cattle in the field, and those ruminating leg-deep in the water turned their heads at my approach, swallowed their unfinished cuds, and scampered off as if they had seen a spectre. I surprised the fish on their spawning-beds and feeding-grounds; they scattered, as my shadow glided down upon them, like chickens when a hawk appears. I surprised an ancient fisherman seated on a spit of gravelly beach, with his back upstream, and leisurely angling in a deep, still eddy, and mumbling to himself. As I slid into the circle of his vision his grip on the pole relaxed, his jaw dropped, and he was too bewildered to reply to my salutation for some moments. As I turned a bend in the river I looked back and saw him hastening away with great precipitation. I presumed he had angled there for forty years without having his privacy thus intruded upon. I surprised hawks and herons and kingfishers. I came suddenly upon muskrats, and raced with them down the rifts, they having no time to take to their holes. At one point, as I rounded an elbow in the stream, a black eagle sprang from the top of a dead tree, and flapped hurriedly away. A kingbird gave chase, and disappeared for some moments in the gulf between the great wings of the eagle, and I imagined him seated upon his back delivering his puny blows upon the royal bird. I interrupted two or three minks fishing and hunting alongshore. They would dart under the bank when they saw me, then presently thrust out their sharp, weasel-like noses, to see if the danger was imminent. At one point, in a little cove behind the willows, I surprised some schoolgirls, with skirts amazingly abbreviated, wading and playing in the water. And as much surprised as any, I am sure, was that hard-worked-looking housewife, when I came up from under the bank in front of her house, and with pail in hand appeared at her door and asked for milk, taking the precaution to intimate that I had no objection to the yellow scum that is supposed to rise on a fresh article of that kind.

"What kind of milk do you want?"

"The best you have. Give me two quarts of it," I replied.

"What do you want to do with it?" with an anxious tone, as if I might want to blow up something or burn her barns with it.

"Oh, drink it," I answered, as if I frequently put milk to that use.

"Well, I suppose I can get you some;" and she presently reappeared with swimming pail, with those little yellow flakes floating about upon it that one likes to see.

I passed several low dams the second day, but had no trouble. I dismounted and stood upon the apron, and the boat, with plenty of line, came over as lightly as a chip, and swung around in the eddy below like a steed that knows its master. In the afternoon, while slowly drifting down a long eddy, the moist southwest wind brought me the welcome odor of strawberries, and running ashore by a meadow, a short distance below, I was soon parting the daisies and filling my cup with the dead-ripe fruit. Berries, be they red, blue, or black, seem like a special providence to the camper-out; they are luxuries he has not counted on, and I prized these accordingly. Later in the day it threatened rain, and I drew up to shore under the shelter of some thick overhanging hemlocks, and proceeded to eat my berries and milk, glad of an excuse not to delay my lunch longer. While tarrying here I heard young voices upstream, and looking in that direction saw two boys coming down the rapids on rude floats. They were racing along at a lively pace, each with a pole in his hand, dexterously avoiding the rocks and the breakers, and schooling themselves thus early in the duties and perils of the raftsmen. As they saw me one observed to the other—

"There is the man we saw go by when we were building our floats. If we had known he was coming so far, maybe we could have got him to give us a ride."

They drew near, guided their crafts to shore beside me, and tied up, their poles answering for hawsers. They proved to be Johnny and Denny Dwire, aged ten and twelve. They were friendly boys, and though not a bit bashful were not a bit impertinent. And Johnny, who did the most of the talking, had such a sweet, musical voice; it was like a bird's. It seems Denny had run away a day or two before, to his uncle's, five miles above, and Johnny had been after him, and was bringing the prisoner home on a float; and it was hard to tell which was enjoying the fun most, the captor or the captured.

"Why did you run away?" said I to Denny.

"Oh, 'cause," replied he, with an air which said plainly, "The reasons are too numerous to mention."

"Boys, you know, will do so, sometimes," said Johnny, and he smiled upon his brother in a way that made me think they had a very good understanding upon the subject.

They could both swim, yet their floats looked very perilous—three pieces of old plank or slabs, with two cross-pieces and a fragment of a board for a rider, and made without nails or withes.

"In some places," said Johnny, "one plank was here and another off there, but we managed, somehow, to keep atop of them."

"Let's leave our floats here, and ride with him the rest of the way," said one to the other.

"All right; may we, mister?"

I assented, and we were soon afloat again. How they enjoyed the passage; how smooth it was; how the boat glided along; how quickly she felt the paddle! They admired her much; they praised my steersmanship; they praised my fish-pole and all my fishings down to my hateful rubber boots. When we stuck on the rifts, as we did several times, they leaped out quickly with their bare feet and legs, and pushed us off.

"I think," said Johnny, "if you keep her straight and let her have her own way, she will find the deepest water. Don't you, Denny?"

"I think she will," replied Denny; and I found the boys were pretty nearly right.

I tried them on a point of natural history. I had observed, coming along, a great many dead eels lying on the bottom of the river, that I supposed had died from spear wounds. "No," said Johnny, "they are lamper eels. They die as soon as they have built their nests and laid their eggs."

"Are you sure?"

"That's what they all say, and I know they are lampers."

So I fished one up out of the deep water with my paddle-blade and examined it; and sure enough it was a lamprey. There was the row of holes along its head, and its ugly suction mouth. I had noticed their nests, too, all along, where the water in the pools shallowed to a few feet and began to hurry toward the rifts: they were low mounds of small stones, as if a bushel or more of large pebbles had been dumped upon the river bottom; occasionally they were so near the surface as to make a big ripple. The eel attaches itself to the stones by its mouth, and thus moves them at will. An old fisherman told me that a strong man could not pull a large lamprey loose from a rock to which it had attached itself.

It fastens to its prey in this way, and sucks the life out. A friend of mine says he once saw in the St. Lawrence a pike as long as his arm with a lamprey eel attached to him. The fish was nearly dead and was quite white, the eel had so sucked out his blood and substance. The fish, when seized, darts against rocks and stones, and tries in vain to rub the eel off, then succumbs to the sucker.

"The lampers do not all die," said Denny, "because they do not all spawn;" and I observed that the dead ones were all of one size and doubtless of the same age.

The lamprey is the octopus, the devil-fish, of these waters, and there is, perhaps, no tragedy enacted here that equals that of one of these vampires slowly sucking the life out of a bass or a trout.

My boys went to school part of the time. Did they have a good teacher?

"Good enough for me," said Johnny.

"Good enough for me," echoed Denny.

Just below Bark-a-boom—the name is worth keeping—they left me. I was loath to part with them; their musical voices and their thorough good fellowship had been very acceptable. With a little persuasion, I think they would have left their home and humble fortunes, and gone a-roving with me.

About four o'clock the warm, vapor-laden southwest wind brought forth the expected thunder shower. I saw the storm rapidly developing behind the mountains in my front. Presently I came in sight of a long covered wooden bridge that spanned the river about a mile ahead, and I put my paddle into the water with all my force to reach this cover before the storm. It was neck and neck most of the way. The storm had the wind, and I had it—in my teeth. The bridge was at Shavertown, and it was by a close shave that I got under it before the rain was upon me. How it poured and rattled and whipped in around the abutment of the bridge to reach me! I looked out well satisfied upon the foaming water, upon the wet, unpainted houses and barns of the Shavertowners, and upon the trees,

"Caught and cuffed by the gale."

Another traveler—the spotted-winged nighthawk—was also roughly used by the storm. He faced it bravely, and beat and beat, but was unable to stem it, or even hold his own; gradually he drifted back till he was lost to sight in the wet obscurity. The water in the river rose an inch while I waited, about three quarters of an hour. Only one man, I reckon, saw me in Shavertown, and he came and gossiped with me from the bank above when the storm had abated.

The second night I stopped at the sign of the elm-tree. The woods were too wet, and I concluded to make my boat my bed. A superb elm, on a smooth grassy plain a few feet from the water's edge, looked hospitable in the twilight, and I drew my boat up beneath it. I hung my clothes on the jagged edges of its rough bark, and went to bed with the moon, "in her third quarter," peeping under the branches upon me. I had been reading Stevenson's amusing "Travels with a Donkey," and the lines he pretends to quote from an old play kept running in my head:—

"The bed was made, the room was fit,
By punctual eve the stars were lit;
The air was sweet, the water ran;
No need was there for maid or man,
When we put up, my ass and I,
At God's green caravanserai."

But the stately elm played me a trick: it slyly and at long intervals let great drops of water down upon me, now with a sharp smack upon my rubber coat; then with a heavy thud upon the seat in the bow or stern of my boat; then plump into my upturned ear, or upon my uncovered arm, or with a ring into my tin cup, or with a splash into my coffee-pail that stood at my side full of water from a spring I had just passed. After two hours' trial I found dropping off to sleep, under such circumstances, was out of the question; so I sprang up, in no very amicable mood toward my host, and drew my boat clean from under the elm. I had refreshing slumber thenceforth, and the birds were astir in the morning long before I was.

There is one way, at least, in which the denuding the country of its forests has lessened the rainfall: in certain conditions of the atmosphere every tree is a great condenser of moisture, as I had just observed in the case of the old elm; little showers are generated in their branches, and in the aggregate the amount of water precipitated in this way is considerable. Of a foggy summer morning one may see little puddles of water standing, on the stones beneath maple-trees, along the street; and in winter, when there is a sudden change from cold to warm, with fog, the water fairly runs down the trunks of the trees, and streams from their naked branches. The temperature of the tree is so much below that of the atmosphere in such cases that the condensation is very rapid. In lieu of these arboreal rains we have the dew upon the grass, but it is doubtful if the grass ever drips as does a tree.

The birds, I say, were astir in the morning before I was, and some of them were more wakeful through the night, unless they sing in their dreams. At this season one may hear at intervals numerous bird voices during the night. The whip-poor-will was piping when I lay down, and I still heard one when I woke up after midnight. I heard the song sparrow and the kingbird also, like watchers calling the hour, and several times I heard the cuckoo. Indeed, I am convinced that our cuckoo is to a considerable extent a night bird, and that he moves about freely from tree to tree. His peculiar guttural note, now here, now there, may be heard almost any summer night, in any part of the country, and occasionally his better known cuckoo call. He is a great recluse by day, but seems to wander abroad freely by night.

The birds do indeed begin with the day. The farmer who is in the field at work while he can yet see stars catches their first matin hymns. In the longest June days the robin strikes up about half-past three o'clock, and is quickly followed by the song sparrow, the oriole, the catbird, the wren, the wood thrush, and all the rest of the tuneful choir. Along the Potomac I have heard the Virginia cardinal whistle so loudly and persistently in the treetops above, that sleeping after four o'clock was out of the question. Just before the sun is up, there is a marked lull, during which, I imagine, the birds are at breakfast. While building their nest, it is very early in the morning that they put in their big strokes; the back of their day's work is broken before you have begun yours.

A lady once asked me if there was any individuality among the birds, or if those of the same kind were as near alike as two peas. I was obliged to answer that to the eye those of the same species were as near alike as two peas, but that in their songs there *were* often marks of originality. Caged or domesticated birds develop notes and traits of their own, and among the more familiar orchard and garden birds one may notice the same tendency. I observe a great variety of songs, and even qualities of voice, among the orioles and among the song sparrows. On this trip my ear was especially attracted to some striking and original sparrow songs. At one point I was half afraid I had let pass an opportunity to identify a new warbler, but finally concluded it was a song sparrow. On another occasion I used to hear day after day a sparrow that appeared to have some organic defect in its voice: part of its song was scarcely above a whisper, as if the bird was suffering from a very bad cold. I have heard a robin with a part of the whistle of the quail in his song. It was out of time and out of tune, but the robin seemed insensible of the incongruity, and sang as loudly and as joyously as any of his mates. A catbird will sometimes show a special genius for mimicry, and I have known one to suggest very plainly some notes of the bobolink.

There are numerous long covered bridges spanning the Delaware, and under some of these I saw the cliff swallow at home, the nests being fastened to the under sides of the timbers—as it were, suspended from the ceiling instead of planted upon the shelving or perpendicular side, as is usual with them. To have laid the foundation, indeed, to have sprung the vault downward and finished it successfully, must have required special engineering skill. I had never before seen or heard of these nests being so placed. But birds are quick to adjust their needs to the exigencies of any case. Not long before, I had seen in a deserted house, on the head of the Rondout, the chimney swallows entering the chamber through a stove-pipe hole in the roof, and gluing their nests to the sides of the rafters, like the barn swallows.

I was now, on the third day, well down in the wilds of Colchester, with a current that made between two and three miles an hour—just a summer idler's pace. The atmosphere of the river had improved much since the first day—was, indeed without taint—and the water was sweet and good. There were farmhouses at intervals of a mile or so; but the amount of tillable land in the river valley or on the adjacent mountains was very small. Occasionally there would be forty or fifty acres of flat, usually in grass or corn, with a thrifty-looking farmhouse. One could see how surely the land made the house and its surrounding;—good land bearing good buildings and poor land poor.

In mid-forenoon I reached the long placid eddy at Downsville and here again fell in with two boys. They were out paddling about in a boat when I drew near, and they evidently regarded me in the light of a rare prize which fortune had wafted them.

"Ain't you glad we come, Benny?" I heard one of them observe to the other, as they were conducting me to the best place to land. They were bright, good boys, of the same piece as my acquaintances of the day before, and about the same ages—differing only in being village boys. With what curiosity they looked me over! Where had I come from; where was I going; how long had I been on the way; who built my boat; was I a carpenter, to build such a neat craft, etc.? They never had seen such a traveler before. Had I had no mishaps? And then they bethought them of the dangerous passes that awaited me, and in good faith began to warn and advise me. They had heard the tales of rafts-men, and had conceived a vivid idea of the perils of the river below, gauging their notions of it from the spring, and fall freshets tossing, about the heavy and cumbrous rafts. There was a whirlpool, a rock eddy, and a binocle within a mile. I might be caught in the binocle, or engulfed in the whirlpool, or smashed up in the eddy. But I felt much reassured when they told me I had already passed several whirlpools and rock eddies; but that terrible binocle—what was that? I had never heard of such a monster. Oh, it was a still, miry place at the head of a big eddy. The current might carry me up there, but I could easily get out again; the rafts did. But there was another place I must beware of, where two eddies faced each other; raftsmen were sometimes swept off there by the oars and drowned. And when I came to rock eddy, which I would know, because the river divided there (a part of the water being afraid to risk the eddy, I suppose), I must go ashore and survey the pass; but in any case it would be prudent to keep to the left. I might stick on the rift, but that was nothing to being wrecked upon those rocks. The boys were quite in earnest, and I told them I would walk up to the village and post some letters to my friends before I braved all these dangers. So they marched me up the street, pointing out to their chums what they had found.

"Going way to Phil— What place is that near where the river goes into the sea?"

"Philadelphia?"

"Yes; thinks he may go way there. Won't he have fun?"

The boys escorted me about the town, then back to the river, and got in their boat and came down to the bend, where they could see me go through the whirlpool and pass the binocle (I am not sure about the orthography of the word, but I suppose it means a double, or a sort of mock eddy). I looked back as I shot over the rough current beside a gentle vortex, and saw them watching me with great interest. Rock eddy, also, was quite harmless, and I passed it without any preliminary survey.

I nooned at Sodom, and found good milk in a humble cottage. In the afternoon I was amused by a great blue heron that kept flying up in advance of me. Every mile or so, as I rounded some point, I would come unexpectedly upon him, till finally he grew disgusted with my silent pursuit, and took a long turn to the left up along the side of the mountain, and passed back up the river, uttering a hoarse, low note.

The wind still boded rain, and about four o'clock, announced by deep-toned thunder and portentous clouds, it began to charge down the mountain-side in front of me. I ran ashore, covered my traps, and took my way up through an orchard to a quaint little farmhouse. But there was not a soul about, outside or in, that I could find, though the door was unfastened; so I went into an open shed with the hens, and lounged upon some straw, while the unloosed floods came down. It was better than boating or fishing. Indeed, there are few summer pleasures to be placed before that of reclining at ease directly under a sloping roof, after toil or travel in the hot sun, and looking out into the rain-drenched air and fields. It is such a vital yet soothing spectacle. We sympathize with the earth. We know how good a bath is, and the unspeakable deliciousness of water to a parched tongue. The office of the sunshine is slow, subtle, occult, unsuspected; but when the clouds do their work, the benefaction is so palpable and copious, so direct and wholesale, that all creatures take note of it, and for the most part rejoice in it. It is a completion, a consummation, a paying of a debt with a royal hand; the measure is heaped and overflowing. It was the simple vapor of water that the clouds borrowed of the earth; now they pay back more than water: the drops are charged with electricity and with the gases of the air, and have new solvent powers. Then, how the slate is sponged off, and left all clean and new again!

In the shed where I was sheltered were many relics and odds and ends of the farm. In juxtaposition with two of the most stalwart wagon or truck wheels I ever looked upon was a cradle of ancient and peculiar make—an aristocratic cradle, with high-turned posts and an elaborately carved and moulded body, that was suspended upon rods and swung from the top. How I should have liked to hear its history and the story of the lives it had rocked, as the rain sang and the boughs tossed without! Above it was the cradle of a phoebe-bird saddled upon a stick that ran behind the rafter; its occupants had not flown, and its story was easy to read.

Soon after the first shock of the storm was over, and before I could see breaking sky, the birds tuned up with new ardor—the robin, the indigo-bird, the purple finch, the song sparrow, and in the meadow below the bobolink. The cockerel near me followed suit, and repented his refrain till my meditations were so disturbed that I was compelled to eject him from the cover, albeit he had the best right there. But he crowed his defiance with drooping tail from the yard in front. I, too, had mentally crowed over the good fortune of the shower; but before I closed my eyes that night my crest was a good deal fallen, and I could have wished the friendly elements had not squared their accounts quite so readily and uproariously.

The one shower did not exhaust the supply a bit; Nature's hand was full of trumps yet—yea, and her sleeve too. I stopped at a trout brook, which came down out of the mountains on the right, and took a few trout for my supper; but its current was too roily from the shower for fly-fishing. Another farmhouse attracted me, but there was no one at home; so I picked a quart of strawberries in the meadow in front, not minding the wet grass, and about six o'clock, thinking another storm that had been threatening on my right had miscarried, I pushed off, and went floating down into the deepening gloom of the river valley. The mountains, densely wooded from base to summit, shut in the view on every hand. They cut in from the right and from the left, one ahead of the other, matching, like the teeth of an enormous trap; the river was caught and bent, but not long detained, by them. Presently I saw the rain creeping slowly over them in my rear, for the wind had changed; but I apprehended nothing but a moderate sundown drizzle, such as we often get from the tail end of a shower, and drew up in the eddy of a big rock under an overhanging tree till it should have passed. But it did not pass; it thickened and deepened, and reached a steady pour by the time I had calculated the sun would be gilding the mountaintops. I had wrapped my rubber coat about my blankets and groceries, and bared my back to the storm. In sullen silence I saw the night settling down and the rain increasing; my roof-tree gave way, and every leaf poured its accumulated drops upon me. There were streams and splashes where before there had been little more than a mist. I was getting well soaked and uncomplimentary in my remarks on the weather. A saucy catbird, near by, flirted and squealed very plainly, "There! there! What did I tell you! what did I tell you! Pretty pickle! pretty pickle! pretty pickle to be in!" But I had been in worse pickles, though if the water had been salt, my pickling had been pretty thorough. Seeing the wind was in the northeast, and that the weather had fairly stolen a march on me, I let go my hold of the tree, and paddled rapidly to the opposite shore, which was low and pebbly, drew my boat up on a little peninsula, turned her over upon a shot which I cleared of its coarser stone, propped up one end with the scat, and crept beneath. I would now test the virtues of my craft as a roof and I found she was without flaw, though she was pretty narrow. The tension of her timber was such that the rain upon her bottom made a low, musical hum.

Crouched on my blankets and boughs—for I had gathered a good supply of the latter before the rain overtook me—and dry only about my middle, I placidly took life as it came. A great blue heron flew by, and let off something like ironical horse laughter. Before it became dark I proceeded to eat my supper—my berries, but not my trout. What a fuss we make about the "hulls" upon strawberries! We are hypercritical; we may yet be glad to dine off the hulls alone. Some people see something to pick and carp at in every good that comes to them; I was thankful that I had the berries, and resolutely ignored their little scalloped ruffles, which I found pleased the eye and did not disturb the palate.

When bedtime arrived, I found undressing a little awkward, my berth was so low; there was plenty of room in the aisle, and the other passengers were nowhere to be seen, but I did not venture out. It rained nearly all night, but the train made good speed, and reached the land of daybreak nearly on time. The water in the river had crept up during the night to within a few inches of my boat, but I rolled over and took another nap, all the same. Then I arose, had a delicious bath in the sweet, swift-running current, and turned my thoughts toward breakfast. The making of the coffee was the only serious problem. With everything soaked and a fine rain still falling, how shall one build a fire? I made my way to a little island above in quest of driftwood. Before I had found the wood I chanced upon another patch of delicious wild strawberries, and took an appetizer of them out of hand. Presently I picked up a yellow birch stick the size of my arm. The wood was decayed, but the bark was perfect. I broke it in two, punched out the rotten wood, and had the bark intact. The fatty or resinous substance in this bark preserves it, and makes it excellent kindling. With some seasoned twigs and a scrap of paper I soon had a fire going that answered my every purpose. More berries were picked while the coffee was brewing, and the breakfast was a success.

The camper-out often finds himself in what seems a distressing predicament to people seated in their snug, well-ordered houses; but there is often a real satisfaction when things come to their worst—a satisfaction in seeing what a small matter it is, after all; that one is really neither sugar nor salt, to be afraid of the wet; and that life is just as well worth living beneath a scow or a dug-out as beneath the highest and broadest roof in Christendom.

By ten o'clock it became necessary to move, on account of the rise of the water, and as the rain had abated, I picked up and continued my journey. Before long, however, the rain increased again, and I took refuge in a barn. The snug, tree-embowered farmhouse looked very inviting, just across the road from the barn; but as no one was about, and no faces appeared at the window that I might judge of the inmates, I contented myself with the hospitality the barn offered, filling my pockets with some dry birch shavings I found there where the farmer had made an ox-yoke, against the needs of the next kindling.

After an hour's detention I was off again. I stopped at Baxter's Brook, which flows hard by the classic hamlet of Harvard, and tried for trout, but with poor success, as I did not think it worth while to go far upstream.

At several points I saw rafts of hemlock lumber tied to the shore, ready to take advantage of the first freshet. Rafting is an important industry for a hundred miles or more along the Delaware. The lumbermen sometimes take their families or friends, and have a jollification all the way to Trenton or to Philadelphia. In some places the speed is very great, almost equaling that of an express train. The passage of such places as Cochecton Falls and "Foul Rift" is attended with no little danger. The raft is guided by two immense oars, one before and one behind. I frequently saw these huge implements in the driftwood alongshore, suggesting some colossal race of men. The raftsmen have names of their own. From the upper Delaware, where I had set in, small rafts are run down which they call "colts." They come frisking down at a lively pace. At Hancock they usually couple two rafts together, then I suppose they have a span of colts; or do two colts make one horse? Some parts of the framework of the raft they call "grubs;" much depends upon these grubs. The lumbermen were and are a hardy, virile race. The Hon. Charles Knapp, of Deposit, now eighty-three years of age, but with the look and step of a man of sixty, told me he had stood nearly all one December day in the water to his waist, reconstructing his raft, which had gone to pieces on the head of an island. Mr. Knapp had passed the first half of his life in Colchester and Hancock, and, although no sportsman, had once taken part in a great bear hunt there. The bear was an enormous one, and was hard pressed by a gang of men and dogs. Their muskets and assaults upon the beast with clubs had made no impression. Mr. Knapp saw where the bear was coming, and he thought he would show them how easy it was to dispatch a bear with a club, if you only knew where to strike. He had seen how quickly the largest hog would wilt beneath a slight blow across the "small of the back." So, armed with an immense hand-spike, he took up a position by a large rock that the bear must pass. On she came, panting and nearly exhausted, and at the right moment down came the club with great force upon the small of her back. "If a fly had alighted upon her," said Mr. Knapp, "I think she would have paid just as much attention to it as she did me."

Early in the afternoon I encountered another boy, Henry Ingersoll, who was so surprised by any sudden and unwanted appearance that he did not know east from west. "Which way is west?" I inquired, to see if my own head was straight on the subject.

"That way," he said, indicating east within a few degrees.

"You are wrong," I replied "Where does the sun rise?"

"There," he said, pointing almost in the direction he had pointed before.

"But does not the sun rise in the east here as well as elsewhere?" I rejoined.

"Well, they call that west, anyhow."

But Henry's needle was subjected to a disturbing influence just then. His house was near the river, and he was its sole guardian and keeper for the time; his father had gone up to the next neighbor's (it was Sunday), and his sister had gone with the schoolmistress down the road to get black birch. He came out in the road, with wide eyes, to view me as I passed, when I drew rein, and demanded the points of the compass, as above. Then I shook my sooty pail at him and asked for milk. Yes, I could have some milk, but I would have to wait till his sister came back; after he had recovered a little, he concluded he could get it. He came for my pail, and then his boyish curiosity appeared. My story interested him immensely. He had seen twelve summers, but he had been only four miles from home up and down the river: he had been down to the East branch, and he had been up to Trout Brook. He took a pecuniary interest in me. What did my pole cost? What my rubber coat, and what my revolver? The latter he must take in his hand; he had never seen such a thing to shoot with before in *his* life, etc. He thought I might make the trip cheaper and easier by stage and by the cars. He went to school: there were six scholars in summer, one or two more in winter. The population is not crowded in the town of Hancock, certainly, and never will be. The people live close to the bone, as Thoreau would say, or rather close to the stump. Many years ago the young men there resolved upon having a ball. They concluded not to go to a hotel, on account of the expense, and so chose a private house. There was a man in the neighborhood who could play the fife; he offered to furnish the music for seventy-five cents. But this was deemed too much, so one of the party agreed to whistle. History does not tell how many beaux there were bent upon this reckless enterprise, but there were three girls. For refreshments they bought a couple of gallons of whiskey and a few pounds of sugar. When the spree was over, and the expenses were reckoned up, there was a shilling—a York shilling—apiece to pay. Some of the revelers were dissatisfied with this charge, and intimated that the managers had not counted themselves in, but taxed the whole expense upon the rest of the party.

As I moved on, I saw Henry's sister and the schoolmistress picking their way along the muddy road near the river's bank. One of them saw me, and, dropping her skirts, said to the other (I could read the motions), "See that man!" The other lowered her flounces and looked up and down the road, then glanced over into the field, and lastly out upon the river. They paused and had a good look at me, though I could see that their impulse to run away, like that of a frightened deer, was strong.

At the East Branch the Big Beaver Kill joins the Delaware, almost doubling its volume. Here I struck the railroad, the forlorn Midland, and here another set of men and manners cropped out—what may be called the railroad conglomerate overlying this mountain freestone.

"Where did you steal that boat?" and "What you running away for?" greeted me from a handcar that went by.

I paused for some time and watched the fish hawks, or ospreys, of which there were nearly a dozen sailing about above the junction of the two streams, squealing and diving, and occasionally striking a fish on the rifts. I am convinced that the fish hawk sometimes feeds on the wing. I saw him do it on this and on another occasion. He raises himself by a peculiar motion, and brings his head and his talons together, and apparently takes a bite of a fish. While doing this his flight presents a sharply undulating line; at the crest of each rise the morsel is taken.

In a long, deep eddy under the west shore I came upon a brood of wild ducks, the hooded merganser. The young were about half grown, but of course entirely destitute of plumage. They started off at great speed, kicking the water into foam behind them, the mother duck keeping upon their flank and rear. Near the outlet of the pool I saw them go ashore, and I expected they would conceal themselves in the woods; but as I drew near the place they came out, and I saw by their motions they were going to make a rush by me upstream. At a signal from the old one, on they came, and passed within a few feet of me. It was almost incredible the speed they made. Their pink feet were like swiftly revolving wheels placed a little to the rear; their breasts just skimmed the surface, and the water was beaten into spray behind them. They had no need of wings; even the mother bird did not use hers; a steamboat could hardly have kept up with them. I dropped my paddle and cheered. They kept the race up for a long distance, and I saw them making a fresh spirt as I entered upon the rift and dropped quickly out of sight. I next disturbed an eagle in his meditations upon a dead treetop, and a cat sprang out of some weeds near the foot of the tree. Was he watching for puss, while she was watching for some smaller prey?

I passed Partridge Island—which is or used to be the name of a post-office—unwittingly, and encamped for the night on an island near Hawk's Point. I slept in my boat on the bench, and in the morning my locks were literally wet with the dews of the night, and my blankets too; so I waited for the sun to dry them. As I was gathering driftwood for a fire, a voice came over from the shadows of the east shore: "Seems to me you lay abed pretty late!"

"I call this early," I rejoined, glancing at the sun.

"Wall, it may be airly in the forenoon, but it ain't very airly in the mornin';" a distinction I was forced to admit. Before I had reëmbarked some cows came down to the shore, and I watched them ford the river to the island. They did it with great ease and precision. I was told they will sometimes, during high water, swim over to the islands, striking in well upstream, and swimming diagonally across. At one point some cattle had crossed the river, and evidently got into mischief, for a large dog rushed them down the bank into the current, and worried them all the way over, part of the time swimming and part of the time leaping very high, as a dog will in deep snow, coming down with a great splash. The cattle were shrouded with spray as they ran, and altogether it was a novel picture.

My voyage ended that forenoon at Hancock, and was crowned by a few idyllic days with some friends in their cottage in the woods by Lake Oquaga, a body of crystal water on the hills near Deposit, and a haven as peaceful and perfect as voyager ever came to port in.

RESPONDING JOURNAL

The volume of Burroughs' essays entitled *Pepacton,* from the name of the selection you have just read, was introduced to public schools and found extremely wide audiences in the early decades of this century. Explain in a sustained journal entry of at least two pages why this essay and others like it would be used for educational purposes. Would it have a similar appeal today? Why or why not?

QUESTIONS FOR CRITICAL THINKING

STRATEGY

1. Think about the course of Burroughs' essay. Does it in any way resemble the flow of a river? Do you detect any points at which you might argue that Burroughs is modeling his writing on the flow of the river? For example, does he offer his deeper, more extensive reflections on specific topics much as the river seems to hesitate and deepen when it pools?

2. Consider Burroughs' comments on the tree and its place within the hydrologic circle. To what extent might such a commentary serve as a means of alerting people to the importance of preservation and ecological awareness? Can you put such a plea in more modern terms that would appeal to a contemporary audience, such as your classmates or members of your community? You might also work in a group to brainstorm means of conveying the lessons of the hydrologic circle to a wider audience or series of audiences.

1. Reflecting on his life and career when he was an old man, Burroughs said, "In boyhood I have had more delight on a hay-mow with two companions and a big dog—delight that came nearer intoxication—than I have ever had in all the subsequent holidays of my life." "The youth enjoys what the man tries to understand." How do these statements help to explain Burroughs' response to the experience of returning to his native stream and thinking about his journey?

2. Think about Burroughs' opening proposition that he build something before he begin his journey. Have you ever felt the need for such an elaborate preparation? How did the venture turn out after your intricate plan?

COLLABORATIVE WRITING ACTIVITY

In another essay Burroughs states, "In fact, winter, like some great calamity, changes the status of most creatures and sets them adrift. Winter, like poverty, makes us acquainted with strange bedfellows." Using the structure of Burroughs' second sentence, substitute each of the other seasons in a similar statement. Once you have completed your three, enter those in a list of those written by all the other class members. Once everyone has heard the others, discuss those most likely to lead to productive or provocative essays.

SUGGESTIONS FOR FORMAL WRITING ASSIGNMENTS

1. Write an essay in which you describe how your developing personality and intellectual growth have modified the way you look at the world and changed your impression of a single landscape. What shifts in your world view—your "mental lenses"—have in turn modified your response to this landscape? You may want to choose a landscape that held a particular significance for you at one time in your life, but that you have now come to view in a different way. You have, in effect, revised your estimate of that landscape.

2. Take a look at an essay you have written previously. What parts of that essay would you change today? Think about your changes in terms of your different mental lenses. Write an essay accounting for the changes in your perspective by providing an autobiography of your mental lenses. What values have you modified as you changed? You may find it helpful to make a chart similar to those in the text as an aid to your brainstorming.

3. Write an essay applying your own mental lenses, or those of a writer other than yourself, to the situation another writer describes. You might choose one

of the writers in this section, such as John Burroughs, and examine his situation from the perspective of a writer such as David Rains Wallace. Would Austin find the great issues along the water trails of the Ceriso that Du Bois examines from his perspective at the rim of the Grand Canyon? How would Austin's mental lenses differ from Du Bois'? How would Annie Dillard see the landscape of the Ceriso or the Pepacton as opposed to Tinker Creek? What different mental lenses would be necessary?

You might also want to approach this essay from the point of view of another character in an essay. Think for example of Mrs. William Sycamore, the widow of the speaker in Benét's poem. How might her mental lenses differ from those of her husband? Be willing to try some imaginative speculation here.

Chapter 4

DRAFTING AND REVISING: RE-SEEING THE WORLD

Randy Lee White, *Custer's Last Stand—Revised,* 1980

1. Examine the point of view from which you view the painting. What is the effect of the perspective? Of your sense of depth?

2. If we look at the squares as used car lots, how do the painting's specific details portray Indian-White relations?

3. How does this painting cause you to revise or modify your ideas about the nature of painting and "Custer's Last Stand"?

Chapter 4

DRAFTING AND REVISING: RE-SEEING THE WORLD

Re-vision means literally a seeing again, taking a fresh look at a concern or problem or subject you considered previously. Just as all athletes have to practice to perform well in their sport, you have to expect to revise to produce successful writing. Virtually no writer produces a piece of writing the first time—the initial effort—that will do an adequate job. As we look at what we have written, we will always see changes we can make that will state our points better, that will help us to make our points more clearly. What follows here, then, is a series of lists that will help you to think about the categories of revision you will consider as you look over your work.

One point you will want to consider right away is the use of a computer for word processing. Since your college almost certainly has facilities set up so you can use a word processing system if you do not have one of your own, you will want to check on this facility. Using a computer for your writing certainly speeds revision and makes it much easier than handwriting or typing. Even if you are not a typist, after a little while on the word processor, you will see how much time you save by not having to retype each draft. The computer allows you to move around blocks of text, to add elements easily, and to take out for just a minute something you are not sure you want to include. If you should decide you want it, you can always re-enter the block. Manipulating your text is wonderfully easy on a word processor; after you get used to using one, you will wonder how you ever existed without one. If you are not already using the computer for your writing, make it your #1 academic priority.

The ideal time to revise is several days after you have completed the first draft. Of course, you cannot always arrange your schedule to accommodate such a long period between drafts, but as a general rule, the longer the layoff the better because you will be able to look at your essay as the work of another person and you will be more objective. Remember, the point is to look at your work critically and ask yourself tough questions about how well your writing works.

At the early stages of revision, you may want to change directions entirely. Sometimes a thesis statement sounds fine as you work it out, but then in practice you find that you cannot or do not want to write an entire essay with it. Often you will learn enough about your subject in the process of writing about it to see that you are really interested in only one part of what you had previously thought was your narrowed subject. In such cases, you are better off revising thoroughly so you can write the essay you really want to write, not something that will fulfill an assignment and leave you lukewarm.

Bring yourself back to the fundamental design of your essay. Recall its basic parts:

- thesis
- audience
- author's stance or point of view

Thesis Does your thesis clearly state exactly the point you want to make, or should you revise that sentence? Remember that this statement is a help to you as well as to your reader because it not only sends a clear message of the point you wish to relay, but it also serves as a baseline to which you can refer back as you write and revise the rest of your essay. If material does not advance your thesis, you should almost certainly delete that material. If you do not delete it, then you have to revise to make its relation to your point—your thesis—crystal clear.

Audience Should you make any adjustment to your thesis in order to make your ideas more accessible to your audience? Should you modify your tone in any way to keep your audience on your side? You should look for a good fit between your audience and the way you have presented your material and yourself as writer.

Author's stance or point of view Are you still satisfied that your attitude toward your subject is appropriate and comes through clearly? What does your peer reviewer think about your point of view? If you take one side or the other of an issue, does your position come logically from your thesis and the material you present?

Once you have gone through these three steps, decide on the best strategy to serve your revision purposes. Will you regard your first draft as a "discovery draft," necessary to your thinking process but not necessary to the final draft you know you should write, or will you begin to revise and reshape the draft you have already written? Adding, deleting, moving around blocks of text, and sharpening are the activities you will carry out here.

If you are satisfied that you have devised a strategy to address these three areas sufficiently, and you have completed that work, move on to these tricks of the trade with your revision:

- Put yourself in the role of your toughest previous English teacher to see how your essay stacks up.
- Try reading your essay out loud to yourself to see how it sounds.

■ Read your essay slowly, paragraph by paragraph, beginning with the last paragraph and working your way forward.

If you have already worked diligently with the earlier stages of revision, these final checks might remind you of what you have already done to modify your essay. If not, then continue your process of revision until you feel satisfied.

Remember that once an idea is written down, it starts to take on a life of its own. Sometimes you will find that this life leads in a direction you had not intended traveling, and so you will have to retrace your steps. Revision is that retracing, that adjusting of direction to be sure you are telling your readers exactly what you want to tell them.

This brings us to the second point. When you revise your work, you take a different attitude toward it than when you generate it. When you are revising, you are evaluating your performance to this point. Both composing and revising modes are necessary, but they have to be executed in the proper order. In the generating and writing stage you need to indulge yourself; write down whatever comes to mind. When you are revising, your critical faculty must operate to help you manage your work and make it as good as it can be. Revision is the process that will teach you the most about writing well.

Writing well means understanding the nature of writing as recursive, or moving back upon itself. Revision and editing are parts of this recursive process: you will sometimes reach the editing stage, only to realize you have to go back and do more revising. Rather than trying to fight it, give in to this situation when you encounter it. Revision of successive drafts leads to the final editing of work that satisfies you; editing is your last shot at the best possible finished product. Remember that your audience knows you through your work: in many situations, all your readers will see of you is your written presentation. The following specific strategies will help you to take maximum advantage of your ideas and abilities.

You will need a dictionary at your fingertips when you do your final editing. Some of the words you used in your earlier drafts may not have been quite the best ones to capture your meaning, and so you will need to check them in the dictionary to be sure. You might also have thought of some words that might fit, but you did not use them because you were uncertain; this is the time to check on those as well. A thesaurus does not serve the same purpose as a dictionary; a thesaurus has the potential to cause as many problems as it will fix. A thesaurus may work for you, but only if you are familiar with the connotations of the words the thesaurus calls to your attention. Better to leave that book alone and stick with your dictionary. Many dictionaries list synonyms, thus eliminating the biggest reason for using a thesaurus.

Your classmates or friends can be a big help to you at the editing stage. It is always best to have someone else look over what you have written. A good, close reading at this point can help significantly because your reader looks at your writing not as a work in progress, but as a *final draft*. This piece of work

represents you to the world. Remember that even though you are getting opinions and responses from your peers, *you* are the final decision-maker. You
alone are responsible for the quality of your work. Therefore, at the editing
stage, you must be first and last: you must take the first look at what you have
written, make it as good as you can, and then—and only then—share it with
your friends. After they have responded, then you take your last shot.
Remember that they can make more useful comments if you have done your
best to make the essay effective. If they have to puzzle over organization or sentence construction or if they get distracted by spelling or typographical errors,
they cannot give your essay as effective a reading as they could if you had elim-
inated difficulties first.

Think of the editing stage as having two parts: your focus on organization
on one hand, mechanics and style on the other. With organization, you look
for the last time at the overall pattern and effect of your writing. You want to
see how well your revisions have worked to achieve your purpose in writing,
and you also have to decide if you need to do any final rethinking. In the second category, you need to move down to the minute level of grammar, sentence structure, spelling, punctuation, and typography to make sure your final
version is mechanically correct. Again, because writing is recursive, your work
at the editing stage may send you back to the revision stage. You may have progressed so far with your writing that you now see how you might have revised
in a different way.

If your writing is unclear to you, make it clear. If you have a complex and
subtle idea, fine. But that complexity and subtlety are out of place in the essay's
organization. Simplicity in organization is always welcome, and always a virtue.
You must provide a clear trail for your reader, and you most need to do so
when your argument or reasoning process is complicated. If something both-
ers you about the essay, that same passage or section will surely bother your
reader as well.

An important principle all writers have to keep in mind is the principle that
"less is more." You often heighten the effect of what you say when you do not
drown it out with unnecessary distractions. This means cutting out unnecessary or repetitive sentences and words. You have to be merciless in cutting
everything that does not contribute directly to your essay and the effect you
want that essay to have on your reader. Again, you can ask someone to read or
listen to a passage to make sure you should take something out. We all hate to
discard what we have written because it is the fruit of our hard effort—we
almost feel like we are killing our child when we kill something we have written—but if the overall essay benefits from the cut, we have to do it. Eliminating
even a phrase such as "the fact that" or reconstructing a sentence to eliminate
a "there is" or "there are" construction is almost always beneficial.

Once you are satisfied that you have finished your revising and editing, read
the paper out loud to yourself to make sure it sounds good to you. Of course, you
can do this, and hopefully you have, at several stages during the composition

process. But here you will do your final reading; think about how your writing will sound to your readers, and how you want to come across to them. Do you like what you have written? Does it sound good? Do the sentences seem too simple, or do you run out of breath before you finish reading them? If either of these is the case, modify your text so that you are comfortable with it.

There are several ways to handle the proofreading of your final copy. If you are using a computer, you have probably run the spell checker at several points during your drafts. Run it a final time, but remember that some words can be correct to the computer (*to* for *too, there* for *their,* pronoun shifts) but incorrect to a careful reader. You want to be the ultimate careful reader at this stage to find these potential trouble spots.

In order to locate these pesky demons, you might want to do first a reading of your essay to look at sentence structure. Treat each sentence as an individual unit and decide if some should be combined, broken down into two or even more separate units, or modified in other ways. Remember that the point of this activity is not another reading of your entire essay, but rather an opportunity to check your work purely in terms of your building of individual sentences. Next, try a line-by-line reading of your paper during which you put a ruler or blank sheet of paper under each line of your text. This has the effect of isolating each line so you can be aware of the flow of your essay, but at the same time you are highlighting each word as you move through the essay.

Here, then, is a brief checklist of editing principles:

1. Use the dictionary to be sure.
2. Less is more, so make sure you do not use what you do not need.
3. Read it aloud to be sure it works.
4. Make sure the thesis sends a clear message to your reader.

The writers in this section display some results of effective editing. They have revisited sites that hold importance in their lives, and they move through their changing responses to this landscape until they reach a stopping place, a suitable vantage point from which to look back over their relation to the human community and to the natural world. Writing with the benefit of age and experience, they reflect on the ultimate power of landscape to bind together humans and their world. Lewis and Clark present in their *Journals* a revised judgment of the threats represented by the grizzly bear; Dayton Duncan, a later traveler on the trail blazed by Lewis and Clark, finds the bears are still formidable. The country is in many ways tamer today than in 1803, but the grizzly remains unafraid of humans. Reid accounts for his changing attitudes towards mountains and the proper attitude to take toward them, just as Wright shows how his experiments with trout feeding have taught him about managing both natural systems and his relation to them. Estes explores some concepts with us that may change the way we look at both women and wolves, while Easterbrook's challenges to received ideas about the environment may make us revise our sense of the danger facing us.

from *The Journals of Lewis and Clark*
Meriwether Lewis and William Clark

*Although the history of the exploration of the west and the subsequent devel-
opment of the United States are synonymous with Lewis and Clark and
Sacajawea, few Americans appreciate the magnitude and significance of the
voyage of the Corps of Discovery, as the expedition was officially chartered
by President Thomas Jefferson in 1803. This daring exploratory trip began
on May 13, 1804, and traveled up the Missouri from its start at Camp
Dubois, near St. Louis. Although Jefferson's hopes of the discovery of a
Northwest Passage were not realized, the expedition succeeded far beyond
anyone's dreams in its discovery and classification of flora, fauna, insects,
reptiles, fishes, geographical formations, and Native American tribes.*

*Both Lewis and Clark wrote entries in their journals, as did other members
of the Corps of Discovery. Clark, born in 1770, and Lewis, in 1774, were both
experienced soldiers and fellow Virginians, as was their patron Jefferson. As
co-commanders of the journey, they balanced each other remarkably well.
Not only did they never seem to have had any differences over how the expedition ought to be governed, but they also demonstrated an amazing ability
to alternate roles of leadership, to take turns at encouraging and uplifting the
spirits of the group. Virtually always in danger and at one time only a day
or two away from death by starvation, Lewis and Clark record an amazing
determination to know what lay to the West. The passages excerpted here
show the spirit not only of adventure, but also of duty to complete the mission and report back to Jefferson just what did lay out there beyond the
Louisiana Purchase. For those of you interested in further work on the significance of this expedition in American history and culture, the definitive
account of the life and role of Sacajawea is still to be written.*

*Meriwether Lewis wrote the excerpts presented here. Although they do not
display the wide range of topics and concerns that the members of the expedition shared, they do show the theme of the party's growing familiarity with
and respect for the American grizzly.*

EXPLORING JOURNAL

Recall what you can remember from other classes, your reading, movies and
television, or accounts told by older people about moving across the landscape.
Do all of your ideas come from history books? What are your mental pictures
of pioneers like? Try a ten-minute freewrite on this subject. If you have any difficulty with it, try thinking back to your earliest experiences and move through
your school career. Did you have strong images of pioneers or pilgrims? How
about Columbus and the Native Americans who helped him and other explorers,

often to their own destruction? Perhaps thinking about certain holidays will help you with this assignment; once you get started, though, you may find that ten minutes is not enough. If so, keep on going.

————⋙◦⋘————

Monday April 29th, 1805

I walked on shore with one man. about 8 A.M. we fell in with two brown or yellow [*white*] bear; both of which we wounded; one of them made his escape, the other after my firing on him pursued me seventy or eighty yards, but fortunately had been so badly wounded that he was unable to pursue so closely as to prevent my charging my gun; we again repeated our fir[e] and killed him. it was a male not fully grown, we estimated his weight at 300 lbs. not having the means of ascertaining it precisely. The legs of this bear are somewhat longer than those of the black, as are it's tallons and tusks incomparably larger and longer. the testicles, which in the black bear are placed pretty well back between the thyes and contained in one pouch like those of the dog and most quadrupeds, are in the yellow or brown bear placed much further forward, and are suspended in separate pouches from two to four inches assunder; it's colour is yellowish brown, the eyes small, black, piercing; the front of the fore legs near the feet is usually black; the fur is finer thicker and deeper than that of the black bear. these are all the particulars in which this anamal appeared to me to differ from the black bear; it is a much more furious and formidable anamal, and will frequently pursue the hunter when wounded. it is asstonishing to see the wounds they will bear before they can be put to death. the Indians may well fear this anamal equiped as they generally are with their bows and arrows or indifferent fuzees, but in the hands of skillfull riflemen they are by no means as formidable or dangerous as they have been represented. . . .

Capt. Clark and Drewyer killed the largest brown bear this evening which we have yet seen. it was a most tremendious looking anamal, and extreemly hard to kill notwithstanding he had five balls through his lungs and five others in various parts he swam more than half the distance across the river to a sandbar, & it was at least twenty minutes before he died; he did not attempt to attack, but fled and made the most tremendous roaring from the moment he was shot. We had no means of weighing this monster; Capt. Clark thought he would weigh 500 lbs. for my own part I think the estimate too small by 100 lbs. he measured 8. Feet 7 1/2 Inches from the nose to the extremety of the hind feet, 5 F. 10 1/2 Ins. arround the breast, 1 F. 11. I. arround the middle of the arm, & 3.F. 11.I. arround the neck; his tallons which were five in number on each foot were 4 3/8 Inches in length. he was in good order, we therefore divided him among the party and made them boil the oil and put it in a cask for future uce; the oil is as hard as hogs lard when cool, much more so than that of the black bear. this bear differs from the common black bear in several respects; it's tallons are much longer and more blont, it's tale shorter, it's hair which is of a redish or bey brown, is longer thicker and finer than that of the black bear; his liver lungs and heart are much larger even in proportion with his size; the heart particularly was as large as that of a large Ox. his maw was also ten times the size of black bear, and was filled with flesh and fish. his testicles were pendant from the belly and placed four inches assunder in separate bags or pouches. this animal also feeds on roots and almost every species of wild fruit.

Monday May 6th, 1805

saw a brown bear swim the river above us, he disappeared before we can get in reach of him; I find that the curiossity of our party is pretty well satisfyed with rispect to this anamal, the formidable appearance of the male bear killed on the 5th. added to the difficulty with which they die when even shot through the vital parts, has staggered the resolution [of] several of them, others however seem keen for action with the bear; I expect these gentlemen will give us some amusement sho[r]rtly as they soon begin now to coppolate. saw a great quantity of game of every species common here. Capt Clark walked on shore and killed two Elk, they were not in very good order, we therefore took a part of the meat only; it is now only amusement for Capt. C. and myself to kill as much meat as the party can consum; I hope it may continue thus through our whole rout, but this I do not much expect. two beaver were taken in traps this mo[r]ning and one since shot by one of the party. saw numbers of these anamals peeping at us as we passed out of their wholes which they form of a cilindric shape, by burrowing in the face of the abbrupt banks of the river. . . .

Friday June 28th, 1805

The White bear have become so troublesome to us that I do not think it prudent to send one man alone on an errand of any kind, particularly where he has to pass through the brush. we have seen two of them on the large Island opposite to us today but are so much engaged that we could not spare the time to hunt them but will make a frolick of it when the party return and drive them from these islands they come close arround our camp every night but have never yet ventured to attack us and our dog gives us timely notice of their visits, he keeps constantly padroling all night. I have made the men sleep with their arms by them as usual for fear of accedents.

RESPONDING JOURNAL

What surprises you about these entries? Do the descriptions of action coincide with your sense of what the pioneers and early explorers had to face? Consider these ideas in your own journal entry.

QUESTIONS FOR CRITICAL THINKING

STRATEGY

1. What seem to you to be the identifying characteristics of this group of journal entries? Jot down a list of these qualities.
2. What objective does Lewis seem to have had in mind when he writes these entries? How effectively do you feel he conveys the sense of danger the members of his party must have felt?

ISSUES

1. Jot down a brief statement about what you judge to be the audience for the journal entries you have read thus far. What clues do you find to suggest who the author has in mind as he or she writes? When a journal has to be kept as part of an obligation—as in an expedition underwritten by a government—to what extent does the obligation seem to affect the content or nature of the journal?
2. As you note in the entries you have read, some journals are actually published as they were written in the field, while others are refined and polished before publication. Where would further exposition of any of the issues in the entries you have read make the journals in which they appear more effective?

COLLABORATIVE WRITING ACTIVITY

Look back over the attitudes toward the land represented in this journal as well as selections you may have read earlier. Record a list of characteristics to share with the group and compile a master list. What do you find here? Discuss these issues among yourselves.

from *Out West*
Dayton Duncan

On several trips between 1983 and 1985, Dayton Duncan set out to negotiate again the trails followed by Lewis and Clark in their expeditions. Others have also retraced the journeys of Lewis and Clark, but Duncan has produced an especially effective account. He follows the famous explorers by forming a narrative, produced from his journal notes, that explains his changing attitudes while he also muses on the adventures of the members of the Corps of Discovery.

Duncan's travels take him to many people across the western half of the
United States, and in his reserved enthusiasm for the various folks he meets,
he parallels William Least Heat Moon, whose Blue Highways *(also represented*
in this book) gives an account of a less specifically planned journey across
America to see its landscapes and talk with its people. Armed with fifteen
pints of New Hampshire maple syrup to pass out as gifts along his travels,
Duncan begins his journey to see what he will find; he has along with him
the eight-volume set of The Journals of Lewis and Clark, *to which he refers fre-*
quently as he takes his trips.

Excerpted here is one section that describes one of the problems the Corps
of Discovery encountered: the grizzly. The explorers finally decided this bear
was too dangerous to deal with, and so they learned to avoid any contact
with the animals. Today, our ability to control much of nature for our own
enjoyment seems to have affected our respect for the bears, as some visitors
to Yellowstone have tragically learned. Duncan's accounts of modern griz-
zly encounters offer an updating of the lessons learned by Lewis and Clark.

EXPLORING JOURNAL

Imagine that you have been asked to retrace the efforts of Lewis and Clark in a
modern setting. How would you have to go about retracing their steps today?
How would you use their original accounts? Develop a set of ground rules to
guide your retracing of their journeys.

Sometimes myths turn out to be based on fact, not fantasy. Such was the case
of the grizzly bear.

The Indians at Fort Mandan had warned Lewis and Clark about this beast,
but the captains discounted much of it as tall tales told around an earth-lodge
fire. The explorers could believe in a Northwest Passage or the existence of
Welsh Indians, but the idea of a bear so big, so ferocious, and so tough to kill
was simply beyond belief, obviously a concoction of overly fertile imaginations.

Near the mouth of the Yellowstone, as they entered what is now Montana,
the Corps of Discovery started seeing the bears' tracks. They were big, all right,
but they only heightened the men's eagerness to see the animal itself. On April 29,
1805, Lewis finally met one. This myth came in the form of three hundred pounds
of yellow-brown fur, with sharp claws and snarling teeth. He wounded it with one
shot, was pursued by the maddened animal, but got the chance to recharge his
muzzleloader and killed the bear with a second round. While previous explorers
had reported sightings of grizzly bears, Lewis's description in his journal that day
was the world's first detailed notation of its size, color, and anatomy.

Flushed with his success, Lewis notes: "the Indians may well fear this anamal equiped as they generally are with their bows and arrows or indifferent

fuzees, but in the hands of skillfull riflemen they are by no means as formidable or dangerous as they have been represented."

About a week later the expedition encountered its second grizzly. It took ten slugs, five of them through the bear's lungs, to bring the grizzly down; even so, the bear swam across the river before expiring on a sandbar.

Myths die hard, especially when they're well earned.

After seeing a third grizzly the next day, Lewis, somewhat chastened after his earlier braggadocio, writes: "I find that the curiossity of our party is pretty well satisfyed with rispect to this anamal."

Satisfied or not, they saw a lot more of these "monsters" throughout Montana. Men were chased across the prairies, chased up trees, and chased over bluffs into the Missouri by grizzlies. Lewis's dog was reduced to "a constant state of alarm with these bear and keeps barking all night," according to the captain. By the end of June he decided that the grizzlies were "so troublesome that I do not think it prudent to send one man alone on an errand of any kind."

The captains called this animal by various names—white bear, grey bear, yellow bear, brown bear—before settling on "grisley" or "grizly." Years later, it would receive the scientific name of *Ursus horribilis* and common name of grizzly. (Teddy Roosevelt preferred "grisly," because it means ghastly, over "grizzly," which refers to the look of its fur.) Whatever, before he left Montana soil, Lewis had been impressed enough by its strength and ferocity to say he "had reather fight two Indians than one bear."

One hundred and eighty years later, I have covered more than 1,500 miles within Montana's borders along the expedition's route without seeing a single grizzly. Not that I'm abjectly disappointed. I'd rather not fight an Indian *or* a grizzly bear. But before leaving Montana, I stop in Missoula (population 33,388) and talk to Charles Jonkel, a biologist and grizzly expert associated with the University of Montana, to find out what happened.

The story of the grizzly has many of the same elements as the stories of other species in the years after Lewis and Clark encountered them during the Voyage of Discovery. The beaver were overtrapped. Whooping cranes and bald eagles were killed and poisoned to the brink of extinction. Passenger pigeons disappeared entirely in 1914. Wolves have all but vanished. Pronghorn sheep and Rocky Mountain goats are scarce. The elk that formerly grazed the open plains have retreated to the mountains. The buffalo, once too many to count, survive now only in protected parks and private ranches.

No one knows how many grizzlies once lived in the American West, but reports of early travelers starting with Lewis and Clark certainly show they were common sights. Estimates range up to the tens of thousands. Compared to the Indians with their bows and arrows, the men of the expedition had an easier, albeit still difficult, time killing the grizzly with their single-shot muzzle-loaders. Later, settlers armed with high-powered repeating rifles found *Ursus horribilis* even less *horribilis.* The bears had existed for centuries without encountering a predator. They were therefore unprepared for the influx of man and rifle. Since they did not run and hide, they were relatively easy pickings as long as the hunter's shot was true. Intelligent though they may be, they still did not learn to fear humans. By the end of the 1970s, they were listed as a threatened species and placed under federal protection. No one knows for sure how many remain in the American West—estimates put their numbers at fewer than a thousand.

The grizzly's story has one big difference from that of, say, the now extinct passenger pigeon. It is this difference that both continues to threaten their survival and attracts a lot of attention: grizzly bears still kill people on occasion. Usually it occurs when they feel provoked or threatened near their cubs or near their feeding grounds. In their ceaseless search to find enough protein to survive their winter hibernation, they will eat anything from roots and insects to moose and elk. A sheep also provides a good meal.

"It's different to live with grizzlies than being just concerned about them in New Orleans," Jonkel says. "It's a different thing if you live in Baltimore and want to save the grizzly bear than if you live on the east front of the Rockies [one of the bears' last refuges]. City people do not understand the problem, the way we do not understand why they cannot stop crime in the cities."

The rub comes, he says, in the fact that "good grizzly habitat is good people habitat." Places with clean water, good fishing, good soil for plants and grazing, and even nice scenery are the places grizzlies like. They're also the places for things like ski resorts, second-home developments, cattle ranches, and even oil exploration.

Jonkel compares the grizzlies' plight to that of the Indians: "The bear has to be where the land is rich. We've taken all of that, given them some of the leftovers, and told them to survive."

Depending on the availability of food sources, the home range of one male grizzly can be anywhere from 189 to 387 square miles. Female "sows" and their cubs roam around in smaller ranges. Not surprisingly, then, despite their dwindling numbers and increasingly remote refuges, grizzlies still keep running into their principal obstacles for survival: man and his ever expanding home range. And so Jonkel is among the handful of biologists and environmentalists trying to educate people on how to live with bears, and bears on how to live with people.

Jonkel heads a project aimed at teaching "problem" bears—bears that have caused trouble in human settlements—to fear humans. The hope is that this "aversive training," combined with the grizzlies' natural intelligence, will result in a wariness that can be passed on to the remnants of the species. Captured bears are deliberately provoked, and when they charge across their cage toward the provoker, they're sprayed in the face with an acrid concoction of red pepper. As the training is repeated, the bear is supposed to start associating humans with unpleasant experiences.

The harder task is teaching humans to leave grizzlies alone. Not that long ago, tourist areas near Yellowstone National Park set their garbage out for the grizzlies to feed on in view of people who had paid to sit on specially erected bleachers and see the West in all its wildness. (Unfortunately, despite advice to the contrary, when the dumps were finally closed it was done so rapidly that the grizzlies, reacting to this "cold turkey" treatment, fanned out to campgrounds and other remaining sources of garbage. Nearly two hundred bears, suddenly "problem" bears, had to be killed.)

In addition to developers and oil drillers who encroach on the bears' dwindling refuges, the "problem" people these days are often hikers and campers who are careless with their food in grizzly territory. (In Glacier, more than two million people—about eleven thousand humans for each grizzly—passed through the park in 1983.) The bears learn that the easiest source of protein is where people are; the next campers encounter a hungry grizzly; complaints are lodged; grizzly is killed.

Jonkel has prepared lists of advice for people in grizzly land. The precautions include having the good sense to avoid places posted by park personnel warning off hikers; keeping campgrounds clean and your food stored in containers away from where you sleep; wearing bells on your boots to prevent your surprising a grizzly on a turn in the trail; and knowing how to react in a confrontation.

"There's no precise formula on how to act if you meet a grizzly," he says. Don't turn and run; it often encourages the bear to chase you. Break eye contact and back away, seeking a tree or rock to hide behind; the bear often accepts this as a signal that you are not challenging its territory and do not want trouble any more than it does. While they can climb trees, grizzlies sometimes accept your climbing as a submissive act and will leave you alone. If the bear's on top of you, playing dead sometimes defuses the situation; if all else fails, fight for your life—both have worked, Jonkel says.

Reports of grizzlies' killing people (about one a year) are oversensationalized, he says. More people die each year being struck by lightning, more children simply disappear or are murdered by their parents.

"If you are going to be done in this summer, it's probably going to be on the highway or by a friend with a butcher knife," Jonkel tells me. "You should be a lot more frightened of Highway 93 than you are of grizzly bears." I've had experience with the highway, so I understand that threat; I'm not sure my friends would appreciate the comparison.

I also have doubts about playing dead beneath a grizzly, I tell him. Seems like playing it might lead to the reality of it. But Jonkel knows several people who have done it and survived. One couple he knows was surprised by a grizzly in their campsite. They both played dead. The grizzly sniffed their bodies and nudged them a couple of times. Then it squatted over the man, defecated on him, and walked off.

"How'd the guy feel?" I ask, seeking some sort of greater symbol in the act, perhaps a grand comment from one species to another.

"He accepted it," Jonkel says. "Better that than being eaten."

———→⬧←———

RESPONDING JOURNAL

How has your sense of the modern version of the "grizzly problem" changed? How have you revised your sense of how well we are able to control bears today? As you write your response, think as well about how this situation reflects a more general trend in our modern dealing with the wild.

QUESTIONS FOR CRITICAL THINKING

STRATEGY

1. How do these writers arouse your interest in their subjects? If you were not familiar with the journeys of the earlier explorers, would these updated journals encourage you to read them? To what writing strategies could you attribute this interest?
2. Think about the detail presented here that does not appear in the original travel journals. Does Duncan match his earlier counterparts in terms of the enthusiasm for his subjects? Explain which one you find more effective as writing.

ISSUES

1. Duncan employs a tone that verges on humor. Does this levity do justice to the genuine dangers the original explorers faced? Could Duncan have taken the comic elements any further?
2. Lewis and Clark clearly felt awed by their experiences with nature. Does Duncan reflect a similar respect for the natural landscape? Explain how you believe they value the physical sites they have described.

COLLABORATIVE WRITING ACTIVITY

Choose any of the journal entries you have read as the basis for your own recreation of the original voyage or for reexperiencing the events or stimuli leading to the original journal entry. Bring your account to the group and read it. The group will then decide on the most promising means of assembling the most effective portions of individual journals to produce a new journal of collected adventures.

"The Mountain of Love and Death"
Robert Leonard Reid

Robert Leonard Reid was trained as an astronomer at Harvard, but he soon realized that his fascination with the heavens was less strictly academic than visionary. He quickly found that his desire to know could not be satisfied by telescopes and the scientific method alone. He became a premier mountain climber and stored up the memories that would furnish the material for his writing about the intense emotional experiences he has had in mountains and in less exotic locations.

Reid brings to the literature of mountain climbing a mature perspective that recalls Wordsworth's phrase "emotion recollected in tranquility." In his recollections, Reid frequently returns to the energy and enthusiasm of his adolescence, where he finds the seeds of his later interests. In addition to Mountains of the Great Blue Dream *(first published in 1991 and from which "The Mountain of Love and Death" is taken), he has also edited* A Treasury of the Sierra Nevada *for Wilderness Press.*

EXPLORING JOURNAL

Freewrite for ten minutes on the concepts of mountain climbing. Imagine a climb to the higher reaches of mountains where snowstorms and sheer rock faces await you. What would be the most frightening parts of a mountain climb for you? The most exhilarating? What argument would you present to a friend to get that person to accompany you on your climb?

⎯⎯►◄⎯⎯

> *To him who in the love of nature holds*
> *Communion with her visible forms, she speaks*
> *A various language.*
>
> —WILLIAM CULLEN BRYANT, "THANATOPSIS"

Mountain climbers speak easily and often eloquently of the beauties of the grand arenas where they practice their dangerous sport. A man with whom I have climbed occasionally, a taciturn soul of not notably refined sensibilities, once said to me matter-of-factly that his first glimpse of Mount McKinley high above the Alaskan tundra reminded him of an enchanted cloud. Where love is the issue, the climber speaks mountains.

Love, yes, but what of death, the other great theme that figures so prominently in the drama of mountaineering? Here the climber is strangely tight-lipped. This seeming indifference to so commanding a presence puzzles those who do not climb, and is sometimes taken as proof of the mountaineer's fool-hardiness, or ignorance, or derangement. A kinder explanation allows climbers their sanity. Their deep and sustained communion with death has left them, I believe, not indifferent to that essential truth but, rather, respectfully speech-less. The insights into the nature of death they have gained while practicing their perilous craft are so complex and so precious that out of fear and wonder they have taken them into their hearts to contemplate in silence. I climbed for twenty years, and while the possibility of dying was something I rarely dis-cussed with my friends in the climbing community—and never with those out-side—I secretly carried on an ardent relationship with death all the years that I climbed. The apprehension I felt so often while climbing was mingled with curiosity and even affection; my fear was blended with bliss. The countenance of death, which once had terrified me, grew ever more alluring the more I stud-ied that not unsightly face. The longer I climbed, the more comfortable I became with the notion that I could die content in the mountains. Death in the city continued to strike me as a bad idea, but death in the mountains seemed fitting and natural. It is worth remembering that Melville's Billy Budd was fear-less in the face of death because he stood nearer to what his creator called "unadulterate Nature" than did his citified counterparts. I don't claim to have achieved the Handsome Sailor's complete acceptance of death, but in the nat-ural surroundings of the mountains I overcame at least the worst of my fears. I think that the same can be said of many mountaineers.

   To nonclimbers' regret, we keep our insights to ourselves. When called upon to explain why we climb, we turn to our more expressive friend love for an answer. Why, it's the beauty of the mountains, we say. It's the adventure, the exhilaration, the challenge, the camaraderie, the quest for our limits. It's . . . the view from the top! We climb because climbing is a joy, we say; we climb because climbing takes us closer to God.

   All of which is true, but none, save possibly the last, suggests the full truth, that more than love animates our affairs with the high peaks.

   One summer past, weary at last of my own twenty-year affair, certain that I had no more to gain from its peculiar glow, and overwhelmed by a series of unhappy and apparently significant events in my life, I laid down my rope and my ice axe and retired from the sport of mountaineering. In the months since, I've had an opportunity to reflect at length on the issue of death in the moun-tains and to cast off some of the climber's customary reserve. Let me, then, take a crack at the ever-popular question I referred to a moment ago. Mountaineers climb because they love the mountains, yes; but they climb too because climb-ing prepares them boldly and tenaciously for death, then guides them faithfully to the edge of another world, a world I now recognize as the world of the dead, and there allows them to dance, mountain after mountain, year after year, as close to death as it is possible to dance; which is to say, within a single step. They go, *not to die*—that is very important—but far from the tumult of the val-ley below to linger in safe communion with death, to feel the exquisite tension that separates it from life, to glimpse its radiant smile and comprehend its peace. Climbing is a way of studying the ultimate unknown. In the curious playgrounds of their sport, mountaineers learn what primitive peoples know instinctively—that mountains are the abode of the dead, and that to travel into the high country is not simply to risk death but to risk understanding it.

Guido Rey, a Swiss who climbed at the turn of the century, wrote beautifully and compellingly of the joys of mountaineering. Seeking to explain why climbers court death in their quest for the summit, Rey made the following frequently quoted observation: "It is important to affirm and prove that we go to the mountains to live and not to die."

What he said is true, I think, but the way he said it is greatly misleading. By postulating a dichotomy between life and death that, through experience, the climber knows to be false, Rey inhibits a full understanding of the reasons people climb. It is precisely because life and death are not disparate conditions but, rather, gentle extensions of one another that the mountaineer moves so easily from one to the edge of the other. Rey meant to say that in the mountains we discover and express the beauty of life; what he failed to say but what is central to his meaning is that we cannot do so without glorying in the quiddity of death. Frequent confrontation with death is essential to mountaineering, not because, as Rey would have it, death can thereby be defied, but because it can thereby be tasted.

Who can admit to so mad an appetite! Countless times I have touched my tongue to the illicit brew, then, pulling back quickly, flopped onto a narrow ledge or trudged exhausted from a storm or felt the wicked angle of the snow beneath my feet lean back gently—and the joy that I knew each time surpassed words. Is it true that this joy at my return to life told only half the story . . . that there had been unspeakable joy, too, as I'd stood wide-eyed at the edge of the abyss, gazing silently and gratefully into the great darkness beyond? . . .

The system is not foolproof: occasionally innocent climbers are done in by circumstances beyond their control. Usually, however, a flawless performance guarantees another sunrise, another brushing of the teeth and packing of the lunch, another trip to the edge. Historically this has held true regardless of the climber's race, color, creed, class, gender, or nationality, making mountaineering perhaps the most democratic of all sports.

How is this flawless performance achieved? Don't fall: that's central. By choosing routes that are within one's level of ability and preparing mentally and physically for the climb, one should be able to get to the top without falling. Don't freeze to death: easy—carry the right clothing and use it properly.

But the list is so long! Each threat can in theory be averted by the prudent climber—but after a long, exhausting, tension-filled day, with steep slopes ahead, the weather turning, and miles to go before camp is reached, will the climber in fact act prudently? A rappel is a doubled rope which one attaches to the mountain side to facilitate descent. One mountaineering humorist has identified two dozen ways in which a rappel can go wrong. Jim Madsen, an out-standing climber of a generation ago, zipped off the end of one on Yosemite's El Capitan and fell half a mile to his death. Marty Hoey was killed on Mount Everest when she failed to secure her waist harness properly and fell free of the rappel rope. The beautiful long red hair of a woman I know was sucked into a rappel braking device just as she started down the ropes on Wyoming's Grand Teton. She dangled in pain and horror a hundred feet above the ground for half an hour, but survived her ordeal by cutting off all her hair with a pocketknife.

A lovely day turns foul, a beautiful chunk of granite sails toward your unhel-meted head, a ten-thousand-year-old handhold that you failed to test snaps off at last. . . . Climbers in Yosemite have completed brilliant multi-day ascents on walls several times the height of the Empire State Building, then been swept to their deaths as they crossed swollen streams during the easy jaunt home. Once on vertical rock in desert country I reached for a crack, inserted my fingers, stepped up, stared into the steely eyes of a rattlesnake coiled two feet from my hand. At such jolly times will one indeed perform flawlessly? I kept moving and the snake inexplicably allowed me to pass. Skilled but startled climbers have fallen to their deaths in just these circumstances.

The cumulative effect of this delirium of hazards is a fatality rate that one recent set of statistics shows to be seventeen times higher than that of moun-taineering's nearest non-equipment-dependent competitor, college football, and forty times higher than that of boxing. Ten percent of those who attempt to climb mountains in the Himalaya die trying. Overall, one in two hundred climbers is killed in the practice of the sport. . . .

A book called *The Conquest of Everest* by Sir John Hunt brought me to the high country. Growing up in Pennsylvania, where mountain climbers were about as common as crocodiles, I knew nothing of the sport until at sixteen, more or less at random I selected Hunt's account of the first ascent of the world's tallest peak for an English-class book report. In those days it was my habit to hang out in the public library—not for the reading matter, mind you, but for several studious and quite charming girls who also hung out there, and to whom I aspired greatly. When my own charm failed, as it often did, I sometimes wandered the stacks in search of strength and courage, and occasionally I located a book instead.

On the evening before my report was due I pulled the Hunt from a shelf and took it to a long oak table where a rabble of whispering teenagers, including two of my inamoratas, were seated. Perhaps I fancied that Janie or Susie would be impressed by my choice. Rather gravely, I expect, after checking to see that both were watching, I opened the book and began to survey its contents.

I was astonished and delighted to discover that, quite by accident, I had made a brilliant choice. Among the many chapters of the book was one entitled "The Summit," written not by the principal author John Hunt but by Edmund Hillary, the man who with Tenzing Norgay had been first to set foot on the top of the world. Clearly Hillary's contribution was the climax of the story. I saw that I could base my report on this very brief memoir (it was only sixteen pages long) and disregard the rest of the book altogether. If I hurried I could finish in time for a cherry Coke with Susie. (I discovered the pleasures of reading entire books only later in life and have since read all of the Conquest. I can attest now, on belatedly complete evidence, that Hunt's part is rather plodding and color-less—I believe I mentioned that in my original report—but that Hillary's is, indeed, worth a read. My mountaineering instincts, it seems, were solid from the beginning.)

My first brush with the high peaks, then, I can now recreate for you with some accuracy. It is a lusty spring evening in 1959 in Titusville, Pennsylvania, altitude 220 feet. Inside the Benson Memorial Library, a probably climbable structure of, as I recall, somewhat craggy construction, the scent of pretty girl hangs like edelweiss in the air, the whispering and flirting have reached Himalayan proportions, and Mrs. Doris Krimble, a librarian icy of demeanor and glacial of stare, has just issued her final warning for the hundredth time. Momentarily abashed, I turn to page 197 of the book and begin to read:

> Early on the morning of May 27th I awoke from an uneasy sleep feeling very cold and miserable. We were on the South Col of Everest. My companions in our Pyramid tent, Lowe, Gregory, and Tenzing, were all tossing and turning in unsuccessful efforts to gain relief from the bitter cold. The relentless wind was blowing in all its fury and the constant loud drumming on the tent made deep sleep impossible. Reluctantly removing my hand from my sleeping bag, I looked at my watch. It was 4 A.M. In the flickering light of a match, the ther-mometer lying against the tent wall read -25° Centigrade.

Neither the library's craggy construction nor Mrs. Krimble's glacial stare could have prepared me for the shock: it was staggering. The events and con-ditions that Hillary described were as Balinese to me. Yet something in his words struck a primal nerve in me. I emerged from my reading on fire. One sen-tence stood out, one that I now know touches on two of mountaineering's most persistent and addicting pleasures: "Our faculties seemed numbed," wrote Hillary of the summit pair's condition during the descent, "and time passed as if in a dream."

Something of that numbing and that dreaminess came over me. Suddenly I longed to be in some faraway place, alone in elemental conditions, surrounded by ice and space and silence. I remember seeing myself wandering through such a place, moving in a thick ether, experiencing an all-consuming peace. Mountains, which I had never seen or even thought about before, were sud-denly mysterious, thrilling places—enchanted clouds. Perhaps they embodied the remembered joys of my childhood, or my fantasies of life on another world. Historian Roderick Nash, who has traced the human attitude toward wilderness over several thousand years, has concluded, quite correctly, I think, that wilder-ness has never existed. Wilderness, says Nash, is a feeling about a place, part of what he calls "the geography of the mind." The geography of my mind, sketched in the distant past and fleshed out in 1959, is a high and snowy place, shrouded in mist and haunted by an ethereal quiet. There are trees and rocks about, and even, somehow, bright flowers and narrow winding streams. High above float mountaintops, their shadowy ramparts just visible where the mist is wearing thin. They seem light years distant, yet so near that I hold my breath at the sight. Standing in that landscape, as I often have since that visionary evening so long ago, I glimpse what poet Eunice Tietjens has called "the white windy presence of eternity." . . .

None of this could I have guessed in 1959. But something in Hillary's words touched me, and not long after, on a slip of paper that I kept that bore the heading of "Life List," I added two words: *Climb mountains.*

Seven years passed before I took my first step toward fulfilling that goal. Then, in the dead of winter 1966, on an impulse available only to the young and foolish, I agreed to join two friends on a four-day snowshoe traverse of the Franconia Ridge in the East's most arctic of ranges, the White Mountains of New Hampshire.

Of that hideous misadventure two remembrances stand out: the brain-curdling cold and the terror. My equipment was ludicrous. I slept in a Sears dacron sleeping bag designed, I believe, for summer in Mississippi. My pack was a Boy Scout knapsack barely large enough to carry my base equipment, let alone an extra sweater or two; the overflow I tied to the outside of the pack with mailing twine. As I stumbled upward through the trees on the first day, nimble-fingered pine and hemlock branches untied my granny knots and sent an avalanche of cooking pots and canteens clanging down the mountainside.

The second night we camped in a blizzard in the saddle between Mount Lafayette and Mount Lincoln. As the wind screamed and the tent shook so violently I feared it and its precious contents would be hurled to the bottom of the mountain, I came to understand one thing—that I was surely going to die.

Now to understand that one is going to die is a useful and profound understanding, not often reached at such an early age. Of course, like Saint Paul, I thought that the end was near and designed my prayers accordingly. That death lay some years in the future was information denied to both Saint Paul and me, and thus in no way diminished the impact of our perceptions or altered our subsequent behavior. When I did not die on the spot, I vowed solemnly to return to the mountains as quickly as possible.

Two summers later I went west with my friends Leon and Henry Bills, and there in the Tetons of Wyoming I saw my first high mountains. I knew I had to climb them. I had no idea how I would accomplish that seemingly impossible task, but, camped at their base, gazing up each day at faraway summits drifting in and out of clouds, I somehow guessed at the intimacy that is available only to the mountaineer, and longed to share in it. I was spellbound by one peak in particular. Teewinot was a perfect snow-draped triangle that swept up and up, ever narrower and more thrilling, to a wondrous pinnacle piercing the sky. By day she quickened my pulse; by night she haunted my dreams. She lay so close to my campsite that I began to know her habits and moods. As though courting her, I wove a sinuous dance at the mountain's foot—exploring her circling paths, dipping at midday into her refreshing streams, pausing in her cool shadows, withdrawing to admire her from afar.

One fine day my friends and I rented ice axes, purchased a length of gold-line rope, and set off to consummate the wild affair. We were quite impetuous all the way, I think. It was our first time; we were hale and determined, but lacked style. Several times in my eagerness I dislodged large, efficient-looking rocks that sailed harmlessly past my companions (but woe to anyone below!) On steep snow I came loose but was saved by the rope, a stroke of luck that cheered us greatly. It became clear that we were not to be thwarted, so excellent was our mission.

As we approached the top I saw to my horror that the summit was a smooth rock the size and shape of a very large almond. On all sides the drop-off was terrible. It was evident that the final act was to be a desperate business. The three of us looked at each other with blank faces. I took a deep breath, said a heart-felt prayer, dropped to my hands and knees, and crawled to the summit like a turtle.

From far below came a thin sound of cheering that failed to encourage me. I lay there hugging that inadequate slab of masonry as though it were my dear mama. To distract myself I thought of other places I had visited, nice places, low places.

Slowly then, after what seemed like an interminable length of time, I began to gain confidence. Following careful and exhaustive planning, I rolled onto my back. Then I sat up and looked around.

My first impression was one of a curious drifting of time and space. I seemed to be moving. I know now that I was. I liked it. Summits, I quickly realized, are not the solid, precisely defined spots that are shown on maps. Rather, they are capricious, meandering places, whose locations, like those of electrons and small children, are incapable of being pinpointed at any given moment.

My senses came to attention. Below I saw my friends snapping pictures. I felt the wind and the dazzle of the nearby sun. I felt the top of my head rubbing the sky. (I am now bald there; many climbers suffer this affliction.) I gazed across a choppy ocean of space at the mountain called the Grand Teton, a resplendent peak that reached even higher into the heavens than the point where I hovered. Suddenly, for the first time, I felt the terrible addiction of the mountaineer: *I wanted that one!* I wanted, more precisely, the east ridge of the Grand Teton, a steep, narrow, utterly bewitching buttress that sliced upward in one magnificent mile-long line from the glacier thousands of feet below me to the very summit of the peak. That long and elegant route introduced me to the aesthetic element of mountaineering: the east ridge of the Grand Teton (or the Grand, as we soon began calling it) was desirable not simply because it led to the top of the peak, but because it was so beautiful. From my lofty perch I studied it raptly: its explosive eruption in the glacier below, its urgent coalescence into great towers and gleaming black ice and rock tumbled on snow, its dark aprons and white cornices and inexorable upward thrust, now narrowing, now rushing, now rock, now only snow, now only sky—

My eyes ran swiftly over the ridge, taking in more and more with each pass. I grew more familiar with my newfound beauty by the minute, more needful, and all I could think was . . . *to be there! to be there!*

It was beyond my present powers. But I vowed that one day, when I had perfected my skills, I would return to climb the ridge. . . .

In the years to come I took many. I became a typical mountaineer—modest ability, limited achievement, an amateur all the way but *possessed* by the mountains. That is to say, the backbone of the sport. I climbed year round, several days a month and usually several weeks during the summer. I began collecting books on mountaineering, and stamps and posters that depicted the peaks that had come to mean so much. Nearly every day for twenty years I studied maps and guidebooks, planned journeys to the far reaches of the universe. I climbed, or attempted to climb, some two hundred peaks throughout the United States and Canada; and on neighborhood crags—the Pinnacles, Wallface, Crow Hill, Goat Rock—I ascended hundreds of routes that only a rock climber could love: Twinkle Toes Traverse, Dick's Prick, Highway To Heaven Portent. In the accustomed manner I had my brushes with death: a long fall in New York's Shawangunks that ended when the rope came tight around my waist; a slip unroped on a steep, exposed slab high on California's Mount Robinson. There I instinctively threw up an arm as I spun off; miraculously my hand flew into a crack above my head and jammed there, stopping me cold. The conviction grew that I was dancing ever closer to the edge. I dreamed of the east ridge of the Grand. Five times I climbed in the Tetons, ascending dozens of routes but never the ridge; it always felt beyond me. Several times I quit the sport, but I was always back within months.

Late one winter I returned to Pennsylvania from my home in California to be with my mother for the last time. She was eighty-one, and for more than a decade had suffered terribly as a series of increasingly debilitating illnesses had robbed her of her faculties one by one—her ability to walk, to feed herself, to turn over in bed, to see. Despite all of these hardships, she remained decidedly alive, something regarded as a miracle by most who knew her, and an added hardship as well. Even more amazing, she retained the capacity to settle into her wheelchair each morning with a measure of excitement about the day to come that was unfathomable.

In her final years her voice, never robust, left her almost entirely. Because my father was hearing-impaired, this led to frequent episodes of high comedy, to say nothing of maddening frustration, as she tried to make her needs known to him. During the day she did usually succeed, relying as much as anything on a kind of mental telepathy perfected over more than fifty years of marriage.

At night, however, something else was needed. My father, a sleeper of legendary soundness, tuned out the moment his head hit the pillow. To rouse him if she needed to get up in the middle of the night, my mother, who slept in a separate bed, required something more vigorous than mental telepathy.

So it was that during their final years together, my parents began joining their wrists with a length of cord each night when they retired. By giving a slight tug at her end—about all she was capable of—my mother was able to wake her partner at the other end of the cord.

Naturally, I saw this lifeline not only as a symbol of my parents' commitment to each other through half a century of marriage, but as an echo of the rope that connects two climbers as they work their way slowly and surely toward the summit of their mountain. The mountaineering technique is perfectly equitable, never favoring neither climber; each protects the other during the ascent. The same was true of my parents, though I failed to recognize this fact at the time. Clearly my mother needed and received the protection of my father; what I didn't see was my father, healthy and vital at his end of the lifeline, was equally dependent on the frail creature at the opposite end of the rope.

At last the difficulties of caring for his wife at home became too great, and my father was forced to find a place for her in a local nursing home. She moved into a sunny room with a sloe-eyed woman who sat rigidly upright in her bed, silent except for the blood-curdling moan she emitted every hour or so. Even now my parents remained intimately connected. With time out only to bolt down meals and hurriedly run errands, my father spent his days and evenings with my mother for more than a year. Seated beside her bed, he would tell her the latest news of my brother or me, or the torn-up sidewalk on Chestnut Street, or the dripping faucet in the kitchen. Often he read to her from the *Reader's Digest*. As the hours lengthened he would grow silent, sometimes falling asleep in his chair or gazing vaguely out the window at a small bed of flowers that was planted there as a reminder.

When I visited my mother last, she lay motionless in her bed, navigating some wonderful dream and preparing easily for what lay ahead. It was clear to me that, despite appearances, she understood everything I said. It was time for me to speak mountains. What I wanted most was to tell her it was all right for her to die, but the words would not come out. Like climbers, we had never much discussed death. Have I told you, she would whisper, pulling me close, that I want you to have Grandma's love seat? But that was circling the issue. And when the time came for me to sever the line that connected us, to ease the strain at her end through an act of selflessness and generosity, I could not do it. She lingered a month longer, and when I saw her next and kissed her cheek for the last time, it was cold as ice.

My father felt the loss of his companion of fifty-five years as a terrible blow. He seemed healthy, but during the next few months, each time we talked he told me that he would be joining her soon. I continued to fight: that's God's decision, I told him, not yours; you've got to take care of yourself, to do your best to go on. Reluctantly, he would agree, but with each call he sounded wearier, more despondent, and ever more certain of a fate he seemed to have taken firmly into his own hands.

I worried about him endlessly. With equal fervor I began to rejoice more and more in the wonder of life. I had recently emerged from the nightmare of divorce after losing to apparently greener pastures a wife of a dozen years. As that bad dream had progressed it had grown darker by the day. With little to show after ten years of struggle, I had given up hope of being a writer. My spirits, my self-esteem, and, not incidentally, life savings had dwindled to zero. I had lost a dear friend in a car accident and a beloved cat to old age. My closest friend had developed cancer. One morning he called me. "Bobby," he began in a faltering voice, and then he blurted it out: "I just tried to kill myself." So dreadful was that year that my friend's bungled suicide attempt emerged as one of the real high points.

And yet . . .

And yet I had survived. Bleeding and gasping, I had flopped onto a very narrow ledge, rolled over, checked my pulse, and found it to be ticking along as merrily as ever. Inexplicably, spring had blossomed once again in my life. I had met and fallen crazily in love with the woman whom music, the mountains, the moon, and the stars were created to celebrate. What's more, I had married her. I had left the job I'd taken during my earlier disillusionment and was recommitted to my writing. And to provide a fresh and exciting context for our astonishing new lives together, Carol and I had moved from suddenly tired California to young and hopeful New Mexico. Such was the spring when I began again, battered and patched but still warm at forty-three.

Came summer, the time to grapple with the mountains of my December dreams. I joined three friends, Kai Wiedman, Art Calkins, and John Flinn, for another go at the Tetons. When we arrived at our trailhead the knot in my stomach was tighter than it had ever been before. Never had I pulled on a pack and headed up a trail toward the high peaks with greater misgivings about my purpose than I did now. Beguiling mountains rose before me, beckoning as they always had, but I no longer craved union through the sacrament of climbing. My soul and my heart were not here, where they had to be if climbing was to have meaning; they were with my father in his grief; they were with Carol at home, where I longed to be, reveling in the beauties of life with her. As I headed into the high country I understood as I never had before how utterly stupid and unforgivable it would be for me to perish in the mountains.

On Mount Owen we climbed steep hard snow unroped, and in a gathering storm a few hundred feet from the summit traversed high over kingdom come. In a clouded vision I saw a body sailing silently into the void, saw three friends turning to watch the plunge. My boot skipped from a shallow footstep. I grabbed frantically at the snow, for a suspended moment felt the consummate peace of falling, the taking up into the blue, the serene absorption into interstellar matter. . . .

I stopped, paralyzed with fear. Kai and Art went on; John huddled with me in the lee of a wall of rocks to await their return. As the wind sharpened and the first drops of rain slashed down on us, we foretold in hushed tones the epic that was unfolding: the cold and the wet, the awful wind, the icy ropes, the slick rock and scary rappels, the down climbing in the dark, the miles and miles, the bivouac in the freezing rain. I faced it all with stark detachment. Already I was off the mountain; already I was in the place where I belonged.

Our friends arrived at the summit in a beehive of electricity. Kai's hardware buzzed at his waist as he touched the mountain's high point, a touch he knew might register as a microsecond of pleasure before dissolving in a white blindness of heat and light. On the way down Art slipped on steep snow but quickly arrested himself with his axe. I did not expect my friends to return but they did, quite suddenly. "We knocked the bastard off," announced Kai, echoing Hillary's words. Quickly we strung a rappel and slid off for home.

In sight throughout the descent, slanting up through the deepening light of slowly falling day, was the route I now knew I would never climb, the east ridge of the Grand Teton. I was too tired to appreciate my tragedy fully. From time to time I glanced up at the ridge, taking it in not as a beautiful woman I had lost but as a welter of problems I was relieved to have escaped. Deep in my heart stirred that irreconcilable sadness of never-to-be-realized dreams—but I would grieve tomorrow, not today. . . .

I went home, content with my decision. I would not stop going to the mountains; I could never do that. But my adventures now would be gentle ones—bright rambles up rounded slopes, where a slip might result in a twisted ankle but never a broken neck. I vowed that I would learn to be content with the tranquillity of the meadows, streams, and forests I had so often hurried past in my quest for the solitude of the summits.

A few weeks later I returned again to Pennsylvania, where late one night my father, perhaps carrying out an unconscious decision of his own, had been taken by ambulance to the local hospital. When I arrived he had been moved to the intensive care unit and there connected to the bewildering implements of survival which, unlike the rope of the mountaineer, we have come to fear so much.

Through the long hours at my father's bedside I did not pray that he would recover. I told him in a steady voice that if he wanted to die I understood, and that he had my blessing. As he slowly slipped away I was happy simply to hold his hand and to stroke his head, and to let him know that he was loved. I let go at last, and on the third morning he did the same, setting off on a voyage to a strange and enchanting world that I did not fear, that I knew to be peaceful and generous and good.

Among the mythologies of the world no theme is more persistent than that of the interdependence of love and death. On the Indonesian island of West Ceram they tell of the murder of the maiden Hainuwele and how, because of her death, both plants and sexual organs came into the world. The Ojibway of Lake Superior believe that the Great Spirit allowed his angel Mondawmin to die in order that Indians might have maize; Christians hold a similar belief involving God's beloved son and a gift of eternal life. So enduring is the motif that we must ask ourselves whether love and death, like Nirvana and the summit, may not be one and the same—neither beginning nor end, neither process nor experience nor condition of any kind, but a means through which we share in that universal and boundless energy we call nature. A year has passed and now the dazzling clarity of late summer has come to the mountains of New Mexico. Under a sky of incandescent blue, Carol and I wander upward among the breathtaking pinnacles of the Sandias, a rock climber's paradise just a stone's throw from where we live. On this quiet afternoon we have given ourselves over to reverie; Carol is lost in thought, while I speculate on pine cones, bumblebees, and, most of all, the baby of whom we have begun to dream.

Across the canyon a tower called the Pulpit rises suddenly into view. The shock is stunning. Unable to resist, I am tossed violently into the past: I stop, rivet my eyes on the face, scan it minutely for the route, the way, the edge, the last dance on thin air . . .

The slip lasts only a moment. Then Carol and I fall into step and resume our gentle stroll.

RESPONDING JOURNAL

Summarize in a paragraph the ideas Reid wants us to absorb from the essay. Which do you feel is the most important? Explain your response in an entry.

QUESTIONS FOR CRITICAL THINKING

STRATEGY

1. Examine Reid's central metaphor of love; in what different ways does he express the meaning of love? For example, does he confine himself to conventional romantic love? How does this line of thinking help him to convey a sense of how a dedicated mountain climber feels?
2. Think about what Reid accomplishes with the sentence structure in which he describes the mountain climber's lack of concern for death "as proof of the mountaineer's foolhardiness, or ignorance, or derangement." Can you duplicate this effect with a sentence of your own construction?

ISSUES

1. List several new ideas you have as a result of reading Reid's essay. Which of these would you like to pursue further? Does the practice of mountain climbing seem more or less intimidating to you now? Explain your response.
2. Look at Reid's narrative in comparison to the journals of Lewis and Clark or Colt. Does the deprivation suggested in these journals parallel that which Reid describes? Have you ever experienced such deprivation, or can you imagine it? How would you describe it?

COLLABORATIVE WRITING ACTIVITY

Experienced mountain climbers often travel to the summit of a mountain in pairs, connected by a rope. Reid says of this practice, "The mountaineering technique is perfectly equitable, favoring neither climber; each protects the other during the ascent." Modify this sentence in whatever way you wish to change the subject from mountain climbing to collaborative writing. Bring your sentence to your group, and read, share, and discuss how to shape these ideas into an essay.

from *Neversink: One Angler's Intense Exploration of a Trout River*
Leonard Wright, Jr.

Leonard M. Wright, Jr., is one of the most noted American writers on fly fishing for trout. Among his books are Fishing the Dry Fly as a Living Insect, First Cast: The Beginner's Guide to Fly Fishing, *and* The Field & Stream Treasury of Trout Fishing. *His essays have appeared in* The New York Times Magazine, Esquire, *and* Travel & Leisure. *Though based in New York City, he spends as much time as possible upstate, in the Catskill Mountains, on his beloved Neversink River. One of the most famous trout streams in the country, the Neversink is the river where the dry fly—and the art of fishing such creations—was developed.*

But Wright's task in Neversink *is not to provide another fishing story. In the introduction, he tells his reader that he will explain his progress of learning about the river; he will duplicate his process of moving back from the fish themselves to the factors and conditions which determine both the river and the fish that inhabit it. Even non-anglers, he feels, will find the natural history of such a stream interesting. Judge for yourself how well he has succeeded.*

The excerpt here is chapter 14; Wright is trying to find ways to increase the food supply for the trout in the river so they will grow both larger and more plentiful.

EXPLORING JOURNAL

What do you know about fly fishing? What does it mean to you? Could you ever imagine yourself doing it? In a ten-minute freewrite, explore your ideas about this sport and the people who practice it.

Since it appeared to be beyond my powers to significantly increase the supply of stream-bred insects, I started looking for a parallel, or supplementary, food source. The solution seemed obvious, but the obvious wasn't the solution.

For decades, hatcheries have been raising trout on artificial pellets, and the latest versions of these small, dry cylinders are said to provide the perfect mix of a trout's nutritional requirements. I wouldn't think of challenging the biologists' analysis of the ideal trout diet nor their ability to recreate it. The only problem I have with their product is that I can't get wild trout to eat it.

Over the years, I'd tossed enough pellets into the river to risk tennis elbow, and yet my fish would not feed on them. Only once did I see one of my trout take a pellet into his mouth—and he spat it out instantly. Apparently, a wild trout's gums don't find hard, unyielding objects anything like squishy insect bodies, and pellets probably flunk the taste test as well. There's no point in yelling "Eat—it will make you big and strong when you grow up." Trout are even less persuadable than small children.

I must admit I have seen wild trout take pellets, but only under special circumstances. I've watched a caretaker feed a heavily stocked stretch of the Neversink, and all the domestic fish pounced on the pellets as if it were feeding time back at the hatchery. Once they'd been boiling regularly for a few minutes, I saw a few wild trout join in. It seems that feeding activity among trout is as social and contagious as it is among chickens. However, loading up my water with large, stocked shills seemed a self-defeating way to cram more food into native trout.

Some time after my inadequate feeding attempts with dace and Japanese beetles, I did hit upon a method of supplying natural insect food that actually dispensed itself and really worked. Well, almost.

Before I sank my garden fence two feet underground, woodchucks kept tunneling under the wire while I was away during the week. As soon as my .22 claimed one, another seemed to show up. To avoid a total loss from these depredations, I decided to convert my romaine and broccoli into nutritious trout food.

I'd read somewhere that British poachers used to hang a sheep's head, or the like, from a tree over a trout pool; when the dropping maggots had attracted a cluster of trout, the fish were easy marks for a maggot impaled on a small hook. Guessing that a plump woodchuck was as good as or better than a gristly sheep's head, I strung up my victims from tree limbs overhanging one of my lower, deeper pools.

In a few days, a steady supply of maggots began trickling out of the carcasses—especially during the heat of the day. A school of trout assembled below each one, and I could almost see them putting on heft with each passing day. This feeding system offered natural food, was fully automated, kept on dispensing while I was away in the city, and was, I felt, the ultimate solution. Then one hot day my wife and kids dropped down to the pool for a swim. As they ran back to the house shrieking and holding their noses, I knew that was the end of an inspired fishery-improvement scheme.

Casting about for another handy source of trout food, I dimly remembered reading in a book by Ed Hewitt (yes, there's that man again) that the best supplemental trout food was ground-up lungs, or lights. He claimed that, while perhaps not as rich a diet as the chopped liver that hatcheries fed in his day, lights were superior river food for two reasons. First, since lung tissue is made up of tiny air sacs that make it float, it trains trout to look for their food on the surface, which is what every fly-fisher wants. Second, floating food has little or no waste. Chopped liver quickly sinks to the bottom and is lost in rock interstices, whereas the floating lights that weren't eaten in the first pool bobbed on downcurrent to be cleaned up by the trout below.

While the great man had the reputation of being always positive and sometimes right, this seemed too good a lead to pass up. I went to my local butcher and asked for twenty pounds of lungs. After convincing him that I was serious, he told me that only a few shops in heavily German neighborhoods ever carried them and that a slaughterhouse was my best bet.

After some research I finally located a small, nearby abatoir that slaughtered and halved veal for wholesalers. The head man seemed willing enough to sell me whole lungs at the going rate (pet-food canners purchase seldom-eaten spare parts in bulk), but explained that he would have to discolor them first with charcoal dust. Apparently, this procedure had been enacted into state law to prevent unscrupulous retailers from adulterating their hamburger with a cheap substitute. It took me a half hour to convince him that I wasn't a crooked butcher, merely a harmless eccentric who had a crush on trout.

After I'd cut away the cartilaginous tracheae, I lugged the lungs down to my local butcher, who'd agreed to run the lot through his grinder for a fee if I appeared at closing time, when he had to clean the machine anyway. The resulting lungburger was stuffed into plastic bags in five-pound lots and parked in my freezer for future use.

It had been quite an exercise, but it turned out to have been worth it. The very first time I ladled out a couple of pounds into a riffle above a good pool I was rewarded by the sight of dozens of fish swirling and splashing down in the slow water. This show was as humbling as it was gratifying because I'd never been able to raise anything like that number of trout when fishing my way up the pool.

Hewitt had been 101 percent right, after all. Here was a relatively cheap food in endless supply that even wild trout would gorge on. These tiny blobs of meat may not present the silhouette of any known insect, but trout wallop them the first time they see them. Apparently, the smell of lights, or of the blood that trickles out of them, identifies them instantly as lean red meat. Our gentle, shy trout seem to have the noses and appetites of piranhas.

The two- to three-pound dosage I usually mete out during a feeding may not sound like much, but when the particles separate in the turbulence of a pool's inflow currents, they number in the many thousands. After ten or so handfuls have been scattered, they'll cover a pool fifty feet wide and two hundred feet long with over a dozen red specks per square foot of surface, creating an artificial red tide. I have never, at any time on any river, seen that many natural insects floating down a pool.

This synthetic super-hatch usually lasts only two or three minutes, but the trout, brookies in particular, make the most of it. Some will hold just under the surface and grab the overhead particles with a rapid, rocking-horse motion. A few gluttons will eat their way steadily up the pool until they're within a few feet of me and, when the food supply peters out, charge down to the tail and start gulping their way up again with their dorsal fins cutting the surface.

Such performances go a long way toward explaining the surprising mobility of pool-dwelling trout. They may be bunched up in the deep fast water at the head of the pool all day only to drift down to the tail end at dusk for easier feeding on dead or dying insects. Apparently, this migration can take place extremely quickly when a food source sends out the right signals.

Trout living in fast runs and pockets will nail these meat particles readily enough as they speed by, but they rarely follow the food stream down into the slow flat or pool below. It seems that homesteading and defending the choice lie they've won is higher on their list of priorities than a few extra mouthfuls of food.

The largest fish in a pool, almost always brown trout, seldom wallow on the surface, but they, too, get their share. About ten percent of the lungburger sinks slowly because its air sacs have been ruptured in grinding and, as a result, becomes easy pickings for bigger, more cautious fish. When the light is right, I can make out the golden flashes from their sides as they twist and turn to intercept the red blobs drifting just above the streambed.

Feeding trout this way is almost as much fun as fishing. There may be an element of playing God to it since trout appear and perform at my bidding, but that's not the major attraction. What fascinates most is the insight this gives into where trout live and how and when they feed.

Through supplemental feeding I have located trout lies in places I never suspected. Some will settle in surprisingly shallow water and well away from the current as long as there's the overriding amenity of overhead cover. Spotting trout this way seems to me a form of cheating. After all, trout have a right to some privacy, too. However, the greater good of the extra food squelches my misgivings about such prying.

The most exciting and revealing trout feeding requires an accomplice. I like to position myself one-third to halfway down a pool, preferably ten to fifteen feet up a steep bank, while a volunteer stands in the run above with bucket in hand. When I say, "Okay, let her go," dispensing begins and soon the trout start feeding, at first hesitantly and then with abandon. From this perspective, I can tell not only the size but also the species of the feeders, both of which are impossible from upriver. Many of my sightings are of fish I've never raised, or even suspected the existence of, in that section of stream.

One trout secret that still eludes me is where these fish hide when they're not on the feed. From my perch on the hillside I can scrutinize the stream bottom of at least one-third of the pool, and I can usually make out only three or four small trout hovering just above the rocks. Then, when the lungburger blankets the surface, a dozen or two will appear out of nowhere. Try as I will, I can't catch these fish emerging from their hides or returning to them, nor can I count enough large, undercut rocks on the bottom to house them all. A trip through the area in a wet suit with mask and snorkel doesn't turn up the fish or their sanctuaries, either.

The most valuable lesson I've learned from this feeding and spying is when—or, more accurately, under which conditions—trout will feed greedily and when they'll abstain. If I scatter lights at a poor time or under adverse conditions only two or three small fish will respond, rising only once or twice each. On the other hand, at the best of times, the same dosage in the same pool will induce forty to fifty trout to feed voraciously for several minutes. The contrast is astonishing.

After years of feeding, I can now predict with 99 percent accuracy how trout will respond to my handouts long before I reach the feeding area. All I need to do is check the cloud cover and river level and reconfirm what they tell me with a stream thermometer.

If the water level is up half a foot or more due to a recent rain, feeding response will be poor to marginal. I know I'll get the same reaction if the day is, or has been, cloudy, regardless of river height. On the other hand, when the sun is out and the river is at normal flow, I can look forward to a full show of eager splashers *if* I choose the right time of day. The triggering factors are water temperature and its rate of change, and my thermometer can readily pinpoint the best feeding time, or times, for the day.

All this is not nearly as mysterious as it may sound. Trout feeding activity is dictated by their metabolism rate and this, in turn, is a function of body temperature. Since trout are cold-blooded creatures, their bodies are essentially the same temperature as the water they're swimming in. When the river heats up or cools down, they do, too.

Biologists have determined that brown trout metabolism peaks at 63 degrees. The figure may be slightly lower for brook trout and a shade higher for rainbows, but 63 degrees is close enough for our purposes. At that reading, a trout's gills are most efficient in extracting oxygen from the water, its digestion rate is most rapid, and all bodily functions are performing at top efficiency.

Obviously, then, trout feeding should be most active when water temperature is nearest to the magic number 63—and it almost always is. But there's a joker in the deck. A flat, unchanging 63 degrees always means that lights-feeding and fly-fishing will be only mediocre at best. It is only when the rate of change toward and through 63 degrees is rapid that fish will feed like gluttons.

You get these rapid changes only in fair weather when the sun's rays warm up stream-bottom rocks, which then transmit this heat to the water flowing over them. The lower and slower the water, and the more hours of sunshine, the faster the heat-up. Similarly, in late afternoon and evening, heat will radiate out of both rocks and stream water fastest when clear skies abet this escape.

During sunny, high-pressure weather in July, the spread of temperatures I record on the lower Neversink may surprise you. At daybreak after a clear night during average flows I usually get 54 degrees. By 11:00 A.M. it will be up to 60 degrees, by noon 62 degrees, and by 1:00 P.M., 64 degrees. Temperature will continue to climb until 4:30 or 5:00 P.M. when it will peak at about 70 degrees. After that, readings start to drop off and should be down to 64 or 65 degrees just before dark. There has been a variation of 16 degrees during that eighteen-hour period, and I can expect active feeding to take place only from 11:00 A.M. to 1:00 P.M. and from 8:00 P.M. until dark.

If the day is cloudy after a cloudy night, on the other hand, temperature variation will be minimum: 60 degrees at 8:00 A.M., 61 degrees at noon, 63 degrees at 4:00 P.M., and 62 degrees at dusk. Trout may take an occasional mouthful of food at any time during that period, but at no time will they feed voraciously.

Why doesn't a flat 63 degrees produce aggressive feeding all day long? I'm convinced that the reason for this is that the supply of oxygen in the trout's bloodstream, though high, remains constant when temperatures are stable. On the other hand, when temperatures are racing up or down toward and through the magic 63 mark, trout are getting a sudden surge of oxygen through rapidly increasing gill efficiency. Whether or not this gives the fish the "high" that we experience when we hyperventilate though forced breathing, I have no clue. I only know from years of observation that fly-fishing success and active feeding on lights are both at their peak under these rapid-change conditions. I have written in a previous book a lengthy documentation with complex charts explaining how and why this works so I won't bore you with the complicated calculations here. Just take my word for it: When temperatures race through the lower to mid-sixties from either direction, trout will feed with a near frenzy, and they simply never do when the thermometer stalls at that mark.

Having discovered this, I could then both fish and feed with far greater efficiency than before. In sunny, summer weather, I'd hit the stream with rod and lights bucket at 11:00 A.M. Leaving the food at the head of a pool, I'd walk down to the tail and fish my way back up. When I'd covered the water to my satisfaction, I'd distribute half the lungburger, saving the rest for the pool above, which would receive the identical treatment. Evenings, I'd go out at 8:00 P.M. and follow the same procedure on a different stretch.

I couldn't resist showing off my fish-feeding prowess to the other fly-fishers in the Valley and they were duly impressed with the spectacle my fish put on. One donated an old but working freezer, which we installed in an empty shed behind the general store. I kept it stocked with bags of frozen lights and the other users put a check mark beside their names every time they picked up a bag. All accounts were settled at the end of the season.

This program, I felt, was nearly ideal. Fish were getting extra rations all up and down the Valley rather than just in my limited stretch. Then too, this feeding was actually cost-efficient. If four pounds of lights would, indeed, build a pound of wild-trout flesh, as biologists said, that quantity of lungs could be bought, ground, and stored at half the price of a pound of inferior hatchery trout. I was convinced that the Neversink would soon regain its position as one of the most productive trout fisheries in the East.

RESPONDING JOURNAL

What do you think of Wright's program to fatten up his fish? Even if he were to release all the fish he caught, does his practice of feeding the fish bother you? Comment on your estimate of his feeding, and try to think of other ways in which people feed animals in the wild. How does his action differ from feeding your dog or cat? Reflect on these issues in a substantial journal entry.

QUESTIONS FOR CRITICAL THINKING

STRATEGY

1. Map out the process Wright describes here. Do you find the order of events he describes to be effective? Can you see any other details he might have included to make the process of trout feeding clearer?
2. Account for Wright's explanations—which occur early in the selection—of unsuccessful attempts at trout feeding. How do these failures help to show how the final process works?

ISSUES

1. Explain what you think the most important lessons for fisherpeople might be from Wright's experiments. Would nature enthusiasts or natural science students find his data equally interesting? Do you? Why?
2. Look at Wright's last paragraph. Does he give you a sense of something to come that will be less than perfect? Do you hear a "but" coming? What drawbacks might this great scheme ultimately have?

COLLABORATIVE WRITING ACTIVITY

If you are not already familiar with a local body of water, get a map and find the closest river, stream, lake, pond, or ocean. Learn what rules govern its use: you will need to check fishing, boating, swimming, access, and seasonal use regulations. After you have assembled as much information as you can, bring it to your group and combine what you have discovered. After you discuss what you have, shift the topic to what you have learned in the process of acquiring this material. What categories of use seem most important to you? How can you best insure the proper use of this resource for future generations? Working as a group, draft a policy statement that reflects your consensus priorities to insure protection of the body of water.

from *Women Who Run with the Wolves*
Clarissa Pinkola Estes

Clarissa Pinkola Estes is a Jungian analyst, a psychiatrist who employs the concepts of therapy achieved through examination of a patient's dreams and subconscious experiences. Estes works primarily with women, and her experiences with women from numerous cultures have led her to the asser-tions she makes in her best-selling book, Women Who Run with the Wolves.

Surprisingly, Estes shopped her book around to publishers for twenty years before she found a house willing to publish it. The wide acclaim and enthusiasm that greeted the book are evident in its publication history: released in July 1992, it was in its 18th printing by December. As you read the selection here, you might ask yourself why Estes' ideas encountered such initial opposition, and then generated such wide interest. She presents 19 folktales of women who did not suppress their energies, but who were rejected or outcast by their societies because of their display of character. The selection below presents Estes' thesis and one folktale.

EXPLORING JOURNAL

What possible links could there be between a woman and a wolf? Apart from Jack Nicholson's recent venture into wolfhood, what more logical connection might exist? Freewrite for ten minutes on the idea of women and wolves. What potentially beneficial characteristics of wolves do you see in women?

SINGING OVER THE BONES

Wildlife and the Wild Woman are both endangered species.

Over time, we have seen the feminine instinctive nature looted, driven back, and overbuilt. For long periods it has been mismanaged like the wildlife and the wildlands. For several thousand years, as soon and as often as we turn our backs, it is relegated to the poorest land in the psyche. The spiritual lands of Wild Woman have, throughout history, been plundered or burnt, dens bull-dozed, and natural cycles forced into unnatural rhythms to please others.

It's not by accident that the pristine wilderness of our planet disappears as the understanding of our own inner wild natures fades. It is not so difficult to comprehend why old forests and old women are viewed as not very important resources. It is not such a mystery. It is not so coincidental that wolves and coyotes, bears and wildish women have similar reputations. They all share related instinctual archetypes, and as such, both are erroneously reputed to be ingracious, wholly and innately dangerous, and ravenous. . . .

Healthy wolves and healthy women share certain psychic characteristics: keen sensing, playful spirit, and a heightened capacity for devotion. Wolves and women are relational by nature, inquiring, possessed of great endurance and strength. They are deeply intuitive, intensely concerned with their young, their mate and their pack. They are experienced in adapting to constantly changing circumstances; they are fiercely stalwart and very brave.

Yet both have been hounded, harassed, and falsely imputed to be devouring and devious, overly aggressive, of less value than those who are their detractors. They have been the targets of those who would clean up the wilds as well as the wildish environs of the psyche, extincting the instinctual, and leaving no trace of it behind. The predation of wolves and women by those who misunderstand them is strikingly similar.

So that is where the concept of the Wild Woman archetype first crystallized for me, in the study of wolves. I've studied other creatures as well, such as bear, elephant, and the soul-birds—butterflies. The characteristics of each species give abundant hints into what is knowable about the feminine instinctual psyche. . . .

Although I did not call her by that name then, my love for Wild Woman began when I was a little child. I was an aesthete rather than an athlete, and my only wish was to be an ecstatic wanderer. Rather than chairs and tables, I preferred the ground, trees, and caves, for in those places I felt I could lean against the cheek of God.

The river *always* called to be visited after dark, the fields *needed* to be walked in so they could make their rustle-talk. Fires *needed* to be built in the forest at night, and stories *needed* to be told outside the hearing of grown-ups.

I was lucky to be brought up in Nature. There, lightning strikes taught me about sudden death and the evanescence of life. Mice litters showed that death was softened by new life. When I unearthed "Indian beads," trilobites from the loam, I understood that humans have been here a long, long time. I learned about the sacred art of self-decoration with monarch butterflies perched atop my head, lightning bugs as my night jewelry, and emerald-green frogs as bracelets.

A wolf mother killed one of her mortally injured pups; this taught a hard compassion and the necessity of allowing death to come to the dying. The fuzzy caterpillars which fell from their branches and crawled back up again taught single-mindedness. Their tickle-walking on my arm taught how skin can come alive. Climbing to the tops of trees taught what sex would someday feel like.

My own post–World War II generation grew up in a time when women were infantilized and treated as property. They were kept as fallow gardens . . . but thankfully there was always wild seed which arrived on the wind. Though what they wrote was unauthorized, women blazed away anyway. Though what they painted went unrecognized, it fed the soul anyway. Women had to beg for the instruments and the spaces needed for their arts, and if none were forthcoming, they made space in trees, caves, woods, and closets.

Dancing was barely tolerated, if at all, so they danced in the forest where no one could see them, or in the basement, or on the way out to empty the trash. Self-decoration caused suspicion. Joyful body or dress increased the danger of being harmed or sexually assaulted. The very clothes on one's shoulders could not be called one's own.

It was a time when parents who abused their children were simply called "strict," when the spiritual lacerations of profoundly exploited women were referred to as "nervous breakdowns," when girls and women who were tightly girdled, tightly reined, and tightly muzzled were called "nice," and those other females who managed to slip the collar for a moment or two of life were branded "bad."

So like many women before and after me, I lived my life as a disguised *criatura,* creature. Like my kith and kin before me, I swagger-staggered in high heels, and I wore a dress and hat to church. But my fabulous tail often fell below my hemline, and my ears twitched until my hat pitched, at the very least, down over both my eyes, and sometimes clear across the room.

I've not forgotten the song of those dark years, *hambre del alma,* the song of the starved soul. But neither have I forgotten the joyous *canto hondo,* the deep song, the words of which come back to us when we do the work of soulful reclamation.

Like a trail through a forest which becomes more and more faint and finally seems to diminish to a nothing, traditional psychological theory too soon runs out for the creative, the gifted, the deep woman. Traditional psychology is often spare or entirely silent about deeper issues important to women: the archetypal, the intuitive, the sexual and cyclical, the ages of women, a woman's way, a woman's knowing, her creative fire. This is what has driven my work on the Wild Woman archetype for the better part of two decades.

A woman's issues of soul cannot be treated by carving her into a more acceptable form as defined by an unconscious culture, nor can she be bent into a more intellectually acceptable shape by those who claim to be the sole bearers of consciousness. No, that is what has already caused millions of women who began as strong and natural powers to become outsiders in their own cultures. Instead, the goal must be the retrieval and succor of women's beauteous and natural psychic form.

Fairy tales, myths, and stories provide understandings which sharpen our sight so that we can pick out and pick up the path left by the wildish nature. The instruction found in story reassures us that the path has not run out, but still leads women deeper, and more deeply still, into their own knowing. The tracks which we all are following are those of the Wild Woman archetype, the innate instinctual Self.

I call her Wild Woman, for those very words, *wild* and *woman,* create *llamar o tocar a la puerta,* the fairy-tale knock at the door of the deep female psyche. *Llamar o tocar a la puerta* means literally to play upon the instrument of the name in order to open a door. It means using words that summon up the opening of a passageway. No matter by which culture a woman is influenced, she understands the words *wild* and *woman,* intuitively.

When women hear those words, an old, old memory is stirred and brought back to life. The memory is of our absolute, undeniable, and irrevocable kinship with the wild feminine, a relationship which may have become ghostly from neglect, buried by overdomestication, outlawed by the surrounding culture, or no longer understood anymore. We may have forgotten her names, we may not answer when she calls ours, but in our bones we know her, we yearn toward her; we know she belongs to us and we to her.

It is into this fundamental, elemental, and essential relationship that we were born and that in our essence we are also derived from. The Wild Woman archetype sheaths the alpha matrilineal being. There are times when we experience her, even if only fleetingly, and it makes us mad with wanting to continue. For some women, this vitalizing "taste of the wild" comes during pregnancy, during nursing their young, during the miracle of change in oneself as one raises a child, during attending to a love relationship as one would attend to a beloved garden. . . .

When women reassert their relationship with the wildish nature, they are gifted with a permanent and internal watcher, a knower, a visionary, an oracle, an inspiratrice, an intuitive, a maker, a creator, an inventor, and a listener who guide, suggest, and urge vibrant life in the inner and outer worlds. When women are with the Wild Woman, the fact of that relationship glows through them. This wild teacher, wild mother, wild mentor supports their inner and outer lives, no matter what.

So, the word *wild* here is not used in its modern pejorative sense, meaning out of control, but in its original sense, which means to live a natural life, one in which the *criatura,* creature, has innate integrity and healthy boundaries. These words, *wild* and *woman,* cause women to remember who they are and what they are about. They create a metaphor to describe the force which funds all females. They personify a force that women cannot live without.

The Wild Woman archetype can be expressed in other terms which are equally apt. You can call this powerful psychological nature the instinctive nature, but Wild Woman is the force which lies behind that. You can call it the natural psyche, but the archetype of the Wild Woman stands behind that as well. You can call it the innate, the basic nature of women. You can call it the indigenous, the intrinsic nature of women. In poetry it might be called the "Other," or the "seven oceans of the universe," or "the far woods," or "The Friend." In various psychologies and from various perspectives it would be called the id, the Self, the medial nature. In biology it would be called the typical or fundamental nature.

But because it is tacit, prescient, and visceral, among *cantadoras* it is called the wise or knowing nature. It is sometimes called the "woman who lives at the end of time," or the "woman who lives at the edge of the world." And this *criatura* is always a creator-hag, or a death Goddess, or a maiden in descent, or any number of other personifications. She is both friend and mother to all those who have lost their way, all those who need a learning, all those who have a riddle to solve, all those out in the forest or the desert wandering and searching.

In actuality, in the psychoid unconscious—the layer from which the Wild Woman emanates—Wild Woman has no name, for she is so vast. But, since Wild Woman engenders every important facet of womanliness, here on earth she is named many names, not only in order to peer into the myriad aspects of her nature but also to hold on to her. Because in the beginning of retrieving our relationship with her she can turn to smoke in an instant, by naming her we create for her a territory of thought and feeling within us. Then she will come, and if valued, she will stay.

So, in Spanish she might be called *Río Abajo Río,* the river beneath the river; *La Mujer Grande,* the Great Woman; *Luz del abismo,* the light from the abyss. In Mexico, she is *La Loba,* the wolf woman, and *La Huesera,* the bone woman.

She is called in Hungarian, *Ö, Erdöben,* She of the Woods, and *Rozsomák,* The Wolverine. In Navajo, she is *Na'ashjé'ii Asdzáá,* The Spider Woman, who weaves the fate of humans and animals and plants and rocks. In Guatemala, among many other names, she is *Humana del Niebla,* The Mist Being, the woman who has lived forever. In Japanese, she is *Amaterasu Omikami,* The Numina, who brings all light, all consciousness. In Tibet she is called *Dakini,* the dancing force which has clear-seeing within women. And it goes on. She goes on. . . .

RESPONDING JOURNAL

In a substantial journal entry of at least two pages, explore the ways in which this excerpt has changed how you look at women and their creative abilities. If you feel Estes is correct, explain how your experiences lead you to that conclusion. If you feel she is incorrect, refute her points with evidence from your own experience and reflection.

QUESTIONS FOR CRITICAL THINKING

STRATEGY

1. Explain Estes' comment that "my fabulous tail often fell below my hemline." What does she mean by this metaphor? How extensively does she employ this technique? Explain how you feel this metaphor helps or hinders Estes in arguing successfully her points.

2. How well can Estes argue her points by using folktales? How can folk-tales support a "scientific" argument? Can a psychiatrist actually base treatment on folktale knowledge? Explain your response by referring to your own experience and to your thinking on the subject.

ISSUES

1. How valid do you find Estes' assertion that old women are generally viewed by society as having little value? Perform a content analysis search of a few magazines or a paperback book display. Select at least six magazines and chart the number of women, along with their approximate ages, that appear. Do the same for a rack of paperbacks. (Once you begin this exercise, you may want to extend the field of your investigation. You might want to perform an analysis of the covers of current periodicals in your school library.) What do you find? Does this research seem to bear out or to negate Estes' claim?
2. Estes claims that Wolf Woman lives in various parts of nature. What do such claims suggest for the environmental movement? You may well want to create metaphors of your own in responding to this question.

COLLABORATIVE WRITING ACTIVITY

Estes states, "*Llamar o tocar a la puerta* means literally to play upon the instrument of the name in order to open a door." Explore this idea in writing by developing a metaphor of your own choice and explaining how this idea helps you to articulate a concept related to the environment or to the natural world. After you have completed your work, bring it to the group to read and to seek comments on what the metaphor says to other group members. After you revise your work in response to your colleagues, pair up with another group member and look at each other's writing to confer on how well you each feel the idea of a controlling metaphor has worked in your writing.

from "Everything You Know about the Environment Is Wrong"
Gregg Easterbrook

Gregg Easterbrook is a professional journalist who has been writing on environment-related topics for twenty years. After beginning with the garbage industry magazine Waste Age, *he has moved on to other periodicals; presently he is a contributing editor at* Newsweek *and* The Atlantic Monthly. *His down-to-earth style suits him well for the environmentally-related subjects he normally covers.*

"Everything You Know about the Environment Is Wrong" was published first in The New Republic *in 1990; the magazine has a reputation as an intellectual publication.*

EXPLORING JOURNAL

What would you most like to be mistaken about with regard to the environment? That is, what problem would you most like to have someone tell you does not really exist or has ceased to exist? Freewrite on this topic for ten minutes.

——➤◆≺——

The air and water are getting cleaner, not dirtier. Acid rain may be preventing global warming. Smog protects you from ozone depletion. Family farmers dump more chemicals than toxic waste sites. The "poisoning of America" is already over. Nature kills more species than humanity. The Third World is a greater threat to the ecology than the West. Some environmentalists actually long for the environment to get worse. Some business leaders want it to get better. OK, OK. That last taxes credence, though occasionally it's true. All the others are bang-on actual.

There is a growing sense that the only socially respectable attitude toward the environment is pushing the panic button. Fashionable alarmism may eventually create a Chicken Little backlash: as the years pass and nature does not end, people may stop listening when environmentalists issue warnings. The tough-minded case for environmental protection is ultimately more persuasive than the folk song and flowers approach. For liberals, being tough-minded means shedding some cherished preconceived notions, but it also means creating more rigorous arguments in favor of ecological respect as a human value, and, perhaps, of pointing the way toward finding humankind a constructive role in creation's scheme. Herewith a guide to what's really going on in the environment.

ACID RAIN

After a decade of Reaganesque mumbling about further study and congressional Democratic stalling, action finally seems assured. (Democrats stalled because the high-sulfur coals that are the chief cause of acid rain come from underground mines staffed by the United Mine Workers; low-sulfur coals come mainly from non-union surface mines in the West.) The new Clean Air Act will reduce national sulfur dioxide by about half and cut a related acid rain source, nitrogen oxide. Probably it will include a "cap" mandating that no matter how much future electricity production grows, total acid rain emissions may never exceed the new (reduced) level. This provision, brainchild of Environmental

Protection Agency administrator William Reilly, is a far-reaching barrier against future new pollution.

Little-known note: in the past fifteen years national sulfur dioxide emissions have already fallen by about one-quarter even as coal use increased nearly fifty percent, owing to controls under the old Clean Air Act and to the construction of new power plants with superior technology. Second note: studies, including a ten-year, $500 million federal project, show acid rain effects to be considerably less than theory predicts. Only high-altitude red spruce trees, not forests generally, so far display acid rain damage; and though some enviros projected that a majority of Eastern U.S. lakes would by now be too acidic for most life, the federal study found that only four percent have crossed this threshold. . . .

The "Garbage Crisis"

In the 1970s my first job as a journalist was with *Waste Age* (its real name), the trade magazine of the garbage industry. One of my first stories quoted various luminaries warning that a "garbage crisis" was about to strike. Fifteen years later it's still about to strike.

Except in a few densely populated cities, it's nutty to maintain that a country as vast as America is "running out" of space for landfills. There is room to landfill our trash till the Lord's return. What we are running out of is willingness to tolerate landfills. That's as it should be. Though landfills can be built with reasonable environmental safety, they are fundamentally bad ideas: enablers of an irresponsible attitude toward resource consumption. What are responsible alternatives? Enviros say recycling. Supposedly it's about to take off. Fifteen years ago I wrote story after story saying recycling was about to take off.

Grumps maintain that recycling is a fraud, noting that more recycled newsprint is already available than processors will buy. But don't markets always take time to develop? There's no reason the United States can't eventually recycle a quarter of its trash, considering that the advanced economy of Japan recycles more. Though separation of aluminum cans and newspaper, the two most viable recycling commodities, is a big pain for homeowners, it's a good idea for society.

But recycling won't ever solve all disposal problems; and reductions in the packaging content of products aren't going to make more than a tiny nick in the problem. Municipal waste is probably best managed with a combination of moderate recycling, waste-to-energy plants burning the bulk of the trash, and some landfills (there's no scheme that eliminates them) for ash from the burners. . . .

Global Warming: Is It Happening?

Data are dueling over two aspects of this subject: whether warming has already been detected, and whether computers can project temperatures for the next century. Widescale climate effects are so little understood that there are currently scientific debates on matters as basic as whether warming would cause increased rainfall or drought; whether plants would flourish in more carbon dioxide or gag on the stuff.

There's even a faction that thinks global warming might be acceptable. Under this theory, my hometown of Buffalo would become a vacation paradise. The United Nations Intergovernmental Panel on Climate Change, the same outfit George Bush was blasted for failing to deliver a doomsday speech to, has estimated that a world warming of 3.5 degrees Fahrenheit would increase agricultural output in the Soviet Union by forty percent and in China by twenty percent (currently chill latitudes would suddenly have growing seasons), while aiding reforestation worldwide. One IPCC committee projected that on balance gains in agriculture and forest growth would outweigh losses of coastal areas owing to sea-level rises.

GLOBAL WARMING: VEILED?

Perhaps the most important question, raised by Pat Michaels of the University of Virginia, is why the industrial era's output of greenhouse gas hasn't yet triggered runaway heat.

One possibility is that scientific understanding of the greenhouse effect is fundamentally mistaken. Another is that nature's climate control systems are more resilient than currently assumed. Nature manages greenhouse gases by having carbon dioxide inhaled by plants and absorbed by ocean life; research increasingly suggests that biologically moderated greenhouse cycles help keep Earth's climate steady and temperate. Plant decay, volcanic seepage, and other natural processes annually add about 200 billion tons of carbon dioxide to the atmosphere; human activity accounts for 7 billion tons, perhaps a small enough amount that nature can for the moment handle it.

A third possibility is that we haven't reached the greenhouse tipping point. Environmentalists urgently warn that atmospheric carbon dioxide has increased by twenty-five percent in the last century, which sounds like a guarantee of woe. They don't add that the carbon dioxide background level is about 290 parts per million (0.029 percent), so that even a huge relative increase in carbon dioxide has little absolute effect on the amount in the air.

Next, it's possible a greenhouse would be occurring but is being held in check by some other influence. One candidate is "smog block." Industrial activity has put lots of smudgy pollutants into the air, which may be blocking some of the sunlight that would otherwise reach the ground—a kind of nuclear frost. Evidence to support this view comes from data suggesting that temperatures in the Northern Hemisphere, where the industrial activity is, have not risen, while those in the Southern Hemisphere have. Emissions from volcanic activity, up recently, might also be creating a natural block.

Or maybe warming is being temporarily inhibited by the oceans, whose enormous mass tends to smooth out global temperature swings. Since the middle decades of this century were chilly, this line holds, the oceans cooled; now they are giving up their coolness by drawing extra heat from the air, and once they become warm—look out.

Then there is the incredibly annoying possibility that acid rain prevents global warming. Thomas Wigley and other scientists have produced studies suggesting that aerosol particles of sulfur dioxide increase the "albedo," or reflectivity, of Earth's cloud layer, causing more solar heat to bounce back into space. If this theory is correct, acid rain reductions without greenhouse gas control might backfire.

Finally, it's possible that global temperatures can be skewed in either direction by small variations in the output of the sun. Our star is nearing the peak of an unusual sunspot cycle; whether that has any relationship to its energy production, no one knows. In fact precious little is known about the internal processes of the sun. A paper by the astronomer-SDI [Strategic Defense Initiative] salesman Robert Jastrow suggesting that solar variability is the driving force in climate change has had considerable influence in the Bush White House. . . .

HEALTH SCORECARD

Though environmentalists predicted it would get much, much worse, U.S. public health has shown steady improvement in recent decades. AIDS aside, incidence of stroke, heart disease, and hypertension have declined while the life expectancy continues to lengthen. A revealing statistic is that if the mortality rates from 1940 applied to 1988, 4 million Americans would have passed on in 1989. Instead, 2.2 million died. This represents a spectacular net public health improvement occurring during the very period when the manufacture and use of toxic chemicals, dangerous machines, and nuclear materials expanded exponentially.

Considering that gross industrial pollution was far worse in the past than it is today, if the repeatedly predicted pollution-induced health crisis were going to occur, it should already be in evidence: a huge demographic cohort of Americans exposed to gross pollution from the turn of the century through the 1960s having reached the point in life where disease is typical. Yet the age-adjusted incidence of cancer has increased only slightly. Would public health have improved even more with greater environmental regulation? Almost certainly. This, in fact, is the best argument for substantial new investments in pollution control. . . .

NUCLEAR POWER

No one has ever been killed by a nuclear power plant in the U.S. Using coal for electric power annually condemns to death 101 U.S. miners, the average for the last decade; the mining leaves huge gashes in the Earth. Nuclear generators emit no greenhouse gases, acid rain, or smog; they displace fossil fuels, which we ought to be conserving for our grandchildren.

Nuclear power has many problems. But this is a reason to shut down the bad plants, not to oppose new, technically improved facilities, with passive-safety and waste-reducing features. The cost of nuclear plants may soon be reduced by an overdue shift on the part of American manufacturers to standard modular designs, rather than making every installation one of a kind

Waste remains an indelible concern. Even Energy Secretary James Watkins, an old nuclear hard-liner, has said that because of waste, contemporary fission reactors are only acceptable as a transitional power source, till alternatives like fusion or perhaps solar energy collected in space become available in the next century. In theory, fusion reactors could burn seawater, producing almost no byproducts; space solar power could be entirely benign. So how about this for a compromise liberal position on nuclear power: live with fission reactors till we know for sure about the greenhouse effect, or something better comes along.

OZONE DEPLETION

It's real. Even Margaret Thatcher, whose high-latitude nation lies near the North Pole ozone breach, is worried. Stratospheric ozone screens out ultraviolet radiation, which can cause skin cancer and birth defects. Satellite data show that ozone over Antarctica seasonally declines forty percent or more compared with a decade ago. Unanswered is: What does this mean? First it's far from proven that the current degree of ozone depletion harms anyone. Holes have only been observed over the poles where there's hardly any life anyway; and it's not clear that even the UV [ultraviolet radiation] at the poles reaches a level that is dangerous, though some researchers think this will be proved eventually. *Nature* published a study suggesting that regardless of what the ozone layer is doing, observed UV levels at the ground have declined in recent years.

Second, though CFCs [chlorofluorocarbons] unquestionably create a breakdown product that depletes ozone, it's not yet proved they are solely responsible for the holes. Little is known about natural ozone cycles. The destructive CFC by-product is chlorine; natural emissions of chlorine (mainly from volcanos) far exceed artificial quantities. For technical reasons CFC chlorine is more likely to reach the stratosphere than chlorine from volcanos. But with world volcanic activity up in recent years, there may be some relationship between that and the sudden holes.

Third, there may be a solar ozone cycle. Stratospheric ozone is thought to be made when solar energies (including the very ultraviolet radiation we want blocked) strike normal two-atom oxygen molecules, converting them to three-atom ozone variations. Do sun output changes create ozone depletion and restoration cycles? Nobody knows.

This said, humanity can ban CFCs but it can't plug volcanos. The Western nations are committed to a fifty percent reduction of CFC manufacturing by the year 2000. (A bigger deal across the Atlantic since European countries never banned CFCs as a spray-can propellant, as the United States did twenty years ago.) Last year Bush said he would seek a total CFC halt, though he has yet to introduce legislation to this effect. Du Pont, the principal CFC manufacturer, says it will cease production by 2000, offering a new substitute. If CFCs someday prove not to be the ozone villain, we can always go back to them. . . .

PESTICIDES

Concern about synthetic pesticides is now well out of proportion, though it's easy to see why this happened, since memories remain fresh of DDT and other "hard" pesticides that did enormous damage before being banned. The public seems to have forgotten that most pesticides are there for legit reasons: increasing yields so food is cheap, killing off microbes that otherwise would make consumers sick. Those stories about people getting ill from what they bought at the health food store are no joke.

That said, it's crazy not to ban any suspect food chemical that does not have some strong compensating virtue. Alar was an easy choice—who cares if apples are shiny red or only sorta red? Others may be tricky. In the wake of the Alar flap the EPA also banned EDBC, a common fungicide that causes cancer in lab animals. The lowly fungus produces most of nature's deadliest poisons and carcinogens—aflatoxin and a list of others. The net carcinogen content of the U.S. food supply may increase as a result of the EDBC decision.

One positive sign is that the trend in pesticide chemistry is toward compounds that can be used in small quantities ("microdose" pesticides), have narrow ranges of toxicity, and lose their potency rapidly. Great promise is held by the fledgling science of altering plants so that they do not require pesticides. If you're horrified by the thought of genetic engineering, bear in mind that one of its likely first effects on society is a reduction in food chemicals.

POLLUTION SCORECARD

Since the first Earth Day in 1970 most measures of U.S. ecological quality have improved, not declined. Direct industrial water pollution is down significantly; the Great Lakes and some other water bodies are recovering. (Groundwater pollution is more of a question.) Though old toxic waste dumps identified under the Superfund program continue to bedevil everyone, creation of new dumps has essentially ceased—an easily overlooked accomplishment. Development of wetlands has slowed; growth, in general, is less likely to come at the expense of land or vistas better preserved. Airborne sulfur and particulates, the two leading varieties of air pollution in 1970, have fallen sharply; lead, the worst atmospheric poison, has nearly disappeared from the U.S. sky; though smog, an index of American prosperity as measured by per capita auto ownership, has increased slightly, the rate of increase is below that of the increase in autos and economic output, indicating that with further effort smog can be bested, too. . . .

REVERSIBILITY

Though commentators often claim human abuses cause aspects of the environment to be "destroyed," nearly every environmental mistake is reversible—except extinction. Sometimes reverses can come surprisingly fast. Lake Erie and the Thames River, both pronounced "dead" in the 1960s, are already nearly thriving again merely through moderate pollution control. The bald eagle may be taken off the endangered species list a decade after many said there was no hope of averting extinction. Within ten years it should be impossible to determine where the *Exxon Valdez* spill even occurred. Whenever the ecology is pronounced "destroyed" by a transitory event, bear in mind nature is an elaborately defended fortress that has been repelling assaults for four billion years.

RESPONDING JOURNAL

Explain how hopeful you are after reading Easterbrook's essay. Do his arguments convince you that the environment is actually in pretty good shape? Respond in a journal entry of several paragraphs.

QUESTIONS FOR CRITICAL THINKING

STRATEGY

1. Look at Easterbrook's tone. How does he establish a relationship with his reader? Point out places where he uses a device to set up your sense of him as a writer. Do you respond positively or negatively to such practices? Why?
2. Easterbrook ends two paragraphs with references to "fifteen years ago." What is the effect of this repetition?

ISSUES

1. Easterbrook maintains that "grumps" say recycling is a fraud. Is it only grumps who do not believe that recycling pays off? What other people seem to dismiss recycling? What sorts of reasoning do they offer for their positions?
2. Think about Bandow's final comment suggesting that nature is a four-billion-year-old fortress. Does his choice of image bother you? How effective is it? Explain your response.

COLLABORATIVE WRITING ACTIVITY

Given the nature of Easterbrook's comments, you might be more confused than ever in terms of what to think about the deterioration of the environment. Provide two examples of key areas you feel you need to examine thoroughly in order to satisfy yourself that Easterbrook has represented his case accurately. When you bring this work to your peer group, you can begin the work of deciding what sort of evidence you would require to be satisfied that Easterbrook has argued his point successfully. Your recorder can list those requirements your group has generated for such categories as acid rain and the ozone hole, and then you can decide which of the categories would work best for a test run. Then write a group statement, putting it in the form of a letter to Easterbrook, of what sort of argument he would have to provide to satisfy your group.

SUGGESTIONS FOR FORMAL WRITING ASSIGNMENTS

1. Write an essay in which you outline a trip that will provide significant new knowledge about a particular region. You might think about how the land or the people deserve to be better known or better served than they are at present. As you write, be aware of the need to take a writer's stance that will help your readers to see how they might revise the opinions they presently hold.

2. "But on the life there, the unforgettable life, modern America has laid a greedy, vulgarizing hand." Mary Austin's comment from her autobiography *Earth Horizon* indicates a strong emotion, but how fairly can we apply the idea to any particular region of the country? Can we argue that the progress of civilization in America in the twentieth century has meant a vulgarization of the natural world? In a substantial essay, explore this idea in relation to the region where you live or go to school. Does the development carried out in the twentieth century constitute a vulgarization? Does simple greed characterize the nature of American development? Explore in your essay the sort of life that might seem to be harmed by modern development. You might want to address your parents' generation in this essay to inform them of how you feel; in this way you might help them to see how they might consider revising their ideas about the world they inhabit.

3. Draw a parallel between yourself and an animal of your choice. What are the most important characteristics of that animal? How does the animal's character change your self-image?

PART THREE

Rhetorical Strategies for Writing

Chapter 5

NARRATION AND DESCRIPTION: DIVERSE CONCEPTS OF "HOME"

Dorothea Lange, *Childress County, Texas, June 1938*

1. What is the effect of the positioning of the house within the photograph? How does its position order or arrange the composition?

2. Describe the visual details the picture presents you. Beginning with those closest to you, explain what you see to someone who cannot view the photo.

3. Imagine the family for whom this house was home. What would their lives have been like before the years of drought? What would the process of leaving have been like for them?

Chapter 5

NARRATION AND DESCRIPTION: DIVERSE CONCEPTS OF "HOME"

Narration and description are two of the most frequently used forms of development and organization. Narrative means the telling of a story, an account of events or feelings that has a beginning, middle, and end. In other words, a story is a progression that carries its readers or listeners along toward a conclusion. Description is a mode of writing that depends on the accumulation of detail for the purposes of constructing a strong sense for the reader of something—a person, place, thing, or process—that the writer wishes to explore at some length.

It might help to visualize the two forms. Narration represents a development, a trajectory:

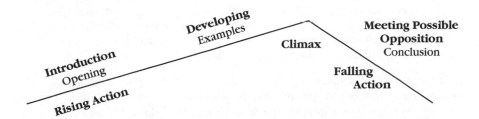

On the other hand, description resembles a painting or photograph, a series of details that the viewer would notice. In the reproduction of Dorothea Lange's photograph "Childress County, Texas, June 1938," see how many of the details you can list. After you have listed all you see, discuss with your classmates or in your collaborative writing group the possible relations among the details you noticed. Which details seem inconsequential at first but later seem to carry significant meaning?

In order to see how expository writing can operate to describe and narrate, look at Pete Hamill's description of his hallway at 378 Seventh Avenue. Hamill's method in this passage reveals his remembered home through the use of narration with the reinforcing power of description. Consider the structure of the first two paragraphs as Hamill moves from precise details toward a more idiosyncratic and personal response as he imagines what the scary details of the hall lead him to believe. Describe his organizational pattern: what sort of narrative line, or "story," does he use to create "Home—378 Seventh Avenue"?

Working individually, with your class, or in collaborative writing groups, discuss first the sequence of events in Hamill's account by applying the diagram from the first page of this chapter. Does he readily signal the different parts of his narration? Next, discuss the detail Hamill provides in order to make his narration more effective. What details work most effectively? You might try the five W's (plus H) as an aid in your discussions. Does Hamill cover all of these, or does he seem to skip some?

Other writers may feel that the detail they require is often more directly descriptive. For illustration, consider the first paragraph of N. Scott Momaday's *The Way to Rainy Mountain.*

> A single knoll rises out of the plain in Oklahoma, north and west of the Wichita Range. For my people, the Kiowas, it is an old landmark, and they gave it the name Rainy Mountain. The hardest weather in the world is there. Winter brings blizzards, hot tornadic winds arise in the spring, and in summer the prairie is an anvil's edge. The grass turns brittle and brown, and it cracks beneath your feet. There are green belts along the rivers and creeks, linear groves of hickory and pecan, willow and witch hazel. At a distance in July or August the steaming foliage seems almost to writhe in fire. Great green and yellow grasshoppers are everywhere in the tall grass, popping up like corn to sting the flesh, and tortoises crawl about on the red earth, going nowhere in the plenty of time. Loneliness is an aspect of the land. All things on the plain are isolate; there is no confusion of objects in the eye, but *one* hill or *one* tree or *one* man. To look upon that landscape in the early morning, with the sun at your back, is to lose the sense of proportion. Your imagination comes to life, and this, you think, is where Creation was begun.

Rather than overwhelming you with an abundance of detail, Momaday offers his reader a sparse landscape, but one in which each concrete item is treated carefully and occupies a definite space in Momaday's plan. He has chosen and arranged his details with great care. How do the descriptions of summer heat prepare you for Momaday's comments on loneliness?

Or look at the definition of loneliness Linda Flowers provides in her selection. Has she helped you—by means of the description in her first paragraph—to understand what "throwed away" means in eastern North Carolina? In describing "good country people," Flowers offers a quick thumbnail sketch of their development across several decades by providing the brand-name detail that suggests their situation in the 1990s. She writes:

> This was a generation the like of which will never be seen again; they were the last of that breed of farmer and hand who had started out during the Great Depression. Their sons and daughters were the first taking up public work, and their grandchildren, the first in an entire lineage never to have known the land

at all. Housewives in the 1960s wore pedalpushers and went to the beauty shop. If they didn't yet work at the sewing plant or Hamilton-Beach, they kept up with "As the World Turns" and "The Edge of Night," and they'd put supper on the table some nights out of a can. The men would wear overall pants and low-cut shoes, wide white belts and slacks when they dressed up, Mennen Skin Bracer maybe, or Old Spice; they'd go when they could to Topsail [a beach on the North Carolina coast] fishing (calling it "Topsl"), or they'd take the family to the drive-in at Mount Olive where, in the 1950s, it had cost a dollar, but a dollar a head in the 1960s.

Notice how Flowers is able to maintain her point of view; she never flinches for a moment as she describes the life she knew as a child, never takes her eye for a moment off the precise details of the people she knows so well. Yet she never condescends, never accuses, simply explains as candidly and clearly as possible how her beloved part of the world is changing. In so doing, she gives her readers a means of measuring the world they know.

Consider for a moment the highly visual nature of the writing in this chapter. Do the photographs in this book help you to imagine what the Oklahoma landscape must look like? Does the Lange photograph at the opening of this chapter suggest the same dryness that Momaday describes? Or does the Lange photo *Westward to the Pacific Coast on U.S. 80* (p. 16) help you to think about the solitary, isolated quality that Momaday evokes for you? Effective description is often as visual as these photos because it creates a mental image in three dimensions which accurately reproduces the picture the author strives to share.

You may recall that Steinbeck in the *Harvest Gypsies* selection clearly tries very hard for most of that essay to maintain an objective point of view, but the horror of what he is experiencing causes him to abandon his objectivity in his final paragraph. In this way, he uses point of view to emphasize his emotion. Wakatsuki Houston also controls point of view carefully in her presentation. How do you account for the resignation with which her family, and particularly her father, regarded the internment at Manzanar? Where do you see instances in which her careful modulation of descriptive details emphasizes the calm her family forced on themselves?

These two examples also suggest how both narrative and descriptive writing seek to make a dominant impression on the reader; long ago, Edgar Allen Poe noted the importance that a single, unified effect could have on a reader. Steinbeck puts up with the deprivation he observes as long as he can, but finally the human overcomes the rational and he suggests in an emotional aside that ill results will follow ill treatment of the migrants. Wakatsuki Houston carries us along with her as she worries about the potential effect of the internment on her family's spirit. In both selections, the writers carefully control the reader's response through their use of concrete description in the narrative.

In Wakatsuki Houston's description and narration, you see the quick history of a newcomer to a place her family is required to make their home. As a writer, you can develop your description of home in a parallel fashion: you have a sense of home built up over an extended period of time, and you have a sense of other people whose presence helps to define exactly what "home" means to you. Some other ideas of home will also help us think about writing.

> Home is the place where, when you have to go there,
> They have to take you in.
>
>   . . . .
>
> I should have called it
> Something you somehow haven't to deserve.

These lines from Robert Frost's poem "The Death of the Hired Man" provide one sense of home. The speaker in Frost's poem is a generous woman who recognizes the psychological importance of home: home means security and some degree of comfort to most people, but it also means a place where you can always count on being taken in. When faced with the possible loss of such a haven, we grow uncertain, fearful, or resentful because we feel threatened by the specter of losing this tie to the physical world. This problem can be compounded if we feel we have been forcibly or unfairly moved from our home place. If we asked Wakatsuki Houston what her dominant image of home from her childhood might be, she might tell us about the rock garden her father arranged, or the image of her mother in her starched yellow hat coming down the dusty road. What would it be for you? Is that image a happy or a sad one for you today?

Allied here to a sense of home is the idea of displacement, of the need to move away from the comfort and security of home to a place that may be far less receptive. Displacement as an American phenomenon in relation to the land began with the earliest interlopers from Europe. When indigenous peoples were forced to find new living spaces as alternatives to persecution or destruction, they began to display the psychological discomforts characteristic of forced removal from an accustomed spot—or home. Such relocations have not stopped. In the selections that follow, many writers locate the qualities necessary to a sense of home in the landscape. As you read these, think about how your own feelings of home and landscape are challenged.

"Home—378 Seventh Avenue"
Pete Hamill

Born in 1935 in Brooklyn, New York, Hamill determined early in life to become a writer and to depict faithfully the gritty, urban New York life he knew from real-world experiences. After various manual labor stints, he became a reporter for the Saturday Evening Post *and won awards for his coverage of the war in South Vietnam. The following selection, from his book*

The Gift, *published in 1973, demonstrates Hamill's ability to render detail in such a way as to make his readers feel as if they are themselves experiencing that scene.*

This talent serves Hamill well in other ways: he has become a highly regarded writer of screenplays, justly known for his ability to evoke the life of New York. His writer's sensibility allows him to focus as easily on everyday human drama as on the most notorious gangland crimes. Hamill's sense of New York life moves beyond mere newspaper reporting, however; he has a keen ear for the inflections of speech among classes of New Yorkers, a well-informed sense of the principles guiding the New York underworld, and the eye for visual detail that makes us feel we ascend the stairs at 378 Seventh Avenue with him.

EXPLORING JOURNAL

Make a list of all the ideas that occur to you when you see or hear the word *home.* After you have brainstormed for at least five minutes, take a look at your list and put a check by those entries that are *general,* or applicable to other people's definitions of home. The others, then, will be details that are particular to you: relatives and family members, details of specific places where you have lived, and so on. Now that you have made this division, perhaps you have thought of other possible entries. Record them as well.

The house was at 378 Seventh Avenue. There was a small butcher shop to the left and Teddy's Sandwich Shop to the right, and when I went in, I saw that the mailbox was still broken and the hall smelled of backed-up sewers and wet garbage. There were, of course, no locks on the doors, and I stood for a while in the yellow light of the thirty-watt bulb, and shifted the sea bag to the other shoulder. Two baby carriages were parked beside the stairs, and in the blackness at the back of the hall, I caught a glimpse of battered garbage cans. A small shudder went through me; the back of that hall had always been a fearful place when I was small, a place where I always felt vulnerable: to sudden attacks from the open door leading to the cellar, to rats feasting on the wet garbage, to unnamed things, specters, icy hands, the vengeance of God. Once, I'd had to go to the cellar late at night. To the right, inside the cellar door, there was a light switch, covered with a ceramic knob. I reached for the knob and it was gone, and there was a raw wire there instead and the shock knocked me over backwards, into the garbage cans, my heart spinning and racing away, and then rushing back again. I thought of that night trip, the strangeness later when I realized for the first time what it must feel like to die, and I started up the stairs.

It was a hall as familiar as anything I've ever known before or since. First floor right, Mae McAvoy; on the left, Poppa Clark; second floor right, Anne Sharkey and Mae Irwin; left, Carrie Woods. Carrie Woods was a tiny sparrow of a woman, who kept dogs and drank whiskey, and the dogs started a ferocious attack on the locked door, trying to get at me—alarmed, I suppose, by a smell they had not sensed for many weeks. All the apartments had the feeling of tossing bodies within, and I remembered fragments of other nights: the scream when a husband punched out a wife, and how he left and never came back; the glasses breaking at some forgotten party and the blood in the hall later; how they all hated one of the women because she was a wine drinker and therefore a snob; the great large silent man in one of those apartments, who played each Christmas with a vast Lionel electric train set, while forcing his only daughter to play at an untuned upright piano, who rooted for the Giants in that neighborhood of Dodger fans, and who had a strange tortured set of eyes. At each landing, there were sealed metal doors where the dumbwaiter once had been, a pit that dropped away, like some bottomless well, to a boarded-over access door in the cellar, and which I thought, when I was eight, was the way to Hell itself, or at the very least, to the secret cave where Shazam granted Billy Batson the magic powers. There were two more baby carriages at the top of the second floor, the floor where my father had so often stopped on his way home, emptied of songs, dry and hoarse, unable to make that one final flight of stairs to bed. Billy Batson. Billy Hamill. Shazam.

There were traces of dinner smells in the hall, as if you could chew the air itself. It was almost three.

The door to our apartment was not locked. I dropped the sea bag, pushed the door open easily, and stepped into the dark kitchen, groping for the light cord in the center of the room. I bumped into a chair, then the table, and then found the light cord. A transformer hummed for a few seconds and then the round fluorescent ceiling light blinked on. The room was as I had remembered it: a white-topped gas range against the far wall where the old coal stove had once stood, a tall white cabinet to the left, and then the sink, high, one side shallow and the other deep, next to the window that never opened. A Servel refrigerator with a broken handle was next to the bathroom door. A closet loomed behind me next to the front door, with a curtain covering the disorder within, and there was a table in the center of the room, linoleum on the floor, and a clothesline running the length of the room because there was no backyard, and in winter the clothes froze on the line on the roof. There was a picture of Franklin Roosevelt on one wall, a map of Ireland from the *Daily News* on another, and beside it I saw some of the drawings I had sent from boot camp. Some of them were cartoons, drawings of soldiers and pilots I had copied from Milton Caniff; the others were something new, drawings of sailors' faces, done in ink washes, the first drawings I had made that didn't look like comic-strip figures. Roaches scurried across the table, panicked by the harshness of the sudden blue-tinged light. I could hear movement at the other end of the railroad flat, the smell of heavy breathing and milk, and then my mother was coming through the rooms.

"Oh Peter," she said "You're home."
And she hugged me.

RESPONDING JOURNAL

Give yourself the chance to play with tone and point of view a little bit. Choose a persona—a social worker, a private detective, a burglar—and rewrite part (or all) of Hamill's piece from this new point of view. Try keeping his sentence structure intact, but modifying it to fit the new speaker.

QUESTIONS FOR CRITICAL THINKING

STRATEGY

1. Explain Hamill's plan in writing this essay. How does he guide you through his relatively brief description? Can you compare this organization to that of a medium other than writing?
2. Since Hamill was trained as a journalist, you can expect him to be well aware of the five W's. Locate the information in this essay that gives you some idea of each category. Which seems most important to Hamill's purpose here?

ISSUES

1. From Hamill's description of his home, what can you say about his neighborhood? Do some of the details he presents seem to indicate how other people in the neighborhood would behave toward one another? Substantiate your response.
2. We know that for most people, home means other people. How does Hamill structure his essay to emphasize the importance of other people in the place that he knows as home? Why do you think he ends with meeting his mother in the kitchen? If you had to write a similar account of home, would you finish in the same way Hamill does? Think of the advantages and disadvantages of such a strategy.

COLLABORATIVE WRITING ACTIVITY

Compare your lists of general terms regarding home that you generated for your exploratory journal. Which terms appear in more than one list? After hearing the lists your peers generated, can you now suggest any new ones? Which seems to be the most promising of those you explored in your group discussion?

"Return to Cannonsville"
Anonymous

Though numerous efforts have failed to locate the anonymous author of this essay, the editor and publisher would be happy to learn the writer's identity so that she may be credited in the next edition of this text. The essay appeared in Catskill Country, *a free magazine running ads for real estate and local recreational services in New York's Catskill Mountain region. The events described in "Return to Cannonsville" took place over a longer period of time than the essay seems to indicate; perhaps this compression of time suggests the young girl's memory and sense of what was important to her at the time. Otherwise the essay is historically accurate—the village of Cannonsville was evacuated by 1968 and the filling of Cannonsville Reservoir was begun soon afterward.*

EXPLORING JOURNAL

Think about how the various media have presented different concepts of home to you. List as many as you can. Generalize about categories that seem to arise from your listing, and do a focused freewrite on these general ideas.

Last night I dreamed I was back in Cannonsville. Imagine. It isn't the first time I had that dream. I dreamed it a lot of times since we left, but I thought I was all over it by now. Sometimes I think I will never forget, though it has been how many years?

I was sixteen when it all started. Seems like only such a short time instead of twenty some years. I used to go with Daddy to the creamery in the old Dodge every morning. I loved to load the empty milk cans on the truck as they came back out after being washed. Dad would go in the creamery and I'd have the cans all loaded and tied up when he came out.

Then one morning George Pepper was telling all the farmers he heard they were going to dam up the West Branch and make it a reservoir like they did at Pepacton a few years back. We laughed. George was a great one for spreading rumors, always getting excited over nothing. But this wasn't nothing, it was something, and things moved fast after that terrible morning that changed all our lives.

First the surveyors, then the lawyers, the politicians, the appraisers. Suddenly the valley was full of strange people in fancy clothes everywhere, in Mr. Nelson's store, in the post office, in the street. And there were official notices on the telephones explaining that New York City was taking our farms in order to increase their water supply.

I guess if anybody had thought about it they would have figured out that some day it would happen. Our long, narrow valley with the Delaware rolling down through the middle, high hills on both sides, just a perfect spot to dam the river and close enough to New York City to be piped down there, but nobody ever thought about it.

I remember the happy times along that old river, daddy plowing the long meadow in the spring and brother Jim and I fishing and swimming in that lovely water. And the happy times we had at the Grange, and in the church, where we used to go every Sunday, the picnics, strawberry festivals, halloween parties, something always happening. We were such a happy family, but I guess you never know till it's gone.

Mom refused to listen to any talk about the dam at all. "I don't want to hear about the damn dam," she used to say. Mom was born on the farm and dad worked for grandpa, and finally married mom, so ma was really the boss about things like that. She said nobody was going to take her farm away and she would get the shotgun and chase them off if she had to. But when the appraisers came with papers to sign, mom ran upstairs and locked herself in the bedroom and bawled like I never heard her before. I didn't think mom ever cried, but she sure let it out that day.

"Mr. Hanson, I think we have been very fair with you, we have given you every dollar your farm is worth."

Dad looked up the stairs to where mom was locked in the bedroom and said, "How much do you pay for a broken heart?"

Things happened fast after that, auctions all around, our beautiful herd of Holsteins sold for almost nothing. There were so many people selling out, and butchers jamming our pet cows with electric prods to load them in their trucks. I ran behind the barn and cried worse than mom and that was the night daddy had his heart attack and he was dead in a week, and we couldn't even bury him in our local churchyard where all our people were buried. They had to take him clear to Walton to some strange cemetery I never saw before, because the church was being torn down and the graves had to be moved.

Well, mom moved to Deposit to an apartment over a store. Imagine, after her whole life on the farm, being cooped up in an apartment. Jim got a job in Pennsylvania, and I hardly ever see him. And me, I married Eddie Mancini, they had the farm just below us on the river and his folks and ours used to change work year after year. I guess that's how we fell in love, working in the hay mow in the big red barn, and sometimes when nobody was around we would forget about mowing and be making out in the back mow. Jim caught us once but he never told.

We moved to Delhi and Eddie got a job on the farm but he finally went to work in the college. It is a pretty good job, but sometimes we'd give anything to be down in Cannonsville Valley. I just wish I could stop dreaming about it.

RESPONDING JOURNAL

Do a ten-minute freewrite beginning with the idea of "happy valley." What does this phrase mean to you? Does it suggest anything in your own experience, or is it the sort of idea that occurs in stories but not in real life?

QUESTIONS FOR CRITICAL THINKING

STRATEGY

1. What effect does this essay have on you as a reader? What effect do you think the author intended? Do you find yourself sympathizing with the author's situation? If you do, explain how you think the writer was able to involve you in this way.
2. Examine the way in which the writer contrasts her situation after Cannonsville with that of her mother. Suggest any possible ways you feel might have shown these differences in an even more striking manner.

ISSUES

1. Explore how this essay uses memory. Both positive and negative memories appear; how does the writer decide to balance the two of these? Are all memories, whether positive or negative, solely functions of what happened and thus random? Explain your response. Or does memory preserve the most intense moments? You may want to discuss this issue with your classmates.
2. Explain the change in Mrs. Hanson. At first she is determined not to hear about the "damn dam," but later she seems to give up rather easily. What might account for such a difference?

COLLABORATIVE WRITING ACTIVITY

Some people who lived in Cannonsville Valley had to move once for the reservoir, and then had to move again when their new residences to the north of the valley were taken so that Interstate 88 could be built. Though this "double jeopardy" is rare, it does indicate a recurrent situation in America. Individually, begin to compile a list of situations in your own community that have led to development and changed patterns of residence. Have such instances of "progress" or uses of eminent domain changed people's lives significantly? Reconvene as a group to share lists and discuss the extent of such displacements. Record in your journal a final summary paragraph of what you have learned.

from *The Way to Rainy Mountain*
N. Scott Momaday

*N. Scott Momaday, of Kiowa and Cherokee ancestry, has achieved consider-
able fame for his prose and fiction. His first novel,* House Made of Dawn, *won
the Pulitzer Prize for fiction in 1969. In the same year, his collection of essays
entitled* The Way to Rainy Mountain *appeared. In all of his work, he combines
Native American myth and history with his own original perspective to present a unique voice in American literature, one that evokes a fascinating, if
elusive, landscape.*

*Momaday frequently combines elements of Native American traditions
with those of the dominant Euro-American population; in areas such as religion and language, he achieves rich blends. In the essay presented here, he
combines autobiography with family and Kiowa history to create a haunting Oklahoma landscape as a backdrop for the account of his pilgrimage to
visit his grandmother's grave.*

EXPLORING JOURNAL

Freewrite for ten minutes on the subject of Native American "Indian reservations." What are the implications of such a descriptive term? What do you know
about such places? Has your concept of these reservations changed as you have
grown older? Look back over what you have written to see if you have suggested any promising leads for further development.

A single knoll rises out of the plain in Oklahoma, north and west of the Wichita
Range. For my people, the Kiowas, it is an old landmark, and they gave it the
name Rainy Mountain. The hardest weather in the world is there. Winter brings
blizzards, hot tornadic winds arise in the spring, and in summer the prairie is
an anvil's edge. The grass turns brittle and brown, and it cracks beneath your
feet. There are green belts along the rivers and creeks, linear groves of hickory
and pecan, willow and witch hazel. At a distance in July or August the steaming foliage seems almost to writhe in fire. Great green and yellow grasshoppers
are everywhere in the tall grass, popping up like corn to sting the flesh, and tortoises crawl about on the red earth, going nowhere in the plenty of time.
Loneliness is an aspect of the land. All things in the plain are isolate; there is no
confusion of objects in the eye, but *one* hill or *one* tree or *one* man. To look
upon that landscape in the early morning, with the sun at your back, is to lose
the sense of proportion. Your imagination comes to life, and this, you think, is
where Creation was begun.

I returned to Rainy Mountain in July. My grandmother had died in the
spring, and I wanted to be at her grave. She had lived to be very old and at last

infirm. Her only living daughter was with her when she died, and I was told that in death her face was that of a child.

I like to think of her as a child. When she was born, the Kiowas were living the last great moment of their history. For more than a hundred years they had controlled the open range from the Smoky Hill River to the Red, from the headwaters of the Canadian to the fork of the Arkansas and Cimarron. In alliance with the Comanches, they had ruled the whole of the southern Plains. War was their sacred business, and they were among the finest horsemen the world has ever known. But warfare for the Kiowas was preeminently a matter of disposition rather than of survival, and they never understood the grim, unrelenting advance of the U.S Cavalry. When at last, divided and ill-provisioned, they were driven onto the Staked Plains in the cold rains of autumn, they fell into panic. In Palo Duro Canyon they abandoned their crucial stores to pillage and had nothing then but their lives. In order to save themselves, they surrendered to the soldiers at Fort Sill and were imprisoned in the old stone corral that now stands as a military museum. My grandmother was spared the humiliation of those high gray walls by eight or ten years, but she must have known from birth the affliction of defeat, the dark brooding of old warriors.

Her name was Aho, and she belonged to the last culture to evolve in North America. Her forebears came down from the high country in western Montana nearly three centuries ago. They were mountain people, a mysterious tribe of hunters whose language has never been positively classified in any major group. In the late seventeenth century they began a long migration to the south and east. It was a journey toward the dawn, and it led to a golden age. Along the way the Kiowas were befriended by the Crows, who gave them a culture and religion of the Plains. They acquired horses, and their ancient nomadic spirit was suddenly free of the ground. They acquired Tai-me, the sacred Sun Dance doll, from that moment the object and symbol of their worship, and so shared in the divinity of the sun. Not least, they acquired the sense of destiny, therefore courage and pride. When they entered upon the southern Plains they had been transformed. No longer were they slaves to the simple necessity of survival; they were a lordly and dangerous society of fighters and thieves, hunters and priests of the sun. According to their origin myth, they entered the world through a hollow log. From one point of view, their migration was the fruit of an old prophecy, for indeed they emerged from a sunless world.

Although my grandmother lived out her long life in the shadow of Rainy Mountain, the immense landscape of the continental interior lay like memory in her blood. She could tell of the Crows, whom she had never seen, and of the Black Hills, where she had never been. I wanted to see in reality what she had seen more perfectly in mind's eye, and travelled fifteen hundred miles to begin my pilgrimage.

Yellowstone, it seemed to me, was the top of the world, a region of deep lakes and dark timber, canyons and waterfalls. But, beautiful as it is, one might have the sense of confinement there. The skyline in all directions is close at hand, the high wall of the woods and deep cleavages of shade. There is a perfect freedom in the mountains, but it belongs to the eagle and the elk, the badger and the bear. The Kiowas reckoned their stature by the distance they could see, and they were bent and blind in the wilderness.

Descending eastward, the highland meadows are a stairway to the plain. In July the inland slope of the Rockies is luxuriant with flax and buckwheat, stonecrop and larkspur. The earth unfolds and the limit of the land recedes. Clusters of trees, and animals grazing far in the distance, cause the vision to reach away and wonder to build upon the mind. The sun follows a longer course in the day, and the sky is immense beyond all comparison. The great billowing clouds that sail upon it are shadows that move upon the grain like water, dividing light. Farther down, in the land of the Crows and Blackfeet, the plain is yellow. Sweet clover takes hold of the hills and bends upon itself to cover and seal the soil. There the Kiowas paused on their way; they had come to the pace where they must change their lives. The sun is at home on the plains. Precisely there does it have the certain character of a god. When the Kiowas come to the land of the Crows, they could see the dark lees of the hills at dawn across the Bighorn River, the profusion of light on the grain shelves, the oldest deity ranging after the solstices. Not yet would they veer southward to the caldron of the land that lay below; they must wean their blood from the northern winter and hold the mountains a while longer in their view. They bore Tai-me in procession to the east.

A dark mist lay over the Black Hills. At the top of a ridge I caught sight of Devil's Tower upthrust against the gray sky as if in the birth of time the core of the earth had broken through its crust and the motion of the world was begun. There are things in nature that engender an awful quiet in the heart of man; Devil's Tower is one of them. Two centuries ago, because they could not do otherwise, the Kiowas made a legend at the base of the rock. My grandmother said:

> Eight children were they at play, seven sisters and their brother. Suddenly the boy was struck dumb; he trembled and began to run upon his hands and feet. His fingers became claws, and his body was covered with fur. Directly there was a bear where the boy had been. The sisters were terrified; they ran, and the bear after them. They came to the stump of a great tree, and the tree spoke to them. It bade them climb upon it, and as they did so it began to rise into the air. The bear came to kill them, but they were just beyond its reach. It reared against the tree and scored the bark all around with its claws. The seven sisters were borne into the sky, and they became the stars of the Big Dipper.

From that moment, and so long as the legend lives, the Kiowas have kinsmen in the night sky. Whatever they were in the mountains, they could be no more. However tenuous their well-being, however much they had suffered and would suffer again, they had found a way out of the wilderness.

My grandmother had a reverence for the sun, a holy regard that now is all but gone out of mankind. There was a wariness in her, and an ancient awe. She was a Christian in her later years, but she had come a long way about, and she never forgot her birthright. As a child she had been to the Sun Dances; she had taken part in those annual rites, and by them she had learned the restoration of her people in the presence of Tai-me. She was about seven when the last Kiowa Sun Dance was held in 1887 on the Washita River above Rainy Mountain Creek. The buffalo were gone. In order to consummate the ancient sacrifice—to impale the head of a buffalo bull upon the medicine tree—a delegation of old men journeyed into Texas, there to beg and barter for an animal from the Goodnight herd. She was ten when the Kiowas came together for the last time as a living Sun Dance culture. They could find no buffalo; they had to hang an old hide from the sacred tree. Before the dance could begin, a company of soldiers rode out from Fort Sill under orders to disperse the tribe. Forbidden without cause the essential act of their faith, having seen the wild herds slaughtered and left to rot upon the ground, the Kiowas backed away forever from the medicine tree. That was July 20, 1890, at the great bend of the Washita. My grandmother was there. Without bitterness, and for as long as she lived, she bore a vision of deicide.

Now that I can have her only in memory, I see my grandmother in the several postures that were peculiar to her: standing at the wood stove on a winter morning and turning meat in a great iron skillet; sitting at the south window, bent above her beadwork, and afterwards, when her vision failed, looking down for a long time into the fold of her hands; going out upon a cane, very slowly as she did when the weight of age came upon her; praying, I remember her most often at prayer. She said long rambling prayers out of suffering and hope, having seen many things. I was never sure that I had the right to hear, so exclusive were they of all mere custom and company. The last time I saw her she prayed standing by the side of her bed at night, naked to the waist, the light of a kerosene lamp moving upon her dark skin. Her long, black hair, always drawn and braided in the day, lay upon her shoulders and against her breasts like a shawl. I do not speak Kiowa, and I never understood her prayers, but there was something inherently sad in the sound, some merest hesitation upon the syllables of sorrow. She began in a high and descending pitch, exhausting her breath to silence; then again and again— and always the same intensity of effort, of something that is, and is not, like urgency in the human voice. Transported so in the dancing light among the shadows of her room, she seemed beyond the reach of time. But that was illusion; I think I knew then that I should not see her again.

Houses are like sentinels in the plain, old keepers of the weather watch. There, in a very little while, wood takes on the appearance of great age. All colors wear soon away in the wind and rain, and then the wood is burned gray and the grain appears and the nails turn red with rust. The windowpanes are black and opaque; you imagine there is nothing within, and indeed there are many ghosts, lives given up to the land. They stand here and there against the sky, and you approach them for a longer time than you expect. They belong in the distance; it is their domain.

Once there was a lot of sound in my grandmother's house, a lot of coming and going, feasting and talk. The summers there were full of excitement and reunion. Kiowas are a summer people; they abide the cold and keep to themselves, but when the season turns and the land becomes warm and vital they cannot hold still; an old love of going returns upon them. The aged visitors who came to my grandfather's house when I was a child were made of lean and leather, and they bore themselves upright. They wore great black hats and bright ample shirts that shook in the wind. They rubbed fat upon their hair and wound their braids with strips of red cloth. Some of them painted their faces and carried the scars of old and cherished enmities. They were an old council of warlords, come to remind and be reminded of who they were. Their wives and daughters served them well. The women might indulge themselves; gossip was at once the mark and compensation of their servitude. They made loud and elaborate talk among themselves, full of jest and gesture, fright and false alarm. They went abroad in fringed and flowered shawls, bright beadwork and German silver. They were at home in the kitchen, and they prepared meals that were banquets.

There were frequent prayer meetings, and great nocturnal feasts. When I was a child I played with my cousins outside, where the lamplight fell upon the ground, the singing of the old people rose up around us and carried away into the darkness. There were a lot of good things to eat, a lot of laughter and surprise. And afterwards, when the quiet returned, I lay down with my grandmother and could hear the frogs away by the river and feel the motion of the air.

Now there is a funeral silence in the rooms, the endless wake of some final word. The walls have closed in upon my grandmother's house. When I returned to it in mourning, I saw for the first time in my life how small it was. It was late at night, and there was a white moon, nearly full. I sat for a long time on the stone steps by the kitchen door. From there I could see out across the land, I could see the long row of trees by the creek, the low light upon the rolling plains, and the stars of the Big Dipper. Once I looked at the moon and caught sight of a strange thing. A cricket had perched upon the handrail, only a few inches away from me. My line of vision was such that the creature filled the moon like a fossil. It had gone there, I thought, to live and die, for there, of all places, was its small definition made whole and eternal. A warm wind rose up and purled like the longing within me.

The next morning I awoke at dawn and went out on the dirt road to Rainy Mountain. It was already hot, and the grasshoppers began to fill the air. Still, it was early in the morning, and the birds sang out of the shadows. The long yellow grass on the mountain shone in the bright light, and a scissortail hied above the land. There, where it ought to be, at the end of a long and legendary way, was my grandmother's grave. Here and there on the dark stones were ancestral names. Looking back once, I saw the mountain and came away.

———

RESPONDING JOURNAL

In a well-developed paragraph or two, explain why Momaday uses the title that he does. Indicate whether you think he deliberately creates an ambiguity in his choice of words as they relate to the events he describes.

QUESTIONS FOR CRITICAL THINKING

STRATEGY

1. Make a list of the words Momaday uses in his opening paragraph to establish a sense of setting for his reader. Which do you find most effective?
2. Look at the paragraph beginning "My grandmother had a reverence for the sun" and decide how Momaday moves through to the last word the paragraph, "deicide." What does this word mean? How has Momaday invested it with special importance? Can you use this paragraph as a model in miniature for the entire selection? Discuss these ideas with your classmates.

ISSUES

1. Explain the various readings we might get from the work's title, *The Way to Rainy Mountain*. Relate your readings to the figure of Momaday's grandmother.
2. Look at the paragraph beginning "Houses are like sentinels in the plain, . . ." Does this paragraph contribute directly to the narrative, or does it appear in Momaday's account of his grandmother's house for another reason? Does this function remind you in any way of other writing about the sense of home?

COLLABORATIVE WRITING ACTIVITY

Try this exercise. On the left side of a sheet of paper, list those details of landscape that Momaday mentions. Do not try to put them in any sort of order unless you want to do so in a second list. On the right side of that sheet, suggest briefly what his reason for mentioning that detail might be. Compare your lists with those of others in your group and create a master list to be shared with the rest of the class.

from *Throwed Away*
Linda Flowers

Linda Flowers is the daughter and granddaughter of tenant farmers in Faison, North Carolina, the eastern North Carolina town in which she finds much of the material for her study Throwed Away: Failures of Progress in Eastern North Carolina. *In this riveting book, she traces the changes in her world that began slowly in the 1950s, intensified through the 1960s, and by the early 1970s transformed forever the world she knew as a child. She looks not only at the changing landscape and the lives of the tenant farmers who depended on that land for their livings, but also at the system of progressive education that simply did not provide adequately for the different world that emerged by the 1970s.*

In her evocative and haunting prose, the reader catches on occasion echoes of the writing of Agee or Faulkner, two of Flowers' favorite writers. She thus continues the tradition of Southern writers who draw strength from their native soil. Upon graduating from high school, she was able to attend college solely because the University of North Carolina at Greensboro provided her a full scholarship. After graduate school and teaching in the midwest and northeast, she returned to eastern North Carolina from a twelve-year absence. Like Antaeus, her renewed contact with her native land reinvigorated her; Throwed Away *appeared in 1990.*

EXPLORING JOURNAL

In a freewriting entry for your journal, explain your ideas about sharecropping and tenant farming. How much do you know about this? Where did you acquire the knowledge you have? How much of your knowledge is accurate? You may want to check in the library if you are unsure about the accuracy of your information.

PREFACE

"Throwed away" is an expression peculiar to eastern North Carolina. If a piece of land or a person or a stretch along the highway looks "throwed away," it can be in no worse shape. Fields left unattended and overcome with cockleburs are "throwed away." Ramshackly houses with boarded-up windows and rotten porches, or country stores that have bitten the dust are "throwed away." A man and woman are "throwed away" when they've outraged morality; a divorcée may "throw herself away." With less force the term can, however, refer to nothing more ominous than simply a feeling of depression; a woman with the blues might say she feels "plumb throwed away." To say that a friend looks "throwed away" is to declare in the plainest yet most solicitous fashion that he or she looks miserable.

The expression is pejorative, though often but mildly—sadly—so. Used to refer to one's self or a loved one, "throwed away" may convey little more than bemused exasperation, perhaps even affection. Tone is everything. By appropriating the phrase I mean to pay tribute to that capacity my mother's generation had of driving home an idea forcefully with colorful language. Not everybody in eastern North Carolina, of course, is throwed away, but for those who are, no other term will serve.

HOME

AN INTRODUCTION

The land lies off to the east of I-95, a far-flung quarter moon or a lazy dogleg if you think of starting up in Northampton and Halifax counties, then bearing south down through Nash and Edgecombe, Wilson and Johnston and Wayne, Duplin and Sampson. Fayetteville and Wilmington, Kinston and Greenville, Scotland Neck and New Bern: all are more or less in the vicinity, though not in these counties. From the north, you enter North Carolina for the most part unaware that you've left Virginia; Richmond is but an hour or so behind, and between Richmond and Roanoke Rapids—Emporia, Virginia and Weldon, North Carolina—there's not much to see from the Interstate. Gas stations, the places selling hams and guns and cigarettes, the outlet or two once you cross the state line, not much more.

But from the south, if you pick up I-95 say out of Savannah, then follow it through Florence and Dillon, the entry into the state takes you through vast reaches of land that hearken to an era you would have thought had long since vanished. You see cotton fields, old tenant houses off in the distance now falling down, and roads that run straight for miles through standing timber thinned now by the logging and pulpwood people. And from late April and May through late September, the sun beats down with merciless abandon, the pavement before you, the tin of barns and housetops, the occasional other car or pickup truck shimmering in the heat.

Coming out of New York and Pennsylvania, the Baltimore–Washington area, then Northern Virginia, the farther south you drive the less heavy industry you see. There's the gradual awareness that farms are smaller, the crops different, and the land flattens out as you get into southern Virginia and North Carolina, tidewater and coastal plain supplanting the higher elevation of Pennsylvania. But change direction and head north from Mississippi and through Tuskegee and Macon and Milledgeville, skirting Atlanta altogether, do this and the perspective shifts: now the Commerce Department's designation of eastern North Carolina as "Gold Leaf Urban" or "Highland Plains Urban" will strike you as much less fanciful.

Firestone, Monsanto, Crown-Zellerbach, Burroughs-Welcome, Ingersoll-Rand, Rockwell International, Campbell Soup, General Signal, Cummins, and other national firms have in the last ten or fifteen or twenty years set up major operations in the region. Alongside the textile mills, DuPont and J.P. Stevens, are the smaller cotton mills that once were the mainstay of the towns growing up around them, and the tobacco companies and warehouses for which everybody knows us. These newer industries do much to give the area a look of having been brought closer within the mainstream of American working life. Now in the mid-1980s even this part of the state—the "Down East" region forever caricatured as backward, irredeemably insular and conservative, redneck—has a look and feel inexplicably both different from and the same as that I remember as a child growing up on a tenant farm outside Faison, in Duplin County, forty and thirty and twenty years ago.

East of I–95 is, however, a long way from the more urban, more heavily industrialized and congested Central Piedmont. If Raleigh and Durham and Greensboro, the Research Triangle Park, and those colleges and universities whose names are recognized nationally are closer now to Faison than they used to be, still they are a good ways off; and for some people still, they are as distant, even as unimaginable, as once they were to most of the area's shirt-tail farmers, tenants, small businessmen, women, and schoolchildren. Nobody I knew back then, in Faison, ever thought of Durham or Chapel Hill unless somebody was in the hospital there, was "bad off," this is to say, or else he would have been at Goldsboro or Clinton or, later, Kenansville. To be taken to Raleigh on a class trip in the seventh or eighth grade was an excursion for many of us comparable to that the seniors took the years they went to Washington, D.C.

If some people shopped at Cameron Village or Crabtree Valley Mall in Raleigh, my mother never did, nor did farm women generally. They would go to Goldsboro in August or September when the last of the tobacco had been sold, and sometimes to Mount Olive or Warsaw in the spring; other than this, whatever else that had to be bought came out of the catalogue: Sears or Montgomery Ward, Aldens or National Bellas Hess. Even biddies the mailman would deliver, their cardboard crates high in the back seat of his car, holes in them the size of quarters.

Except for the places on the Interstate catering to tourists, probably a little less is done here than farther south to play up the clichés still governing outsiders' conception of the South. Confederate flags there are, but fewer than you see in Alabama and Mississippi, and there are magnolia trees but nobody pays them any mind. High-school teams may still call themselves "Rebels," but "Dixie" isn't heard much anymore. Well before the North Carolina line, the red clay of south Georgia has given way to sandy, peach-growing flatlands, and the dirt roads of Mississippi and Georgia and Alabama aren't nearly as numerous here. There is less cotton, as well as fewer Bible colleges, more pine trees, far more tobacco fields, roads that may be a little better. The old tobacco barns for curing are abandoned now; bulk barns, the heat forced around and through the tobacco by a new technology, have supplanted them. There aren't as many dish antennae as in Alabama and Mississippi, but neither do there seem to be as many people living in the worst of inadequate houses.

Except around Wilmington, but also occasionally in fields and towns throughout the region, you won't come across as many mansions of the Civil

War era as you will in South Carolina and beyond, although houses dating from the Revolution can still be visited in Edenton and Tarboro, New Bern and Bath.

Nowadays in eastern North Carolina the finest of brick homes are as likely to be found in the country as in the towns; next to them or down the road may be a house with the porch caved in. Trailer parks are all over, as are single trailers off to themselves or in a backyard. Government housing projects have come even to many of the smaller towns. For long stretches in certain areas, there may be no (or very few) houses at all, or much sign of any habitation. In some pockets, what dwellings you do see are pitiable enough; you could be forgiven for thinking the Depression had come again, they look so much like the photographs taken then. Other places, however, you would think had never known anything else but affluence. Poverty and wealth coexist, if not always literally side by side, at least in such proximity as to startle anybody used to seeing the poor (or the black) contained in big-city neighborhoods; here, where life still is essentially rural, unless they live well inside one of the larger towns, rich and poor—and black and white—aren't likely to escape the sight of one another. . . .

3. THE HANDS AND US

Tenant farmers themselves were often in a way like landlords. They frequently hired "hands" on their own, men and women and children who knew them to be the boss, and who may or may not have known that, in fact, another man owned the land on which they had been brought to work. In Faison all during the 1940s and 1950s and into, but not much beyond, the early 1960s, white tenants in the summer drove twice daily into Juniper . . . or into the black sections of Clinton or Mount Olive or Warsaw, even Goldsboro, picking up workers early in the morning and bringing them back at close of day. In front of one ramshackly house after another, they would gather: mamas and those grown or half-grown daughters still at home, some with babies of their own; little girls barely tall enough to reach into the slides to hand out the green tobacco that big girls tied steadily all day to sticks, which then were hung in barns for curing; and young men who worked in the field, sleepy-eyed in the morning and sullen, not to be bothered. Summer work was setting out and suckering and putting in tobacco; picking beans and cucumbers and pepper and squash; chopping; grading dry tobacco when it came out of the barn, Granddaddy Flowers telling us we had it easy, that nowadays nobody made you separate the different shades of green, one hue of yellow from another, the trash, as they had when he had farmed. Fall meant picking cotton and gathering corn; winter meant killing hogs, for people lucky enough to have them.

It was my father's part, as that of tenants everywhere, to pay the hands out of his pocket. He might have to borrow from the landlord in order to do so, but they both knew that at the year's end, at the time to settle up, whatever had been advanced would have to come out of the tenant's share. Farming on thirds meant that the landlord deducted from the tenant's two-thirds the expenses of making the crop, a usual (but not universal) exception being that the owner paid part of the fertilizer bill. Labor and gasoline or, before tractors were much in use, horses and mules and feed; cultivators and plows, disk harrows and planters; seed and tools; everything, in fact, necessary for farming, as well as what it took to feed him and his family during the winter when there was nothing to sell and he had to go to the landlord for grocery money, all were the tenant's lookout, whether the crop made anything or not. Some owners of course furnished more than others, even as tenants who farmed the same land year after year (some for the second or even third generation) were more likely to own what they had to have than were those others who kept afloat by going from place to place, shedding one landlord for another as seasonally as molting birds lose and acquire feathers. Knowing this to be the case, the hands might or might not have known, also, that in some ways the tenant's living was about as hard to make, and infinitely as precarious, as their own. The money the tenant counted out to each as he came forward on Saturday when his name was called, the others holding back and respectful, was not, or not yet, his.

To the child of such a farmer, however, nothing was more exhilarating than riding in the back of the truck with these familiar but unknown people, than playing around the tobacco barn all day, or between the rows in the field, listening without consciously doing so as they talked among themselves, their voices cast in a different timbre, the words falling into another rhythm, than when they talked to us. Humored and taken up for by them, saved from spankings by black women appropriating by the sheer force of their person the right to scold even your own mother if she offered to correct you (so long as your offenses, though legion, still fell within the scheme of things), a white child of six or seven, too little to work and in command just of the gentler truths of the world, might have been forgiven for thinking the tobacco barn and these people, the house in which she lived, even the white-hot sun itself, existed merely for her own happiness and peace of mind. Nobody told her any different.

GOOD COUNTRY PEOPLE

AN EPILOGUE

You don't see them much anymore. Not in Rocky Mount and Goldsboro, Wilson, Smithfield and Clinton; in Faison, yes, in little towns like that, sometimes. Especially if on Saturday you buy your groceries at one of the less-than-grand supermarkets, your clothes, when you have to have them, at the dry-goods store. They stay out of the shopping malls, away from the stores dazzling as operating rooms. At Christmas time, everybody sees them (but tries not to); they stumble along, slower than other people, more uncertain, as if they're not quite sure where they are. As for the men, you can spot them without too much trouble.

At the tractor places, the filling stations where they go to pass the time of day, the run-down ones; they're driving battered pickup trucks and looking out across the land, poking along at forty and forty-five. But they're not as common as they used to be, these old farmers in faded overalls, in khaki shirts washed thin and almost white, brogans, hats usually: dusty as a March field. And the women, the country women of my childhood are as scarce now almost as hen's teeth.

Oh, but they were something! The beauty they'd had as girls wrung out of them, and in its place another: faces composed, purposeful as iron. A look that went right through you, bottomless and sad. People my mother's age, the blacks who had known her all her life, would stop me on the street sometimes and after getting it right ("Ain'cha Miss Geneva's girl? Ain'cha now? Ain'cha?"), they'd tell me what a fine-looking woman my mama had been; how in the fields chopping, picking cotton maybe, she could outwork anybody, them too, and *did*. They said she hadn't the need for conversation ("all sech as that"), and suffered no fools, gladly or otherwise, that she was all business, as good as her word and meant what she said. They'd tell me they knew my people. "Sho' do! Mist' Jim . . . Miss Annie, all of 'em!" And they did.

To have stood there on the sidewalk with these people, six or seven or eight years old, the year 1950, 1951, 1952, on a Saturday more than likely, and Faison full to beat the band, old black women congregated in the doorways, come to town. And you pulling away even as they told you, yet not wanting to offend, and their high cackle, the start on something else if you weren't quick, backing away: you knew that you were *known,* and exactly who you *were.*

This was a generation the like of which will never be seen again; they were the last of that breed of farmer and hand who had started out during the Great Depression. Their sons and daughters were the first taking up public work, and their grandchildren, the first in an entire lineage never to have known the land at all. Housewives in the 1960s wore pedalpushers and went to the beauty shop. If they didn't yet work at the sewing plant or Hamilton-Beach, they kept up with "As the World Turns" and "The Edge of Night," and they'd put supper on the table some nights out of a can. The men would wear overall pants and low-cut shoes, wide white belts and slacks when they dressed up, Mennen Skin Bracer maybe, or Old Spice; they'd go when they could to Topsail fishing (calling it "Topsl"), or they'd take the family to the drive-in at Mount Olive where, in the 1950s, it had cost a dollar a car, but a dollar a head in the 1960s.

Farm people still, and they'd seldom had much schooling, but the plants and factories then springing up wanted them anyway, and as surely as the sun was going to rise tomorrow, women and men both did need *them.* Tenantry going to hell like it was, and small landowners having a time of it, too, the 1960s and early 1970s saw the virtual disappearance of one way of life, the birth of another: the dying out of good country people and the emergence of a semi-skilled class of laborers, their roots in the land but their future, as that of their children, in—why, whatever job they could pick up; they weren't particular, and they knew how to work.

Of course it was getting to the place where you had to know a little something extra to get on at some plants, education and training coming to matter more and more in the 1970s and 1980s, but nobody had asked them, and they couldn't help it; and if their children hadn't got what they ought to out of school, they couldn't help that, either. They'd done all right so far, and they reckoned the young people just starting out would, too—soon as they settled down some. They were making more than they ever had before, even if they couldn't begin to tell you where it all went, and if the plant where they worked would just stay open, if their hours weren't cut any more, if they didn't get layed off, why they thought they'd fare right well.

For their children a community college or technical school has sometimes made the difference, taking up the slack after high school, bridging the gap between the training they need for certain kinds of employment and the skills they, in fact, have; associate-degree programs provide entry into some of the technical fields, and for those not finishing high school or unable to read, courses in basic education are available. People thrown out of work by changes in the job market, as by layoffs and closings, can sometimes take advantage of the chance to try something else, computers, maybe, or data processing. But not everybody has found retraining to be the answer, or, for that matter, a two-year program. Jobs of any kind, but especially if they hold much long-range promise, still don't grow on trees; wages still are among the lowest in the nation, and the gap between these eastern counties and the Piedmont cities, as between rich and poor, gets wider and wider. Manufacturing is not what it used to be, and, yet, most people looking for work aren't prepared for anything else.

But these are proud people. Throwed away they may be, but it won't do to count them out. Men and women who have seen how, in the 1960s, machines pushed up the demand for land, even as they made farm laborers increasingly obsolescent, who have experienced the breakup of smalltime agriculture, yet who have kept going nonetheless, kept looking ahead—they know they're up against a hard time, but they know, too, they'll make it somehow: they always have. Business is business, and if people still matter less than profits, why they've always known *that*.

It's a hard lesson for the young: the realization that they're not likely to do even as well as their parents; that as many plants are closing as are coming in; that by itself a high-school diploma, as increasingly a college degree, means little to the man doing the hiring. Having sat for twelve years in more modern, more costly schools than any in history, they aren't happier for the experience or scarcely any more prepared for meeting the world head-on; nor are their parents any happier or more financially secure for having taken up public work. Neither schools nor factories have fulfilled the promises inherent in them. The one too often seems irrelevant; the other, willfully capricious if not worse. Fairness is something only little children any longer much expect.

But blood is thicker than water, and in these youngsters as often different from us as night from day, there may yet survive a farmer's cussedness, his equable and solid understanding of what counts and of who really matters. Others have come into less, surely.

—————

RESPONDING JOURNAL

Explain in your own words the different meanings of *throwed away* as Flowers defines the range of possibilities in her first paragraph.

QUESTIONS FOR CRITICAL THINKING

STRATEGY

1. Explain the effect of Flowers' opening section in "Home" where she describes how you would reach her native county by car. What does this technique contribute to her writing? How effective do you find it?
2. Consider Flowers' description of her sense of childhood. What devices or strategies does she use to give you an idea of exactly what it felt like to be a child living in the tenant farming system?

ISSUES

1. Think about the area you consider home; what sorts of changes has it undergone during your lifetime? during the memory of people who have told you what the "old days" were like? Explain any parallels to Flowers' descriptions you can recall.
2. In contrasting ideas of pride and "throwed away," Flowers seems to wait until the end of her essay to suggest how proud the tenant farmers are. But there are numerous clues to the pride and strength of these people throughout the selection. Locate the clues, and decide how Flowers has established these strongly independent characteristics before her concluding section.

COLLABORATIVE WRITING ACTIVITY

Take the writing you did for your journal on tenant farming and share it with your group. After everyone has had the chance to ask questions about tenantry, divide up according to the categories of tenant farming you have identified through your discussion. Do some investigation in the library, with perhaps each member of the group responsible for one issue, and report back to your individual group. This reporting will serve as the basis for a substantial report back to the class on the nature of tenant farming. You may want to learn about its demise, the extent of its practice, the nature of the system during its heyday, and its status today.

from *Farewell to Manzanar*
Jeanne Wakatsuki Houston and James Houston

Jeanne Wakatsuki Houston was moved with her family to the Manzanar Internment in 1942, where they remained for the duration of World War II. While later attending San Jose State, from which she graduated in 1956, she met her husband, James Houston, also a 1956 graduate. Together they tell the story of the landscape in the internment camp called Manzanar; their book Farewell to Manzanar *has won such awards as the Humanities Prize, the Christopher Award, and the National Women's Political Caucus award.*

Despite the horror of the forced relocation, the indomitable spirit of the internees prevails. The Houstons' command of tone and recollection of detail display a tremendous human strength to find a correspondent hope in a landscape, even a "throwed away" one such as the Owens Valley had become by 1942.

EXPLORING JOURNAL

Freewrite for ten minutes on those details from your high school days that mean the most to you. If you run out of time, make a quick list. Later, go back and compose a second entry on why two or three of these items seem most important to you today.

—————◆•◆—————

In Spanish, Manzanar means "apple orchard." Great stretches of Owens Valley were once green with orchards and alfalfa fields. It has been a desert ever since its water started flowing south into Los Angeles, sometime during the twenties. But a few rows of untended pear and apple trees were still growing there when the camp opened, where a shallow water table had kept them alive. In the spring of 1943 we moved to block 28, right up next to one of the old pear orchards. That's where we stayed until the end of the war, and those trees stand in my memory for the turning of our life in camp, from the outrageous to the tolerable.

Papa pruned and cared for the nearest trees. Late that summer we picked the fruit green and stored it in a root cellar he had dug under our new barracks. At night the wind through the leaves would sound like the surf had sounded in Ocean Park, and while drifting off to sleep I could almost imagine we were still living by the beach.

Mama had set up this move. Block 28 was also close to the camp hospital. For the most part, people lived there who had to have easy access to it. Mama's connection was her job as dietician. A whole half of one barracks had fallen empty when another family relocated. Mama hustled us in there almost before they'd snapped their suitcases shut.

For all the pain it caused, the loyalty oath finally did speed up the relocation program. One result was a gradual easing of the congestion in the barracks. A shrewd househunter like Mama could set things up fairly comfortably—by Manzanar standards—if she kept her eyes open. But you had to move fast. As soon as the word got around that so-and-so had been cleared to leave, there would be a kind of tribal restlessness, a nervous rise in the level of neighborhood gossip as wives jockeyed for position to see who would get the empty cubicles.

In Block 28 we doubled our living space—four rooms for the twelve of us. Ray and Woody walled them with sheetrock. We had ceilings this time, and linoleum floors of solid maroon. You had three colors to choose from— maroon, black, and forest green—and there was plenty of it around by this time. Some families would vie with one another for the most elegant floor designs, obtaining a roll of each color from the supply shed, cutting it into diamonds, squares, or triangles, shining it with heating oil, then leaving their doors open so that passers-by could admire the handiwork.

Papa brought his still with him when we moved. He set it up behind the door, where he continued to brew his own sake and brandy. He wasn't drinking as much now, though. He spent a lot of time outdoors. Like many of the older Issei men, he didn't take a regular job in camp. He puttered. He had been working hard for thirty years and, bad as it was for him in some ways, camp did allow him time to dabble with hobbies he would never had found time for otherwise.

Once the first year's turmoil cooled down, the authorities started letting us outside the wire for recreation. Papa used to hike along the creeks that channeled down from the base of the Sierras. He brought back chunks of driftwood, and he would pass long hours sitting on the steps carving myrtle limbs into benches, table legs, and lamps, filling our rooms with bits of gnarled, polished furniture.

He hauled stones in off the desert and built a small rock garden outside our doorway, with succulents and a patch of moss. Near it he laid flat stepping-stones leading to the stairs.

He also painted watercolors. Until this time I had not known he could paint. He loved to sketch the mountains. If anything made that country habitable it was the mountains themselves, purple when the sun dropped and so sharply etched in the morning light the granite dazzled almost more than the bright snow lacing it. The nearest peaks rose ten thousand feet higher than the valley floor, with Whitney, the highest, just off to the south. They were important for all of us, but especially for the Issei. Whitney reminded Papa of Fujiyama, that is, it gave him the same kind of spiritual sustenance. The tremendous beauty of those peaks was inspirational, as so many natural forms are to the Japanese (the rocks outside our doorway could be those mountains in miniature). They also represented those forces in nature, those powerful and inevitable forces that cannot be resisted, reminding a man that sometimes he must simply endure that which cannot be changed.

Subdued, resigned, Papa's life—all our lives—took on a pattern that would hold for the duration of the war. Public shows of resentment pretty much spent themselves over the loyalty oath crises. *Shikata ga nai* again became the motto, but under altered circumstances. What had to be endured was the climate, the confinement, the steady crumbling away of family life. But the camp itself had been made livable. The government provided for our physical needs. My parents and older brothers and sisters, like most of the internees, accepted their lot and did what they could to make the best of a bad situation. "We're here," Woody would say. "We're here, and there's no use moaning about it forever."

Gardens had sprung up everywhere, in the firebreaks, between the rows of barracks—rock gardens, vegetable gardens, cactus and flower gardens. People who lived in Owens Valley during the war still remember the flowers and lush greenery they could see from the highway as they drove past the main gate. The soil around Manzanar is alluvial and very rich. With water siphoned off from the Los Angeles–bound aqueduct, a large farm was under cultivation just outside the camp, providing the mess halls with lettuce, corn, tomatoes, eggplant, string beans, horseradish, and cucumbers. Near Block 28 some of the men who had been professional gardeners built a small park, with mossy nooks, ponds, waterfalls, and curved wooden bridges. Sometimes in the evenings we could walk down the raked gravel paths. You could face away from the barracks, look past a tiny rapids toward the darkening mountains, and for a while not be a prisoner at all. You could hang suspended in some odd, almost lovely land you could not escape from yet almost didn't want to leave.

As the months at Manzanar turned to years, it became a world unto itself, with its own logic and familiar ways. In time, staying there seemed far simpler than moving once again to another, unknown place. It was as if the war were forgotten, our reason for being there forgotten. The present, the little bit of busywork you had right in front of you, became the most urgent thing. In such a narrowed world, in order to survive, you learn to contain your rage and your despair, and you try to re-create, as well as you can, your normality, some sense of things continuing. The fact that America had accused us, or excluded us, or imprisoned us, or whatever it might be called, did not change the kind of world we wanted. Most of us were born in this country; we had no other models. Those parks and gardens lent it an oriental character, but in most ways it was a totally equipped American small town, complete with schools, churches, Boy Scouts, beauty parlors, neighborhood gossip, fire and police departments, glee clubs, softball leagues, Abbott and Costello movies, tennis courts, and traveling shows. (I still remember an Indian who turned up one Saturday billing himself as a Sioux chief, wearing bear claws and head feathers. In the firebreak he sang songs and danced his tribal dances while hundreds of us watched.)

In our family, while Papa puttered, Mama made daily rounds to the mess halls, helping young mothers with their feeding, planning diets for the various ailments people suffered from. She wore a bright yellow, longbilled sun hat she had made herself and always kept stiffly starched. Afternoons I would see her coming from blocks away, heading home, her tiny figure warped by heat waves and that bonnet a yellow flower wavering in the glare.

In their disagreement over serving the country, Woody and Papa had struck a kind of compromise. Papa talked him out of volunteering; Woody waited for the army to induct him. Meanwhile he clerked in the co-op general store. Kiyo, nearly thirteen by this time, looked forward to the heavy winds. They moved the sand around and uncovered obsidian arrowheads he could sell to old men in camp for fifty cents apiece. Ray, a few years older, played in the six-man touch football league, sometimes against Caucasian teams who would come in from Lone Pine or Independence. My sister Lillian was in high school and singing with a hillbilly band called The Sierra Stars—jeans, cowboy hats, two guitars, and a tub bass. And my oldest brother, Bill, led a dance band called The Jive Bombers—brass and rhythm, with cardboard fold-out music stands lettered J.B. Dances were held every weekend in one of the recreation halls. Bill played trumpet and took vocals on Glenn Miller arrangements of such tunes as *In the Mood, String of Pearls,* and *Don't Fence Me In.* He didn't sing *Don't Fence Me In* out of protest, as if trying quietly to mock the authorities. It just happened to be a hit song one year, and they all wanted to be an up-to-date American swing band. They would blast it out into recreation barracks full of bobby-soxed, jitterbugging couples:

> Oh, give me land, lots of land
> Under starry skies above,
> Don't fence me in.
> Let me ride through the wide
> Open country that I love . . .

Pictures of the band, in their bow ties and jackets, appeared in the high school yearbook for 1943–1944, along with pictures of just about everything else in camp that year. It was called *Our World.* In its pages you see school kids with armloads of books, wearing cardigan sweaters and walking past rows of tarpapered shacks. You see chubby girl yell leaders, pompons flying as they leap with glee. You read about the school play, called *Growing Pains* ". . . the story of a typical American home, in this case that of the McIntyres. They see their boy and girl tossed into the normal awkward growing up stage, but can offer little assistance or direction in their turbulent course . . ." with Shoji Katayama as George McIntyre, Takudo Ando as Terry McIntyre, and Mrs. McIntyre played by Kazuko Nagai.

All the class pictures are in there, from the seventh grade through twelfth, with individual head shots of seniors, their names followed by the names of the high schools they would have graduated from on the outside: Theodore Roosevelt, Thomas Jefferson, Herbert Hoover, Sacred Heart. You see pretty girls on bicycles, chicken yards full of fat pullets, patients back-tilted in dental chairs, lines of laundry, and finally, two large blowups, the first of a high tower with a searchlight, against a Sierra backdrop, the next a two-page endsheet showing a wide path that curves among rows of elm trees. White stones border the path. Two dogs are following an old woman in gardening clothes as she strolls along. She is in the middle distance, small beneath the trees, beneath the snowy peaks. It is winter. All the elms are bare. The scene is both stark and comforting. This path leads toward one edge of camp, but the wire is out of sight, or out of focus. The tiny woman seems very much at ease. She and her tiny dogs seem almost swallowed by the landscape, or floating in it.

RESPONDING JOURNAL

Reflect on the natural details that the internees seem to emphasize. What saves them from becoming despondent? Write a journal entry on the resilience and strength you find in them.

QUESTIONS FOR CRITICAL THINKING

STRATEGY

1. Explain the process of the Wakatsukis' move into their home at block 28; next, explain how Houston writes about the process. Do you detect any parallels between moving in and writing?
2. What gives you the most intense feeling about the nature of life in the Manzanar camp? What do you think would have been the most difficult part of internment life for you to have learned to accommodate?

ISSUES

1. Explain how you feel about the listing of high schools from which the Manzanar internees would have graduated; do you think this list is random, or does its progress from Theodore Roosevelt through Sacred Heart carry additional connotations? What do you conclude from this list in Houston's last paragraph?
2. Find a picture, either in this text or elsewhere, that you feel parallels the final image Houston describes. How effective do you find this visual image? Does it make a good conclusion? Think of a story or a narrative that the other picture you found might conclude. What does this exercise lead you to conclude about the power of such images?

COLLABORATIVE WRITING ACTIVITY

Generate a list of situations in America parallel to the interning of Japanese-Americans in World War II; you may use the history of the country from its earliest days through today. Read your list to the group and allow time to discuss the items on your list. Members of your group may be unaware of situations that people have listed, and they will want to ask questions. Together, produce a summary of your discussion for the rest of the class to read.

"XVII" ("I love the wind/ When it blows through my barrio")
Jimmy Santiago Baca

*Poet Jimmy Santiago Baca, abandoned as a child, writes about the nature of community and attachment to land as means of personal fulfillment. The titles of his books—*Immigrants in Our Own Land, The Black Mesa Poems, *and* Martin & Meditations on the South Valley—*indicate his concern with locating human truth on specific landscapes. Blending elements of* mestizo, *Chicano, Catholic, and Anglo culture, Baca shows how we both create and are created by the worlds in which we live.*

"XVII" is from Martin & Meditations on the South Valley, *a reference to the central character and his thoughts in the series of poems leading to final celebration of the building of a new house after his home was destroyed by fire. Baca's characterization of the wind as energizer, linking difficult lives with relentless desert breezes, contrasts with other descriptions of similar winds: Joan Didion's description of the Santa Ana wind hot off the desert, blowing through the city in "Los Angeles Notebook," recalls Raymond Chandler's observation that such winds elicit murder from otherwise meek people. Baca's gospel of hope in the land, however, creates a different landscape that holds promise in spite of the sometimes suffocating depression.*

EXPLORING JOURNAL

Freewrite a journal entry on the ideas and associations the word *barrio* calls up for you. What mental image do you have of this word? How did you develop your concepts? What are their sources?

I love the wind
when it blows through my barrio.
It hisses its snake love
down calles de polvo,
and cracks egg-shell skins
of abandoned homes.
Stray dogs find shelter
along the river,
where great cottonwoods rattle
like old covered wagons,
stuck in stagnant waterholes.
Days when the wind blows
full of sand and grit,
men and women make decisions

that change their whole lives.
Windy days in the barrio
give birth to divorce papers
and squalling separation. The wind tells us
what others refuse to tell us,
informing men and women of a secret,
that they move away to hide from.

RESPONDING JOURNAL

How has your image of the barrio changed after reading Baca's poem? What aspects of the barrio does he emphasize? What surprises you most about his presentation?

QUESTIONS FOR CRITICAL THINKING

STRATEGY

1. Explain the idea of "snake love." What contribution does this image make to the poem? Does it help Baca establish a tone? Why?
2. Compare the first and last lines of the poem. How does the idea expressed in the final line relate to the poet's love of the wind he describes?

ISSUES

1. Explain the reference to "old covered wagons." Why would Baca want to introduce the connotations of this image to his poem?
2. Wind seems to mean a "squalling separation." What sorts of separations does the poem suggest? What sort of direct connection might exist between the blowing of the wind and the separating?

COLLABORATIVE WRITING ACTIVITY

Compose a brief essay on the wind that blows through your neighborhood or the place where you were raised. Bring it to your group and read it. After discussing the entries of the group, decide how best to combine them into a chorus of voices representing the winds blowing through all of your homes. Try both written and choral versions that you can present to the rest of the class.

SUGGESTIONS FOR FORMAL WRITING ASSIGNMENTS

1. The mountaineer's rope is a constant metaphor in Reid's essay, which appears in the previous chapter. He suggested that rope tied him to his sense

of home and family. Write an essay in which you modify this image to suit your own purposes as you describe a mutually supportive relationship acted out against a particular backdrop. This topic is fairly open-ended, so you will be free to choose among a wide range of possibilities.

2. Compose a speculative essay in which you explore the relation between landscape and culture. Since you have not yet read as many accounts of such interrelation as you will by the end of this course, you may find that your essay is somewhat shorter than others you have written; this one may reach only 250–300 words. Of course, you may also find that you have much to say and feel a need to write an essay several times that length.

3. Explain, in a well-detailed essay, how your sense of home is shaped by the people you associate with home. You may wish to include neighbors, as Hamill has done, but you should try to focus on how best to help your reader identify readily the most important people in your home. Does Hamill's dramatic leading up to the appearance of his mother suggest anything to you?

4. Think about the varieties of displacement in your own life. Have you moved, or been forced to move, against your will? Have you had close friends who have suffered more extreme cases of displacement? Did your arrival at college strike you as a displacement in any way? Write a substantial essay in which you explore the psychological costs of having to move from what you consider home to another site. Think about the specific emotions you experienced, and try to describe them fully for your readers.

VISIONS: A LANDSCAPE OF ARTISTIC REPRESENTATIONS

Asher B. Durand, *Kindred Spirits,* 1849

OBSERVATIONS

1. Durand is a member of the "Hudson River School," a group of painters whose shared techniques and interests in the Hudson River and adjoining Catskill Mountains bound them together in the nineteenth century. They often attempted to use sharp detail as a way of drawing the spectator into the new American paradise of forest and field. Explain what you find to be the dominant qualities of this painting. What attracts your attention as you look at it?

2. The title of this painting, *Kindred Spirits,* reflects the relationship between Cole (on the right, pointing with his walking stick) and the poet William Cullen Bryant (on the left, hat in hand). Do you consider the arts of poetry and painting as being "kindred spirits"? Does this painting seem to suggest any ways in which the two might be related or any common subject matter? What sorts of emotions do the two men seem to be experiencing?

3. Describe the specific detail Durand has presented in the foreground and general atmospheric qualities suggested in the background. Exactly what does he include in the foreground? How does he make the background atmosphere appear? Explain how well you think these two categories work together in this painting.

SUGGESTIONS FOR WRITING

1. Think about the associations you bring to the specific details of a scene such as this one. What does this scene cause you to recollect? Write an essay in which you explore the sensations and associations this painting evokes for you.

2. In an essay, explain how the composition of this painting takes advantage of the natural scene. For example, what is the most noticeable part of the foreground? How does the tree at the left work to unify the painting? What is the effect of the projection into the middle of the painting of the rock on which the men stand? Provide your reader with a rationale for your assessment of the effectiveness of this painting's composition.

3. Imagine yourself to be at a particular place in the natural world with a person whose attributes or skills complement your own. This other person may be a close friend of yours, or someone you know solely through reading about him or her. Write an essay in which you explain in detail how you would have to set up the scene in order to depict the particular "kindred spirits" you and your partner share.

Ernest Blumenschein, *Sangre de Christo Mountains,* 1924

OBSERVATIONS

1. Explain exactly what you see in this painting. Indicate what you see in the foreground, the middle distance, and the background. What is your predominant sense of this landscape?

2. How does Blumenschein use color in this painting? Account for the limited range of shades you see here.

3. On a sheet of paper sketch the forms that Blumenschein uses here. What are the four dominant shapes in this painting? Explain why you chose these four. Why do you think Blumenschein might have chosen these to emphasize?

4. How does the season of the year affect this painting? What other details might we see if the season were changed?

SUGGESTIONS FOR WRITING

1. Write an essay in which you explain how each detail of the painting is necessary to achieve the overall effect. You are taking over the role of the artist in making sure that your audience understands why each detail, each color, appears exactly as it does. Your ultimate goal is to explain the unity of effect so that your reader understands how the painting works.

2. What stories have you read or heard that might have taken place in a village like the one depicted here? What sorts of lives do its inhabitants live? Write an essay in which you construct your version of life in this village. Offer as much detail in your essay as you can and feel free to indulge your imagination.

3. Some art critics have credited Blumenschein with the ability to see the natural world as the Indians did: in terms of symbolic or abstract representation of the forces of the earth rather than directly representational narrative accounts. Write an essay in which you support or argue against such a reading of Blumenschein's painting. You may wish to do some further research to view Blumenschein's other works as you plan your essay.

Georgia O'Keeffe, *Red Hills, Lake George,* 1927

OBSERVATIONS

1. Lake George is a large natural lake surrounded by mountains in upstate New York. What is the artist depicting in this painting? To what extent do you have to use your imagination to connect the painting with its title?

2. How does the use of color inform this painting? What emotions does the painting arouse for you? What other colors might have worked as well or better?

3. Does this painting make you think of other images? Explain how the simplicity of these images might be paralleled in the natural world.

4. Explain how this image might help its viewers to think of a lake. Is there any direct connection?

SUGGESTIONS FOR WRITING

1. Think of the emotions associated with either sunrise or sunset. Do you think this painting depicts one or the other, or some combination? Write an essay in which you explain how this painting represents or does not represent a sense of the time of day.

2. Go to the library and look at some of O'Keeffe's other paintings. Write an essay in which you explain how looking at some of her other depictions of place, shape, and color have helped you to understand more fully some of the ideas or techniques she uses in this painting.

3. After consulting a dictionary, form in your own words a definition of "abstract art." Write an essay in which you explain how abstract art might relate to the landscape. In your essay, use this painting and any others that will help you to illustrate your definition.

4. Some art critics find that O'Keeffe depicts a tension between the delicacy of natural forms on the one hand and the desolation, both natural and man-made, of the American land on the other. Write an essay in which you agree or disagree with this idea; support your points with this painting and any of her other works you may have seen.

George Inness, *The Lackawanna Valley,* 1855

OBSERVATIONS

1. You might have noticed several puffs of smoke in the painting. Why might we see them as being related to one another? Is the artist drawing our attention to these by painting them in a similar fashion? What similar elements, if any, do you see in the painting?

2. What effect does the tree on the left have on the painting? Does it direct our attention in a certain way as we look at the picture? How does the human figure near the tree affect the scene? What seems to be the relation between that figure and the rest of the painting?

3. Think about the way stumps are depicted in the painting. What differences do you find between those that are close to the viewer and those that are a considerable distance away? What possible reasons might Innes have had for treating the stumps in this manner?

4. Innes has said that an artist should "represent a living motion." Discuss the ways in which he has or has not represented living motion in this painting. What elements do you find most effective?

SUGGESTIONS FOR WRITING

1. The Delaware, Lackawanna, and Western Railroad commissioned Innes and paid him $75 for this painting in 1855. He was required, however, to include the double tracks and the roundhouse. Select a place in your community with special significance to the economy, history, or culture of your area. Write an essay in which you explain the guidelines that should be observed when depicting that place as part of the natural world in painting, writing, or other means of representing the landscape.

2. Many other visual representations of nature include some sort of mechanized element or activity as a central focus; automobile advertisements are one type of such representation. Think about what you consider to be the characteristics of such representation, and write an essay in which you explain your viewpoint. You may want to include several visual texts in addition to this one in your essay.

3. The effects of industrialization in the nineteenth century are evident in this painting, from the train to the stumps in the cleared field. In an essay, explain how the ideas of landscape and industry are held together in this painting. How well do the colors of the growing world relate to the smoke tones and the lines of mechanization?

Jerry Bywaters, *Oil Field Girls,* **1940**

OBSERVATIONS

1. What are the important features of this painting? What are its most unusual characteristics? Does this painting cause you to respond in any new or different ways? Explain which details seem most important to you.

2. What is the effect of the brand names in this painting? What ideas do they add to the painting that it could not otherwise convey?

3. What features of this painting seem different from other landscapes you have viewed? Could any of these differences be in any sense "political"? How do these differences affect the way you view *Oil Field Girls?*

4. What can you tell about the two women in the painting? Which one seems older? How do you know? What can you say about their emotional states? Are they talking with each other? What possible reasons might Bywaters have for painting these two particular women? Why not men?

SUGGESTIONS FOR WRITING

1. Write an essay in which you compare and contrast the treatment of people in relation to the land as demonstrated in Durand's *Kindred Spirits* and Bywaters's *Oil Field Girls.* What does the relationship between the two people in each painting suggest to you about the relationship to the land examined in each of the paintings? To what extent do the ideas demonstrated in these paintings apply today? If similar paintings were created today, what would change?

2. Write an essay in which you explain the development of the particular plot of land depicted in the painting. What might the area have looked like before the oil wells were drilled? What sorts of changes have taken place? How have these changes affected the people on the land?

3. What murals are on display in your region? Many murals were painted as part of the Works Projects Administration (WPA), a New Deal program of the 1930s designed to employ out-of-work or underemployed artists on public projects. WPA murals were frequently painted in post offices and other federal buildings, where they remain prominently displayed today. Bywaters painted such murals in the 1930s. Write an essay in which you compare and contrast *Oil Field Girls* with the other murals you have been able to find either in your region of the country or in other sources that reproduce American mural art.

Grandma Moses, *Hoosick Falls in Winter,* 1944

Anna Mary Robertson (Grandma) Moses, the most popular American folk artist of this century, began painting while in her seventies. During the next two decades she produced numerous works that celebrated American rural traditions and community life.

OBSERVATIONS

1. What do you think of first when you see this painting? What images have you seen before? Look at each of the dwellings in the painting; make a list of the different types you see here.

2. Identify the elements—trees, water, landforms, machines—in this painting that you find in other paintings. How does their treatment here parallel or differ from those in the other visual texts?

3. What would you compare to Grandma Moses' painting? What else have you seen that resembles this depiction of a New England village? Make a list of the activities you see depicted here. What is their overall effect?

4. Anna May Robertson Moses began painting in 1927 and continued until her death at the age of 101. She took up painting when arthritis forced her to give up needlework. What traces of her previous medium seem to be apparent in her painting?

SUGGESTIONS FOR WRITING

1. Write an essay in which you compare and contrast the train in the landscapes you have seen or read with those trains in your own experience. What has formed your personal sense of trains? How does this one confirm or deny your previous ideas? Which is more realistic? More emotional? Explain your responses as fully as you can.

2. The work of Grandma Moses has been called "folk art." Think of other examples of folk art in the area where you live now or where you previously lived. Write an essay in which you offer your own definition of folk art by explaining how people create images or texts of the daily life around them. What do these illustrations of the everyday lives of people say about the land on which they are lived?

3. Write an essay in which you suggest the benefits and drawbacks of life in a village like Hoosick Falls. Do such communities still exist? What relation do they bear to such imaginary places as the Bedford Falls of the Christmas movie *It's a Wonderful Life?*

Karl Bodmer, *Mih-Tutta-Hang-Kusch, Mandan Village,* 1833–34

OBSERVATIONS

1. Above Bismark, North Dakota, on the Missouri River is Fort Clark, visible on the upper left bank. To the right is the Mandan village of Mih-Tutta-Hang-Kusch. What does this painting suggest to you about the selection of this particular spot for both communities?

2. This watercolor sketch was completed by the Swiss artist Karl Bodmer, who, with his patron the German Prince Maximilian, voyaged on the American frontier from 1832–1834. As you look at the painting, explain what you think Bodmer wanted to convey to his audiences in both the United States and Europe.

3. The Mandans are depicted moving across the frozen Missouri River from their summer village on the high bluffs to their winter camp on the forested, protected lowlands partially shown in the foreground. What does the painting convey about the nature of this journeying back and forth between the camps which went on during the winter? How does Bodmer establish a sense of this process?

4. Explain the effects of color in this painting. Look, for example, at the group of three Mandans, the horse and the dog in the center; what conclusions can you draw from Bodmer's treatment of them?

SUGGESTIONS FOR WRITING

1. Write an essay in which you explain how this painting confirms or challenges your ideas of how American Indians lived during the winter. What does this image suggest to you about their lives?

2. The *Hoosick Falls* painting by Grandma Moses also depicts a winter scene. Write an essay in which you compare and contrast the use of winter details and the level of sophistication in the two paintings. Would the same audiences respond enthusiastically to both paintings? Explain your points by referring to the details of the two paintings.

3. Write an essay in which you explain the relationship between composition and meaning in this painting. Why does Bodmer, for example, place three Mandans in the center of the painting against the white snow? How do these three stand in relation to the other Mandans? What do you make of the different sizes Bodmer uses to depict the Mandans? What does this question of perspective tell you about the Mandans' activities?

Maynard Dixon, *Open Range*, 1942

OBSERVATIONS

1. The sky is often a main concern for Dixon's paintings. What function does the sky serve here? How does it direct your response to the rest of the painting?

2. Explain how the shadows in the painting affect the mood it creates for you. What would change if the shadows were not present?

3. Look at the lines of the yellowish mesa in the middle distance. What purpose do these landscape features seem to serve in the overall picture?

4. Put yourself in the place of the cowboy in the painting. What sorts of thoughts do you think he experiences as he rides through this image?

SUGGESTIONS FOR WRITING

1. Dixon often said that he felt painting must demonstrate the "eternal fitness of things," which he saw in the best examples of mural art. Write an essay in which you explain how this painting exemplifies the "eternal fitness of things" in the particular landscape you find represented here.

2. Write an essay in which you indicate what you feel about the image of the cowboy. How has it been influenced by television, movies, country music, advertising, or fiction? Explain how you would fit Dixon's image of the cowboy into your sense of what that term means.

3. Think about the mood the painting creates for you. How might you relate your response to the colors and lines of the painting? Write an essay in which you explain your response as best you can; you may find some assistance in thinking about similar or opposed experiences you have had as you responded to other visual compositions.

4. Compose an essay in which you relate the painting's title, *Open Range,* to both the image the painting creates and your sense of the meaning of the phrase. Does "open range" suggest a way of life that no longer exists in the United States, or have we found other sorts of "open ranges"?

Arthur Rothstein, *Displaced Sharecropper,* 1939

1. Explain the human qualities this image defines for you. Link your response to specific visual details.

2. Account for the effect of this image by analyzing the elements of its composition. How do horizontal and vertical lines relate to the overall image?

3. Describe the glances of the man, his child, and yourself as you look at the two. How do these perspectives define the world of this photograph?

Chapter 6

DEFINITION: DEBATES ON
ENVIRONMENT AND ECOLOGY

The very idea of *definition,* of what a word *means,* is a standard, accepted way of conveying information. Think about how you have a sense of the word definition itself. We talk about a word's "dictionary definition," "legal definition," "formal definition." You might ask yourself what all of these mean because why should there be, or how can there be, more than one sense of what we mean by a word?

For example, your sense of *home, neighborhood,* or *community* might well be different from mine. Think about your reading on these terms up to this point in the course; how has your concept of these words changed? In much the same way, subtleties of meaning grow around other words until they assume a new shape. This chapter focuses on several words related to the natural world—such as *homeless, wildness, wilderness, ecology*—and explores the networks of meaning they hold. To take one example, the word *homeless* had a formal definition—as being without a home—for many years. Although the word always carried this meaning, at certain points in history it carried other connotations. For example, the selection in this text from Steinbeck's *The Harvest Gypsies* and his major novel *The Grapes of Wrath* deal with the problem of homelessness that came about as a result of the Great Depression of the 1930s. To be homeless in this period was to have fallen victim to larger forces of crop failure or bank failure; hundreds of thousands of Americans found themselves homeless. Although the word had this connotation for a period, it may have lost its meaning during the relative economic prosperity that followed World War II and peaked in the 1960s; when homeless again began appearing in the popular mind during the 1980s, the term changed to still another meaning. Since some people released from mental institutions by new interpretation of legislation in the early 1980s had no homes to which to return after release, the homeless often were erroneously branded as potentially dangerous mental defectives.

Another way to understand the sense of *homeless* is to consider the parallel with *houseless.* Both mean being without a building in which to live some sort of a regular life, but the connotations of support attached to *home* are missing in the case of *house.*

During the past ten years, however, print and visual media have provided enough examples of people thrown out into the streets by desperate economic circumstances that *homeless* once again carries a positive connotation. The homeless, we all know, might reflect our complex contemporary world where financial ruin could be only a hospital stay away. Loss of a job can mean a family forced out of a home and, without relatives or a similar support system, that family may have to live on the streets. In terms of human ecology, homelessness carries different meanings in the 1990s than it did in the 1930s. If we were to diagram the different connotations for homeless today, we might find the following range of meanings.

FORMAL DEFINITION	SPECIFIC MEANING	CIRCUMSTANCES OF MEANING
"being without a home"	"people unfortunately forced by unavoidable circumstances to live on the streets"	Great Depression; Public perceptions of those without permanent homes; 1980s economic conditions
DENOTATION	CONNOTATION	DETERMINING CONDITIONS

Given this range of definitions, you can see that a word can not only take on a powerful array of meanings, but that the act of explaining just what a word means, or how we want to use it in a specific situation, can be an effective means of writing for a variety of purposes.

For example, we can demonstrate how something is special, or how it differs from other, closely related objects or qualities. We can show how the meaning changes over time, and which of the different possible definitions of the word we are using as we write. In doing such tracing of etymology, or the history of a word's meaning, we are using the formal properties of definition as a way of organizing our writing. In an essay on homelessness, we might want to show how the word carries several connotations, explore how these connotations may have come about, and suggest how changing connotations reflect different social values. Suppose we were to find from our work that America seems now to value its homeless population more than it did in the 1980s. We might find that society has more sympathy for them when it can perceive specific reasons or circumstances that directly explain their homelessness. In this case, the essay might move from the most elementary sense of a neutral point of view toward people in the category of "being without a home," to a sympathetic response toward others who are forced out of their homes for reasons over which they have no control, to a reasoned plea that society try to find ways to help these homeless reestablish themselves in homes. Such a persuasive effort would include several meanings in an essay of definition, such as:

- formal definition
- range of meanings

- original meaning
- relation between current definition and other factors, such as time, history, and social conditions
- argument for a particular connotation or set of connotations

To try this concept on another word, let's consider *wildness*. This seems to be a fairly clear word meaning "having qualities of that which is untamed, uncivilized." In one of his famous quotes, Thoreau states that "In wildness is the preservation of the world." What does he mean? Is he thinking of wilderness preservation areas? Does wilderness mean the same thing as wildness? Take a look at the full context of his remark, which comes in the essay "Walking," in which Thoreau describes an inclination toward the west, away from the civilizations of cities and toward the unknown. After telling us that the west "is but another name for the wild," Thoreau goes on to say:

> . . . in Wildness is the preservation of the World. Every tree sends its fibers forth in search of the Wild. The cities import it at any price. Men plow and sail for it. From the forest and wilderness come the tonics and barks which brace mankind. Our ancestors were savages. The story of Romulus and Remus being suckled by a wolf is not a meaningless fable. The founders of every state which has risen to eminence have drawn their nourishment and vigor from a similar wild source. It was because the children of the Empire were not suckled by the wolf that they were conquered and displaced by the children of the northern forests, who were.
>
> I believe in the forest, and in the meadow, and in the night in which the corn grows. We require an infusion of hemlock, spruce or arbor vitae in our tea. There is a difference between eating and drinking for strength and from mere gluttony. The Hottentots eagerly devour the marrow of the koodoo and other antelopes raw, as a matter of course. Some of our northern Indians eat raw the marrow of the Arctic reindeer . . . [which] is probably better than stall-fed beef and slaughterhouse pork to make a man of. Give me a wildness whose glance no civilization can endure—as if we lived on the marrow of koodoos devoured raw.

Thoreau is working with an idea here which underlies all of his writing; perhaps this idea of the wild is the single most important concept he would like to leave future generations. Therefore, he is careful to establish a sense of the history of the word ("our ancestors were savages") and to encourage our sense of the word's importance by focusing on its different qualities in almost the same way as a riddle would ("The cities import it at any price. Men plow and sail for it"). He provides an extended description using first a familiar story—that of Romulus and Remus—and then other, more obscure accounts of Hottentots and Northern Indians. In this way, he uses definition to provide a mini-essay within a far larger essay.

His last remarks defining wildness suggest his attitude toward civilization as well; he defines wildness by showing its opposites ("the children of the empire" vs. "the children of the northern forests," the tame food of civilization vs. "the marrow of koodoos devoured raw"). If he is still deliberately vague about just what wildness is, he has given us a sufficient number of striking examples to insure that we have a good sense of what he means by this complex word.

Yet another perspective on the word *wild* and its meaning comes from Chief Luther Standing Bear of the Oglala, who says:

> We did not think of the great open plains, the beautiful rolling hills, and winding streams with tangled growth, as "wild." Only to the white man was nature a "wilderness" and only to him was the land "infested" with "wild" animals and "savage" people. To us it was tame. Earth was bountiful and we were surrounded with the blessings of the Great Mystery. Not until the hairy man from the east came and with brutal frenzy heaped injustices upon us and the families we loved was it "wild" for us. When the very animals of the forest began fleeing from his approach, then it was that for us the "Wild West" began.

Though Standing Bear's definition parallels Thoreau's in some important ways, it also presents a new connotation of *wild* and suggests the relation of wildness to wilderness.

In his study *Wilderness and the American Mind,* Roderick Nash takes a look at the larger context of Thoreau's definition: he looks at what we mean by our modern term *wilderness.* Nash points out that etymologically the word as an adjective means uncontrolled, savage, disordered; as a noun it suggests a place of wild beasts, a place where humans are normally not found. But there is so much difference among people's definitions of the word that "One man's wilderness may be another's roadside picnic ground. The Yukon trapper would consider a trip to northern Minnesota a return to civilization while for the vacationer from Chicago it is a wilderness adventure indeed" (p. 1).

So, if we want to present an effective definition of what <u>we</u> mean when we use the word *wilderness,* we will have to provide a clear path through those ideas that the concept embraces. We will probably use some specific examples to help locate the idea for our readers, as Nash does when he cites the old English poem *Beowulf* as an early example of the occurrence of the word. But then more modern meanings will help, too; we will want to point out, as Nash does, that the concept of wilderness has many meanings today because we can talk about a wilderness when we are thinking of outer space as well as the remotest parts of Alaska. "Any place in which a person feels stripped of guidance, lost, and perplexed may be called a wilderness" (p. 3). Such a definition allows us to take into account ideas like *The Neon Wilderness,* Nelson Algren's account of the franchised roadside strips full of fast food and standardized retail outlets.

For our purposes here, however, we want to capture the two opposing qualities of the word. On the one hand it threatens; on the other, it invites us to participate in its energy, its positive values. For Thoreau, wildness had only this positive sense of a new energy coming into the world. The purpose of Nash's book-length study of a definition is to show how the word developed its complex and multi-faceted meaning, and in the process we see how Thoreau's definition can be updated and related to our world.

Another means of using definition to explain important concepts is to see how a word has been or can be manipulated, how its meaning can change to suit circumstances. We have seen that a word can go along with an accepted definition that may be very different from its original intention. For example, the name by which an Indian tribe is known may not have anything to do with the tribe itself. The Navajos were given their name by Spanish explorers; many members of this tribe want to return to the more traditional name for their tribe: the Dinetah, or People of the Earth. Similarly, the Sioux tribe (white man's name) calls itself the Dakota (or Lakota for the western tribes) for "alliance of friends." The "Sioux" label was given by French explorers in imitation of the Ojibwa word for the Dakotas, which means "cutthroat," "treacherous." As you can tell, the Ojibwas and Dakotas were mortal enemies. The name Adirondacks, which to most people conveys a sense of the beauty of upstate New York, is actually a pejorative term the Iroquois used for the Algonquins.

What must these names given to a group of people mean to the self-image of the groups? How does such a misnomer affect the people who are stereotyped in this way? Does this renaming seem to be an ancestor of contemporary racial or ethnic slurs? Exploring the answers to any of these questions involves using the techniques of definition as a writing strategy.

The essays in this section offer various means of defining a term or concept. Luci Tapahonso shows "what she is" by defining her matriarchal line and her home. Proper environmental action is another concept that writers work hard to define. In order to convince us of their position, writers such as Doug Bandow and Dave Foreman discuss *ecoterrorism* and *ecowarriors;* the connotations of each are clear. Their common ground on a term such as *ecotage* suggests that proper definition plays a significant role in their debate over how much environmental dissent and activity on behalf of the earth can be tolerated in our society. The Heritage Foundation and Bill Devall offer other definitions of related terms as they seek to support their ideas of how best to orient the reading public to the crucial question of how best to respond to the environment. In doing so, they provide an opportunity to define *ecology* in more accurate ways because of the new ideas they offer us.

"What I Am"
Luci Tapahonso

Born in Shiprock, New Mexico, Luci Tapahonso is a member of the Navajo Nation. She publishes fiction and nonfiction essays as well as her poetry, which has been collected in One More Shiprock Night, Seasonal Woman, *and* A Breeze Swept Through. *In her writing she is concerned with the relationships among members of families and communities across time and space;*

much of her work centers on arrivals and departures from her native New Mexico. She has taught at Northern Arizona University and the University of Kansas.

"What I Am," first published in 1988, displays the sense of generational spirit influencing the present. She relays to her readers a sense of how the spirit of place can follow people virtually anywhere, if they will only seek it. Corn pollen ceremonies appear frequently in Navajo culture, and they parallel forms of communion in other religious contexts.

EXPLORING JOURNAL

How important is a sense of your place of origin? Do you feel that your place of birth and community define you as a person? Freewrite for at least ten minutes on the idea of "What I Am" by referring to the place that has made you.

Nineteen hundred thirty-five. Kinlichíi'nii Bitsí waited, looking across the snow-covered desert stretching out before her. Snow was falling lightly and the desert was flat and white. From where she stood at the foothills of the Carriso Mountains, she could see for miles.

She would see him when he approached—a small, dark speck on the vast whiteness, moving slowly, but closer. Her son, Prettyboy, tall and lanky on the surefooted horse. She would see him.

All evening, she kept watch, stepping out every once in a while. He had gone to visit some relatives at Little Shiprock and should have returned by now. It had begun snowing early and continued into the evening. She knew Prettyboy had started home before the storm and would be arriving soon. She kept watch, looking at the horizon.

In those days, hogans had no windows so she stood at the front door, a shawl around her shoulders. Only her eyes were uncovered as she squinted, looking out into the desert night. "Nihimá, deesk'aaz. Our mother, it's cold," her children called her inside. She would come in for a while, then go back out to watch for him again. All evening she waited, and her children urged her not to worry. He would be home soon they said. She continued watching for him, insisting that she wasn't cold.

Finally she saw him, a dark speck on the horizon. She rushed in and stirred up the fire, heated up the stew, and put on a fresh pot of coffee. She heated up the grease for fry bread. He came in, damp and cold with snow. They laughed because his eyebrows were frozen white. "Tell us everything about your trip." While he ate, he talked about the relatives he visited and the news that he had heard. He said the horse seemed to know the way by itself through the snow and wind. He kept his head down most of the way, he said. The snow was blowing and it was hard to see. The family finally went to bed, relieved that Prettyboy was safely home. Outside the wind blew and the snow formed drifts around the hogan.

In the morning, Kinlichíi'nii Bitsí was sick—feverish and dizzy. She didn't get up and they fed her blue corn meal mush and weak Navajo tea to drink. She slept most of the day and felt very warm. The nearest medical doctor was in Shiprock, fifty miles to the east. On horseback, it was a full day's journey. Even then, the doctor was at the agency only two days a week, and they couldn't remember which days he was there. A medicineman lived nearby on the mountainside, and they decided to wait until morning to go over there and alert him, if they had to. She would get better, they said, and they prayed and sang songs for strength and for the children. Very late that night, she became very ill and talked incessantly about her children and grandchildren.

She died before morning, and Prettyboy went out into the snow and blowing wind to tell other relatives who lived at distances from the hogan of Kinlichíi'nii Bitsí. People gathered quickly despite the snow; they came from all around to help out for the next four days.

Nineteen hundred sixty-eight. The granddaughter of Kinlichíi'nii Bitsí said:

My Uncle Prettyboy died today, and we went over to his house. His aunt, my grandma, was sweeping out her hogan next door and scolding the young people for not helping out. They were listening to their radios in their pickups and holding hands. You know, they are teenagers. My grandma is 104 years old. My real grandma, Kinlichíi'nii Bitsí, would have been 106 if she had not died in the 1930s. I know a lot about her through stories they told me. I know how she was. I think I'm like her in some ways.

At Prettyboy's house, his wife and children were sitting in the front room, and people came in and spoke to them quietly. They were crying and crying; sometimes loudly, sometimes sobbing. In the kitchen, and outside over open fires, we were cooking and preparing food for everyone who had come following his death.

Prettyboy was a tall man and he died of cancer. It was awful because he didn't even smoke. But he had worked in the uranium mines near Red Valley, like many other Navajo men, and was exposed to radioactive materials.

Last week when we were hoeing in the field, my mother told me, "Having a mother is everything. Your mother is your home. When children come home, the mother is always ready with food, stories, and songs for the little ones. She's always happy to see her children and grandchildren."

She had always told me this as I was growing up. That day when we were hoeing corn, I asked her, "Tell me about my grandmother and how you knew something was wrong that time. Tell me the story, shima."

She told me, saying:

That night Prettyboy was coming home. I knew something was wrong. The wind blew hard and roared through the tall pine trees. We lived in the mountains at Oak Springs, ten miles above where my mother lived. We were just married then. Our first baby, your oldest sister, was a month old.

That night the dogs started barking wildly and loudly; they were afraid of something. Then they stopped suddenly. Your father and I looked at each other across the room. Then we heard the coyotes barking and yelping outside. He opened the door and they were circling the hogan, running around and around, yelping the whole time. Your father grabbed the rifle and shot at the coyotes, but he missed each time. He missed. He had been a sharpshooter in the army, and he couldn't shoot them. Finally, they ran off, and we were both afraid. We talked and prayed into the night. We couldn't go anywhere. The snow was deep, and even the horses would have a hard time.

In the morning, I went out to pray and I saw my brother, Prettyboy, riding up to our hogan. He was still far away, but even then I knew something had happened. I tried not to cry, but I knew in my bones something had happened. My brother would not ride out in that weather just to visit. Even though the sun was out, the snow was frozen, and the wind blew steadily. I held the baby and prayed, hoping I was wrong.

Finally, as he came up to our hogan, I went out and I could see that he was crying. He wasn't watching where he was going. The horse led my brother who was crying. I watched him, and then he saw me. I cried out, "Shínaaí, my older brother!" He got off the horse and ran to me, crying out, "Shideezhí, nihimá adin. My younger sister, our mother's gone!" My heart fell. We cried. The wind stopped blowing and we went inside.

I held my baby girl and told her she would not see her grandmother. Neither would our other children. My mother died, and I realized that she was my home. She had always welcomed us, and since I was the youngest, she called me "baby." Even if you are a grandmother, you will be my baby always," she often said to me.

When my mother told this story, we always cried. Even if I had known Kinlichíi'nii Bitsí, I could not love her more than I do now—knowing her only through stories and my mother's memory.

My grandmother had talked to my father about a week before she died. She told him, "Take care of her. She is my youngest, my baby. I trust you and I have faith that you will care for her as I have all these years. She is my baby, but she knows what to do. Listen to her and remember that a woman's wisdom is not foolish. She knows a lot, because I have raised her to be a good and kind person."

My father listened, and he treats my mother well. He listens to her and abides by her wishes.

Nineteen hundred eighty-seven. The great-granddaughter of Kinlichíi'nii Bitsí said:

Early in the morning, we went out to pray. The corn pollen drifted into the swimming pool, becoming little specks of yellow on the blue water. The water lapped quietly against the edges. We prayed and asked the holy ones and my ancestors who died before to watch over me.

I was going so far away to Europe. What a trip it would be. My grandma called the evening before I left and said, "Remember who you are. You're from Oak Springs, and all your relatives are thinking about you and praying that you will come back safely. Do well on your trip, my little one." I was nervous and couldn't sleep. I felt like changing my mind, but my mother had already spent all that money. She promised she wouldn't cry at the airport, then she did. I know that my little sister teased her about it.

I put the bag of pollen in my purse. At La Guardia Airport, I went to the bathroom and tasted some. Everything was confusing and loud, so many people smoking and talking loudly. I wanted my mother's soft, slow voice more than anything. I was the only Indian in the group, and no one knew how I felt. The other girls were looking at the boys and giggling.

At least I had the corn pollen. I was afraid they would arrest me at customs for carrying an unknown substance, but they didn't. I knew I was meant to go to Paris.

I prayed on top of the Eiffel Tower, and the pollen floated down to the brick plaza below. I was so far away from home, so far above everything. The tower swayed a bit in the wind. I never missed Indians until I went abroad. I was lonely to see an Indian the whole time. People thought I was "neat"—being a "real" Indian. They asked all kinds of questions and wanted to learn Navajo. It was weird to be a "real" Indian. All along, I was just regular, one of the bunch, laughing with relatives and friends, mixing Navajo and English. We were always telling jokes about cowboys, computer warriors and stuff.

It was while I stood on top of the Eiffel Tower that I understood that who I am is my mother, her mother, and my great-grandmother, Kinlichíi'nii Bitsí. It was she who made sure I got through customs and wasn't mugged in Paris. When I returned, my grandmother was at the airport. She hugged me tightly. My mother stood back, then came forward and held me. I was home.

RESPONDING JOURNAL

In her poetry Tapahonso refers to "Dinetah," which means "The People's homeland." The narrator's mother says, "Your mother is your home." Compare your experience of such concepts of home to the events presented in Tapahonso's essay. Are such relationships still possible in the 1990s? Explain your answer.

QUESTIONS FOR CRITICAL THINKING

STRATEGY

1. Look at the style of Tapahonso's writing. What strikes you about the way she constructs many of her sentences? What does this style contribute to the ideas of the essay?

2. The divisions of Tapahonso's story correspond to events in the lives of her family members. How else might she have set up her writing? What are the particular strengths of the choice she has made?

ISSUES

1. Tapahonso sees her writing not so much as her own personal production but rather as "a collection of many voices that range from centuries ago and continue into the future." Explain the implications of this idea to her story; what benefits might a person derive from such a view of writing?
2. Is the question of Tapahonso's identity as an Indian always a source of comfort? Are there times when her ethnicity causes problems for her? What in her background seems to help her deal with these potential difficulties?

COLLABORATIVE WRITING ACTIVITY

Think about the times you have had to leave your closest friends and family in order to take a trip, or the times someone else has had to leave you. Explain in an informal essay of several paragraphs how you managed to cope with this departure. What particular devices or activities did you use to make the parting easier? Which of these practices are tied to the place you call home? After you have read your presentation to the group, discuss the sorts of mechanisms you and your group have used. Working together, devise a description of the common ground you find in your efforts to ease the pain of separation.

from "Environment"
The Heritage Foundation

The Heritage Foundation, under the editorship of Mark Liedl, published "Environment" in their Issues '88. *A conservative "think tank," or a body which studies current issues and offers detailed position papers in the hope of influencing legislation through compelling arguments, the Heritage Foundation is located, like most similar bodies, in Washington, D.C.*

In this excerpt, the Heritage Foundation tries to define the environment as an entity extensively affected by state and federal legislation. In recommending specific policy goals, the foundation suggests a specific connotation for the larg-er concept of environment.

EXPLORING JOURNAL

Provide your own definition of *environment*. If you were to define the word from the standpoint of a land developer, would your definition then be different from the definition you think Dave Foreman or Henry David Thoreau would provide?

America's wilderness areas and sensitive animal life are threatened not by big business, but by government bureaucrats and pernicious economic incentives created by government. The reason is that "public ownership" of national parks, forests, and wilderness areas in reality means ownership by nobody. Instead, control is vested in the hands of federal agencies that respond to bureaucratic incentives. Often these are destructive of environmental goals. Anti-pollution policy suffers from a similar disregard of incentives, resulting in policies costly to business and ineffectual in reaching policy goals.

Federal policy should shift to create positive incentives to reach environmental goals effectively and efficiently. To do this the fiction of public ownership should be replaced with the reality of responsible ownership by nongovernmental enti-ties. Example: environmental organizations could manage wilderness areas. The private sector, meanwhile, could be required to pay the full cost of gaining access to forest and range lands. The responsibility for hazardous waste programs could be moved to the states and local communities as much as possible, to encourage innovative approaches and to discourage communities from merely passing the buck, and the bill, to Uncle Sam. Finally, privatization and other market-based strategies could introduce strong incentives for efficient pollution policy.

Create a Wilderness Board to administer wilderness areas. For decades the management of wilderness areas in the U.S. has provoked bitter political disputes. Wilderness advocacy organizations take every opportunity to get Congress to classify sensitive tracts as wilderness areas. This blocks virtually any commercial use of these tracts. Those who oppose such efforts invariably are denounced as antienvironment. Yet it is possible to balance use and con-servation in many sensitive lands. Environmentalists would understand this if they were given the responsibility for running wilderness lands. If they effec-tively "owned" the land, through a long lease arrangement with the federal gov-ernment, they would have the incentive to raise revenues from less sensitive lands to purchase more sensitive areas. As with a private museum, the envi-ronmental groups could buy and sell tracts for their "collection," and augment their revenues with commercial activities that did not conflict with their pri-mary purpose. In fact, several organizations, such as The Nature Conservancy, already own land privately and obtain revenue by permitting carefully con-trolled mineral exploration.

A Wilderness Board should be established, consisting of environmental organizations. This Board would manage public lands, making day-to-day decisions over operation of the lands and decisions regarding sales and purchases of land. The Board would be required to report to Congress, which would have oversight responsibility.

End Forest Service destruction of forests. The U.S. Forest Service is the nation's largest road builder. The Forest Service road network is more than eight times as long as the federal interstate system and has scarred America's woodlands. Having spent billions of taxpayers' dollars to build these roads, to give access to commercial timber companies, it then sells wood below cost. And in its efforts to give away timber, the Forest Service encourages both the "clear cutting" of forests, which leads to huge gaping holes on thousands of otherwise scenic hillsides, and also the deforestation of sensitive high-elevation timberlands.

The more roads that the Forest Service builds and the more clear-cut areas that must be replanted, the bigger the Forest Service's budget and workforce. Since the Forest Service does not in any sense own forests, the long term damage to the value of forests from its actions, and the artificial economics of timber sales do not concern it.

More productive management of the nation's forests would result from transferring ownership to conservation organizations, or selling essentially commercial forests—with assurances for public access—to commercial timber companies. In both cases, the new owners would have a direct interest in preserving the value of the forests to protect their investment, and in ending uneconomic destruction of trees.

Increase state and local control of hazardous waste policy. The cleanup of hazardous waste dumps is rapidly becoming the nation's most expensive public works program. Billions of dollars are committed to dealing with the results of toxic emissions. Few Americans would dispute the need to deal with the hazardous waste problem, but it should be done by states and local governments, not the federal government and its Superfund program. Federal responsibility simply allows local officials to escape the obligation to take quick and decisive action before a problem reaches crisis proportions. Meanwhile, paying for cleanup through taxes on all firms, as now is the policy, imposes a double cost on responsible firms which already control their toxic pollution, while allowing the worst polluters to evade the cost of their actions.

POLICY CHANGES NECESSARY

Two steps are needed to make hazardous waste policy effective, fair, and economical. First, it must be made clear that it is the responsibility of state and local governments close to the scene to take the initiative in controlling toxic dumping and dealing with its consequences. Second, future hazardous dumps could be limited by instituting a tax on dangerous waste products rather than on all firms in industries producing toxic waste. This would encourage firms to find the least costly way of reducing or safely disposing of waste, and it would place the heaviest tax burden on the worst offenders.

Use market mechanisms, such as a production rights market, to achieve pollution goals. Virtually all Americans want pollution reduced. And virtually all Americans want to keep their jobs. Yet laws forcing businesses to reduce pollution increase business costs and thus reduce employment. Government policy, moreover, discourages firms from finding the least costly method of reducing pollution. Typically, regulations require a firm to meet a particular emission standard or to install a particular anti-pollution device, irrespective of immediate or long-term cost.

A better anti-pollution policy would mandate industry-wide reductions in pollution, but allow firms more flexibility to meet these standards. The whole industry could reach the standard by encouraging firms that could reduce pollution substantially, but inexpensively, to cut pollution the most. Meanwhile, some other firms, for whom meeting the standard would mean heavy costs and job losses, would be able to exceed the standard. In this way, pollution could be reduced at the lowest possible industry-wide cost. And individual firms would be permitted to find the least costly way of reducing pollution, rather than installing particular equipment.

The way to achieve industry-wide standards while minimizing harm to some firms would be by a "production rights market." The total permissible pollution would be set for an industry, or a group of firms in a particular area, and licenses would be sold for the right to contribute to that total. The revenues collected would be used by government to deal with such effects of pollution as health problems or corrosion of buildings. Polluting firms would have to purchase the licenses but could trade them between each other. The result would be that firms easily able to reduce pollution would do so to avoid licensing costs, while other firms would prefer to pay license fees, rather than facing heavy costs and layoffs. The overall standard would be achieved.

Review Environmental Protection Agency grant incentives to encourage private sector financing and operation of wastewater plants. The privatization of municipal wastewater treatment plants, which clean and treat sewage, is a recent success story. Cities such as Chandler, Arizona, and Auburn, Alabama, which have contracted with private firms to construct and operate wastewater treatment plants, have reduced costs by about 20 to 30 percent. Because these competitive private firms are more innovative and less costly, the integrity of the water supply is better protected. As such, wastewater treatment privatization has won accolades from environmental groups and local taxpayers.

The federal grant program, however, discourages privatization. Cities must forfeit their federal funds for wastewater plant construction if they seek private financing. This can make privatization unattractive to cities even when the cost of a privately built wastewater plant is less than the specifications required by an EPA grant. To reverse these perverse incentives the federal government should allow cities to receive 20 percent of a project's cost if it is privately owned and financed. The cost to the federal government would be far less than the 55 percent federal funding that cities can receive for publicly owned plants. But because of the significant savings that can be achieved through private design and financing, even with the smaller grant many cities would find it more economic to use the private sector.

STRATEGY

Environmental protection policy and procedures need to be improved. This can be achieved through innovative policies which replace bureaucratic plodding with aggressive free market incentives. While government has a role to play in environmental protection, so does the marketplace. And while pollution control is an important goal, there is a cost in achieving it. This cost is in jobs. Consequently, policymakers should focus on the least costly way of achieving reasonable standards. Such an approach would improve environmental protection, resulting in more cleanup for current expenditures or the same cleanup for less cost.

Environmental reform in the 1990s offers the welcome opportunity of building a partnership between conservatives and traditional liberals. Recently, environmental groups such as the Audubon Society have recognized the value of partnership with the private sector to achieve conservation goals. And increasingly, such environmentalists are admitting that government is not always the right answer.

OBSTACLES

One obstacle to innovative environmental solutions is the mindset that the private sector is, by definition, the enemy of a clean environment. This, of course, has served as the rationale for a massive buildup of public sector regulatory industries. Although this mindset has moderated somewhat among certain conservation groups, it remains a significant obstacle to reform. Another obstacle is the federal bureaucracy. A more free market approach to conservation and environmental protection will mean less government spending, and thus fewer bureaucrats.

Congress is another barrier. Federal control over the environment means power for members of Congress. Whether it comes in the form of grass roots support from environmental groups or campaign contributions from industries seeking favors, these benefits are something that politicians do not want to relinquish. And finally, special interests would oppose changes in the status quo. These are of two types: First is the environmentalist on the leftist fringe of the movement who opposes anything that would lessen government control. Second are the lobbyists and industries which benefit from regulatory policies. Many firms, for example, have learned that environmental policies provide an excellent opportunity to tilt the competitive playing field to favor their interests. Rust Belt industries lobbied hard for "grandfather clauses" in the Clean Air Act which create a bias against new plants in the South and Southeast. Similarly, the restriction on the use of low sulfur coal results from the effective use of environmental laws to benefit the "dirty coal" sector and its union allies. Conservatives should use such examples to explain the environmental risks of relying so heavily on federal environmental regulations. . . .

TACTICS

The problems of current environmental programs must be understood. The current effort to reform federal water policies, to reduce forestry subsidies, and to rethink the sewage grant program was preceded by research projects and articles documenting how these programs were wasting money and endangering the environment. Similar studies are needed for other environmental programs. The public must learn how the Superfund has become a vast public works boondoggle, and how the Clean Air Act encourages the use and continued operation of older, more polluting plants. Also more studies are needed detailing the ways in which environmental regulators discourage the development and introduction of less polluting technologies. . . .

RESPONDING JOURNAL

Think of the destruction of environmental resources you have witnessed in your lifetime. Explain what you feel to be the causes of this destruction. Indicate the extent to which you feel government contributes to or causes this destruction.

QUESTIONS FOR CRITICAL THINKING

STRATEGY

1. Look at the headings for the major divisions of the essay. To what extent do you find these helpful? Why are they included in this essay and not in all writing? What, then, is their particular function?
2. Look at the language of the Heritage Foundation's proposed solution to our environmentally related dilemmas. They wish to see implemented "innovative policies which replace bureaucratic plodding with aggressive free market incentives." Explain the implications of this assertion for other possible viewpoints; how compelling do you find this presentation of the argument? How convincing is the presentation of government inefficiency?

ISSUES

1. Evaluate the claim that free market interests will be the salvation of our environmental woes. How complete do you find this assertion to be? Will it work equally well in all parts of the country? Why or why not?
2. If the federal government is to decrease its ownership of environmental resources, then some other body must take over ownership. What other bodies do the Heritage Foundation recommend take over these resources? To what extent do you find the Heritage Foundation arguments convincing? Are there other alternatives to ownership? If so, what might they be?

COLLABORATIVE WRITING ACTIVITY

Think about how your definition of *environment* has changed as a result of reading this essay and perhaps reflecting on what you have read in other works. List some of the ways you have changed your views on the environment or learned new facts about it. Exchange your list with your group members and discuss the most important points you feel came from reviewing your definitions.

"Ecotourism," from *Living Richly in an Age of Limits*
Bill Devall

Bill Devall has been at the front of the aggressive environmental movement in the United States. A professor of ecology at Humboldt State University in the northern California redwood country, he has written for numerous publications and taken field trips to participate in the "green revolution" that he chronicles in his writing. Devall has also written Simple in Means, Rich in Ends: Practicing Deep Ecology *as a follow-up to* Deep Ecology: Living as If Nature Mattered. *In* Simple in Means, *he explores the implications of the deep ecology philosophy.*

One of the interesting features of Deep Ecology: Living as If Nature Mattered *is that it presents substantial excerpts from the work of other writers. Rather than paraphrasing the ideas of other writers, Devall presents lengthy quotes from poets, philosophers, and writers. In this way, the quotations are able to suggest that the very spirit of another work is worth considering. For a scholarly book, this concerted effort to present intact the ideas of others is a welcome variation on the usual attempt to make one's work seem as if it is the newest thing on the planet; but could the true spirit of ecology offer anything less?*

EXPLORING JOURNAL

In a freewriting entry, explore what deep ecology means to you as a concept. How can something like ecology be "deep"? Is it like "Deep Thoughts" on "Saturday Night Live"? Is this a concept that some take seriously and others might ridicule? What might be the possible causes of such ridicule? Allow yourself to consider these ideas freely.

———※◦※———

Dogen, the Buddhist teacher credited with bringing Zen Buddhism to Japan in the twelfth century, advised his students to walk in the mountains. "The mountains are walking," he wrote in his famous "Mountains and Rivers Sutra," and "He who does not know his own walking does not know the blue mountains are walking." John Muir, remembering his mountain walks, "ramblings," he called them, told his readers, "Go to the mountains and get their glad tidings." Dogen was talking about a pathway, an opening to enlightenment. Muir's enlightenment in the Sierra Nevada, his "range of light," was strengthened by an emphatic understanding for the mountains' feeling.

Dogen, John Muir, and Henry David Thoreau were a few travelers among many on journeys to further reaches of transpersonal awareness. They were not looking for just another roadside attraction on their journey. They were engaged in a journey of spiritual growth. Journey into wild nature on spiritual quests is an ancient practice in many religious and cultural traditions. Tibetans journey to specific holy mountains of Tibet. After ritual preparation, Yuroks in training to "make medicine" journey to the "high country" in the Siskiyou mountains in northwestern California. Australian aboriginals follow "dream lines" that connect the landscape to a cosmological framework. Such types of journeys take careful preparation, skilled guides, and willingness to take physical and psychological risks.

Joan Halifax, an anthropologist and Buddhist teacher, in her book *The Fruitful Darkness: On the Ecology of Initiation—Notes on Crossing the Threshold* (1993) describes her personal journey, including many travels in Africa, among American Indians, and among indigenous traditional shamans in Asia, to find what she calls "root-truth." She "sought fresh answers in ancient fields." Based on the ancient tradition of initiation, Halifax used her travels to learn truth about the practice of ecology, "an ecology of mind and spirit in relation to the earth." For Halifax, the practice of ecology is "based on the experience of engagement and the mystery of participation."

Contemporary travelers who are engaged in the practice of ecology need new skills and clear intentions, as well as skilled guides, when they enter wild

areas or areas of special biological and cultural sensitivity. These contemporary travelers are labeled in this chapter as "ecotourists." Ecotourists are both
open to new experience and, paradoxically, carry with them an ideological
framework which includes the premise that the traveler should both participate in the ongoing life of the community in which he or she is a visitor and
attempt to protect, nurture, help, and defend the ecological integrity of the
places visited.

In the following pages I focus on some of the attitudes we carry with us
when we begin a vacation or make a short trip into a nature reserve. I offer
what I hope are practical suggestions for preparing our attitude and changing
our behavior to conform to the needs of the area we are visiting. I am defining
nature tourism, or "ecotourism," as it is most frequently called in travel literature, as a visit to an area of biological interest or geological, biological and/or
Native cultural interest. Examples of participation in commercial trips featuring
themes of ecotourism include boat trips to visit areas frequented by whales,
trips designed to help the visitor explore some of the rich biological diversity
of the Amazon rainforest, trips to the Galapagos Islands of Ecuador, and visits
to the Kodiak area of Alaska to view grizzly bears in their native habitat. For purposes of this chapter I am also including under the term "ecotourism" certain
types of adventure—such as organized trips on whitewater sections of rivers,
mountain trekking in the Himalayan region, organized diving trips to waters off
Belize, and walking trips in mountain regions—travel that focuses on personal
growth for the "practice of ecology."

By some estimates of researchers for the travel industry, only about 10 percent of the two trillion dollars a year devoted to tourist travel is spent on ecotourism, but this form of tourism is growing by 20 to 30 percent a year. Even the
most dedicated bioregionalist sometimes has the opportunity to travel away from
home to enjoy a different area of biological, geological, or ecological interest.

INTENTION OF TRIPS TO NATURE RESERVES AND SPECTACULAR SITES

Ecotourism involves different attitudes than those found in other types of travel.
The general theme of tourist travel advertising is basically hedonistic—that is,
"enjoy yourself!" This seductive slogan is broadcast in thousands of brochures
and advertisements distributed by the tourist industry and by Chambers of
Commerce and local governments. Throughout this book I have emphasized
the importance of finding multiple sources of joy in our daily experiences; however joyful experiences are not hedonistic. In my definition, hedonism involves
unthoughtful indulgence. However, our general desire to find joy should not
distract us from our responsibilities as visitors.

For example, millions of people visit national parks and monuments in the
United States and Canada each year. The environmental impact of tourist development—including hotels, roads, golf courses, and shopping malls to serve the

desires of some tourists and increase profits for concessionaires—on wilderness values in many of these parks, as well as on biodiversity and biological habitats of specific creatures in national parks, has been documented by numerous researchers. The situation in national parks—decline in species diversity, pollution, overcrowded park facilities, lack of restoration programs—has led one sympathetic critic of national parks to call for a "regreening" of national parks (Michael Frome, *Regreening the National Parks,* 1992).

Unless we travel in an "environmental bubble" in poverty-stricken third world nations and poor regions of North America—taking all the conveniences of home with us or staying in hotels built to serve tourists from wealthy nations with conventional conveniences such as hot water and air conditioning— unless we close our eyes to the suffering around us, we must be aware of the wounds to forests, pollution in rivers, and hunger of people in local regions through which we travel.

In my experience, some people who embark on a trip to a nature reserve, wildlife sanctuary, or national park have an intention that is too narrowly focused. They want to see the "most spectacular," whether that be a special rock formation, a famous mountain, or a famous seasonal gathering of wildlife of some species— grizzly bears, for example, gathering to feed on the salmon in certain rivers in Alaska. An intention that is too narrowly focused may blind visitors to the possibility of learning something new about themselves and something new about the people in different cultures that they encounter on their journey. On virtually all kinds of nature trips, visitors must come in contact with local people, residents of the region. It seems to me that only by engaging in some form of denial can a visitor be unaware of political controversy over local environmental issues. The openness to new experience, for ecotourists, is not without a "filter." The ideological position of an ecotourist is to minimize his or her own impact on the processes of nature and to encourage socially responsible protection of the integrity of nature.

Exploring ways to protect the integrity, beauty, and life of a region visited on a specific trip to a specific location can be seen as part of the discussion within the group. While it is possible to visit some mountains, some deserts, some arctic regions by planning one's own trip, by chartering a boat or plane for a small group of people, or by hiring guides or trackers to find wildlife, tourists more commonly use a tour company.

Let me offer an example of some of the difficulties of intent that I discovered during a nature trip organized by a commercial tour company a few years ago. The trip was advertised by the tour company as a whale-watching trip to winter resting areas for California gray whales in lagoons on the west coast of Baja, California. I flew to San Diego and boarded a one-hundred-foot boat with thirty other intrepid whale watchers—teachers, retired people, a single mother with her teenaged daughter, several businessmen with their wives.

The itinerary included stops on nearly a dozen islands off the coast of Baja, looking at native plants and exploring some of the impact of exotic, introduced plants and animals on island systems. Some of these islands are occupied year-round by local fishermen and their families; others are occupied only during certain months when local fishing cooperatives send small groups of people to fish for specific species. On one island we observed burros grazing hillsides to bare dirt. We were told the burros were brought to the island a decade before to transport material to build a new lighthouse on the island for fishers. When the lighthouse was completed, the burros were left to fend for themselves.

The passengers on this trip had the opportunity to read scientific and nat-uralistic reports of the areas in an extensive library on board. Each evening the naturalist on board gave an illustrated lecture on the island or natural area we would visit the next day.

When our boat reached Magdalena Bay, we anchored in the zone pre-scribed by Mexican authorities, outside the area of the lagoon where most of the whales engage in amorous activities and bear their young. Each day while anchored in the lagoon we took small boats across the bay, watching whales playing in the waters. On a few occasions a friendly whale passed under our small boat close enough for us to see the hairs on its back and even to allow some of us to reach our arms into the waters and touch the back of the whale.

I was a young ecophilosopher and environmental activist at the time, and in evening conversations, I kept asking questions about our relationship with whales and about how we could help protect them (at the time there was no worldwide ban on commercial killing of whales). I asked the naturalist on board during conversations after his evening lectures, "Can we write letters to the Mexican government asking the government to take more stringent pre-cautions to protect the whales? About the burros we saw on the island—they are destroying native vegetation. Could we demand that they he transported back to mainland Baja?" After several attempts to ask such questions at dinner time, I was told by other passengers, in polite terms, to shut up. They were on this trip for pleasure, to enjoy their visit with whales, to commune with whales, not to discuss boring political questions or engage in political activism.

At the conclusion of our trip our leader admonished us to "keep in touch with each other" and provided us with the names and addresses of other pas-sengers. I did write each passenger soon after returning home, enclosing a copy of a whale action newsletter from some environmental group and suggesting that we write to a specific politician to get his support for whale protection leg-islation. I did not receive any responses to my letters.

After that experience I realized that I personally cannot separate my polit-ical activism from my recreation. I feel an obligation to do something to help when I visit a natural area. I discovered that my intention for traveling was not only to learn new information, experience different cultures and situations, but also to use my time as a traveler to engage in activism. Having clarified my intention, I have devoted my vacation and recreational travel during the past few years to exploring regions of my own backyard—in the Klamath-Siskiyou Mountains. I have hiked drainages that the U.S. Forest Service planned to clearcut. I have visited sites where massive forest fires burned through tree plantations planted by the Forest Service after ancient forests were clearcut. I have taken trips on the Trinity River by canoe with groups of students seeking body-mind-spirit awareness. We worked with the river as our teacher, carefully listening to birds calling before dawn, diving deep into the pools to watch fish, watching miners dredge in spawning gravels along the river. When I tell my friends, even some environmentalists, that I plan to travel to this watershed or that mountain trail this weekend or for a week long trip, I sometimes receive an incredulous "I've never heard of that place. Why would you want to go there?"

This leads me to discuss another issue concerning our intention when planning trips to natural reserves, wildlife sanctuaries, or wild areas. In some social circles of well-educated, well-traveled people in our culture, it seems that one can score points in conversations on some not-clearly-defined scale of worth with friends or coworkers by visiting a spectacular "named" place, especially a world-class named place.

During the 1960s and 1970s trekking in the Himalayas seemed to be in vogue. In the 1980s the rainforests of the Amazon became more popular with North Americans. In the early 1990s my hip friends are going to Lake Baikal and the farther reaches of Siberia.

Tourists, we are told in news accounts in 1992, are returning to Tibet after the worldwide expression of horror over the Tianamen Square massacre and repression of human rights in China subsided. Western visitors going to Tibet seek not only the spectacular scenery of the mountains and sight of the wildlife but also the beautiful culture of traditional Tibetans. Visitors, however, cannot be blind to the contemporary situation. The Tibetan people have suffered greatly at the hands of Chinese oppression. Tourists in Tibet, and many other regions of the world, may feel compelled to speak out against human rights violations, even—if they happen to be in a public place when the events happen—to film police attacks on protesters, public executions, or damage from government destruction of sacred sites, or damage due to warfare or ethnic violence.

If we do not clarify our intentions as visitors and if we are not psychologically open to the wide range of experiences available during our journey, we may overlook the richness of experience and possibilities for social activism based on our experiences as tourists.

Arne Naess led several expeditions to the Himalayan mountains from his home in Norway and, while looking for a new route up a previously unclimbed mountain, he came to the conclusion that the moral action in that situation would be to leave the mountain unclimbed. He concluded that the intent to conquer mountains can lead to hubris. Going to a mountain and not climbing can encourage modesty. Naess suggests that modesty has little value unless motivated by deep feelings, feelings of how we understand ourselves as part of nature. "The smaller we come to feel ourselves compared to the mountain, the nearer we come to participating in its greatness. I do not know why this is so" ("Modesty and the Conquest of Mountains," in Tobias, *The Mountain Spirit*). *Modesty* is an alien word for some visitors, who take pride in telling other people about their exploits in finding wild grizzlies or kayaking previously unrun whitewater stretches of wild rivers, but Naess suggests that modesty includes maintaining an attitude that emphasizes protecting the integrity of a place we have visited even when talking about and reflecting upon our visit after we return home.

Naess expresses a concern that some nature enthusiasts will increase human impacts on beautiful places by incessantly writing, lecturing, showing pictures about certain places, implying "you *must* visit this area." The implication is that some spectacular places are more important to visit than other places. After he visited the canyonlands of the American Southwest, some people asked Naess to compare them to the fjords of Norway. He admitted that in comparison, the fjords seemed small and cramped. But the comparison had no useful purpose. The fjords of Norway have their own beauty, their own integrity, and their own natural history that visitors can appreciate and learn from. Naess suggests that visitors appreciate the intrinsic value of their experiences in each place and not be caught up in a game of false comparison. This game has many variations: for example, "I've counted X number of species of birds at such-and-such sanctuary," implying that more species are better than the few species that live around one's home.

After experiencing the spectacular, that is the biggest whales, the deepest canyon, the highest mountain, the largest herd of wild rhinos—the tourist is inclined to say, "When will I see again the most spectacular?" The danger is clear, as Naess says, "In the long run such a person mostly will develop an urge and need for the spectacular and a decrease of sensitivity. Using a long range perspective [on our experiences] nothing is gained and something is lost." What can be lost is deep appreciation for the biological diversity, the beauty, indeed the suffering in our own home bioregion—watching the flowering and withering of native plants, caring for the changes in the scenery of our own backyards. Naess concludes that it is wise to consider carefully ways to make our tourist travels relevant to greening our lifestyles and helping to protect the areas that we visit.

GREENING TOURIST TRAVEL

We can incorporate our tourist travels as part of greening our lifestyles. While we do not need to stay in our home bioregions for our whole lives, we should recognize that there are limits on ecotourism that, if transgressed, can reduce planetary and personal quality of life. In an age of easier access to most regions of the earth, the social system as well as ecological processes of vast areas can be damaged by too many tourists traveling through the area within a period of a few years. For example, after the discovery of the tropical rainforests of the Amazon by North American tourists during the 1980s, within a few years the government of Brazil recognized the environmental damage of the tourist boom to certain areas. More port facilities and airports were built. Fossil fuel was imported to run tourist boats, and tour operators dumped empty oil barrels outside of local villages. More wildlife was killed to make artifacts for sale to

tourists. Cultural erosion began immediately as tourists began to transfer artifacts they brought with them—such as transistor radios—to indigenous people.
In a report published by World Resources Institute, Kreg Lindberg concludes
that mass tourism can upset the ecological balance in a region, disrupt the
economy of the region, and overwhelm the local culture.

Ecotourism is frequently portrayed as a benign industry, encouraging
preservation of wildlife and wild habitat while contributing to the local economy.
However, the ecotourist industry—transportation, food supply, accommodations, sales of souvenirs to tourists, wastes created by tourists, introduction of
diseases to local human and wildlife populations by tourists—can lead to
environmental and social problems in local areas within very short periods of
time. Even small groups of tourists regularly visiting the ice-free areas of
Antarctica, for example, can contribute to destruction of fragile lifeforms that
a visitor might not even notice unless educated by a knowledgeable tour
guide. Scientists tell us that only one percent of Antarctica is ice-free, and that
one percent supports the greatest variety of organisms on the continent.
Facilities for scientists have been constructed on the ice-free area.
Construction of tourist facilities in Antarctica in this same area continues the
trend toward more road building, runway construction, and creation of dumps
by the major signatories to the Antarctic treaty—the USSR, U.S.A., France,
Chile, Argentina, New Zealand, and Australia. Reports compiled by
Greenpeace indicate that oil drums are rarely taken back to the countries of
origin. They are left by the tarmac where their contents are used to fuel the
planes used by tourists. Facilities for traveling scientists and tourists are far
from the simple huts and sled dogs that the explorers took with them to
Antarctica in the early decades of this century.

How can ecotourists address the problematic moral issues and environmental and social impacts of their own activities as visitors? Recognizing that
the interest we have in ecotourism is part of our general concern for environmental quality, protection of endangered habitat, and endangered ecosystems,
how can we travel in ways that encourage healing rather disruption of the
places we love? How can we avoid, as some commentators have noted, "loving
the place to death?" One general principle is to choose activities of intrinsic
value rather than activities valued for social prestige. We can avoid traveling to
sensitive habitats just to escape from depression, loneliness, and poor human
relationships at home. We can place more emphasis on the traveling—enrich-
ing our lives along the way—rather than on focusing arrival at some destination.

We can avoid activities that local environmentalists, scientific studies, and
government agencies suggest are disruptive to local habitat of wildlife and flora.
For example, extensive research has documented the habitat needs of the
desert tortoise in certain regions of the American Southwest. It seems irresponsible, therefore, to participate in a mass, cross-country motorcycle race
across the California desert or to go offroading in areas of critical habitat in
desert regions. Even though many people enjoy the challenge of offroad racing,
"the needs of nature come first."

Some visitors have discovered that they find joy in restoring areas that were disrupted by their own trips and by other visitors who followed them. For example, regions of the Himalayan Mountains, favored by climbers and other visitors during the 1960s and 1970s, began showing signs of overuse during the 1980s. In 1992 Sir Edmund Hillary, the first man to reach the summit of Mount Everest, called for a moratorium on climbing the world's highest mountain so that it can recover from tourism. He said that his team, during their 1953 expedition to climb Everest, contributed to the junk heap on the mountain. In their exuberance in climbing a mountain uncontaminated by humans, they unthinkingly began littering the mountain. Peter Stone, a spokesperson for a group of mountaineers attending the Earth Summit in Rio in June 1992, launched "An Appeal for the Mountains." whose aim is to preserve mountain regions of the earth for small populations of people living in the mountains and for environmentally responsible visitors.

One of the most general guidelines for ecotourism is to develop sensitivity— to the needs of local residents and to the situation of flora, fauna, and whole ecosystems. Stories brought back from tourists in some Himalayan regions indicate, for example, that tourists trekking in the mountains have increased pressure on local wood supplies. Tourists may be required to carry fossil-fuel stoves, and their own fuel supply, in some areas.

A general rule for visitors in designated wilderness areas administered by the U.S. Forest Service is "Pack it in, pack it out." Leave no trash, bury no trash. That rule is useful for ecotourists traveling in many regions of the earth. If you don't take home the personal trash you generated when you fly back from a tourist trip, work with the tour guide to insure that wastes will not create problems for local inhabitants, including nonhuman inhabitants. On trips by canoe or raft on many rivers in the American West, for example, the U.S. Forest Service and Bureau of Land Management require that defecation by humans be carried in containers and transported out of the river canyon at the end of the trip to an authorized organic waste disposal site. Recyclable materials, including bottles and cans, are taken to recycling centers. Part of the cost of the trip is the cost of handling wastes generated on the trip.

There are many guidebooks on how to act like a native in Paris or how to order a meal in Madrid without offending local sensibilities. We need similar guidance on how to relate to local flora and fauna in their native habitats. In the mountain gorilla park in the Central African nation of Rwandi, for example, small groups of tourists under the direction of an experienced guide can take day hikes in the mountain rainforest. If they encounter a group of gorillas, the tourists must stay a designated distance from the gorillas and can observe them only for one hour because humans can transmit respiratory diseases to gorillas and thus endanger their health. When tourists visit the Galapagos Islands reserve, they are required to stay on designated paths, and not attempt to touch any of the native animals which show no fear of humans (sailors in the nineteenth century easily captured giant turtles by turning them on their backs; the turtles were thus kept alive on the ships until the cook butchered them for the crew).

Ecologists use terms like *fragile, threatened,* and *sensitive* to describe regions that are increasingly popular with adventurous ecotourists. These include tundra regions, tropical coral reefs, islands, marine mammal breeding and resting areas, and waterholes in the deserts. In part of North America, some areas that are particularly sensitive to human impact are being zoned as "natural areas" or "botanical reserves," and only authorized scientists or visitors in small groups under the supervision of a qualified naturalist are allowed to visit the area.

Managers of some nature sanctuaries allow limited numbers of visitors who agree to follow guidelines on how to approach, when to approach, and what to wear when approaching wildlife. When contemplating a trip to a park, reserve, or animal breeding area, a visitor can anticipate some of the potential impact of his or her activities. For example, if one is interested in fishing, consider how heavily the area to be visited might be impacted by sports fishing. (Some nature conservancy groups encourage, or even require, catch and release flyfishing with the use of barbless hooks in the areas they manage.)

Hunters can avoid taking trophy specimens—and certainly hunting any threatened or endangered species. While some people object to any hunting on ethical grounds, some ecologists argue for eradication programs of certain introduced or non-native species in fragile and disturbed ecosystems. For example, some government agencies encourage hunting for wild goats and feral pigs on some lands they manage on the Big Island of Hawaii. Animals were introduced by European explorers and they bred prolifically, destroying native wildlife and flora. Killing wild goats, pigs, and burros—or, as an alternative, live-trapping the animals and transporting them out of sensitive areas and fencing them out—is one possibility for returning some of the habitat to native species.

Another general rule to remember in contemplating nature tourism is that visitors should not expect to impose their culture on the native people of the area visited. Unlike religious missionaries or military authorities, ecotourists are in the region to learn about it and to experience natural and cultural diversity. Openness of spirit, openness of mind, willingness to be flexible in daily routines allow a visitor to appreciate the joys of each day.

When a visitor arrives with an open mind, ready to participate as much as possible in local daily routines, ready to be transformed in a spirit of solidarity, and ready to accommodate rather than be accommodated by local residents, an exchange is possible that is beyond an exchange of commodities or exchange of money.

Travels to many areas of interest to ecotourists means travels to areas heavily impacted by industrial civilization or travels through areas of extreme poverty and suffering in third world nations. Buddhist traditions teach us not to avoid suffering but to attempt to understand the causes of the suffering. One implication of this principle is that ecotourists do not avoid visiting damaged ecosystems, areas of massive clearcut forests, toxic waste dumps, nuclear testing sites, and mining districts. Joanna Macy, one of the originators of the Nuclear Guardianship Project, has taken groups of people on pilgrimages to Kiev, Ukraine, to sit with the beings who suffered from the Chernobyl disaster.

Just as some people visited the sites of Nazi death camps, which are included on the list of World Heritage Sites, to remember the suffering of the victims of these camps and resolve that this horrible action should not happen again, so some people visit Bhopal, or Three Mile Island, or the Hanford nuclear site, or the Nevada test site, or Alamogordo, or Hiroshima, or Bikini Island to witness the horrors of the atomic age and resolve that this knowledge that humanity now has will not be used to destroy the earth.

An ecotourist who is both socially and environmentally conscious does not overlook injustice, poverty, unequal distributions of power, or irresponsible behavior by government officials who are administering and managing national parks, wilderness areas, nature reserves, or other designated sanctuary areas. They should ask if officials are managing the area for the primary benefit of the flora and fauna, of the ecosystem, or managing it for the aesthetic and personal comfort of visitors. Ecotourists can resolve to be active in political and educational efforts to protect the integrity of the area visited after returning home.

Some people travel specifically to show solidarity with local environmentalists. A former student of mine, a resident of Hawaii, owns a small parcel in the Puna district of the Big Island. He invites groups of students and anyone else who is interested to visit his home and camp on his property. He leads visitors on hikes in the geothermal district. For several years he has been active in exposing the threats of geothermal development on the Big Island. Visitors are encouraged to contact government officials concerning geothermal development when they return home. They are encouraged to take photos, make videos, describe their experiences and, when they return home, to help the effort of protecting endangered rainforests on the Big Island.

Showing solidarity can also mean participating in demonstrations with local environmental groups. One time while visiting Mexico City to attend a conference on a different topic, I learned that a march opposing the opening of the first nuclear power plant in Mexico was being held in the city. On Sunday morning the main streets leading to the Presidential Palace were closed to traffic for several hours, and I and several other American tourists joined tens of thousands of Mexicans—farmers, students, women with children—in the demonstration against the nuclear power reactor. (I was opposed to repeating the mistake made in my home bioregion where we must now be perpetual guardians of the decommissioned, and contaminated, Humboldt Bay Nuclear Reactor site after less than fifteen years of service.) While traveling in a region, visitors can check with local environmental groups on planned actions. A friend recently returned from a southern state where she marched in solidarity with people who were opposing the siting of a toxic waste dump in their neighborhood. Just as visitors to Tibet and other parts of China bring home reports of human rights abuses, so ecotourists, on occasion, have the opportunity to bring back firsthand accounts of abuses to native vegetation, forests, flora, and fauna.

SERVICE TRIPS

Groups of recreationists focused around many different kinds of activities—kayaking, skiing, backpacking, surfing, rockclimbing, sailing, and diving, for example—are increasing the number and variety of service trips designed to engage in work projects in some of their favorite recreation sites. In sporting activities on the land, or in the water, people can see degradation occurring and sometimes suspect that they are suffering possible health hazards from industrial or farming operations near their recreation site.

For example, surfers in California saw, during the 1960s and 1970s, degradation to favorite surfing breaks—from breakwaters constructed by the Corps of Engineers, from dredging operations, and raw sewage from outfall pipes in bays and near surfing reefs, as well as from closure of historic public access to beaches by people building housing projects on beachfront property. A group of surfers in California formed the Surfrider Foundation, a nonprofit organization that engages in educational activities, and discussions with public officials, and serves as an information clearinghouse. Upon hearing the stories of surfers who surfed near the outfall of two pulp mills on the Samoa peninsula near the mouth of Humboldt Bay—stories of dizziness after surfing, eruptions on the skin and other ailments—and after discovering thousands of Clean Water Act violations, the Surfrider Foundation took Simpson Timber Corporation and Louisiana Pacific Corporation, owners of the pulp mills, to court. With the support of the EPA, the surfers won a settlement whereby the corporation agreed to install new equipment designed to reduce the danger of dioxin and to pay the surfers to create a new management plan for the beach area where the outfalls from the pulp mills are located.

The Sierra Club national outings department offers a variety of types of service trips. Trip projects include rebuilding sections of trail in designated wilderness areas, cleaning up campsites in wilderness areas, and other activities designed to restore human-damaged areas.

A group working through Earth Island Institute, located in San Francisco, offers trips to the Lake Baikal region of Siberia designed to assist the Russian national park service in their efforts to develop management plans for two national parks, one on either side of the lake. Among those recruited for these annual trips are social scientists, recreation planners, retired U.S. park administrators, and others who could provide some advice to their Russian counterparts while at the same time learning about and enjoying the Lake Baikal region.

Private conservancy groups sponsor work projects on areas under their management. In my own bioregion, the Friends of Dunes, a membership group that supports a project of the Nature Conservancy called the Lansphere-Christiansen Dunes, a unique coastal dune forest system between Humboldt Bay and the ocean, sponsors an annual trip to the dunes for supporters of the dunes projects. Friends of the Dunes, some of whom live hundreds of miles away, come to take walks with naturalists, watch birds, and help with the annual "lupin bashing": An exotic species of lupin (a shrub) was introduced by timber companies in the last century in an attempt to stabilize the shifting dunes so railroad tracks could be laid over them. As with many exotics, this exotic species of lupin outcompeted native vegetation, including a species of wallflower that is currently listed as threatened and endangered. Each year the Friends of the Dunes engage in manual labor to cut down and tear out lupin and restore the natural conditions that allowed wallflowers and other native plants to flourish on the dune system.

In another example, in southern California waters containing shallow reefs in the Santa Barbara Channel, divers for many years took abalone, a prized shell creature that, when prepared properly, provides a gourmet meal. The state Fish and Game Department noted a rapid decline in the number of mature abalone and put restrictions on the taking of abalone. Fearing that the abalone population would become threatened, scientists began an experimental program to raise abalone seeds—small offspring of abalone. Sport diving clubs were enlisted to help with placing these seed abalone in their natural habitat on the reefs. Sport divers, combining their interest in abalone with a sense of responsibility for the rapid decline in numbers of mature abalone, volunteered to put back abalone and to watch over them by reporting poaching activity.

In the Central American nation of Costa Rica, Americans and Canadians have volunteered to spend their vacations in Monteverde, a tropical forest reserve along the continental divide, working on a variety of projects—scientific studies, development of nature interpretation centers, and restoration of areas now included in the reserve that had been previously logged.

A group organized in Humboldt County, California, traveled to southern Chile to hike through and document on film the ancient temperate forests in that nation. After seeing the invasion of American logging corporations into Chile, this group came home resolved to help protect the forests of Chile and formed a new organization, Ancient Forests International.

Each of these examples shows how people can engage in adventure travel mindfully. These are not "fun trips" with endless parties, luxury hotels, and rich food. These are trips for people who are open to new experience, who want adventure but also recognize that they can help heal the wounds of the earth. They have a commitment to live life richly, rich in experience, compassion, and sharing of knowledge they have acquired in life.

A growing number of people forgo conventional tourist travel all together and devote their vacations, holidays, time away from school or jobs to join with others who are directly addressing gross destruction of nature. During the summer of 1990, hundreds of people—students, retirees, annual vacationers, and some unemployed—came to Humboldt County to participate in Redwood Summer, a series of demonstrations seeking protection for the remaining ancient forests held by private timber corporations and asking for radical reforms in timber harvest practices. A base camp and communication center was set up by local activists, and visitors helped to organize and conduct media campaigns, street theater, demonstrations at the gates of pulp mills, and educational efforts of many kinds. Many of these visitors returned to their homes with renewed commitment to work for reform of forest practices and to speak out for the long-term integrity of forests.

Some local environmental groups in northwestern California have produced information fact sheets to distribute to visitors coming through the redwood region. Environmentally conscious and concerned visitors want to take responsibility, to question the abuses of the forests by the timber corporations. People who love the forests, who are visiting regions experiencing gross deforestation encouraged by government agencies—regions such as British Columbia, Alberta, Maine, Washington, Oregon, California, Montana, Burma, Thailand, Malaysia, Indonesia, Victoria state of Australia, Chile, and Siberia, as well as Brazil—sense they have an obligation to tell those responsible for such deforestation their views on these issues.

Visitors to any areas of deforestation have an obligation, it seems to me, to look behind the "scenic strips" along the roads and to express their grief, concern, even outrage at irresponsible logging practices—nonsustainable harvest rates, clearcuts, destruction of the habitat of threatened and endangered species, and aerial spraying of herbicides. Some environmental groups encourage tourists to boycott certain areas until governments and corporations cease environmentally destructive practices. For example, some environmental groups in British Columbia encouraged tourists to boycott their province and inform the provincial government that they were boycotting the province because of the destructive forestry policy of the provincial government.

One option, which increasing numbers of people are taking, is to join an expedition sponsored by a group such as Greenpeace or The Sea Shepherd Society, or the Costeau Society, to sail to areas that are threatened by inappropriate actions and peacefully witness for the whales and other marine mammals, for the ocean ecosystem threatened by French nuclear testing in the South Pacific, for the life in the oceans threatened by massive drift nets in the North Pacific. Just as visitors go on pilgrimages to sites of nuclear disasters, so they go to sites where massive exploitation of forests, oceans, deserts, and coral reefs is occurring, to witness for the life of these beings.

Some groups specialize in helping volunteers participate in scientific studies, ongoing studies, of ecology, habitat change, wildlife behavior. Ecotourists can become involved in local conservation programs. Some of my former students are working in a valley in Ecuador with an Indian tribe, buying land, mapping reservation boundaries, working in community development.

The ecotourist is a new breed of traveler, both environmentally and socially conscious. As with other lifestyle decisions discussed in this book, decisions concerning whether to travel, when to travel, and the intent of travel open up many questions concerning who we are, our relationship with nature, and how we can engage the process of healing ourselves while helping to heal some of the wounds in the Age of Ecology. We want to experience wild places and wild beings joyfully. We resolve that our presence will not contribute to more degradation of the landscape, more misunderstanding, more suffering.

QUESTIONS TO ASK TOUR COMPANIES

For many people desiring to visit a natural area, wildlife area, or scenic attraction, the first and most important question is which tour company to choose. Seek a responsible tour company, one willing to answer questions about the impacts of tourist travel and willing to discuss all aspects of the trip, including the political actions that might help further protection of native ecosystems. Keep the ecotourist group small.

Questions to ask the tour operator might include the following:

- Does the operator take local people along on tours, that is, people from that nation or region of that nation?
- Does the operator hire competent local guides who know the difference between observing wildlife and harassing wildlife?
- How much of the money spent by tourists on this trip goes directly to the local economy as opposed to leakage—money that goes back to the United States or Canada because the goods bought in the local area were transported to the tourist site—for example film, shampoo, aspirin, even canned foodstuffs that are sold in local stores but made in America?
- Does the tour operator know local customs, sacred sites, what birds or animals are on the threatened list, and what wood products are from endangered local trees?
- When a company says that X dollars out of the total fee for the tour will be donated to "protect the rainforest" or "protect the whales," ask specifically if the donation will go to a local grassroots environmental group or to a large, national organization that has a big public relations department but perhaps is not as effective as a local, and frequently underfunded, group.
- The socially conscious response, "You are helping the local economy," might be explored by asking what segments of the local population are "helped" by my travel? How are women workers treated in the hotels and other public facilities that I use?
- How are trash and garbage managed on the tour? Small villages near a site of nature observation may have trouble disposing of trash brought by visitors. On some desert islands I have visited, there are only a few places to defecate, and tourists from previous tour boats have discarded trash that does not decompose rapidly.
- Does the tour operator provide simpler, rather than more luxurious accommodations? Ecotourists who are living a simpler lifestyle at home may feel uncomfortable staying in luxury hotels while on vacation just because the host government wants "tourist development" by multinational hotel chains, auto rental agencies, and national tourist development boards. Tourists who are ecologically conscious will demand facilities that have less impact on—and are more representative of—natural, social, and cultural environments.

RESPONDING JOURNAL

Devall echoes Aldo Leopold's claim that humans should not consider themselves superior to other forms of life on this earth, but rather that they are "plain citizens " of the world. Is this a concept that you endorse, or do you want to qualify the idea to shape it to your own beliefs? Explain in a journal entry of several paragraphs.

QUESTIONS FOR CRITICAL THINKING

STRATEGY

1. Do you think this essay requires a genuine interest in the subject on the part of its readers? That is, does a reader have to be a confirmed ecologist to find Devall's essay interesting and persuasive? If, as a writer, you feel that your audience will be deeply interested in the ideas you are presenting, are you excused from making your writing as interesting or engaging as possible? Reflect on these questions as you examine the facts and references Devall mentions.
2. Explain how effective you find the headings for the sections to be. Would you adopt similar structures in your own writing? Does such dividing help you as a reader? What other divisions do you see in the first part of his essay? Would he have gained anything by indicating any of these divisions? If yes, what? Similarly, explain any benefits you see in his final set of questions.

ISSUES

1. Consider the topic of dominance Devall examines. Do you find yourself disagreeing with any of Devall's points or implications? Explain why you believe his arguments should be qualified or why you think he argues effectively.
2. Are you satisfied that you understand the concept of ecotourism? If you do, what parts of the excerpt particularly helped you? If you do not, what additional sorts of information would you need to feel comfortable?

COLLABORATIVE WRITING ACTIVITY

Jot down some notes for a statement about your personal convictions regarding the way people should regard the natural world. Support your arguments with references to sources that you respect. If you cannot cite them accurately, do the best you can for now. After you have read your ideas to your group, discuss with the other group members the sorts of authorities that you used to make your claims. Draft a statement of "ecology" and appropriate "ecotourism" that represents your group's consensus.

from "Ecoterrorism: The Dangerous Fringe of the Environmental Movement"
Doug Bandow

An editor, columnist, attorney, and policy maker, Doug Bandow has written extensively on ecological and environmental topics. Born in 1957, he served during the Reagan administration as a Senior Policy Analyst and also performed other duties at the White House. Presently he is a Senior Fellow at the Cato Institute in Washington, D.C., a public policy research foundation examining and publicizing "traditional American principles of limited government, individual liberty, and peace."

A columnist for the Copley News Service since 1983, Bandow has also edited the journal Inquiry *and worked with the various Cato Institute publications.*

EXPLORING JOURNAL

Explain what *ecoterrorism* means to you; what difficulties do you have in providing a definition with which you feel comfortable? Have you heard or read about any activities that might come under the heading of ecoterrorism?

———⟶•◀———

As the twentieth anniversary of Earth Day approaches, environmental activists and private citizens alike are reflecting on the state of the earth's ecology and what policies best can make the world cleaner. One environmental matter, however, is receiving little attention. Individuals and scattered bands of environmental or ecological radicals, usually called ecoterrorists, have been sabotaging industrial facilities, logging operations, construction projects, and other economic targets around the country. They have inflicted millions of dollars in damage and have maimed innocent people.

These ecoterrorists are a tiny, fringe group. They in no way represent America's broad environmental movement. Yet, mainstream environmentalists and the press remain strangely silent about the atrocities committed by the ecoterrorists. By failing to police their own movement, and by failing to denounce loudly and openly the ecoterrorists, mainstream environmentalists risk bringing their entire movement into disrepute. It thus is time for mainstream environmental groups and their supporters in Congress to disassociate themselves from those who use violence in the name of the environment and to see that they are brought to justice.

THE ROOTS OF ECOTERRORISM

In the early 1970s a lone environmental activist, identified only as "The Fox," engaged in a sustained campaign of eco-sabotage, also termed ecotage, against

Chicago-area firms. For three years he committed acts ranging from vandalizing the offices of corporations to more serious and dangerous crimes such as plugging industrial drains and smokestacks. Around the same time, a group in Minnesota called the "Bolt Weevils" and one in Arizona called the "Ecoraiders" carried out similar activities.

The concept of ecoterrorism gained some attention in the book *Ecotage!*, a "do-it-yourself" guide published in 1972 with the support of Environmental Action. Based on the results of a contest soliciting eco-sabotage ideas, this book extolled the activities of "The Fox," who, it argued, "deserves special credit because he has put his ideas into action, whereas for many, ecotage will remain a fantasy." The book also praised "the Billboard Bandits in Michigan, the Eco-Commandoes in Florida," who carried out their own disruptive activities, and contended that "if Thomas Jefferson, Patrick Henry and George Washington were alive today they'd be ecoteurs by night."

While authors Sam Love and David Obst explained that "we are not advocating that those who buy this book go out and try each one of the tactics included," they added that "it is important for readers to become aware that such ideas do exist and that there are already groups actively involved in implementing some of them."

A few years later, environmental activist Edward Abbey romanticized ecotage in his novel, *The Monkey Wrench Gang*. In this story, four people roam the West wreaking havoc, destroying power poles, railroad lines, billboards, and any other signs of civilization that mar the landscape. The book concludes with the blowing up of a bridge over the Colorado River. The book's message: those genuinely concerned about the environment are entitled to use virtually any tactic, perhaps excluding murder, to stop development. Abbey, who died in 1989, became the spiritual adviser and symbol for activists who turned to outlaw resistance. "If opposition is not enough, we must resist. And if resistance is not enough, then subvert," he said.

In 1981, Dave Foreman, a former lobbyist for the Wilderness Society, founded "Earth First!" This group, Foreman admits, was formed "to inspire others to carry out activities straight from the pages of *The Monkey Wrench Gang* even though Earth First!, we agreed, would itself be ostensibly law-abiding." Strictly speaking, Foreman calls "Earth First!" a movement rather than an organization; there are no membership lists nor officers, for instance. But the group, with about 10,000 people receiving its newsletter, provides a focal point for those interested in destructive and violent forms of protest. "Earth First! as an organization does not support or condone illegal or violent activities" runs a disclaimer in the newsletter. However, it adds: "what an individual does autonomously is his or her own business."

Yet Foreman joined environmental activist Bill Haywood to write *Ecodefense: A Field Guide to Monkeywrenching,* a book that has sold more than 10,000 copies. While purporting to be for "entertainment purposes only," its 311 pages offer detailed advice on how, illegally and violently, to sabotage attempts to develop land and other resources. It describes how to drive spikes into trees to shatter chainsaws and saw mill blades when these cut the trees and logs. This "tree spiking" can injure lumberjacks and mill workers severely. Road spikes are recommended to flatten tires. Methods for destroying roads, disabling construction equipment, and cutting down power lines are discussed. In one chapter, the authors explain that power lines "are highly vulnerable to monkeywrenching from individuals or small groups."

During an Earth First! demonstration at the Arches National Park in mid-1981, power lines in nearby Moab, Utah, were cut. Foreman said that Earth First! was not directly responsible for such acts, but he added that "Other people in Earth First! have *done* things, not as Earth First! though . . . Earth First!, a group, is not going to do any monkey-wrenching. But if people who get the Earth First! newsletter do that, that's fine."

In a later interview he went even further, arguing that monkeywrenching "is morally *required* as self-defense on the part of the Earth."

DEEP ECOLOGY

Underlying the activities of many members of Earth First! and probably most ecoterrorists is the ideology of "Deep Ecology," which places the protection of nature above the promotion of humankind. The principles of Deep Ecology were first enunciated in 1972 by Norwegian philosopher Arne Naess. California sociologist Bill Devall and philosopher George Sessions of Sierra College in California are among the more prominent American Deep Ecologists. Naess advocates "a long range, humane reduction [in the world's population] through mild but tenacious political and economic measures. This will make possible, as a result of increased habitat, population growth for thousands of species which are now constrained by human pressures." According to environmentalist Alston Chase, a newspaper columnist and chairman of the Yellowstone National Park Library and Museum Association, who does not support Naess's views, "poets, philosophers, economists, and physicists joined the ecologists in a search for a new beginning." Through what Chase describes as a "swirl of chaotic, primeval theorizing, patterns began to form, and themes resonated," particularly the notions that nature is sacred and everything within the universe is interconnected.

ECOTERRISM: A PRESENT DANGER?

While most of the actions of ecoteurs to date mainly have destroyed property, injury of innocent people is now becoming part of the ecoterrorist record. Spiking trees with metal or ceramic spikes, the latter of which are not detected by metal detectors, is common in the western U.S. Incidents have also occurred in Canada and Australia. In May 1987, a young California sawmill operator was severely injured when a spike shattered a band saw. A local Earth First! official blamed the sawmill for jeopardizing its workers' lives. Earth First! leader

Foreman said workers fearing injury could quit and that to him "the old-growth forest in North Idaho is a hell of a lot more important than Joe Six-pack." Loggers in California and Oregon since have been injured.

Northwest Forestry Association spokesman Mike Sullivan of Portland, Oregon, says that spiking incidents have been reported throughout the Northwest. After the injury of the California mill worker, the Forest Service said it planned to step up efforts to prevent spiking, but argued that the practice was "not a great epidemic." Though spiking has increased during the mid-1980s, explains Forest Service spokesman Jay Humphries, "there is still less than 100 incidents a year. Most of the illegal activity and threats to Forest Service land are related to marijuana growing, not environmental ecotage."

Many loggers remained unconvinced. In 1988, one Washington lumber mill lost $20,000 worth of blades from cutting spiked trees.

In another incident involving personal injury, demonstrators, some armed with knives and clubs, attacked Forest Service personnel involved in herbicide spraying in the Siskiyou National Forest.

REACTIONS TO ECOTERRORISM

Increased enforcement has been the traditional response to ecoterrorism. Companies are more vigilant in protecting their equipment; the Forest Service tries to watch more closely for saboteurs of trees, roads, and equipment. In 1988, Congress passed a bill offered by Senator James McClure, the Idaho Republican, making tree spiking a federal offense. In 1989, Representative Charles (Chip) Pashayan, the California Republican, introduced legislation to stiffen penalties and create a reward program for informers against tree spikers.

[The next year] the Washington Contract Loggers Association created a Field Intelligence Report to track the activities of ecoteurs and has established a reward program for information leading to the apprehension of such criminals. Similarly, the Mountain States Legal Foundation, based in Denver, Colorado, established an ecotage hotline. In the first two months of hotline operation, Foundation President William Perry Pendley received reports of ecotage from California, Colorado, Idaho, Nevada, Oregon, and Washington. Mountain States also established a clearinghouse to file civil damage actions against saboteurs and to assist the government in prosecuting violators.

THE ENVIRONMENTALISTS' RESPONSE

Adequate penalties are a necessary part of any effort to combat ecoterrorism. Yet western forestland and deserts are too sparsely populated to be patrolled and defended effectively against the determined ecoterrorists. The best defense against ecotage is for mainstream environmentalist community and political leaders and for businessmen to speak out frequently on the issue.

The message should be twofold: 1) violence is not justified as a response to perceived wrongs to the environment, and 2) the protection of human life remains society's paramount responsibility.

Particularly important is the role of the major environmental groups. Though none of them endorse ecotage, few have shown much enthusiasm for publicly criticizing the practice. Some even aid violent ecoteurs. David Brower, past executive director of the Sierra Club and current chairman of Friends of the Earth, gives Earth First! office space and has defended the organization's activities. "I think the environmental movement has room for lots of different views broadcasting on many channels," said Brower. "I'm certainly not going to be against civil disobedience."

Brower has said that "Earth First! makes Friends of the Earth look reasonable. What we need now is an outfit to make Earth First! look reasonable." When challenged to disavow ecoterrorists in 1983, the Sierra Club's then-executive director and now chairman Michael McCloskey responded that "we no more have an obligation to run around denouncing extremists using the environmental movement than Republicans and Democrats have an obligation to go around spending most of their time condemning the views of left or right wing extremists."

McCloskey ignores the fact the Republicans and Democrats have done just that. They overwhelmingly reject the use of violence to achieve their goals. They never have supported the use of tactics that may maim and even kill. And when such cases occur, these political movements have acted to disassociate themselves with the culprits. In the 1950s, the American labor movement purged itself of most communist members and influence. In 1989, George Bush and Republican Party Chairman Lee Atwater denounced the election of former Ku Klux Klan leader David Duke as a Republican to the Louisiana State Legislature and expelled him from the national party.

The political organizations closest to the terrorist group's ideological views should separate themselves from its activities and to help mold a broad social consensus against its activities. The Sierra Club and other organizations, because they are committed to many of the goals of Earth First!, have a special duty to discourage violence committed in the name of the environment. . . .

PEACEFUL CHANGES IN GOVERNMENT POLICY

Environmental destruction underwritten by the federal government certainly should be the target of reformers. But this does not justify extremist tactics, civil disobedience, and violence. Nor does this justify ignoring the balance that must be struck between ecological concerns and economic development. It is neither humane nor does it serve the public good to shut businesses needlessly, to restrict the supply of housing by prohibiting construction of new homes, or to drive up the costs of energy by reducing electrical generating capacity. There are ways to protect the environment without paying those prices. Some of these ways include privatization and ending of federal development subsidies. Environmental policies must be designed around natural market forces which would deliver more ecological amenities at lower cost.

Americans want to preserve a clean world—to conserve their environment. Americans too want an economy that offers them increasing economic opportunities. How to balance these two goals all too often splits Washington between myopic conservationists and equally myopic developers. Out of this split comes the ecoterrorists, who believe that anything short of complete victory for "the environment" is a moral as well as a practical disaster.

Their extremist philosophy is leading to a guerrilla movement that is destroying property and injuring the innocent and one day will kill innocent workers or park employees.

To prevent this, policy makers and particularly establishment environmental groups must respond to the ecoterrorists by rebuilding the moral consensus against the use of violence. The environmental movement has a special responsibility. It must no longer tolerate, let alone encourage, the ecoteurs. In particular environmental groups should publicize the fact that the ecoteurs' violence sabotages legitimate environmental groups. These mainstream groups thus should speak out forcefully to encourage their members to distance themselves from violent and destructive activities.

RESPONDING JOURNAL

Describe what you feel Doug Bandow's position in relation to the environment must be. What definition of his position can you provide ?

QUESTIONS FOR CRITICAL THINKING

STRATEGY

1. Think about the structure of Bandow's first sentence; how does it reflect his position on the issues he covers? Think particularly about his statement that "environmental activists and private citizens alike are reflecting on the state of the earth's ecology."

2. In his conclusion, Bandow does not offer the kind of specific detail that he uses to support his points earlier in the essay. Comment on the effectiveness of his conclusion and decide if he could have presented material in his last remarks to strengthen his case.

1. On page 280, Bandow draws a parallel between the environmental movement on the one hand and the Democratic and Republican parties on the other. He does so by extending and commenting on Michael McCloskey's original suggestion that the three groups have parallels. Which position, Bandow's or McCloskey's, do you find more convincing? Why?

2. In his last section, Bandow refers to "the balance that must be struck between ecological concerns and economic development." To what extent do you find this to be a valid or necessary claim?

COLLABORATIVE WRITING ACTIVITY

After writing a definition of ecoterrorism, explain in a few sentences how you feel the United States should deal with the threat of ecoterrorism. Do you in fact feel that it is a threat? As you read your work to the rest of your group and listen to their responses, think about how your personal definition of ecoterrorism is changing or developing.

from *Confessions of an Eco-Warrior*
Dave Foreman

Dave Foreman has worked as a ranch hand and lobbyist; he has worked in Utah, New Mexico, Washington, D.C., and other locations where ecology and commercial interests have been at odds. As Doug Bandow points out in his essay, Foreman is the founder of Earth First! His most famous (or infamous) book is Ecodefense: A Field Guide to Monkeywrenching, *which he wrote with Bill Haywood. Probably no one except Edward Abbey, who in* The Monkey Wrench Gang *laid out a scenario for an ecological terrorist group, enjoys a greater reputation among extreme champions of the environmental movement than Foreman.*

Confessions of an Eco-Warrior *is Foreman's most recent book, appearing in 1992. Here he attempts to look back over his roles in the environmental movement and to suggest where he feels the movement needs to move next. In setting up a plan of action, he argues for a special kind of environmentalist.*

EXPLORING JOURNAL

Explain what *eco-warrior* might mean. Think about the connotations of both parts of the word; what associations do you bring to these shades of meaning? Think also about the denotations of the words; how does this aspect help? Comment on what other information you feel you need to complete the definition.

———◆◆◆———

If opposition is not enough, we must resist. And if resistance is not enough, then subvert.

—EDWARD ABBEY

The early conservation movement in the United States was a child—and no bastard child—of the Establishment. The founders of the Sierra Club, the National Audubon Society, The Wilderness Society, and the wildlife conservation groups were, as a rule, pillars of American society. They were an elite band—sportsmen of the Teddy Roosevelt variety, naturalists like John Burroughs, outdoorsmen in the mold of John Muir, pioneer foresters and ecologists on the order of Aldo Leopold, and wealthy social reformers like Gifford Pinchot and Robert Marshall. No anarchistic Luddites, these.

When the Sierra Club grew into the politically effective force that blocked Echo Park Dam in 1956 and got the Wilderness Act passed in 1964, its members (and members of like-minded organizations) were likely to be physicians, mathematicians, and nuclear physicists. To be sure, refugees from the mainstream joined the conservation outfits in the 1950s and 1960s, and David Brower, executive director of the Sierra Club during that period, and the man most responsible for the creation of the modern environmental movement, was beginning to ask serious questions about the assumptions and direction of industrial society by the time the Club's board of directors fired him in 1969. But it was not until Earth Day in 1970 that the environmental movement received its first real influx of antiestablishment radicals, as Vietnam War protesters found a new cause—the environment. Suddenly, beards appeared alongside crewcuts in conservation group meetings—and the rhetoric quickened.

The militancy was short-lived. Eco-anarchist groups like Black Mesa Defense, which provided a cutting edge for the movement, peaked at the United Nations' 1972 Stockholm Conference on the Human Environment, but then faded from the scene. Along with dozens of other products of the 1960s who went to work for conservation organizations in the early 1970s, I discovered that compromise seemed to work best. A suit and a tie gained access to regional heads of the U.S. Forest Service and to members of Congress. We learned to moderate our opinions along with our dress. We learned that extremists were ignored in the councils of government, that the way to get a senator to put his arm around your shoulders and drop a Wilderness bill in the hopper was to consider the conflicts—mining, timber, grazing—and pare back the proposal accordingly. *Of course* we were good, patriotic Americans. *Of course* we were concerned with the production of red meat, timber, and minerals. We tried to demonstrate that preserving wilderness did not conflict all that much with the gross national product, and that clean air actually helped the economy. We argued that we could have our booming industry and still not sink oil wells in pristine areas.

This moderate stance appeared to pay off when Jimmy Carter, the first President who was an avowed conservationist since Teddy Roosevelt, took the helm at the White House in 1977. Self-professed conservationists were given decisive positions in Carter's administration. Editorials proclaimed that environmentalism had been enshrined in the Establishment, that conservation was here to stay. A new ethic was at hand: Environmental Quality and Continued Economic Progress.

Yet, although we had access to and influence in high places, something seemed amiss. When the chips were down, conservation still lost out to industry. But these were our friends turning us down. We tried to understand the problems they faced in the real political world. We gave them the benefit of the doubt. We failed to sue when we should have. . . .

I wondered about all this on a gray day in January 1979 as I sat in my office in the headquarters of The Wilderness Society, only three blocks from the White House. I had just returned from a news conference at the South Agriculture Building, where the Forest Service had announced a disappointing decision on the second Roadless Area Review and Evaluation, a twenty-month exercise by the Forest Service to determine which National Forest lands should be protected in their natural condition. As I loosened my tie, propped my cowboy boots up on my desk, and popped the top on another Stroh's, I thought about RARE II and why it had gone so wrong. Jimmy Carter was supposedly a great friend of wilderness. Dr. M. Rupert Cutler, a former assistant executive director of The Wilderness Society, was Assistant Secretary of Agriculture over the Forest Service and had conceived the RARE II program. But we had lost to the timber, mining, and cattle interests on every point. Of 80 million acres still roadless and undeveloped in the 190 million acres of National Forests, the Department of Agriculture recommended that only 15 million receive protection from road building and timber cutting.* Moreover, damn it, we—the conservationists—had been moderate. The antienvironmental side had been extreme, radical, emotional, their arguments full of holes. We had been factual, rational. We had provided more—and better—serious public comment. But we had lost, and now we were worried that some local wilderness group might go off the reservation and sue the Forest Service over the clearly inadequate environmental impact statement for RARE II. We didn't want a lawsuit because we knew we could win and were afraid of the political consequences of such a victory. We might make some powerful senators and representatives angry. So those of us in Washington were plotting how to keep the grassroots in line. Something about all this seemed wrong to me.

---

* Only 62 million acres were actually considered by the Forest Service in RARE II. Another 18 million acres that were also roadless and undeveloped were not considered because of sloppy inventory procedures, political pressure, or because areas had already gone through land-use plans that had supposedly considered their wilderness potential.

After RARE II, I left my position as issues coordinator for The Wilderness Society in Washington to return to New Mexico and my old job as the Society's Southwest representative. I was particularly concerned with overgrazing on the 180 million acres of public lands in the West managed by the Department of the Interior's Bureau of Land Management. For years, these lands—rich in wildlife, scenery, recreation, and wilderness—had been the private preserve of stock growers in the West. BLM had done little to manage the public lands or to control the serious overgrazing that was sending millions of tons of topsoil down the Colorado, the Rio Grande, and other rivers, wiping out wildlife habitat, and generally beating the land to hell.

Prodded by a Natural Resources Defense Council lawsuit, BLM began to address the overgrazing problem through a series of environmental impact statements. These confirmed that most BLM lands were seriously overgrazed, and recommended cuts in livestock numbers. But after the expected outcry from ranchers and their political cronies in Congress and in state capitals, BLM backtracked so quickly that the Department of the Interior building suffered structural damage. Why were BLM and the Department of the Interior so gutless?

While that question gnawed at my innards, I was growing increasingly disturbed about trends in the conservation organizations themselves. When I originally went to work for The Wilderness Society in 1973, the way to get a job with a conservation group was to prove yourself first as a volunteer. It helped to have the right academic background, but experience as a capable grassroots conservation activist was more important.

We realized that we would not receive the salaries we could earn in government or private industry, but we didn't expect them. We were working for nonprofit groups funded by the contributions of concerned people. Give us enough to keep food on the table, pay rent, buy a six-pack—we didn't want to get rich. But a change occurred after the mid-1970s: people seeking to work for conservation groups were career-oriented; they had relevant degrees (science or law, not history or English); they saw jobs in environmental organizations in the same light as jobs in government or industry. One was a steppingstone to another, more powerful position later on. They were less part of a cause than members of a profession.

A gulf began to grow between staff and volunteers. We also began to squabble over salaries. We were no longer content with subsistence wages, and the figures on our paychecks came to mark our status in the movement. Perrier and Brie replaced Bud and beans at gatherings.

Within The Wilderness Society, executive director Celia Hunter, a prominent Alaskan conservationist and outfitter, World War II pilot, and feminist, was replaced in 1978 by Bill Turnage, an eager young businessman who had made his mark by marketing Ansel Adams. Within two years Turnage had replaced virtually all those on the staff under Celia with professional organization people. The governing council also worked to bring millionaires with a vague environmental interest on board. We were, it seemed to some of us, becoming indistinguishable from those we were ostensibly fighting.

I resigned my position in June 1980.

The same dynamics seemed to affect the rest of the movement. Were there any radicals anywhere? Anyone to take the hard stands? Sadly, no. The national groups—Sierra Club, Friends of the Earth, National Audubon Society, Natural Resources Defense Council, and the rest—took almost identical middle-of-the-road positions on most issues. And then those half-a-loaf demands were readily compromised further. The staffs of these groups fretted about keeping local conservationists (and some of their field representatives) in line, keeping them from becoming extreme or unreasonable, keeping them from blowing moderate national strategy. Even Friends of the Earth, which had started out radical back in the heady Earth Day era, had gravitated to the center and, as a rule, was a comfortable member of the informal coalition of big environmental organizations.

For years I advocated this approach. We could, I believed, gain more wilderness by taking a moderate tack. We would stir up less opposition by keeping a low profile. We could inculcate conservation in the Establishment by using rational economic arguments. We needed to present a solid front.

A major crack in my moderate ideas appeared early in 1979, when I returned from Washington to the small ranching community of Glenwood, New Mexico. I had lived there earlier for six years, and although I was a known conservationist, I was fairly well accepted. Shortly after my return, *The New York Times* published an article on RARE II, with the Gila National Forest around Glenwood as chief exhibit. To my amazement, the article quoted a rancher who I considered to be a friend as threatening *my* life because of local fears about the consequences of wilderness designations. A couple of days later I was accosted on the street by four men, one of whom ran the town cafe where I had eaten many a chicken-fried steak. They threatened my life because of RARE II.

I was not afraid, but I was irritated—and surprised. I had been a leading moderate among New Mexico conservationists. I had successfully persuaded New Mexico conservation groups to propose fewer RARE II areas on the Gila National Forest as Wilderness. What had backfired? I thought again about the different approaches to RARE II: the moderate, subdued one advanced by the major conservation groups; the howling, impassioned, extreme stand set forth by off-road-vehicle zealots, many ranchers, local boosters, loggers, and miners. They looked like fools. We looked like statesmen. They won.

The last straw fell on the Fourth of July, 1980, in Moab, Utah. There the local county commission sent a flag-flying bulldozer into an area the Bureau of Land Management had identified as a possible study area for Wilderness designation. The bulldozer incursion was an opening salvo for the so-called Sagebrush Rebellion, a move by chambers of commerce, ranchers, and right-wing fanatics in the West to claim federal public land for the states and eventual transfer to private hands. The Rebellion was clearly an extremist effort, lacking the support of even many conservative members of Congress in the West, yet BLM was afraid to stop the county commission.

What have we really accomplished? I thought. *Are we any better off as far as saving the Earth now than we were ten years ago?* I ticked off the real problems: world population growth, destruction of tropical forests, expanding slaughter of African wildlife, oil pollution of the oceans, acid rain, carbon dioxide buildup in the atmosphere, spreading deserts on every continent, destruction of native peoples and the imposition of a single culture (European) on the entire world, plans to carve up Antarctica, planned deep seabed mining, nuclear proliferation, recombinant DNA research, toxic wastes. . . . It was staggering. And I feared we had done nothing to reverse the tide. Indeed, it had accelerated.

And then: Ronald Reagan. James "Rape 'n' Ruin" Watt became Secretary of the Interior. The Forest Service was Louisiana Pacific's. Interior was Exxon's. The Environmental Protection Agency was Dow's. Quickly, the Reagan administration and the Republican Senate spoke of gutting the already gutless Alaska Lands bill. The Clean Air Act, up for renewal, faced a government more interested in corporate black ink than human black lungs. The lands of the Bureau of Land Management appeared to the Interior Department obscenely naked without the garb of oil wells. Concurrently, the Agriculture Department directed the Forest Service to rid the National Forests of decadent and diseased old-growth trees. The cowboys had the grazing lands, and God help the hiker, Coyote, or blade of grass that got in their way.

Maybe, some of us began to feel, even before Reagan's election, it was time for a new joker in the deck: a militant, uncompromising group unafraid to say what needed to be said or to back it up with stronger actions than the established organizations were willing to take. This idea had been kicking around for a couple of years. Finally, in 1980, several disgruntled conservationists— including Susan Morgan, formerly educational director for The Wilderness Society; Howie Wolke, former Wyoming representative for Friends of the Earth; Bart Koehler, former Wyoming representative for The Wilderness Society; Ron Kezar, a longtime Sierra Club activist; and I—decided that the time for talk was past. We formed a new national group, which we called Earth First! We set out to be radical in style, positions, philosophy, and organization in order to be effective and to avoid the pitfalls of co-option and moderation that we had already experienced.

What, we asked ourselves as we sat around a campfire in the Wyoming mountains, were the reasons and purposes for environmental radicalism?

- To state honestly the views held by many conservationists.
- To demonstrate that the Sierra Club and its allies were raging moderates, believers in the system, and to refute the Reagan/Watt contention that they were "environmental extremists."
- To balance such antienvironmental radicals as the Grand County commission and provide a broader spectrum of viewpoints.
- To return vigor, joy, and enthusiasm to the tired, unimaginative environmental movement.

- To keep the established groups honest. By stating a pure, no-compromise, pro-Earth position, we felt that Earth First! could help keep the other groups from straying too far from their original philosophical base.
- To give an outlet to many hard-line conservationists who were no longer active because of disenchantment with compromise politics and the co-option of environmental organizations.
- To provide a productive fringe, since ideas, creativity, and energy tend to spring up on the edge and later spread into the center.
- To inspire others to carry out activities straight from the pages of *The Monkey Wrench Gang* (a novel of environmental sabotage by Edward Abbey), even though Earth First!, we agreed, would itself be ostensibly law-abiding.
- To help develop a new worldview, a biocentric paradigm, an Earth philosophy. To fight, with uncompromising passion, for Earth.

The name Earth First! was chosen because it succinctly summed up the one thing on which we could all agree: That in *any* decision, consideration for the health of the Earth must come first.

In a true Earth-radical group, concern for wilderness preservation must be the keystone. The idea of wilderness, after all, is the most radical in human thought—more radical than Paine, than Marx, than Mao. Wilderness says: Human beings are not paramount, Earth is not for *Homo sapiens* alone, human life is but one life form on the planet and has no right to take exclusive possession. Yes, wilderness for its own sake, without any need to justify it for human benefit. Wilderness for wilderness. For bears and whales and titmice and rattlesnakes and stink bugs. And . . . wilderness for human beings. Because it is the laboratory of human evolution, and because it is home.

It is not enough to protect our few remaining bits of wilderness. The only hope for Earth (including humanity) is to withdraw huge areas as inviolate natural sanctuaries from the depredations of modern industry and technology. Keep Cleveland, Los Angeles. Contain them. Try to make them habitable. But identify big areas that can be restored to a semblance of natural conditions, reintroduce the Grizzly Bear and wolf and prairie grasses, and declare them off limits to modern civilization.

In the United States, pick an area for each of our major ecosystems and recreate the American wilderness—not in little pieces of a thousand acres, but in chunks of a million or ten million. Move out the people and cars. Reclaim the roads and plowed land. It is not enough any longer to say no more dams on our wild rivers. We must begin tearing down some dams already built—beginning with Glen Canyon on the Colorado River in Arizona, Tellico in Tennessee, Hetch Hetchy and New Melones in California—and freeing shackled rivers.

This emphasis on wilderness does not require ignoring other environmental issues or abandoning social issues. In the United States, blacks and Chicanos of the inner cities are the ones most affected by air and water pollution, the ones most trapped by the unnatural confines of urbanity. So we decided that not only should eco-militants be concerned with these human environmental problems, we should also make common ground with other progressive elements of society whenever possible.

Obviously, for a group more committed to Gila Monsters and Mountain Lions than to people, there will not be a total alliance with other social movements. But there are issues in which Earth radicals can cooperate with feminist, Native American, anti-nuke, peace, civil-rights, and civil-liberties groups. The inherent conservatism of the conservation community has made it wary of snuggling too close to these leftist organizations. We hoped to pave the way for better cooperation from the entire conservation movement.

We believed that new tactics were needed—something more than commenting on dreary environmental-impact statements and writing letters to members of Congress. Politics in the streets. Civil disobedience. Media stunts. Holding the villains up to ridicule. Using music to charge the cause.

Action is the key. Action is more important than philosophical hairsplitting or endless refining of dogma (for which radicals are so well known). Let our actions set the finer points of our philosophy. And let us recognize that diversity is not only the spice of life, but also the strength. All that would be required to join Earth First!, we decided, was a belief in Earth first. Apart from that, Earth First! would be big enough to contain street poets and cowboy bar bouncers, agnostics and pagans, vegetarians and raw-steak eaters, pacifists and those who think that turning the other cheek is a good way to get a sore face.

Radicals frequently verge on a righteous seriousness. But we felt that if we couldn't laugh at ourselves we would be merely another bunch of dangerous fanatics who should be locked up—like oil company executives. Not only does humor preserve individual and group sanity; it retards hubris, a major cause of environmental rape, and it is also an effective weapon. Fire, passion, courage, and emotionalism are also needed. We have been too reasonable, too calm, too understanding. It's time to get angry, to cry, to let rage flow at what the human cancer is doing to Earth, to be uncompromising. For Earth First! there is no truce or cease-fire. No surrender. No partitioning of the territory.

Ever since the Earth goddesses of ancient Greece were supplanted by the macho Olympians, repression of women and Earth has gone hand in hand with imperial organization. Earth First! decided to be nonorganizational: no officers, no bylaws or constitution, no incorporation, no tax status, just a collection of women and men committed to the Earth. At the turn of the century, William Graham Sumner wrote a famous essay titled "The Conquest of the United States by Spain." His thesis was that Spain had ultimately won the Spanish-American War because the United States took on the imperialism and totalitarianism of Spain. We felt that if we took on the organization of the industrial state, we would soon accept their anthropocentric paradigm, much as Audubon and the Sierra Club already had.

And when we are inspired, we *act.*

Massive, powerful, like some creation of Darth Vader, Glen Canyon Dam squats in the canyon of the Colorado River on the Arizona-Utah border and backs the cold, dead waters of "Lake" Powell some 180 miles upstream, drowning the most awesome and magical canyon on Earth. More than any other single entity, Glen Canyon Dam is the symbol of the destruction of wilderness, of the technological ravishment of the West. The finest fantasy of eco-warriors in the West is the destruction of the dam and the liberation of the Colorado. So it was only proper that on March 21, 1981—at the spring equinox, the traditional time of rebirth—Earth First! held its first national gathering at Glen Canyon Dam.

On that morning, seventy-five members of Earth First! lined the walkway of the Colorado River Bridge, seven hundred feet above the once-free river, and watched five compatriots busy at work with an awkward black bundle on the massive dam just upstream. Those on the bridge carried placards reading "Damn Watt, Not Rivers," "Free the Colorado," and "Let it Flow." The five of us on the dam attached ropes to a grille, shouted out "Earth First!" and let three hundred feet of black plastic unfurl down the side of the dam, creating the impression of a growing crack. Those on the bridge returned the cheer.

Then Edward Abbey, author of *The Monkey Wrench Gang*, told the protesters of the "green and living wilderness" that was Glen Canyon only nineteen years ago:

> And they took it away from us. The politicians of Arizona, Utah, New Mexico, and Colorado, in cahoots with the land developers, city developers, industrial developers of the Southwest, stole this treasure from us in order to pursue and promote their crackpot ideology of growth, profit, and power—growth for the sake of power, power for the sake of growth.

Speaking toward the future, Abbey offered this advice: "Oppose. Oppose the destruction of our homeland by these alien forces from Houston, Tokyo, Manhattan, Washington, D.C., and the Pentagon. And if opposition is not enough, we must resist. And if resistance is not enough, then subvert."

Hardly had he finished speaking when Park Service police and Coconino County sheriff's deputies arrived on the scene. While they questioned Howie Wolke and me, and tried to disperse the illegal assembly, outlaw country singer Johnny Sagebrush led the demonstrators in song for another twenty minutes.

The Glen Canyon Dam caper brought Earth First! an unexpected amount of media attention. Membership quickly spiraled to more than a thousand, with members from Maine to Hawaii. Even the government became interested. According to reports from friendly park rangers, the FBI dusted the entire Glen Canyon Dam crack for fingerprints!

When a few of us kicked off Earth First!, we sensed a growing environmental radicalism in the country, but we did not expect the response we received. Maybe Earth First! is in the right place at the right time.

The cynical may smirk. "But what can you really accomplish? How can you fight Exxon, Coors, the World Bank, Japan, and the other great corporate giants of the Earth? How, indeed, can you fight the dominant dogmas of Western civilization?"

Perhaps it *is* a hopeless quest. But one who loves Earth can do no less. Maybe a species will be saved or a forest will go uncut or a dam will be torn down. Maybe not. A monkeywrench thrown into the gears of the machine may not stop it. But it might delay it, make it cost more. And it feels good to put it there.

———————

RESPONDING JOURNAL

Now that you have read Foreman's description of what his role in the environmental movement requires, think about how your attitude toward the environment parallels his. List your attitudes toward some of the organizations he opposes and draft a position statement that you think is defensible and convincing.

QUESTIONS FOR CRITICAL THINKING

STRATEGY

1. Even though Foreman is deadly serious, he manages to take a tone that is often humorous. Explain how he achieves this tone in his first paragraph; you will probably need to look up "Luddites" in a decent dictionary or encyclopedia.
2. Foreman describes what he calls "true" environmental radicalism on page 287 by listing the "reasons and purposes" for Earth First! Decide what you would have to add to this list—perhaps how you would have to change the grammatical structure of the items listed—to make this a clear definition of "environmental radicalism."

ISSUES

1. Evaluate the importance to Foreman's argument the claim that many of the people who went to work for conservation groups after the mid-1970s "were less part of a cause than members of a profession." Assuming the statement is true, what are its implications?
2. Explain to what extent you agree with the proposition "That in *any* decision, consideration for the health of the Earth must come first."

COLLABORATIVE WRITING ACTIVITY

As you look over your response to the Issues question #2, think about how you would modify your response to reshape it as a definition of a just plan for monitoring the health of the earth. What would such a plan include? Remember to think of your definition as something that could appear in a guide or encyclopedia. When you discuss your individual responses in your group, you can decide how effective each one would be as a definition.

"Night and Day"
Linda Hogan

In widely known and highly regarded poems such as "The Truth Is," Linda Hogan explores the tensions and demarcations between her White and Chickasaw ancestries. A noted poet, essayist, novelist, and playwright, she has published extensively; her collection of poetry entitled Seeing Through the Sun, *published in 1985, demonstrates convincingly her sense of the inter-connectedness of her family traditions, the land, and her own unique poetic voice. She now teaches and writes in her native Colorado.*

"Night and Day" is typical of Hogan's work in that it uses the natural world as its setting and finds significance in seemingly trivial details. As you read this poem, think about how the poet defines the oppositions of night and day. You may find meaning here that moves far away from the vague setting Hogan constructs. But after a few moments' reflection, you may realize that the ideas that seem so farfetched are after all very close to the literal meaning of the words of the poem.

EXPLORING JOURNAL

Freewrite on the idea of night and day; how would you explain most effectively what each means to you in terms of the natural world? Use details that are as specific as possible; think about the striking images in poems you have read or songs you have heard.

At night, alone,
the world is a river in me.
Sweet rain falls in the drought.
Leaves grow from lightning-struck trees.

I am across the world from daylight
and know the inside of everything
like the black corn dolls
unearthed in the south.

Near this river
the large female ears of corn listen and open.
Stalks rise up the layers of the world
the way it is said some people emerged
bathed in the black pollen of poppies.

In the darkness, I say,
my face is silent.
Like the corn dolls
my mouth has no more need to smile.

At midnight,
there is an eye in each of my palms.

I said, I have secret powers at night,
dark as the center of poppies,
rich as the rain.

But by morning I am filled up
with some stranger's lies
like those little corn dolls.
Unearthed after a hundred years
they have forgotten everything
in the husk of sunlight
and business
and all they can do is smile.

RESPONDING JOURNAL

How does Hogan explain or suggest the difference between night and day? To
what extent does she use these words with positive connotations? Which of
her images do you find most effective?

QUESTIONS FOR CRITICAL THINKING

STRATEGY

1. Explain why Hogan might begin her poem with the idea of lying in bed
   "At night, alone." What associations does she set up here?
2. List the number of natural processes Hogan either mentions or suggests in her poem. Do all of these have similar effects? How well do
   they work?

ISSUES

1. Suggest why Hogan might be making such a sustained reference to the
   ears of corn. Traditional wisdom holds that corn grows at night; does
   Hogan seem to be thinking about this idea?

2. What might Hogan have in mind by suggesting that "all they can do is smile"? Who does she mean here? And why would they be smiling? Give yourself some freedom to explore the possibilities.

COLLABORATIVE WRITING ACTIVITY

Divide up the poem as fairly as possible; allow each member of the group to take a section. Write an interpretation of what that part of the poem means, and then work together as a group to produce a reading of the entire poem. You will have to discuss your ideas carefully and be open to new ideas or interpretations. As a group, draft a statement summarizing your discussions.

SUGGESTIONS FOR FORMAL WRITING ASSIGNMENTS

1. Evaluate Doug Bandow's assertion that there exists "the balance that must be struck between ecological concerns and economic development" by carefully defining these two conflicting demands. Then argue for a resolution to the dilemma by using your definitions to reinforce your position. Present your case in the form of an essay of at least 500 words.

2. Define a term that is important to you. Explain how this environmentally related term depends on certain circumstances for its meaning to you and your readers. Think about how best to convince your readers as you write your essay. For example, you have read enough in this section to construct a definition of "environmental activist" and explore that definition in a substantial essay.

3. Now that you have explored the concept of definition, you may want to return to Thoreau's definition of *wildness*. Looking at the larger context in which he uses the word by reading his essay "Walking" might suggest additional ideas to you. Other possibilities will help you in developing a sense of how you would provide an extended definition of the word in a substantial essay. For example, brainstorming all the different senses of wild you can think of will help; what does wild mean in relation to cars, to baseball, to people? How do these ideas relate to the natural world and to Thoreau's exploration? He feels that *west* is a synonym for *wild;* do you agree? Does the sense of west as frontier help you with the writing of your essay of extended definition?

Chapter 7

CLASSIFICATION: CATEGORIES IN THE NATURAL WORLD

Alexandre Hogue, *Erosion #2 Mother Earth Laid Bare,* 1938

1. Explain the combined effects of the different distances—foreground, middle distance, background—of the painting. You might want to pay particular attention to the plow in the extreme front of the painting. What relationship to "Mother Earth" might it be suggesting?

2. What are the implications of "Mother Earth Laid Bare"? How do you feel about the idea of a naked woman spread out on the earth? How do you think your response matches up with the ideas of the painter?

3. Explain how you have modified your sense of the earth after seeing the photograph of this painting. Since Hogue was a muralist whose other works often adorned public places, explain how this painting might work in a location where it could be viewed by many people.

Chapter 7

CLASSIFICATION: CATEGORIES IN THE NATURAL WORLD

With a field as diverse as nature, the environment, and human ecology, the ability to sort through material and ideas in a meaningful way is crucial. We will have to look at forests—of material as well as real forests—to see which trees we will want to consider first, which ones we will need to consider most extensively, and which ones we can mention briefly. The process of classification will be a major tool in helping us to negotiate our progress through the mass of data that confronts us. For example, we hear that there *is* a hole in the ozone layer, and that there is *not* a hole in the ozone layer; we hear that global warming will destroy our way of life, and that it is a misconception on the part of some alarmist scientists; we hear that acid rain is killing our forests and fish, and we hear that acid rain is an invention of media. We have to find ways to organize the data in order to build a foundation we can use to evaluate competing claims. In this way, the material to be mastered is like the terrain on a map; classification is the process of rendering that material understandable or navigable.

The process of classifying requires dividing material into multiple groups based on clearly understood principles. Classification is related to simple division, which requires dividing the material into only two groups, based on a simple dividing principle. For example, Edward Abbey offers a pair of extremes in the behavior of trekkers in isolated western deserts:

> If I hike with another person it's usually the same; most of my friends have indolent and melancholy natures too. A cursed lot, all of them. I think of my comrade John De Puy, for example, sloping along for mile after mile like a goddamned camel—indefatigable—with those J.C. Penny hightops on his feet and that plastic pack on his back he got with five books of Green Stamps and nothing inside it but a sketchbook, some homemade jerky and a few cans of green chilies. Or Douglas Peacock, ex-Green Beret, just the opposite. Built like a buffalo, he loads a ninety-pound canvas pannier on his back at trailhead, loaded with guns, ammunition, bayonet, pitons and carabineers, cameras, field books, a 150-foot rope, geologist's sledge, rock samples, assay kit, field glasses, two gallons of water in steel canteens, jungle boots, a case of C-rations, rope

hammock, pharmaceuticals in a pig-iron box, raincoat, overcoat, two-man mountain tent, Dutch oven, hibachi, shovel, ax, inflatable boat, and near the top of the
load and distributed through side and back pockets, easily accessible, a case of
beer. Not because he enjoys or needs all that weight—he may never get to the
bottom of that cargo on a ten-day outing—but simply because Douglas uses his
packbag for general storage both at home and on the trail and prefers not to have
to rearrange everything from time to time merely for the purposes of a hike.
Thus my friends De Puy and Peacock; you may wish to avoid such extremes.

Abbey's advice about avoiding extremes is sound, as anyone who has done any
hiking in arid areas well knows. But how would you help him to classify the
information he presents here? How would you divide Peacock's load into categories? What would you name these categories? You may well wish to discuss
the divisions you recommend with those that other members of your class have
determined. But as hints to help you get started, you may wish to think about
different forms of utility, items for comfort, and foodstuffs as potential categories for classification.

What you have done here is to classify on the basis of *purpose* or *function*,
one of the main reasons for establishing a system of classification in the first place.
In explaining the characteristics of each category and proving that each item fits
the intended category, you convey essential information not only about the items
themselves, but also about the larger patterns into which those individual items
fit. In some areas you might find some overlap in purpose or function. For example, is the camera included to provide a record of plants encountered on the trip
or is it included just in case some scene strikes Peacock so dramatically that he
feels he needs to record it? In other words, does the camera fulfill a utilitarian need
or merely serve as a potential source of pleasure? Offering your reasons for including such an item in one group or another will not only strengthen your writing,
but it will also demonstrate to your reader why you are writing in the first place.

What you are doing here is determining an *overriding or determining principle* for your classification system. This principle works much like a thesis statement, for it explains to your reader what your point will be. In classification, you
advance an idea of how divisions are formed, of why distinctions should be
made. In classifying the contents of Peacock's pack, you determine that there
are several categories of items that are useful to Peacock on his hikes, and you
have sorted those items according to the overriding or determining principle of
the utility each class of items holds for Peacock when he is out hiking.

Often, the categories of classification we choose will in turn direct the
ways we think, both as readers and as authors. As a writer, once you determine
categories, you will eliminate any evidence or material that does not fall into
one of the categories you have established. As a reader, you will be attempting
to assimilate the new information according to the categories your author has
offered you; if the material does not readily fit into the expressed categories,
you will be frustrated as a reader. For example, think about the categories
Abbey proposes elsewhere in his essay: "It has been said, and truly, that everything in the desert either stings, stabs, stinks, or sticks." As a reader you may

be amused, and you are primed to expect an unusual array of examples to exemplify a humorous perspective on desert hiking. As a writer, you would not attempt to follow through on such a classification unless your role as writer made you comfortable with the level of both humor and information necessary to execute such a piece of writing successfully. In order to handle this well, you have to be as comfortable as Abbey in forsaking traditional categories and devising your own.

Establishing traditional categories, however, is a primary means for humans to see and understand relationships. We begin by identifying properties of things, and then sorting these things into groups based on their properties. In classifying people, for example, we find some characteristics or properties (two arms, two legs, a nose, etc.) are common to all members of the classification, while other characteristics are unique to ourselves (reddish brown hair, freckles, mole on left cheek, triangular scar on right knee) and help distinguish us from other people. In setting up the main terms of this text, the overriding principle of classifying writing was ecology, or the study of living things and how they interact with each other and with their environment. When we narrow our classification system by preceding the word *ecology* with *human,* we indicate that we will be looking at living things from the standpoint of how they interact with humans. Even more particularly, we will be interested in seeing how human activity affects the world of living things and the environment—the physical planet—in which they exist.

Simple dividing in two can be the first step in classification, but we have to be sure that we define our terms clearly at each step so that we will be able to rely on the accuracy of our divisions or classifications. For example, we could classify restaurants into such basic categories as "fast food" and "fine dining"; but what do these categories mean? If we define fast food as a restaurant where we order from a counter, where we get our own trays and plastic silverware, where at least 50% of food sold is taken away from the premises, and where most items cost less than $3.00, we have a good sense of what we mean. The other extreme might be more complex, however; do we mean restaurants where we have a tablecloth, where we have real silverware, where someone comes to our table, takes our order, and serves it to us? Can a fine dining restaurant be a franchise operation, like Bennigan's? Does a restaurant have to have a four- or five-star rating to be considered fine dining?

Finally, do you feel limited by the single division into two as a means of classifying restaurants? Would a much more logical division be made on some other categories that would move away from the two extremes to cover restaurants in the middle, such as cafeterias where you have real silverware but must serve yourself? Our earlier examinations of different categories of people who respond to the environment in different ways, from tree-hugger to eco-warrior to environmental activist to hunter to logger, all require some definition of terms to become useful. Indicating such a series of definitions would be your responsibility as a writer. In order to convey what you feel is meaningful about the differences among the different categories, you must make careful distinctions in setting up your classifications.

For example, scientists divide the earth's surface into several regions which share common qualities of topography, climate, and life forms; to regions having similar characteristics in the above categories, scientists have applied the term *biome.* They find several biomes across the globe: prairie grassland, desert, coniferous forest, tundra, deciduous forest, tropical rain forest. Biomes can be found in different parts of the world; for example, desert can be found in North America, Africa, and Asia. Tundra is found, on the other hand, only around the Arctic Circle, while the tropical rain forest straddles the equator. Plant life tends to be short and stubby on the tundra but huge and lush in the rain forest because of the differences in rainfall and climate between the two biomes. Such a system of classification provides scientists with the means to distinguish the important characteristics of each area.

Often scientists will further classify biomes into smaller areas to make study more manageable. They use the term *ecosystem* to denote a part of a biome with its own environment and life forms which interact harmoniously. Your backyard at home may well be an ecosystem with soil, grasses, worms, weeds, birds, insects, and perhaps even moles that together form a harmoniously existing colony. You could take one of these categories out and transplant it to a similar biome without much difficulty or harm to the organism; however, if you were to transplant it to a very different biome, such as tundra or tropical rain forest, it might well not survive.

The selections in this chapter offer various systems of classification. Dave Barry shows us how to choose a form of nature activity that will not put undue stress on us. Both Lopez and Anzaldúa explore how edges, of land and the mind, emphasize the categories and tensions held within borders. In examining native plant cultures, Gary Nabhan offers a means of organizing data about plant life, water resources, and human interaction with the earth. In terms of both length and difficulty, these selections show how the establishing and careful consideration of categories underlie effective writing. Maxine Kumin's poem shows how other systems of division work. Sorting and classifying the particular qualities of a single dog explain the dog's habits in summer, though these distinctions also tell us much more than that. As in all good classification writing, the classification itself is a tool for further thinking as well as for clarifying the ideas we already have.

"Choose Your Nature Well Before Hiking"
Dave Barry

Syndicated columnist Dave Barry is well known to readers who see his work daily in hundreds of newspapers across not only the United States but other countries as well. While he does not usually treat nature-related topics, nothing is safe from his particular brand of analytic humor. In this column, he explains how he has revised his opinion of his editor after said editor invited Barry on a hiking trip to the Swamp of Doom, a.k.a. the Big Cypress National Preserve in Florida.

EXPLORING JOURNAL

Freewrite a journal entry on the idea of visiting the "Swamp of Doom." Explore the possible difficulties you imagine as a consequence of hiking in an area with lush growth and varied animal life—some of which may find you appetizing!

If you look at any list of great modern writers such as Ernest Hemingway, William Faulkner and F. Scott Fitzgerald, you'll notice two things about them:

1. They all had editors.
2. They are all dead.

Thus we can draw the scientific conclusion that editors are fatal. I was made intensely aware of this recently when, as the direct result of an idea conceived of by my editor, I wound up in the Swamp of Doom.

That is not its technical name. Its technical name is the Big Cypress National Preserve, which is part of the Everglades ecosystem, an enormous, wet, nature-intensive area that at one time was considered useless, but which is now recognized as a vital ecological resource providing Florida with an estimated 93 percent of its bloodsucking insects.

No, really, the Everglades are very important. Tragically, they have been tampered with by man, an ecological moron who is always blundering into sensitive areas and befouling them with beer cans, used condoms, golf courses, etc. Only lately has man realized that the best thing for him to do is stay out of the Everglades. This was certainly my policy.

So it never occurred to me to set actual foot in the Everglades until my editor, Tom Shroder, suggested that I go hiking with him out there.

"It's real interesting," he said, never once mentioning alligators, let alone poison trees.

TRIP BEGINS

So one Saturday morning we went. On the edge of the Everglades we stopped for supplies at a combination truck stop/sporting-goods store. I bought the survival basics: a safari-style helmet, a machete, beef jerky, a bottle of Evian water, a snakebite kit and Certs.

I used the machete to cut the tag off the safari-style helmet, so the wildlife creatures would not think I was some easily edible swamp rookie. But I was still nervous. And I did not feel better when we met our guide, John Kalafarski, a Park Service ranger who is extremely knowledgeable about wildlife.

"See this tree?" he said, pointing to a tree that looked, to me, exactly like every other tree in the Everglades. "This is a poisonwood tree. You don't want to touch it."

"I'm not touching anything," I said.

Then we began our hike. At first it was fine. There was an actual path, with little signs to identify the plants. But suddenly John, having apparently brushed up against a lunaticwood tree, plunged right into the swamp! Soon we were up to our knees in murky, festering soup, walking on one of those squishy muck bottoms, surrounded by dense growth and the smell of rotting vegetation. Deeper and deeper we went. I was fighting my way through big snarls of vines, stumbling over logs, falling into hidden holes, while up ahead, John, oblivious to the aura of menace all around us, was delivering a cheerful nonstop commentary on the flora and fauna, pointing out rare mushrooms, tree snails, etc. I wanted to scream: "There could be giant snakes hiding in this water, and you're looking at tree snails?"

But I did not want to act like a weenie. I saved that until the water started getting deeper, and deeper, until finally we were up to our armpits, our feet sinking in goo, and John, pointing right in front of us, said, "This is an alligator hole."

"You mean there's a (bad word) alligator in there?" I asked.

"Yes," said John, "and it's appropriate that you should use that word to describe him, because this is mating season."

"We don't want your women!" I shouted at the hole.

"That might offend him," Tom pointed out.

"Not that we don't find your women attractive!" I shouted at the hole.

Fortunately we got out of there without having any important limbs chomped off. Although the Certs were ruined.

When we got back onto the dry trail, I opened the beef jerky package with my machete and passed it around, and we enjoyed a pleasant sense of fellowship and accomplishment. If you enjoy nature, I strongly recommend that you, too, take a hike in the Everglades. I'll wave to you from the car.

RESPONDING JOURNAL

Think about how Barry achieves his comic effect. What can you do to imitate him? Try writing a journal entry that has the style and shape of a similar column on some outdoor trip you took that did not turn out so well.

QUESTIONS FOR CRITICAL THINKING

STRATEGY

1. Look at several points where Barry's writing made you smile. How does he achieve this effect? Try to generalize about the categories that describe most effectively the different kinds of humorous situations.

2. Barry opens his column with a logical proposition and then goes on to a "scientific conclusion." How well does this work as an opening to his column?

Issues

1. What assumptions does Barry make about what it takes to become an experienced nature person? How much of what he says would you judge to be accurate?
2. Do you think any hikers who enjoy trips to places such as the Big Cypress National Preserve would be offended by Barry's remarks? Explain your response.

Collaborative Writing Activity

Highlight Barry's comic lines and discuss with your classmates how each joke is created. Use this activity to see how you can make your own writing more humorous. Try at least a paragraph of comedy on the outdoors and bring it to class, where you can combine it into a group presentation, perhaps like a monologue. Your group may even get its own byline, just like Barry!

"Perimeter"
Barry Lopez

Barry Lopez, one of America's most widely read contemporary nature writers, is the author of Wolves and Men *(1978) and* Arctic Dreams *(1986). Writing both short fiction and nonfiction essays gives Lopez the opportunity to work with new ways of combining the power and appeal of both forms, as "Perimeter" indicates. The quote from Lopez' essay "Mapping the Real Geography" that opens this book (page 1) indicates how he feels about establishing a relationship with the land, a consistent concern throughout his work.*

"Perimeter" appears in Desert Notes: Reflections in the Eye of a Raven *(1976); the prologue to that book features Charles Darwin's comment from* The Voyage of the Beagle *that the vast deserts of Patagonia, which have existed virtually unchanged for thousands of years, have taken far stronger hold of his mind than more fertile and green areas. Lopez begins his book with an account of driving his van over the vast alkaline desert of the American southwest and realizing that since he is not on any road, he does not have to drive, and so he moves over to the passenger's seat (a practice not sanctioned*

by the American Automobile Association). He can "see the sheen where I'd sat for years" when he looks at the vacant driver's seat and imagines the driver; he moves around the van and ultimately opens the doors, jumps out, runs with and around the van, takes out his bicycle and pedals around the van before finally getting back in. He imagines both the desert and himself in new ways.

The following excerpt explores a new system of classification, one of color and direction. The system, while emotional and even idiosyncratic, is still formal.

EXPLORING JOURNAL

Imagine seeing yourself in a desert. Describe the way you would be moving across it, in whatever mode of transportation you think would be appropriate. Try to imagine yourself from the perspective of a low-flying plane or bird gliding above you as you move across the vast flats. Freewrite on this topic for at least ten minutes, longer if the spirit moves you.

I.

In the west, in the blue mountains, there are creeks of grey water. They angle out of the canyons, come across the brown scratched earth to the edge of the desert and run into nothing. When these creeks are running they make a terrific noise.

No one to my knowledge has ever counted the number, but I think there are more than twenty; it is difficult to be precise. For example, some of the creeks have been given names that, over the years, have had to be given up because a creek has run three or four times and then the channel has been abandoned.

You can easily find the old beds, where the dust has been washed out to reveal a level of rock rubble—cinnabar laced with mercury, fool's gold, clear quartz powder, and fire opal; but it is another thing to find one of the creeks, even when they are full. I have had some success by going at night and listening for the noise.

There is some vegetation in this area; it does not seem to depend on water. The rattlesnakes live here along with the rabbits. When there is any thunder it is coming from this direction. During the day the wind is here. The smells include the hellebore, vallo weed and punchen; each plant puts out its own smell and together they make a sort of pillow that floats a few feet off the ground where they are not as likely to be torn up by the wind.

II.

To the north the blue mountains go white and the creeks become more dependable though there are fewer of them. There is a sort of swamp here at the edge of the desert where the creeks pool and where grasses and sedges grow and the water takes a considerable time to evaporate and seep into the earth. There are some ducks here, but I do not know where they come from or where they go when the swamp dries up in the summer. I have never seen them flying. They are always hiding, slipping away; you will see their tail feathers disappearing in the screens of wire grass. They never quack.

There are four cottonwood trees here and two black locusts. The cottonwoods smell of balsam, send out seeds airborne in a mesh of exceedingly fine white hair, and produce a glue which the bees use to cement their honeycombs. Only one of the cottonwoods, the oldest one, is a female. The leaf stem meets the leaf at right angles and this allows the leaves to twitter and flash in the slightest breeze. The underside of the leaf is a silver green. I enjoy watching this windflash of leaves in strong moonlight.

The black locusts are smaller, younger trees and grow off by themselves a little. They were planted by immigrants and bear sweet smelling pea-like flowers with short, rose-like thorns at the leaf nodes. There are a few chokecherry bushes and also a juniper tree. You can get out of the sun here at noon and sleep. The wind runs down the sides of the cottonwoods like water and cools you.

An old tawny long-haired dog lives here. Sometimes you will see him, walking along and always leaning to one side. There is also part of a cabin made with finished lumber lying on its back; the dark brown boards are dotted with red and yellow lichen and dry as sun-baked, long forgotten shoes.

III.

To the east the white mountains drop off and there is a flat place on the horizon and then the red mountains start. There is almost nothing growing in these mountains, just a little sagebrush. At the base, where they come to the desert, there are dunes, white like gypsum.

Inside the mountains are old creeks that run in circles over the floors of low-ceilinged caves. The fish in these waters are white and translucent; you can see a pink haze of organs beneath the skin. Where there should be eyes there are grey bulges that do not move. On the walls are white spiders like tight buttons of surgical cotton suspended on long hairy legs. There are white beetles, too, scurrying through the hills of black bat dung.

I have always been suspicious of these caves because the walls crumble easily under your fingertips; there is no moisture in the air and it smells like balloons. The water smells like oranges but has no taste. Nothing you do here makes any sound.

You have to squeeze through these red mountains to get around them; you can't walk over them. You have to wedge yourself in somewhere at the base and go in. There is always a moment of panic before you slip in when you are stuck. Your eyes are pinched shut and the heels of your shoes wedge and make you feel foolish.

At night the wind lies in a trough at the base of the red mountains, sprawled asleep over the white sand dunes like a caterpillar. The edge of the desert is most indistinct in this place where the white sand and the alkaline dust blow back and forth in eddies of the wind's breath while it sleeps.

IV.

In the south the red mountains fall away and yellow mountains rise up, full of silver and turquoise rock. There are plenty of rabbits here, a little rain in the middle of the summer, fine clouds tethered on the highest peaks. If you are out in the middle of the desert, this is the way you always end up facing.

In the south twelve buckskin horses are living along the edge of the yellow mountains. The creeks here are weak; the horses have to go off somewhere for water but they always come back. There is a little grass but the horses do not seem to eat it. They seem to be waiting, or finished. Ten miles away you can hear the clack of their hooves against the rocks. In the afternoon they are motionless, with their heads staring down at the ground, at the little stones.

At night they go into the canyons to sleep standing up.

From the middle of the desert even on a dark night you can look out at the mountains and perceive the differences in direction. From the middle of the desert you can see everything well, even in the black dark of a new moon. You know where everything is coming from.

Responding Journal

In a focused freewriting entry, speculate on what Lopez' essay makes you think of. Does it recall to you any book you have read, music you have heard, or television or movie production? What does this sort of writing make you think about?

Questions for Critical Thinking

Strategy

1. Draw a map of the area Lopez describes. You may well want to use colors to indicate the different directions. What has this experience of charting visually elements which are described verbally shown you?
2. What effect do you think Lopez is trying to achieve here? What do you feel he is trying to do to you as a reader? How well does he succeed?

Issues

1. How do you imagine Lopez sees himself in relation to the desert, or, to phrase this question another way, what do you imagine his relation to the desert to be?

2. What is the effect of Lopez' style in this excerpt? Compare it to the scientific precision of Nabhan or Thomas; which do you find more effective? Why?

COLLABORATIVE WRITING ACTIVITY

List what you feel to be the main ideas Lopez wishes to convey in "Perimeter." Next, indicate what you feel Lopez did in terms of structure, organization, diction, or style to illustrate his ideas. Bring your lists to your group and share them; after you have heard all members' lists, write three statements in which you explain how Lopez used a certain technique or techniques to develop a point. Choose three that you can explain to your satisfaction.

from *Borderlands/La Frontera: The New Mestiza*
Gloria Anzaldúa

Poet and fiction writer Gloria Anzaldúa has taught at the University of Texas, San Francisco State University, and Norwich University. She is the author of This Bridge Called My Back: Writings by Radical Women of Color, *which won the Before Columbus Foundation American Book Award. In addition to writing about the problems facing women of color in the United States, she has been active in the migrant farm workers movement.*

Growing up in the region of the Texas-Mexico border, Anzaldúa learned firsthand what it is like to live between two cultures. She intersperses Spanish with English in her work in order to show the borders between languages as well as cultures; she presents the Spanish text so that English language readers can understand it in context. As the book cover explains, "Borderlands/La Frontera *is a meeting ground for all people who, like herself, realize that the work of the 21st century will be about the coming together of diverse cultures." Lands of borders are crucial places.*

EXPLORING JOURNAL

What is your impression of the border between the United States and Mexico? How does this border differ from other borders you know or have heard about? Do a ten-minute focused freewrite on this question of borders.

PREFACE

The actual physical borderland that I'm dealing with in this book is the Texas-U.S. Southwest/Mexican border. The psychological borderlands, the sexual borderlands and the spiritual borderlands are not particular to the Southwest. In fact, the Borderlands are physically present wherever two or more cultures edge each other, where people of different races occupy the same territory, where under, lower, middle and upper classes touch, where the space between two individuals shrinks with intimacy.

I am a border woman. I grew up between two cultures, the Mexican (with a heavy Indian influence) and the Anglo (as a member of a colonized people in our own territory). I have been straddling that *tejas*-Mexican border, and others, all my life. It's not a comfortable territory to live in, this place of contradictions. Hatred, anger and exploitation are the prominent features of this landscape.

However, there have been compensations for this *mestiza,* and certain joys. Living on borders and in margins, keeping intact one's shifting and multiple identity and integrity, is like trying to swim in a new element, an "alien" element. There is an exhilaration in being a participant in the further evolution of humankind, in being "worked" on. I have the sense that certain "faculties"—not just in me but in every border resident, colored and non-colored—and dormant areas of consciousness are being activated, awakened. Strange, huh? And yes, the "alien" element has become familiar—never comfortable, not with society's clamor to uphold the old, to rejoin the flock, to go with the herd. No, not comfortable but home.

This book, then, speaks of my existence. My preoccupations with the inner life of the Self, and with the struggle of that Self amidst adversity and violation; with the confluence of primordial images, with the unique positionings consciousness takes at these confluent streams; and with my almost instinctive urge to communicate, to speak, to write about life on the borders, life in the shadows.

Books saved my sanity, knowledge opened the locked places in me and taught me first how to survive and then how to soar. *La madre naturaleza* succored me, allowed me to grow roots that anchored me to the earth. My lover of images—mesquite flowering, the wind, *Ehécatl,* whispering its secret knowledge, the fleeting images of the soul in fantasy—and words, my passion for the daily struggle to render them concrete in the world and on paper, to render them flesh, keeps me alive.

The switching of "codes" in this book from English to Castillian Spanish to the North Mexican dialect to Tex-Mex to a sprinkling of Nahuatl to a mixture of all of these, reflects my language, a new language—the language of the Borderlands. There, at the juncture of cultures, languages cross-pollinate and are revitalized; they die and are born. Presently this infant language, this bastard language, Chicano Spanish, is not approved by any society. But we Chicanos no longer feel that we need to beg entrance, that we need always to make the first overture—to translate to Anglos, Mexicans and Latinos, apology blurting out of our mouths with every step. Today we ask to be met halfway. This book is our invitation to you—from the new mestizas.

THE HOMELAND, AZTLÁN

EL OTRO MÉXICO

> *El otro México que acá hemos construido*
> *el espacio es lo que ha sido*
> *territorio nacional.*
> *Esté el esfuerzo de todos nuestros hermanos*
> *y latinoamericanos que han sabido*
> *progressar.*
>
> —LOS TIGRES DEL NORTE[1]

"The *Aztecas del norte* . . . compose the largest single tribe or nation of Anishinabeg (Indians) found in the United States today. . . . Some call themselves Chicanos and see themselves as people whose true homeland is Aztlán [the U.S. Southwest]."[2]

Wind tugging at my sleeve
feet sinking into the sand
I stand at the edge where earth touches ocean
where the two overlap
a gentle coming together
at other times and places a violent clash.

Across the border in Mexico
   stark silhouette of houses gutted by waves,
      cliffs crumbling into the sea,
         silver waves marbled with spume
            gashing a hole under the border fence.
            Miro el mar atacar
            la cerca en Border Field Park
         con sus buchones de agua,
an Easter Sunday resurrection
of the brown blood in my veins.

Oigo el llorido del mar, el respiro del aire,
   my heart surges to the beat of the sea.

In the gray haze of the sun
　　the gulls' shrill cry of hunger,
　　　　the tangy smell of the sea seeping into me.

　　　　I walk   through the hole in the fence
　　　　　　to the other side.
Under my fingers I feel the gritty wire
　　rusted by 139 years
　　　　of the salty breath of the sea.

Beneath the iron sky
Mexican children kick their soccer ball across,
run after it, entering the U.S.

　　　　I press my hand to the steel curtain—
　　chainlink fence crowned with rolled barbed wire—
rippling from the sea where Tijuana touches San Diego
　　unrolling over mountains
　　　　and plains
　　　　　　and deserts,
this "Tortilla Curtain" turning into *el río Grande*
　　flowing down to the flatlands
　　　　of the Magic Valley of South Texas
　　its mouth emptying into the Gulf.

1,950 mile-long open wound
　　　　　　dividing a *pueblo,* a culture,
　　　　　　running down the length of my body,
　　　　　　　staking fence rods in my flesh,
　　　　　　splits me   splits me
　　　　　　　　me raja   me raja

　　This is my home
　　this thin edge of
　　　barbwire.

　　But the skin of the earth is seamless.
　　The sea cannot be fenced,
　　el mar does not stop at borders.
To show the white man what she thought of his
　　　　　　　arrogance,
　　　Yemaya blew that wire fence down.

　　This land was Mexican once,
　　　was Indian always

and is.
And   will be again.

Yo soy un puente tendido
   del mundo gabacho al del mojado,
lo pasado me estirá pa' 'trás
   y lo presente pa' 'delante.
Que la Virgen de Guadalupe me cuide
Ay ay ay, soy mexicana de este lado.

The U.S.-Mexican border *es una herida abierta* where the Third World grates against the first and bleeds. And before a scab forms it hemorrhages again, the lifeblood of two worlds merging to form a third country—a border culture. Borders are set up to define the places that are safe and unsafe, to distinguish *us* from *them*. A border is a dividing line, a narrow strip along a steep edge. A borderland is a vague and undetermined place created by the emotional residue of an unnatural boundary. It is in a constant state of transition. The prohibited and forbidden are its inhabitants. *Los atravesados* live here: the squint-eyed, the perverse, the queer, the troublesome, the mongrel, the mulato, the half-breed, the half dead; in short, those who cross over, pass over, or go through the confines of the "normal." Gringos in the U.S. Southwest consider the inhabitants of the borderlands transgressors, aliens— whether they possess documents or not, whether they're Chicanos, Indians or Blacks. Do not enter, trespassers will be raped, maimed, strangled, gassed, shot. The only "legitimate" inhabitants are those in power, the whites and those who align themselves with whites. Tension grips the inhabitants of the borderlands like a virus. Ambivalence and unrest reside there and death is no stranger.

> In the fields, *la migra.* My aunt saying, *"No corran,* don't run. They'll think you're *del otro lao."* In the confusion, Pedro ran, terrified of being caught. He couldn't speak English, couldn't tell them he was fifth generation American. *Sin papeles*—he did not carry his birth certificate to work in the fields. *La migra* took him away while we watched. *Se lo llevaron.* He tried to smile when he looked back at us, to raise his fist. But I saw the shame pushing his head down, I saw the terrible weight of shame hunch his shoulders. They deported him to Guadalajara by plane. The furthest he'd ever been to Mexico was Reynosa, a small border town opposite Hidalgo, Texas, not far from McAllen. Pedro walked all the way to the Valley. *Se lo llevaron sin un centavo al pobre. Se vino andando desde Guadalajara.*

During the original peopling of the Americas, the first inhabitants migrated across the Bering Straits and walked south across the continent. The oldest evidence of humankind in the U.S.—the Chicanos' ancient Indian ancestors—was found in Texas and has been dated to 35000 B.C.[3] In the Southwest United States archaeologists have found 20,000-year-old campsites of the Indians who migrated through, or permanently occupied, the Southwest, Aztlán—land of the herons, land of whiteness, the Edenic place of origin of the Azteca.

In 1000 B.C., descendants of the original Cochise people migrated into what is now Mexico and Central America and became the direct ancestors of many of the Mexican people. (The Cochise culture of the Southwest is the parent culture of the Aztecs. The Uto-Aztecan languages stemmed from the language of the Cochise people.)[4] The Aztecs (the Nahuatl word for people of Aztlán) left the Southwest in 1168 A.D.

Now let us go.
   Tihueque, tihueque,
Vámonos, vámonos.
   Un pájaro cantó.
Con sus ocho tribus salieron
    de la "cueva del origen."
los aztecas siguieron al dios
    Huitzilopochtli.

Huitzilopochtli, the God of War, guided them to the place (that later became Mexico City) where an eagle with a writhing serpent in its beak perched on a cactus. The eagle symbolizes the spirit (as the sun, the father); the serpent symbolizes the soul (as the earth, the mother). Together, they symbolize the struggle between the spiritual/celestial/male and the underworld/earth/feminine. The symbolic sacrifice of the serpent to the "higher" masculine powers indicates that the patriarchal order had already vanquished the feminine and matriarchal order in pre-Columbian America.

At the beginning of the 16th century, the Spaniards and Hernán Cortés invaded Mexico and, with the help of tribes that the Aztecs had subjugated, conquered it. Before the Conquest, there were twenty-five million Indian people in Mexico and the Yucatán. Immediately after the Conquest, the Indian population had been reduced to under seven million. By 1650, only one-and-a-half-million pure-blooded Indians remained. The *mestizos* who were genetically equipped to survive small pox, measles, and typhus (Old World diseases to which the natives had no immunity), founded a new hybrid race and inherited Central and South America.[5] *En 1521 nació una nueva raza, el mestizo, el mexicano* (people of mixed Indian and Spanish blood), a race that had never existed before. Chicanos, Mexican-Americans, are the offspring of those first matings.

Our Spanish, Indian, and *mestizo* ancestors explored and settled parts of the U.S. Southwest as early as the sixteenth century. For every gold-hungry *conquistador* and soul-hungry missionary who came north from Mexico, ten to twenty Indians and *mestizos* went along as porters or in other capacities.[6] For the Indians, this constituted a return to the place of origin, Aztlán, thus making Chicanos originally and secondarily indigenous to the Southwest. Indians and *mestizos* from central Mexico intermarried with North American Indians. The continual intermarriage between Mexican and American Indians and Spaniards formed an even greater *mestizaje.*

NOTES

1. Los Tigres del Norte is a *conjunto* band.
2. Jack D. Forbes, *Aztecas del Norte: The Chicanos of Aztlán.* (Greenwich, CT: Fawcett Publications, Premier Books, 1973), 13, 183; Eric R. Wolf, *Sons of Shaking Earth* (Chicago, IL: University of Chicago Press, Phoenix Books, 1959), 32.
3. John R. Chávez, *The Lost Land: The Chicano Images of the Southwest* (Albuquerque, NM: University of New Mexico Press, 1984), 9.
4. Chávez, 9. Besides the Aztecs, the Ute, Gabrillino of California, Pima of Arizona, some Pueblo of New Mexico, Comanche of Texas, Opata of Sonora, Tarahumara of Sinaloa and Durango, and the Huichol of Jalisco speak Uto-Aztecan languages and are descended from the Cochise people.
5. Reay Tannahill, *Sex in History* (Briarcliff Manor, NY: Stein and Day Publishers/Scarborough House, 1980), 308.
6. Chávez, 21.

RESPONDING JOURNAL

What new ideas come to you as a result of reading Anzaldua's account of border difficulties? Is it really possible for a piece of land to carry so much meaning? Write several paragraphs of response to your feeling about how legal borders can have significant emotional meaning for people on their sides.

QUESTIONS FOR CRITICAL THINKING

STRATEGY

1. Explain how effective you find the interspersed Spanish and English text. How does this strategy relate to what you have seen before of bilingual texts? What, if any, difficulties does it present for you?
2. Writers sometimes include their poetry in their critical essays; Alice Walker is another writer represented in this text who likes to use her poetry to intensify a point in her nonfiction prose. Explain how well you think Anzaldua uses the poetry to emphasize her points here.

ISSUES

1. Explain the different borders Anzaldúa is exploring. Which of these borders is a significant issue in other parts of the country without literal borders (as the United States Immigration Service observes them)?

2. Explain the classification system Anzaldúa presents here. What does she seem to be doing in using the concept of borders in relation to humans?

COLLABORATIVE WRITING ACTIVITY

Consider the Spanish portions of Anzaldúa's text. Write a substantial paragraph or two to explain your reaction to her work. When you get together to share your work with your group, begin by taking turns indicating your degree of familiarity with Spanish. Allow those in the group who have more experience with Spanish to suggest meanings of text that may have given a little trouble to those who do not have experience with Spanish. After your discussion, draft together a one-page statement of the ways in which the Spanish text helps or hinders Anzaldua's text.

"Plants Which Coyote Steals, Spoils, and Shits On," from *The Desert Smells Like Rain*
Gary Paul Nabban

Winner of a MacArthur fellowship, Gary Paul Nabban is an ethnobotanist whose specialty is native plant strains. The titles of his other works indicate his focus: Gathering the Desert *and* Enduring Seeds *show his focus on dry locales and the plant varieties best suited to them. He often writes about his experiences in tracking down rare plants and tracing the development of seed strains. Sometimes, as in the selection presented here, he explains how folklore and the practices of indigenous people explain the importance of various plant forms. Coyote is the premier trickster figure in Native American culture; particularly on the desert country of the southwest, Coyote is an animal to be watched carefully, as he is in Nabban's account.*

Nabban's work also has an extremely important element whose practicality is more immediately apparent. His efforts at identifying endangered native plants help him to discover traditional Indian ways of coexisting with and protecting the environment so that more traditional scientists can profit from his work and help more effectively to preserve and protect endangered ecosystems. Nabban serves as a bridge between little-known and unpublicized local knowledge and practices on the one hand, and influential members of scientific communities on the other. These scientists directly influence the legislation that affects the physical world. As a student, Nabban was torn between his desire to become a writer and his fascination with botany; he has combined the two effectively.

EXPLORING JOURNAL

What do you know about the seeds you plant in your garden or flower pot? How would you explain the different varieties of plant or flower seeds—for example, tomato or marigold seeds? Spend fifteen minutes in a focused freewriting exercise to explore your ideas of how and why seeds might be "developed," or why seed catalogs and nurseries offer so many different varieties of the same species.

> *We were picking watermelons all day, taking them to the shed and stacking them. The next morning as the sun was coming, I was just sitting there waking up, when I looked out and there he was— Coyote—carrying away this watermelon. He was running along, carrying it between his front legs. I went over there to the shed and all those melons were just about gone.*
>
> —REMEDIO CRUZ, *BIG FIELDS*

Coyote—he's a hard one to write about and get away with it.

I tried once before. One winter, I went around to the villages asking the Desert People about him. By the end of the cold season, I had a whole notebook full of stories.

The *old* stories. How, after the Flood, Elder Brother let Coyote help him make a new batch of people to start the world over again. Coyote fooled around and made a bunch of misshapen creatures, with eyes on their knees, with only one leg, or with their sexual organs in the wrong places. Elder Brother just had to throw them away, far across the ocean.

And dirty stories. Like the time he volunteered to carry a pretty girl across a river, but told her to throw her skirts up over her eyes so that they wouldn't get wet. Pretending to help her across, he helped her get pregnant instead.

Yet nothing became of those stories I wrote down. It seemed I had misplaced the notebook. For weeks, I searched without luck.

The following summer, I was walking near the village they call *Ban Dak*, Where Coyote Sat Down, when I came upon a *charco* that had just filled up with floodwaters. There, floating in the pond, was a notebook that looked familiar to me, except it had pawprints smudging the pages, and whole sections ripped out by the teeth.

You have to watch what you say about this one they call Coyote.

For a long while, I wasn't sure of the difference between the legendary Coyote and the wild dog-like creature I'd catch glimpses of out in the desert now and then. Is Coyote a special coyote? And if so, would I recognize him if I saw him up close? I spoke with trappers and zoologists, and memorized all the identifying characteristics of ordinary coyotes. If I saw anything peculiar, then, I'd know who it was.

My time came one dusk as I was driving home to Esperero Canyon after a day of observing birds near Sonata. Just as I pulled off the pavement onto the dirt road, I spotted a coyote up ahead of me. I braked the car, then glided slowly toward the middle of the road where he stood. He moved to the roadside on my left. I killed the engine.

He paused there for a moment, so I quickly grabbed the binoculars out of my knapsack, and took a good look at him: the pale pattern of fur around his neck; the hang of his tail; the taper of his snout; the shape of his eyes. He looked *muy típico.*

He stood there for a couple of minutes, still but attentive. Satisfied that he was just a plain old coyote, I finally put the binoculars down.

Then a funny thing happened.

He started to trot off, but stopped, and turned again toward me. Fixing his eyes on me, he slowly walked a complete ring around the car. When he had come full circle, he stopped, tossed his head back, and *yawned,* then walked off.

By the time it hit me that there's a little of Coyote in every coyote, I had realized that there's some Coyote in a lot of humans too. That wasn't news to my Papago friends—their stories are full of Coyotes in men's clothing.

Ban—coyote and Coyote go by the same simple name in the Papago tongue. The Papago call the legendary Coyote by laudatory euphemisms in certain tales—Our Furry Friend, Gray Partner, Burning-Eyed Buddy—but most of the time the word *Ban* does the trick. On the other hand, there is a rich array of words to describe not-so-praiseworthy Coyote-like attributes in humans (or vice versa). Linguist Madeline Mathiot put a few of them into print before Coyote could swipe her notebook.

Bankaj refers to any coyote-like quality, such as "yelling like a coyote." *Banma* describes one who is being greedy. *Banmad* is a verb meaning "to cheat somebody." *Banmakam* is a glutton. *S-banow* is the superlative for the bad breath of someone who "sure stinks like a coyote."

The name for a Pulaski is *ban wuhiosa*—Coyote's face. Every time I ask a friend why those tools are called that, he breaks out laughing and says, "Well, don't they look just like me?" The *ban wuhiosa* is the main tool used by the crew of "community beautifiers" that get paid by the government to keep the village clean of all weeds and debris. The nickname of the work crew?

Ba:ban-Pioñ—the Coyote Workers. The Papago phrase used for nicknames? *Ban'ĭ Kuadc*—"Coyote peeked in."

My first lesson about Coyote's plants left a bad taste in my mouth, to say the least. I had brought some wild desert gourds out to a village with me, curious

to find if the Papago used them in any way. An elderly woman looked at the little gourds in the bed of my pickup.

"What you got there? Oh, that's what they call *a:d!* Long time ago they used
to go out, and when those fruit got ripe and turned yellow, they would eat it
just like a sweet apple."

Before she had a chance to finish her story. I grabbed one tender, yellow
gourd and took a bite into it. She yelled "DON'T" but it was too late—that taste
was so terrifically bitter that my tongue muscles went into shock. I spat the
pulp and ran for water.

When I returned to where the woman was, she was grinning.

"It *used* to taste just like an apple, they say. Then Coyote came along and
he *shit* on it. I guess ever since then it has had that taste that is in your mouth
right now. . . ."

Over the next couple of years, I learned a lot about the two gourd species
that different Papago refer to as *a:d—Cucurbita foetidissima,* and
Apodanthera undulata. In Spanish as well as English, these gourds are
called coyote melons. Both contain bitter substances called *cucurbitacins,*
that are found in all of the wild relatives of cultivated squashes and pumpkins. In fact, an important change during the evolution of domesticated
squashes was the loss of bitterness in the fruit. Some ancient Indian probably tasted an ancient gourd and discovered a mutant that was *not* bitter!
That was fortunate for him, and fortunate for us too—the seeds of it that he
saved and grew were the start of a line of sweet-tasting fruit. All squashes
with edible pulp have been derived from a few rare mutants found by
chance. Prior to that, gourd-like squashes and pumpkins were grown primarily for their edible seed and the containers that could be made from their
hard rinds.

A few months after I had the Coyote taste-test, I told the story to Don Bahr, who
was deeply involved in learning Pima and Papago songs at the time. "Oh, there
are a number of songs and legends referring to plants that are specifically
named Coyote's this or that. You should find out about the other ones too."

So I did. And while Coyote hadn't exactly dumped on all of them, he hadn't
left any of them in very good shape either. There was *Ban Tokĭ,* or Coyote
Cotton, which grows in the canyons of the Baboquivaris, and occasionally in
Papago fields. Its leaves and flowers look like regular cotton, but the bolls lack
a crucial ingredient—spinnable lint. Thanks to Coyote, they got the short end
of the deal when it came to cotton fiber and are worthless to weavers.

Then there's *Ban Bawĭ*—Coyote Tepary Beans. They, too, frequent canyons, sometimes twining around the stems of Coyote Cotton. The Papago used
to try harvesting them, but it became too much work—the seeds explode
out of the ripe pods when you touch them, scattering over the ground. You
could harvest them by picking them up if they were easy to see like the big
white and red-brown domesticated teparies that Papago grow in their fields.

But no, the Coyote Teparies look just like *gravel,* and are easily lost in the shuffle. Today Papago only grow the larger, brighter-colored tepary bean varieties, having given up the wild harvest by the late 1940s.

The kicker for me was in finding out about *Ban l'hug-ga,* or Coyote's Devil's Claw. Because its yellow flower is shaped like a shoe, one wild devil's claw species is also called *Ban Suisk,* meaning Coyote shoe, sandal, or tire. Another species of devil's claw has been cultivated by the Papago for the fiber produced from its dry, bony capsule—these black fibrous strips are woven into patterns in yucca and beargrass baskets that Papago women make.

The interesting thing is that botanists have been arguing for years as to whether or not the Papago had genetically altered the wild annual devil's claw species into a truly domesticated plant. This domesticated *l'hug* has longer fibers, pale flowera, and white (instead of black) seeds that germinate more quickly. Papago folk taxonomy clearly treats wild devil's claw species just as it does the wild relatives of other domesticated plants.

"Those other ones are *Ban l'hug-ga* because Coyote left them out in the desert. Now they are no good for making baskets with—those fibers are too small, too brittle. They just snap. You can't make anything out of them."

On a sunny winter day a Papago elder from Topowa sat with me in her field and told me the story of Coyote's Devil's Claw.

"One day Coyote was walking all over the desert, trying to find something to eat. He couldn't find anything, and he was too lazy to grow anything himself. So he walked and walked until he found what looked like a bone in the sand.

"He tasted it. It had no taste. It was too dry. So he sat down, thinking. Then he started to jump up and down, yelling, 'I think this bone wants to tell me that I will find something to eat around here.'

"So he ran around. All he saw was desert, no food. Then he came to a wide wash. He tried to jump across, but he landed in the middle of it, on top of a little green plant half-buried in the sand.

"'This looks like it would be good to eat,' Coyote said, and he gobbled down the whole plant—root, bony fruit, seeds and all.

"Glad that he didn't have to work to get his food, he decided to lie down and sleep. But after a while, he woke up with a big pain in his stomach. He got so sick that he had to get the plant out of his insides. He buried it in the sand and hoped he would not see it again. 'They don't like me and I don't like them.'

"But every year when the rain comes, those plants come up again. The floods carry the bony fruit and bury more of them in the sand where they can grow. Pretty soon, Coyote sees those plants he doesn't like all around. When the Desert People learn that they make him sick, they decide to say it is his plant. His devil's claw. *Ban l'hug-ga.*"

Around 1911, Papago Juan Dolores recorded the story of a time when Coyote *did* try to grow his own food. Well, sort of, in a way. He was given some good corn seed after the fall harvest, but instead of saving it for the next planting, he

ate nearly all of it. When the summer rains finally came, he had forgotten to prepare some good land. He finally just threw the seeds along the bad ground
around a wash.

Then Coyote slept through the growing season. He didn't learn the right
songs to sing to the corn when it did come up. Knowing that he had to sing
something to make it grow, he just made up a song. It was terrible.

The corn grew anyway. But it didn't grow up to be corn, because it never
heard the corn's songs. In a poor place like the rough edge of a wash, only
another kind of plant would grow. The plants grew up to be *Ban Wiw-ga,*
Coyote's Tobacco.

Another time, Coyote stole real tobacco for his own. Real tobacco was
grown in secret places. If a man other than its keeper saw it sprouting, the plant
would sink into the ground. It was used in the sacred smoke houses by tribes-
men, and could bring enlightenment: the ability to see in the dark, to talk with
the dead, to sense the source of a companion's disease, and to realize its cure.

Coyote saw this magic plant where it sprouted from the grave of a powerful woman. Before anyone could stop him, the Furry Thief ran along, snatched
some up, and went into the Smoke House as if he had some business there. He
rolled himself a cigarette, then smoked it by himself, not even passing it
around. It didn't matter. By that time it was Coyote's Tobacco, and wasn't too
good for curing and seeing. He just saw the world like he always did. Like a
Coyote.

Nearly all Coyote's plants are closely related to domesticated crops, seeds
which the Papago say they have grown "since the beginning." These wild seed-
stocks, as Coyote's plants, are considered by Papago to be genetic retrogrades
rather than possible progenitors of the crops. This is because Papago life *prior
to* having these cultivated crops is now unthinkable; without these domesticated plants, their culture would not exist in the way it has for centuries.
Hence, these wild plants associated with Coyote must be degenerated from
their original, useful forms. It's as if Coyote snatched sweet apples from a
Papago Garden of Eden, only to watch them turn sour and shriveled.

In contrasting Coyote's plants with *O'odham* domestic plants, Papago storytellers are making their people aware of two matters: that these plants are, in
fact, related; and that they should do their best to care for and improve the quality of their crops, lest they deteriorate into less useful forms.

But it turns out that these wild relatives of crops *aren't* worthless; it has
simply taken scientists time to recognize their relatedness to crops, and how to
make use of it. Plant breeders are now using such wild relatives to improve the
crops in our fields and protect them from pestilence. For it has been discovered
that Coyote's plants are often hardy and resistant.

In the 1940s, plant breeders used wild Arizona cotton in a triple hybrid
including cultivated upland cotton. They found a Coyote trick hidden therein.
Although wild cotton is nearly lintless, its genes contributed fiber strength to

cultivated upland cotton. Thus a better-quality cotton for weaving was developed using a plant that was "worthless" to weavers. Hybrids which included these wild cotton genes were also found to have pink boll worm resistance.

More recently, scientists have attempted to transfer other wild cotton genes to cultivated varieties to diminish the size of the bracts surrounding their flowers. In cotton processing, these bracts crumble into dust-like particles which have in the past been a cause of respiratory disease among workers.

For years, scientists have been interested in transferring the heat, drought, and blight resistance of tepary beans to other kinds of cultivated beans. Crosses between teparies and Great Northerns have been made, but always with considerable difficulty. Dr. Howard Scott Gentry, who made many collections of wild beans over the last half century, then suggested that these wild relatives should be used as "genetic bridges" to facilitate easier crosses between more distantly related cultivated varieties.

At Riverside, California, Claire Thomas and Giles Waines have recently had some success with this approach. By crossing cultivated teparies with Coyote's teparies, and doing the same with wild and domesticated common beans, they built a "bean bridge." These two hybrids can then be crossed with less difficulty, resulting in the transfer of tepary genes to other beans.

Wild cucurbits are now being used to improve the disease resistance of squashes. Collections of Coyote-like gourd species from Mexico have proved to have high levels of resistance to powdery mildew and to certain cucumber and watermelon mosaic viruses.

So I wonder what other tricks Coyote is preparing. Year after year, he steals watermelons from my Papago friends and goes down to hide them in the wash. He eats the big, cool fruit, spitting and shitting out the seed. Now and then, downstream from villages, I see small, ugly watermelons growing by themselves in the dry wash. What has he done to them? How bad do they taste?

RESPONDING JOURNAL

Explain how much Nabhan's characterization of Coyote adds to his explanation of wild seed propagation practices. What particular benefits does Nabhan derive from emphasizing the folkloric aspects of this animal?

QUESTIONS FOR CRITICAL THINKING

STRATEGY

1. Think about the opening of Nabhan's essay. Why might he have decided to start his narration with the idea of the "*old* stories"? What might such a technique contribute to his essay?

2. *Ban* as a Papago word for *coyote* has several shades of meaning. See how many different senses of the word you can find and explain in this selection. In what ways does such a system of classification deepen your understanding of Nabhan's topic?

ISSUES

1. What relations between native culture and seed development do you see? What possible problems do you envision in developing seed strains apart from the cultures that have used them for centuries?
2. The Papago storytellers make much of Coyote's corn, tobacco, and other plants. What information or knowledge do they convey in these stories? What possible use of such knowledge would you hope to see develop in the future, assuming that these stories can be told to a wider audience?

COLLABORATIVE WRITING ACTIVITY

Think of a word that has multiple meanings, particularly in a colloquial sense. You might look at a word drawn from the animal or plant world, such as "dog" or "vegetable," or a more unusual word. Explain in a statement of two pages or more how these different connotations of the word might draw strength from their source. After you have done this brainstorming and composing, present your findings to the group. After you have discussed the group's responses, revise your original and come back to the group with your work. Working as a group, decide on the characteristics you have identified as commonalities— those things that signify your group as a language community. Working from your revised material, compose a statement of the commonalities.

"Spring" from *A Country Year*
Sue Hubbell

After a divorce and a move from Michigan to the Missouri Ozarks in the mid-1980s, former bookstore manager and librarian Sue Hubbell found the emotional climate necessary to develop her great skill as a naturalist and writer. When A Country Year: Living the Questions *appeared in 1986, audiences recognized a writer of great ability. Her specialty is bees and beekeeping; she farms honey on her Ozark retreat.*

More than simply orchestrating the life cycles of her hives, though, her activities center on patient observation of the slowly changing farm and rural community. The book both begins and ends with spring, an annual cycle completed and the promise of the second one begun. The selection printed here is from the first spring the book describes.

EXPLORING JOURNAL

What do you know about beekeeping besides the images of the hives and their caretakers with the protective netting hung around their helmets? What might prompt people to take up beekeeping as a hobby? Try to correlate this activity with your own sense of bees as you explore these ideas in a ten-minute freewrite.

———————

Anyone who has kept bees is a pushover for a swarm of them. We always drop whatever we are doing and go off to pick one up when asked to do so. It doesn't make sense because, from a standpoint of serious beekeeping and honey production a swarm isn't much good. Swarms are headed up by old queens with not much vitality or egg-laying potential left, and so a beekeeper should replace her with a new queen from a queen breeder. He will probably have to feed and coddle the swarm through its first year; it will seldom produce any extra honey the first season. And yet we always hive them.

There is something really odd about swarms, and I notice that beekeepers don't talk about it much, probably because it is the sort of thing we don't feel comfortable about trying to put into words, something the other side of rationality.

The second year I kept bees, I picked up my first swarm. I was in the middle of the spring beework, putting in ten to twelve hours a day, and very attuned to what the bees were doing out there in their hives. That day had begun with a heavy rainstorm, and so rather than working out in the beeyards, I was in the honey house making new equipment. By afternoon the rain had stopped, but the air was warm and heavy, charged and expectant. I began to feel odd, tense and anticipatory, and when the back of my neck began to prickle I decided to take a walk out to the new hives I had started. Near them, hanging pendulously from the branch of an apple tree, was a swarm of bees. Individual bees were still flying in from all directions, adding their numbers to those clinging around their queen.

In the springtime some colonies of bees, for reasons not well understood, obey an impulse to split in two and thus multiply by swarming. The worker bees thoughtfully raise a new queen bee for the parent colony, and then a portion of the bees gather with the old queen, gorge themselves with honey and fly out of the hive, never to return, leaving all memory of their old home behind. They cluster somewhere temporarily, such as on the branch of my apple tree. If a beekeeper doesn't hive them, scout bees fly from the cluster and investigate nearby holes and spaces and report back to the cluster on the suitability of new quarters.

We know about two forms of honeybee communication. One is chemical: information about food sources and the wellbeing of the queen and colony is exchanged as bees continually feed one another with droplets of nectar which they have begun to process and chemically tag. The other form of communication is tactile: bees tell other bees about good things such as food or the location of a new home by patterned motions. These elaborate movements which amount to a highly stylized map of landmarks, direction and the sun's position, are called the bee dance.

Different scout bees may find different locations for the swarm and return to dance about their finds. Eventually, sometimes after several days, an agreement is reached, rather like the arrival of the Sense of the Meeting among Quakers, and all the bees in the cluster fly off to their new home.

I watched the bees on my apple tree for a while with delight and pleasure, and then returned to the barn to gather up enough equipment to hive them. As I did so, I glanced up at the sky. It was still dark from the receding thunderstorm, but a perfect and dazzling rainbow arched shimmering against the deep blue sky, its curve making a stunning and pleasing contrast with the sharp inverted V of the barn roof. I returned to the apple tree and shook the bees into the new beehive, noticing that I was singing snatches of one of Handel's coronation anthems. It seemed as appropriate music to hive a swarm by as any I knew.

Since then, I have learned to pay attention in the springtime when the air feels electric and full of excitement. It was just so one day last week. I had been working quietly along the row of twelve hives in an outyard when the hair on the back of my neck began to stand on end. I looked up to see the air thick with bees flying in toward me from the north. The swarm was not from any of my hives, but for some reason bees often cluster near existing hives while they scout a new location. I closed up the hive I was working on and stood back to watch. I was near a slender post oak sapling, and the bees began to light on one of its lower limbs right next to my elbow. They came flying in, swirling as they descended, spiraling around me and the post oak until I was enveloped by the swarm, the air moving gently from the beat of their wings. I am not sure how long I stood there. I lost all sense of time and felt only elation, a kind of human emotional counterpart of the springlike, optimistic, burgeoning state that the bees were in. I stood quietly; I was nothing more to the bees than an object to be encircled on their way to the spot where they had decided, in a way I could not know, to cluster. In another sense, I was not remote from them at all, but was receiving all sorts of meaningful messages in the strongest way imaginable outside of human mental process and language. My skin was tingling as the bees brushed past and I felt almost a part of the swarm.

Eventually the bees settled down in the cluster. Regaining a more suitable sense of my human condition and responsibilities, I went over to my pickup and got the empty hive that I always carry with me during swarming season. I propped it up so that its entrance was just under the swarm. A frame of comb from another hive was inside and the bees in the cluster could smell it, so they began to walk up into the entrance. I watched, looking for the queen, for without her the swarm would die. It took perhaps twenty minutes for all of them to file in, and the queen, a long elegant bee, was one of the last to enter.

I screened up the entrance and put the hive in the back of the pickup. After I was finished with my work with the other hives in the beeyard, I drove back home with my new swarm.

I should have ordered a new queen bee, killed the old one and replaced her, but in doing that I would have destroyed the identity of the swarm. Every colony of bees takes its essence, character and personality from the queen who is mother to all its members. As a commercial beekeeper, it was certainly my business to kill the old queen and replace her with a vigorous new one so that the colony would become a good honey producer.

But I did not.

RESPONDING JOURNAL

Hubbell gives us just enough of a sense of similarity between bee and human behavior to make us think there are some interesting parallels. In an entry of several paragraphs, explore the idea that the behavior of bees can suggest some patterns of human nature and community.

QUESTIONS FOR CRITICAL THINKING

STRATEGY

1. Think about the pattern of this chapter, this mini-essay, from Hubbell's book. Why does she begin the chapter with the particular details she chooses? Why does she end the chapter in the terse fashion she does? What does she gain by such a structure?

2. Explain the significance of the final sentence in the first paragraph, "And yet we always hive them." Particularly in a short essay, the last sentence in the first paragraph can carry special weight; it often functions as the thesis sentence for the essay. What does Hubbell accomplish with her short sentence?

ISSUES

1. In the foreword to *A Country Year,* Hubbell states that she has learned on her farm that "there are more questions than answers." What are some of the questions her essay raises but does not answer?

2. Explain why Hubbell evidently does not share the fear that most of her readers would experience if they were as close to a million bees as she is in her essay.

COLLABORATIVE WRITING ACTIVITY

Write several paragraphs to explain why beekeepers, according to Hubbell, are able to rest comfortably with knowledge that lies on "the other side of rationality." What other situations that lie on that far side of the rational come to mind for you? Do people like farmers seem to have a special inside track on the ability to achieve such a situation of specialized knowledge? Bring your comments and read them to the rest of the group; after you have had a chance to discuss your ideas, work together to draft a general statement of how this "other side of rationality" can be seen and understood.

"Custodian"
Maxine Kumin

One of the most widely read contemporary poets who celebrate the world of external nature, Maxine Kumin won the Pulitzer Prize for Poetry in 1973. Educated at Radcliffe, she has taught at Columbia, Princeton, and Washington University. She presently lives in New Hampshire, where she finds the setting for much of her poetry.

"Custodian" appears in the collection Nurture, *first published in 1989. After you have had opportunity to read the poem enough times to feel comfortable with its topics and ideas, locate the stanza that provides the classification. If you wish, you may use the poem's title as a guide to help you determine the overriding principle on which the classification is based. Think about how the final stanza deepens or complicates the ideas upon which the classification rests.*

EXPLORING JOURNAL

Compose an entry on the idea of animals as custodians. What do they protect or guard? In what sense can animals be said to look after humans or other species of animals?

Every spring when the ice goes out
black commas come scribbling across the shallows.
Soon they sprout forelegs.
Slowly they absorb their tails
and by mid-June, full-voiced, announce themselves.

Enter our spotted dog.
Every summer, tense with the scent of them,
tail arced like a pointer's but wagging
in anticipation, he stalks his frogs
two hundred yards clockwise around
the perimeter of this mucky pond,
then counterclockwise, an old pensioner
happy in his work.

Once every ten or so pounces
he succeeds, carries his captive north
in his soft mouth, uncorks him on the grass,
and then sits, head cocked, watching the slightly
dazed amphibian hop back to sanctuary.

Over the years the pond's inhabitants
seem to have grown accustomed
to this ritual of capture and release.
They ride untroubled in the wet pocket
of the dog's mouth, disembark in the meadow
like hitchhikers, and strike out again for home.

I have seen others of his species kill
and swallow their catch and then be seized
with violent retchings. I have seen children
corner polliwogs in the sun-flecked hollow
by the green rock and lovingly squeeze
the life out of them in their small fists.
I have seen the great blue heron swoop in
time after wing-slapping time to carry
frogs back to the fledglings in the rookery.

Nothing is to be said here
of need or desire. No moral arises
nor is this, probably, purgatory.
We have this old dog,
custodian of an ancient race of frogs,
doing what he knows how to do
and we too, taking and letting go,
that same story.

—————◆—————

RESPONDING JOURNAL

Write an entry that outlines the classifications Kumin mentions. If her dog is the custodian, what are the other people she mentions? What sorts of roles do they have?

QUESTIONS FOR CRITICAL THINKING

STRATEGY

1. What is the image presented in the first two lines? How might this image be particularly effective for a poem?
2. What point might Kumin have in mind when she mentions dogs other than her own, dogs who eat the frogs, or children who find the pollywogs?

ISSUES

1. To what extent do the frogs seem to be disturbed by their trips in the dog's mouth? From what perspective could they be said to be in "custody"?
2. Explain how the poem's final image of "that same story" might be applied to the principle of classification.

COLLABORATIVE WRITING ACTIVITY

Think of different situations in the natural world that might parallel Kumin's idea of the animal in its habits as "an old pensioner/happy in his work." Bring your example back to your group, where you can share your responses and together compose an introductory statement of how animals work happily with one another.

SUGGESTIONS FOR FORMAL WRITING ASSIGNMENTS

1. Devise a system of classification of your own choosing based on differing attitudes toward the environment. You may want to look at the list of possible roles, sketched in the introduction, that people might take toward the natural world and thus base your determining principle on characteristic stances or attitudes.

2. The writers represented in this section have used widely varying sorts of classification systems. Evaluate at least three different sorts of classification systems you have encountered, and write an essay in which you set forth the merits and potential drawbacks of each; decide which are likely to be most useful in which situations. Your conclusion will give you an opportunity for some original writing and argument.

3. Compose a humorous classification much in the spirit of the nature walk described by Dave Barry. You may well want to refer to Nabhan's taxonomies to help you generate ideas, but you are free to use all or any part of the natural world as your field of concentration for this essay.

Chapter 8

PROCESS ANALYSIS: NATURAL AND MECHANICAL SYSTEMS

John K. Hilliers, *Hopi Pueblo of Walpi,* **1876**

1. Analyze your first impressions of this photograph. Why do you
   think the viewer needs time to look at the image carefully in order
   to comprehend it?

2. What is the effect of the composition of this photograph? Why
   might the photographer have chosen to present the image from
   this particular perspective?

3. Discuss the similarities between natural processes and patterns of
   human activity that have shaped the landscape shown here. What
   evidence of the two processes can you detect? How effective a
   medium is black-and-white photography to show such processes?

Chapter 8

PROCESS ANALYSIS: NATURAL AND MECHANICAL SYSTEMS

You see process at work all the time, although you may not notice it. Whenever you perform a series of tasks related to the achievement of a specific goal, you are carrying out a process. When you get on a plane and a flight attendant explains the emergency exit procedure, he is explaining a process designed to get everyone off the plane safely in case of an accident. Similarly, when that person explains what to do if the cabin should lose pressure and the emergency oxygen masks should drop down, he is again explaining a process. If you are wearing shoes with laces, then you completed the process of tying your shoes (assuming the laces are of course tied). Actually, describing in an essay the process of tying your shoes and then having a partner actually tie your shoe by following your directions has long been used in composition classes as a good way to determine how effective a process essay actually is: can your partner tie your shoes according to your process description?

If you look at the shelves of the "how-to" section in your library or bookstore, you will find numerous examples of process analysis. The title of Stephen R. Covey's recent bestseller, *The Seven Habits of Highly Effective People*, promises that it will outline seven forms of process which will in turn provide its readers a guide to success in various areas of life. By patterning our own behavior on that of people who have impressively demonstrated their own personal effectiveness, we can imitate the process by which they achieved their success. If you look at any fishing or hunting magazine, you will find numerous articles on how to achieve success with a certain species by following the plans—the process—spelled out in the article.

In such essays, the plan is clear. You as writer provide a thesis—for example, that you can catch large saltwater fish with a fly rod by following proper tackle techniques—and then outline the equipment and procedures by taking your reader through the actual process. Have you ever seen anyone tie a fly, or have you tied one yourself? All fly tying instructions are examples of a complex process that is broken down into its component parts for analysis; such instructions are also often called process analysis, since they break down and analyze each step in the process so the reader can see why the feathers and yarn should be tied to the hook in certain prescribed manners. To complete a scientific analysis of an artificial fly, we might dissect it as carefully as we would a real insect; such a process would reveal to us each successive bit of material attached to the hook.

Tying a fly is an example of a mechanical process; even though the finished creation is supposed to imitate a natural fly, the means whereby the artificial is created is clearly not natural. An example of the natural process would be the movement through life cycles of the actual fly. The mechanical and natural are two parallel versions of the process description; both require a clear statement of purpose, often provided in an opening thesis, and clearly defined steps in the process so the reader will be able to see quickly the important points along the way.

A short description of process can also be used to begin an essay, as in "Last of the Wild Salmon" where Marie De Santis describes the spawning run. If she were asked to explain why she uses process, she might say: "I decided that in using process as my organizing principle, I could show effectively the power and determination of the salmon by describing its physical condition and its actions. I will convey a sense of the salmon's instincts by explaining the process of the salmon's upstream movement."

Such a technique of process is a useful way to begin an essay in which your strategy of development will benefit from a description of a process used as an example, an illustration, or background for a particular point. For example, in "Thinking Like a Mountain" Aldo Leopold explains how the scream of a wolf initiates a process of response. He begins his essay with gripping detail: "A deep chesty bawl echoes from rimrock to rimrock, rolls down the mountain, and fades into the far blackness of the night." By describing the responses of things both living and dead to this howl, he helps his reader understand the subtleties of the process he is describing. He believes that "only the mountain has lived long enough to listen objectively to the howl of a wolf." In the remainder of his essay, he explains the process of thought which leads him to such a conclusion.

The concept of process description as technique can also work well for writers at other points in an essay. In his essay "Coon Valley: An Adventure in Cooperative Conservation," Leopold sets out to explain the significance of the Coon Valley Erosion Project as a model for careful, responsible land stewardship. In order to do so, he first compares and contrasts different views of land conservation; he next proposes a hybrid model. After listing the benefits of such a model, he provides background for Coon Valley by sketching the process by which the land changed from what it was before white settlers arrived to what it was at the beginning of the project. Leopold writes:

> Coon Valley is one of those innumerable little units of the Mississippi Valley which collectively fill the national dinner pail. Its particular contribution is butterfat, tobacco, and scenery.
>   When the cows which make the butter were first turned out upon the hills which comprise the scenery, everything was all right because there were more

hills than cows, and because the soil still retained the humus which the wilderness vegetation through centuries had built up. The trout streams ran clear, deep, narrow, and full. They seldom overflowed. This is proven by the fact that the first settlers stacked their hay on the creekbanks, a procedure now quite unthinkable. The deep loam of even the steepest fields and pastures showed never a gully, being able to take on any rain as it came, and turn it either upward into crops, or downward into perennial springs. It was a land to please everyone, be he an empire-builder or a poet.

But pastoral poems had no place in the competitive industrialization of prewar America, least of all in Coon Valley with its thrifty and ambitious Norse farmers. More cows, more silos to feed them, then machines to milk them, and then more pasture to graze them—this is the epic cycle which tells in one sentence the history of the modern Wisconsin dairy farm. More pasture was obtainable only on the steep upper slopes, which were timber to begin with, and should have remained so. But pasture they now are, and gone is the humus of the old prairie which until recently enabled the upland ridges to take on the rains as they came.

Result: Every rain pours off the ridges as from a roof. The ravines of the grazed slopes are the gutters. In their pastured condition they cannot resist the abrasion of the silt-laden torrents. Great gashing gullies are torn out of the hillside. Each gully dumps its load of hillside rocks upon the fields of the creek bottom, and its muddy waters into the already swollen streams. Coon Valley, in short, is one of the thousand farm communities which, through the abuse of its originally rich soil, has not only filled the navigational dinner pail, but has created the Mississippi flood problem, the navigation problem, the overproduction problem, and the problem of its own future continuity.

In describing the process by which Coon Valley's present problems have come into existence, Leopold also offers a clear example of cause and effect as a theme and structuring principle in his writing.

Process, however, is even more likely to be used as the organizational basis for an entire essay. Most often, you will want to think of some task to be accomplished. The examples we have just seen include the salmon's instinctual progress upstream, the effect of a wild animal's cry echoing in a wilderness, and remedial actions necessary to reverse long-standing patterns of erosion and loss of local resources. In each of these, the step-by-step nature of the sequence—the nuts and bolts of getting each job done—comes through clearly. So it will be with your own writing.

You will find clues to situations in which process will work very well by looking for statements like: "What steps should we take to protect our water supply?," or "How does the topography of our region affect rainfall?," or "Describe the route our water travels from its source to our faucets." Supplying answers in essay form to these or similar questions will often result in an example of a process essay.

We are likely to find two kinds of process in much of our reading: natural and mechanical. In matters related to the natural world, we will encounter numerous examples of both. The essays in this chapter feature different versions of the process approach. Marie De Santis explains how the biological process of salmon migration is affected by man-made changes in the salmon's environment. Aldo Leopold describes a process of relating to the land through which a person attempts to "think like a mountain," to find a way of looking at the world that would be in harmony with the great natural forms. Alice Walker reverses the process and suggests that parts of the natural world have the same qualities that we value in humans. Tim Cahill takes a walk across Death Valley in midsummer and lives to tell about it. Throughout these varied accounts of the processes operating in the natural world, we see writers attempting to develop processes of their own to describe their responses.

"Last of the Wild Salmon"
Marie De Santis

Marie De Santis began her professional career in a fairly traditional way by pursuing a graduate degree, but on the way to the doctorate she changed direction radically: she became a commercial fisherwoman. After working on other captains' boats, she herself became a captain and recorded her experience in her 1984 book, Neptune's Apprentice.

Her second book, California Currents, *from which "Last of the Wild Salmon" is taken, treats animals that have been removed from their natural habitats and forced to live under the domination of man in new environments. In describing the life cycle of the Pacific salmon, she explains how dams and other water management programs have seriously eroded the salmon's traditional means of spawning and have led to hatchery-bred salmon that have little in common with their wild cousins.*

EXPLORING JOURNAL

Explain what you know about hatchery-reared fish. Have you heard distinctions made before between "wild" or "native" fish and "hatchery" fish? What groups do you think would care about the differences that might exist? Does such a distinction have any relevance to you in your life? Do any of the fish you eat come from one source or the other?

———⊷⊶———

In a stream so shallow that its full body is no longer submerged in the water, the salmon twists on its side to get a better grip with its tail. Its gillplate is torn, big hunks of skin hang off its sides from collisions with rocks, there are deep gouges in its body, and all around for miles to go there is only the cruelty of more jagged rocks and less and less water to sustain the swim. Surely the animal is dying!

And then the salmon leaps like an arrow shot from a bow; some urge and will and passion ignores the animal body and focuses on the stream.

Of all the extremes of adaptation to the ocean's awful toll on the young, none is more mythic in proportion than the salmon's mighty journey to the mountain streams: a journey that brings life to meet death at a point on a perfect circle, a return through miles of narrowing waters to the exact gravel-bedded streamlet of its birth. A journey to spawning and death, so clear in its resemblance to the migrations of the sperm to the egg as to entwine their meanings in a single reflection.

On every continent of the northern hemisphere, from the temperate zone to the arctic, there is hardly a river that hasn't teemed with salmon's spawn: the Thames, the Rhine, the rivers of France and Spain, Kamchatka and Siberia, Japan (which alone has more than 200 salmon rivers) and the arctic streams of Greenland. From the Aleutians to Monterey Bay, through the broadest byways to the most rugged and narrow gorge, the salmon have made their way home. There are many journeys for which the salmon endure more than 1000 miles.

As soon as the ice melts on the Yukon, the king salmon enter the river's mouth, and for a month, the fish swim against the current, 50 miles a day for a total of 1500. And like every other salmon on its run, the king salmon fasts completely along the way. In other rivers, salmon scale vertical rocks up to 60 feet high, against hurtling waterfalls.

The salmon gets to spawn once in life, and maybe that's reason enough. The salmon's instinct to return to the place of its birth is so unmodifiable and of such purity as to have inspired hundreds of spiritual rites in as many societies of human beings.

The salmon arrives battered and starved, with a mate chosen along the way, and never has passion seemed less likely from two more wretched-looking beings. But, there in the gravel of the streamlet, the female fans out a nest with the sweep of her powerful tail and the male fends off intruders. The nest done, the two fish lie next to each other suspended in the water over the nest; their bodies quiver with intense vibrations, and simultaneously they throw the eggs and the sperm. Compared with the millions of eggs thrown by a cod in a stream, the salmon need throw only 2000 to 5000. Despite the predators and other hazards of the stream, these cold mountain waters are a sanctuary compared with the sea. For the next two or three days, the pair continue nesting and spawning until all the eggs are laid. Then the salmon, whose journey has spanned the ocean and the stream, lies by the nest and dies.

Soon the banks of the streams are stacked with ragged carcasses, and the animals of the woods come down for a feast. The stream lies quiet in the winter's deepening cold. But within a month two black eyes appear through the skin of each egg. And two weeks later, the water is again alive with the pulsing of millions of small fish feeling the first clumsy kicks of their tails. The fingerlings stay for a while, growing on the insects and larvae that have been nurtured by the forest. Then, one day, they realize what that tail is for and begin their descent to the sea, a journey mapped in their genes by the parents they left behind.

The young salmon arrive in the estuary facing the sea, where they linger again and learn to feed on shrimp, small crustaceans and other creatures of the brine. Here, also, their bodies complete an upheaval of internal and external changes that allow them to move on to the saltier sea. These adaptations require such extraordinary body transformations that when the same events occur on the stage of evolution they take millions and millions of years. In the life of the salmon, the changes take place in only a matter of months. One of life's most prohibitive barriers—that between fresh and salt water—is crossed, and the salmon swim back and forth, in and out of the sea, trying it on for size.

Then one day, the youngsters do not return. The stream is only a distant memory drifting further and further back in the wake of time, only different— a memory that will resurrect and demand that its path be retraced.

So accessible is the salmon's life in the stream that more is known about the reproduction of this fish than any other ocean animal. With the ease of placing cameras underwater, there isn't any aspect of this dramatic cycle that hasn't been captured in full color in some of the most spectacular film footage ever made.

But once the salmon enters the sea, the story of its life is a secret as deep and dark as the farthest reaches of the ocean it roams. The human eye with its most sophisticated aids, from satellite to sonar, has never caught more than a glance of the salmon at sea. Extensive tagging programs have been carried out, but they tell us little more than that the salmon is likely to be found anywhere within thousands of miles of its origins, and even this is only a sliver of the picture because the tags are recovered only when the salmon is caught by fishermen, who work solely within the narrow coastal zone. Along with a few other pelagic fishes, like the tuna, that claim vast stretches of sea for their pasture, the salmon's life remains one of the most mysterious on earth.

—————>•◦•<—————

RESPONDING JOURNAL

Explain in a ten-minute freewrite how well you think the idea of process works to describe the plight of the salmon. Might any other strategy that we have considered so far work better than the process approach?

QUESTIONS FOR CRITICAL THINKING

STRATEGY

1. Think about the effect of the first three paragraphs. How does De Santis dramatize her reader's sense of the process she is describing?
2. Look at De Santis' description of the mating of the salmon. Does she convey to you a sense of their determination? Explore her success with this quality and explain the devices she uses to achieve her objective. Might any other strategies have helped her?

ISSUES

1. De Santis mentions "one of life's most prohibitive barriers" as the line "between fresh and salt water." Think of other instances you can recall that suggest the importance of this dividing line. Can you think of any other parallel barriers in the natural world? How do these compare to the barriers the salmon face?

2. Are you familiar with any of the "hundreds of spiritual rites" that De Santis claims humans have devised in imitation of the single mindedness of the salmon in their journey? Is she making a claim for something that *actually* exists, or for something that she *hopes* exists? Try to think of compelling reasons to support your assertions.

COLLABORATIVE WRITING ACTIVITY

Based on what you know about the life cycle of salmon, from the possibility of actual experience through films you might have seen of bears hunting salmon, speculate on what might occur when the salmon are away at sea. Concentrate on these two questions: where do they go, and why do they move so far away? Does your recollection of knowledge gained from your science classes help you with this in any way? Record your ideas in the form of two lists.

Next, bring your lists to your group. After the group members have read their responses, discuss the ideas presented and compile a list of three or four responses that make the most sense to the entire group. When you have finished this, draft as a group a statement of what information you would need to answer the questions your group has raised. Think about the sorts of authorities whose information you would need to have to complete your research.

"Thinking Like a Mountain"
Aldo Leopold

Aldo Leopold (1887–1948) is widely regarded as America's premier ecologist in the first half of the twentieth century. Beginning as a forestry student at Yale, he worked as a ranger in the Southwest and Mexico in the early 1920s when this land was truly a wilderness. When he began his long teaching career at the University of Wisconsin, he turned his efforts to working for the promotion of biodiversity in land use at the same time the sanctity of each form of life is respected. Many of his essays and speeches were presented for public or scholarly audiences; his 1930 study, Game Management, *was the only book published during his lifetime. He died of a heart attack while help-ing a neighbor put out a brush fire near his second home, a scraggly tract of reclaimed Wisconsin farmland that Leopold was trying to restore to its pre-settler status.*

"Thinking Like a Mountain" was first published in A Sand County Almanac *in 1949, the first and the most popular collection of Leopold's essays. In it, we see some of the reasons for Leopold's growing popularity: he is able to take a personal experience and suggest how it has guided his thinking and taught him something about the natural world. Leopold shows how a pro-cess of thought can be as dramatic as a life-or-death struggle in the wild.*

Exploring Journal

Explain what you think Leopold's title might mean. In what possible way could you "think like a mountain"? What sort of mental process would you have to undergo so that your pattern of thinking would approximate that of a mountain?

A deep chesty bawl echoes from rimrock to rimrock, rolls down the mountain, and fades into the far blackness of the night. It is an outburst of wild defiant sor-row, and of contempt for all the adversities of the world.

Every living thing (and perhaps many a dead one as well) pays heed to that call. To the deer it is a reminder of the way of all flesh, to the pine a forecast of midnight scuffles and of blood upon the snow, to the coyote a promise of glean-ings to come, to the cowman a threat of red ink at the bank, to the hunter a challenge of fang against bullet. Yet behind these obvious and immediate hopes and fears there lies a deeper meaning, known only to the mountain itself. Only the mountain has lived long enough to listen objectively to the howl of a wolf.

Those unable to decipher the hidden meaning know nevertheless that it is there, for it is felt in all wolf country, and distinguishes that country from all other land. It tingles in the spine of all who hear wolves by night, or who scan their tracks by day. Even without sight or sound of wolf, it is implicit in a hun-dred small events: the midnight whinny of a pack horse, the rattle of rolling rocks, the bound of a fleeing deer, the way shadows lie under the spruces. Only the ineducable tyro can fail to sense the presence or absence of wolves, or the fact that mountains have a secret opinion about them.

My own conviction on this score dates from the day I saw a wolf die. We were eating lunch on a high rimrock, at the foot of which a turbulent river elbowed its way. We saw what we thought was a doe fording the torrent, her breast awash in white water. When she climbed the bank toward us and shook out her tail, we realized our error: it was a wolf. A half-dozen others, evidently grown pups, sprang from the willows and all joined in a welcoming mâlée of wagging tails and playful maulings. What was literally a pile of wolves writhed and tumbled in the center of an open flat at the foot of our rimrock.

In those days we had never heard of passing up a chance to kill a wolf. In a second we were pumping lead into the pack, but with more excitement than accuracy: how to aim a steep downhill shot is always confusing. When our rifles were empty, the old wolf was down, and a pup was dragging a leg into impassable slide-rocks.

We reached the old wolf in time to watch a fierce green fire dying in her eyes. I realized then, and have known ever since, that there was something new to me in those eyes—something known only to her and to the mountain. I was young then, and full of trigger-itch; I thought that because fewer wolves meant more deer, that no wolves would mean hunters' paradise. But after seeing the green fire die, I sensed that neither the wolf nor the mountain agreed with such a view.

Since then I have lived to see state after state extirpate its wolves. I have watched the face of many a newly wolfless mountain, and seen the south-facing slopes wrinkle with a maze of new deer trails. I have seen every edible bush and seedling browsed, first to anaemic desuetude, and then to death. I have seen every edible tree defoliated to the height of a saddlehorn. Such a mountain looks as if someone had given God a new pruning shears, and forbidden Him all other exercise. In the end the starved bones of the hoped-for deer herd, dead of its own too-much, bleach with the bones of the dead sage, or molder under the high-lined junipers.

I now suspect that just as a deer herd lives in mortal fear of its wolves, so does a mountain live in mortal fear of its deer. And perhaps with better cause, for while a buck pulled down by wolves can be replaced in two or three years, a range pulled down by too many deer may fail of replacement in as many decades.

So also with cows. The cowman who cleans his range of wolves does not realize that he is taking over the wolf's job of trimming the herd to fit the range. He has not learned to think like a mountain. Hence we have dustbowls, and rivers washing the future into the sea.

We all strive for safety, prosperity, comfort, long life, and dullness. The deer strives with his supple legs, the cowman with trap and poison, the statesman with pen, the most of us with machines, votes, and dollars, but it all comes to the same thing: peace in our time. A measure of success in this is all well enough, and perhaps is a requisite to objective thinking, but too much safety seems to yield only danger in the long run. Perhaps this is behind Thoreau's dictum: In wildness is the [preservation] of the world. Perhaps this is the hidden meaning in the howl of the wolf, long known among mountains, but seldom perceived among men.

——➤◦◄——

RESPONDING JOURNAL

How realistic do you believe Leopold's position to be? Could we actually take his suggestion and allow packs of wolves to live? What could be the costs of such an action? Justify your position on this issue; what other species might provide parallel examples?

QUESTIONS FOR CRITICAL THINKING

STRATEGY

1. Think about the construction of Leopold's second paragraph. Does he adequately prepare you to accept the idea of his final sentence in that paragraph?
2. Does Leopold explain sufficiently what he means by "peace in our time"? Has he provided sufficient examples for you to feel compelled by his argument? How effectively does he establish significant points of the process?

ISSUES

1. Leopold's process of thought has evolved, as he tells us, over the years to a point where he questions the wisdom of our traditional practices regarding wolves. Can you think of other instances where your experience has led you to question traditional or accepted ways of doing things? Can you offer some key points in your experience that led you to question that "business as usual"?
2. Does Leopold's essay give you any further support for your feelings about hunting? Whether you hunt or not, whether you approve of hunting or not, has his essay caused you to change your position? If so, what points in the essay encourage you to modify your thinking about hunting?

COLLABORATIVE WRITING ACTIVITY

Generate a list of your reasons to agree or disagree with Leopold's statement that "We all strive for safety, prosperity, comfort, long life, and dullness." After you bring your list to your group, see if the group as a whole agrees or disagrees with Leopold. Work together to generate a statement of the utility of Leopold's idea for the 1990s. Can our world afford to act on his recommendation? Can we adopt such a process of thinking as the one he describes?

"Everything Is a Human Being"
Alice Walker

Alice Walker is a contemporary author of fiction and essays, probably best known for her 1982 novel The Color Purple. *In the following year, she published the collection of essays titled* In Search of Our Mothers' Gardens.

In all of her writing, Walker has long argued vigorously for the emancipation of African American women from the various forces that seek to dominate them.

Her work during the 1960s and 1970s for voter registration rights and welfare rights prefigured her strong insistence on the necessary protection of human rights that occurs so frequently in her writing. In her noted essay "In Search of Our Mothers' Gardens," she uses the wide range of examples and texts that typifies her writing. Like Momaday with his strong interest in his Native American roots, she frequently arranges collisions between the dominant Euro-centric culture and her own African American cultural traditions. She then enters to see where the pieces will fall. "Everything Is a Human Being" comes from her collection of essays, Living by the Word.

EXPLORING JOURNAL

Speculate on the possible meaning of Walker's title. Surely not everything could be *human,* could it? Try to stretch your thinking in a ten-minute freewrite to come up with possible ways in which objects such as trees that we do not normally think of as human might in some way be considered as people.

Some years ago a friend and I walked out into the countryside to listen to what the Earth was saying, and to better hear our own thoughts. We had prepared ourselves to experience what in the old days would have been called a vision, and what today probably has no name that is not found somewhat amusing by many. Because there is no longer countryside that is not owned by someone, we stopped at the entrance to a large park, many miles distant from the city. By the time we had walked a hundred yards, I felt I could go no farther and lay myself down where I was, across the path in a grove of trees. For several hours I lay there, and other people entering the park had to walk around me. But I was hardly aware of them. I was in intense dialogue with the trees.

As I was lying there, really across their feet, I felt or "heard" with my feelings the distinct request from them that I remove myself. But these are not feet, I thought, peering at them closely, but roots. Roots do not tell you to go away. It was then that I looked up and around me into the "faces." These "faces" were all middle-aged to old conifers, and they were all suffering from some kind of disease, the most obvious sign of which was a light green fungus, resembling moss and lichen, that nearly covered them, giving them—in spite of the bright spring sunlight—an eerie, fantastical aspect. Beneath this greenish envelopment, the limbs of the trees, the "arms," were bent in hundreds of shapes in a profusion of deformity. Indeed, the trees reminded me of nothing so much as badly rheumatoid elderly people, as I began to realize how difficult, given their bent shapes, it would be for their limbs to move freely in the breeze. Clearly these were sick people, or trees; irritable, angry, and growing old in pain. And they did not want me lying on their gnarled and no doubt aching feet.

Looking again at their feet, or roots—which stuck up all over the ground and directly beneath my cheek—I saw that the ground from which they emerged was gray and dead-looking, as if it had been poisoned. Aha, I thought, this is obviously a place where chemicals were dumped. The soil has been poisoned, the trees afflicted, slowly dying, and they do not like it. I hastily communicated this deduction to the trees and asked that they understand it was not *I* who had done this. I just moved to this part of the country, I said. But they were not appeased. Get up. Go away, they replied. But I refused to move. Nor could I. I needed to make them agree to my innocence.

The summer before this encounter I lived in the northern hills of California, where much logging is done. Each day on the highway, as I went to buy groceries or to the river to swim, I saw the loggers' trucks, like enormous hearses, carrying the battered bodies of the old sisters and brothers, as I thought of them, down to the lumberyards in the valley. In fact, this sight, in an otherwise peaceful setting, distressed me—as if I lived in a beautiful neighborhood that daily lost hundreds of its finest members, while I sat mournful but impotent beside the avenue that carried them away.

It was of this endless funeral procession that I thought as I lay across the feet of the sick old relatives whose "safe" existence in a public park (away from the logging trucks) had not kept them safe at all.

I *love* trees, I said.

Human, *please*, they replied.

But, I do not cut you down in the prime of life. I do not haul your mutilated and stripped bodies shamelessly down the highway. It is the lumber companies, I said.

Just go away, said the trees.

All my life you have meant a lot to me, I said. I love your grace, your dignity, your serenity, your generosity . . .

Well, said the trees, before I actually finished this list, we find you without grace, without dignity, without serenity, and there is no generosity in you either—just ask any tree. You butcher us, you burn us, you grow us only to destroy us. Even when we grow ourselves, you kill us, or cut off our limbs. That we are alive and have feelings means nothing to you.

But *I,* as an individual, am innocent, I said. Though it did occur to me that I live in a wood house, I eat on a wood table, I sleep on a wood bed.

My uses of wood are modest, I said, and always tailored to my needs. I do not slash through whole forests, destroying hundreds of trees in the process of "harvesting" a few.

But finally, after much discourse, I understood what the trees were telling me: Being an individual doesn't matter. Just as human beings perceive all trees as one (didn't a U.S. official say recently that "when you've seen one tree, you've seen 'em all"?), all human beings, to the trees, are one. We are judged by our worst collective behavior, since it is so vast; not by our singular best. The Earth holds us responsible for our crimes against it, not as individuals, but as a species—this was the message of the trees. I found it to be a terrifying thought. For I had assumed that the Earth, the spirit of the Earth, noticed exceptions—those who wantonly damage it and those who do not. But the Earth is wise. It has given itself into the keeping of all, and all are therefore accountable.

And how hard it will be to change our worst behavior!

Last spring I moved even deeper into the country, and went eagerly up the hill from my cabin to start a new garden. As I was patting the soil around the root of a new tomato plant, I awakened a small garden snake who lived in the tomato bed. Though panicked and not knowing at the time what kind of snake it was, I tried calmly to direct it out of the garden, now that I, a human being, had arrived to take possession of it. It went. The next day, however, because the tomato bed *was* its home, the snake came back. Once more I directed it away. The third time it came back, I called a friend—who thought I was badly frightened, from my nervous behavior—and he killed it. It looked very small and harmless, hanging from the end of his hoe.

Everything I was ever taught about snakes—that they are dangerous, frightful, repulsive, sinister—went into the murder of this snake person, who was only, after all, trying to remain in his or her home, perhaps the only home he or she had ever known. Even my ladylike "nervousness" in its presence was learned behavior. I knew at once that killing the snake was not the first act that should have occurred in my new garden, and I grieved that I had apparently learned nothing, as a human being, since the days of Adam and Eve.

Even on a practical level, killing this small, no doubt bewildered and disoriented creature made poor sense, because throughout the summer, snakes just like it regularly visited the garden (and deer, by the way, ate all the tomatoes), so that it appeared to me that the little snake I killed was always with me. Occasionally a very large mama or papa snake wandered into the cabin yard, as if to let me know its child had been murdered, and it knew who was responsible for it.

These garden snakes, said my neighbors, are harmless; they eat mice and other pests that invade the garden. In this respect, they are even helpful to humans. And yet, I am still afraid of them, because that is how I was taught to be. Deep in the psyche of most of us there is this fear—and long ago, I do not doubt, in the psyche of ancient peoples, there was a similar fear of trees. And of course a fear of other human beings, for that is where all fear of natural things leads us: to fear of ourselves, fear of each other, and fear even of the spirit of the Universe, because out of fear we often greet its outrageousness with murder.

That fall, they say, the last of the bison herds was slaughtered by the Wasichus.* I can remember when the bison were so many that they could not

---

*Wasichu was a term used by the Oglala Sioux to designate the white man, but it had no reference to the color of his skin. It means: He who takes the fat. It is possible to be white and not a Wasichu or to be a Wasichu and not white. In the United States, historically speaking, Wasichus of color have usually been in the employ of the military, which is the essence of Wasichu.

be counted, but more and more Wasichus came to kill them until there were only heaps of bones scattered where they used to be. The Wasichus did not kill them to eat; they killed them for the metal that makes them crazy, and they took only the hides to sell. Sometimes they did not even take the hides, only the tongues; and I have heard that fire-boats came down the Missouri River loaded with dried bison tongues. You can see that the men who did this were crazy. Sometimes they did not even take the tongues; they just killed and killed because they liked to do that. When we hunted bison, we killed only what we needed. And when there was nothing left but heaps of bones, the Wasichus came and gathered up even the bones and sold them. —BLACK ELK SPEAKS†

In this way, the Wasichus starved the Indians into submission, and forced them to live on impoverished "reservations" in their own land. Like the little snake in my garden, many of the Indians returned again and again to their ancient homes and hunting grounds, only to be driven off with greater and greater brutality until they were broken or killed.

The Wasichus in Washington who ordered the slaughter of bison and Indian and those on the prairies who did the deed are frequently thought of, by some of us, as "fathers of our country," along with the Indian killers and slave owners Washington and Jefferson and the like.

Yet what "father" would needlessly exterminate any of his children?

Are not the "fathers," rather, those Native Americans, those "wild Indians" like Black Elk, who said, "It is the story of all life that is holy and is good to tell, and of us two-leggeds sharing in it with the four-leggeds and the wings of the air and all green things; for these are children of one mother and their father is one Spirit"?

Indeed, America, the country, acts so badly, so much like a spoiled adolescent boy, because it has never acknowledged the "fathers" that existed before the "fathers" of its own creation. It has been led instead—in every period of its brief and troubled history—by someone who might be called Younger Brother (after the character in E. L. Doctorow's novel *Ragtime*, set in turn-of-the-century America), who occasionally blunders into good and useful deeds, but on the whole never escapes from the white Victorian house of racist and sexist repression, puritanism, and greed.

The Wasichu speaks, in all his U.S. history books, of "opening up virgin lands." Yet there were people living here, on "Turtle Island," as the Indians called it, for thousands of years; but living so gently on the land that to Wasichu eyes it looked untouched. Yes, it was "still," as they wrote over and over again, with lust, "virginal." If it were a bride, the Wasichus would have permitted it to wear a white dress. For centuries on end Native Americans lived on the land, making love to it through worship and praise, without once raping or defiling it. The Wasichus—who might have chosen to imitate the Indians, but didn't because to them the *Indians* were savages—have been raping and defiling it

†By John G. Neihardt (New York: William Morrow, 1932).

since the day they came. It is ironic to think that if the Indians who were here then "discovered" America as it is now, they would find little reason to want to stay. This is a fabulous *land,* not because it is a country, but because it is soaked in so many years of love. And though the Native Americans fought as much as any other people among themselves (much to their loss!), never did they fight against the earth, which they correctly perceived as their mother, or against their father, the sky, now thought of mainly as "outer space," where primarily bigger and "better" wars have a projected future.

The Wasichus may be fathers of the country, but the Native Americans, the Indians, are the parents ("guardians," as they've always said they are) of the land.‡ And, in my opinion, as Earthling above all, we must get to know these parents "from our mother's side" before it is too late. It has been proved that the land can exist without the country—and be better for it; it has not been proved (though some space enthusiasts appear to think so) that the country can exist without the land. And the land is being killed.

Sometimes when I teach, I try to help my students understand what it must feel like to be a slave. Not many of them can go to South Africa and ask the black people enslaved by the Wasichus there, or visit the migrant-labor camps kept hidden from their neighborhoods, so we talk about slavery as it existed in America, a little over a hundred years ago. One day I asked if any of them felt they had been treated "like dirt." No; many of them felt they had been treated badly at some time in their lives (they were largely middle class and white) but no one felt he or she had been treated like dirt. Yet what pollution you breathe, I pointed out, which the atmosphere also breathes; what a vast number of poisons you eat with your food, which the Earth has eaten just before you. How unexpectedly many of you will fall in and die from cancer because the very ground on which you build your homes will be carcinogenic. As the Earth is treated "like dirt"—its dignity demeaned by wanton dumpings of lethal materials all across its proud face and in its crystal seas—so are we all treated.

Some of us have become used to thinking that woman is the nigger of the world, that a person of color is the nigger of the world, that a poor person is the nigger of the world. But, in truth, Earth itself has become the nigger of the world. It is perceived, ironically, as other, alien, evil, and threatening by those who are finding they cannot draw a healthful breath without its cooperation. While the Earth is poisoned, everything it supports is poisoned. While the Earth is enslaved, none of us is free. While the Earth is "a nigger," it has no choice but to think of us all as Wasichus. While it is "treated like dirt," so are we.

---

‡Though much of what we know of our Indian ancestors concerns the male, it is good to remember who produced him; that women in some tribes were shamans, could vote, and among the Onondaga still elect the men who lead the tribe. And, inasmuch as "women's work" has always involved cleaning up after the young, as well as teaching them principles by which to live, we have our Indian female parent to thank for her care of Turtle Island, as well as the better documented male who took her instructions so utterly to heart.

In this time, when human life—because of human greed, avarice, igno-rance, and fear—hangs by a thread, it is of disarmament that every thoughtful person thinks; for regardless of whether we all agree that we deserve to live, or not, as a species, most of us have the desire. But disarmament must also occur in the heart and in the spirit. We must absolutely reject the way of the Wasichu that we are so disastrously traveling, the way that respects most (above nature, obviously above life itself, above even the spirit of the Universe) the "metal that makes men crazy." The United States, the country, has no doubt damned its soul because of how it has treated others, and if it is true that we reap what we sow, as a country we have only to recognize the poison inside us as the poison we forced others to drink. But the land is innocent. It is still Turtle Island, and more connected to the rest of the Universe than to the United States govern-ment. It is beginning to throw up the poisons it has been forced to drink, and we must help it by letting go of our own; for until it is healthy and well, we can-not be.

Our primary connection is to the Earth, our mother and father; regardless of who "owns" pieces and parts, we, as sister and brother beings to the "four-leggeds (and the fishes) and the wings of the air," share the whole. No one should be permitted to buy a part of our Earth to dump poisons in, just as we would not sell one of our legs to be used as a trash can.

Many of us are afraid to abandon the way of the Wasichu because we have become addicted to his way of death. The Wasichu has promised us so many good things, and has actually delivered several. But "progress," once claimed by the present chief of the Wasichus to be their "most important product," has meant hunger, misery, enslavement, unemployment, and worse to millions of people on the globe. The many time-saving devices we have become addicted to, because of our "progress," have freed us to watch endless reruns of com-mercials, sitcoms, and murders.

Our thoughts must be on how to restore to the Earth its dignity as a living being; how to stop raping and plundering it as a matter of course. We must begin to develop the consciousness that everything has equal rights because existence itself is equal. In other words, we are all here: trees, people, snakes, alike. We must realize that even tiny insects in the South American jungle know how to make plastic, for instance; they have simply chosen not to cover the Earth with it. The Wasichu's uniqueness is not his ability to "think" and "invent"—from the evidence, almost everything does this in some fashion or other—it is his profound unnaturalness. His lack of harmony with other peo-ples and places, and with the very environment to which he owes his life.

In James Mooney's *Myths of the Cherokee and Sacred Formulas of the Cherokees,* collected between 1887 and 1890, he relates many interesting prac-tices of the original inhabitants of this land, among them the custom of asking pardon of slain or offended animals. And in writing about the needless murder of the snake who inhabited our garden—the snake's and mine—I ask its pardon and, in the telling of its death, hope to save the lives of many of its kin.

The missionary Washbum [says Mooney] tells how among the Cherokees of Arkansas, he was once riding along, accompanied by an Indian on foot, when they discovered a poisonous snake coiled beside the path. "I observed Blanket turned aside to avoid the serpent, but made no sign of attack, and I requested the interpreter to get down and kill it. He did so, and I then inquired of Blanket why he did not kill the serpent. He answered, 'I never kill snakes and so snakes never kill me.'"

The trader Henry [Mooney observes elsewhere] tells of similar behavior among the Objibwa of Lake Superior in 1764. While gathering wood he was startled by a sudden rattle. . . . "I no sooner saw the snake, than I hastened to the canoe, in order to procure my gun; but, the Indians observing what I was doing, inquired the occasion, and being informed, *begged* me to desist. At the same time, they followed me to the spot, with their pipes and tobacco pouches in their hands. On returning, I found the snake still coiled.

"The Indians, on their part, surrounded it, all addressing it by turns, and call-ing it their *grandfather*; but yet keeping at some distance. During this part of the ceremony, they filled their pipes; and now each blew the smoke toward the snake, who, as it appeared to me, really received it with pleasure. In a word, after remaining coiled, and receiving incense, for the space of half an hour, it stretched itself along the ground, in visible good humor. Its length was between four and five feet. Having remained outstretched for some time, at last it moved slowly away, the Indians following it, and still addressing it by the title of grandfather, beseeching it to take care of their families during their absence, and to be pleased to open the heart of Sir William Johnson (the British Indian Agent, whom they were about to visit) so that he might *show them charity,* and fill their canoe with rum. One of the chiefs added a petition, that the snake would take no notice of the insult which had been offered by the Englishman, who would even have put him to death, but for the interference of the Indians, to whom it was hoped he would impute no part of the offense. They further requested, that he would remain, and inhabit their country, and not return among the English. . . ."

What makes this remarkable tale more so is that the "bite" of the English-man's rum was to afflict the Indians far more severely than the bite of any tremendous number of poisonous snakes.

That the Indians were often sexist, prone to war, humanly flawed, I do not dispute. It is their light step upon the Earth that I admire and would have us emulate. The new way to exist on the Earth may well be the ancient way of the steadfast lovers of this particular land. No one has better appreciated Earth than the Native American. Whereas to the Wasichus only the white male attains full human status, everything to the Indian was a relative. Everything was a human being.

As I finish writing this, I notice a large spider sleeping underneath my desk. It does not look like me. It is a different size. But that it loves life as I do, I have no doubt. It is something to think about as I study its many strange but oddly beautiful dozen or so legs, its glowing coral-and-amber coloring, its thick web, whose intricate pattern I would never be able to duplicate. Imagine building your house from your own spit!

In its modesty, its fine artistry and self-respecting competency, is it not like some gay, independent person many of us have known? Perhaps a rule for permissible murder should be that beyond feeding and clothing and sheltering ourselves, even abundantly, we should be allowed to destroy only what we ourselves can re-create. We cannot re-create this world. We cannot re-create "wilderness." We cannot even, truly, re-create ourselves. Only our behavior can we re-create, or create anew.

> Hear me, four quarters of the world—a relative I am! Give me the strength to walk the soft earth, a relative to all that is! Give me the eyes to see and the strength to understand, that I may be like you. . . .
>
>   Great Spirit, Great Spirit, my Grandfather, all over the earth the faces of living things are all alike. With tenderness have these come up out of the ground. Look upon these faces of children without number and with children in their arms, that they may face the winds and walk the good road to the day of quiet.
> —BLACK ELK SPEAKS

Note

The Onondagas are the "Keepers of the Fire" of the Six Nation Confederacy in New York state. The Confederacy (originally composed of five nations) is perhaps the oldest democratic union of nations in the Western world, dating back roughly to the time of the Magna Carta. It is governed under an ancient set of principles known as the "Gayaneshakgowa," or Great Law of Peace, which in written form is the constitution of the Six Nation Confederacy.

This remarkable document contains what well may have been the first detailed pronouncements on democratic popular elections, the consent of the governed, the need to monitor and approve the behavior of governmental leaders, the importance of public opinion, the rights of women, guarantees of free speech and religion, and the equitable distribution of wealth.

Benjamin Franklin and Thomas Jefferson acknowledged in the mid-18th century that their own ideas for a democratic confederacy were based largely on what they had learned from the Six Nations. A century later Friedrich Engels paid a similar tribute to the Great Law of Peace while making his contribution to the theory of Marxism. —*Jon Stewart,* Pacific News Service

RESPONDING JOURNAL

Respond to Walker's comment that "Many of us are afraid to abandon the way of the Wasichu because we have become addicted to his way of death." To what extent do you agree with her position? How would you distinguish your feelings and behaviors from those you see around you? What would Walker probably say about such attempts to separate ourselves from others? Explore these ideas in a substantial journal entry.

QUESTIONS FOR CRITICAL THINKING

STRATEGY

1. Consider the process of thought that Alice Walker explains for her readers. List the significant steps that take Walker through to her conclusion. How does she indicate when she takes another step, when she reaches another significant point in the development of her thinking?
2. Do you find that you have to work harder to appreciate Walker's sense of organization and paragraph development than you do Leopold's? What benefits do you think Walker derives from making you as reader work hard, as you do when you come to the first quote from Black Elk (p. 343) and she has not told you the quote is coming?

ISSUES

1. If you have read Richard Wright's short story "Silt," you may wish to compare the experience of reading that story to Walker's account of helping her students to feel what slavery must have been like. She asks if they have ever felt that they had been treated like dirt in order to show how the dirt, the earth, of the planet has been treated. Explain how effective you find such an imagined experience to be; have you yourself ever felt like dirt or like silt?
2. Explain Walker's title. How convincingly does she argue that every thing upon the earth should be accorded human status?

COLLABORATIVE WRITING ACTIVITY

Walker presented this essay as a speech at a ceremony honoring the birth of Dr. Martin Luther King, Jr., held at the University of California at Davis. Dr. King's birthday was not yet a national holiday when Walker delivered the speech in 1983. Explain in a few substantial paragraphs—an address of two or three minutes—why you feel Walker's essay would be an effective speech to present on King's birthday. Suggest in your address why Walker did not mention King at all in her speech. What might she have in mind by forcing the listener or reader to make the connection?

After you have made your address to the group, decide on the most compelling points made by members of your group. Decide on a speaker to represent your collective views. Once you have determined the consensus choice, the entire group should collaboratively extend the address to five minutes. After completing the address in collaboration, the designated speaker can then present it to the rest of the class.

"Fear of Frying"
Tim Cahill

One of the founding editors of Outdoors *magazine and a frequent contributor to* Rolling Stone, *Cahill has had as many and as varied adventures as any outdoor writer. From an on-the-scene report of the Jim Jones massacre at Jonestown in Guyana to the subtleties of ice fishing in Wisconsin, Cahill has tried just about everything and lived to tell about it. His style is irreverent and almost always humorous, even when he is describing brushes with death.*

His trip through Death Valley in summertime is one such brush with death, an experience he describes in "Fear of Frying." The title is a humorous allusion to Fear of Flying, *a novel by Erica Jong that enjoyed a reputation in the 1970s as racy soft porn. There is no soft porn in Cahill's essay, but there is a breezy account of his misadventures. The essay comes from a collection of his works entitled* A Wolverine Is Eating My Leg; *his other works are* Jaguars Ripped My Flesh *and* Buried Dreams: Inside the Mind of a Serial Killer. *Despite the tabloid quality of these titles, Cahill does have a fine sense of how to describe a process.*

EXPLORING JOURNAL

What does the term *Death Valley* mean to you? What sorts of images does it arouse? Freewrite for ten minutes on the ideas that occur to you as you consider this term.

———❖———

We have always depended on the kindness of rangers.

DEATH VALLEY HIKERS FOUND SAFE BUT TIRED, read the front-page headline in the *Death Valley Gateway Gazette*. Since the article was about me and a good pal—and it's the only time I've been front-page news, anywhere, ever—I found the unstated assumptions, uh, distressing. Even in the *Death Valley Gateway Gazette* a man tends to read his clippings compulsively, and this story, by implication, might have been titled HALF-DEAD DUMBSHITS IN THE DESERT.

According to the article,

> An aerial search of the rugged desert and mountain areas between Badwater in Death Valley National Monument and Mount Whitney in the eastern Sierras located two overdue hikers on Wednesday (June 20th).
>
> Rangers found the pair, a writer and photographer from *Rolling Stone* magazine, to be in "good but fatigued" condition due to exhaustion from the rugged hike in temperatures well over the 100-degree-Fahrenheit mark. . . . [The journalists] started their rugged trek from the lowest elevation in the continental

United States, approximately 282 feet below sea level, and plan to conclude their journey to reach Mount Whitney, the highest point, at an elevation of 14,375 feet above sea level.

[Chief Ranger Dick] Rayner considered the rescued hikers to be "extremely fortunate" and cautioned monument visitors on the hazards of summer hiking and backpacking in extreme Death Valley temperatures.

THE ABSOLUTE PIT

I am as low as a man can get in the United States, and I am slowly sinking lower. Death Valley, 550 square miles of it below sea level—all scalding salt flats and dunes—is surrounded by mountains: by the Amargoasa, the Panamint and the Last Chance ranges, which rise from four to eleven thousand feet above the valley floor. These mountains catch what rain the westerly air current didn't drop on the Sierras, and water rolls down the mountainsides into the valley, where it immediately evaporates, leaving the accumulated mineral residue of chlorides, sulphates, and carbonates.

Not all of the water is lost, however. Some of it skulks in a steaming, muddy bog that lies just under a brittle salt crust out toward the center of the valley. Somewhere near the lowest point in the continental United States, the salt crust refuses to support the weight of a man; it takes the boot to the ankle, then the leg to the calf, the knee. Walking becomes a crack-splash affair, and the sharp, crystalline salt crust scrapes and cuts the shins. The bog below is a musty-gray combination of hot mud and salt that clings to boots and legs like hot clay. First the valley chews up your legs then it rubs salt in your wounds.

There are rumors that "in some places in the middle of the bog, the soft salty area in the bottom of the valley, a team of horses or a man walking have been instantly sucked down out of sight." This bit of cheerful information comes from Daniel Cronkhite's well-researched book *Death Valley's Victims.* The author acknowledges that the story may be apocryphal and goes on to quote Old Johnnie, who told of "finding a dead man's face looking up at him out of the ground. 'He was a Swede with yellow hair, and he stared at the sun. He sank standing up.'"

This is a report to brood upon when walking across Death Valley around two in the morning with a photographer from *Rolling Stone.* You want to crack-splash through the steaming mud about thirty feet apart, so that if one should go down, the other can more efficiently panic and go lurching off into the desert night, hands in the air, screaming and gibbering.

When the photographer, Nick Nichols, and I reached what we supposed was the nadir of life in the United States, the absolute pit, Nick also discovered that he had dropped his strobe, "back there." He began trudging along our back trail, muttering malign imprecations and leaving me standing knee-deep in hot, salty mud. There was no place to sit down, unless I wanted to take a scalding mud bath, and the Van Gogh stars spun madly overhead. The desert sky was impossibly clear, and I could make out the colors of various stars and planets, so that, glancing up, I felt as if I were stranded in space.

Thick, muddy water was draining back into our posthole footprints—it was a sick sort of squishing sound—and the dead of a Death Valley night swallowed up Nick's receding light. Alone, in the darkness, I stared down at the unbroken salt crust of the valley floor. There were innumerable pillars of salt standing in inch-high clusters. Some formations, like certain tropical corals, took on the shapes of crystalline flowers, and they wound about in baroque curlicues, snaking across the floor like endless meandering rivers.

The mountains—waiting to reveal themselves in the light of the rising moon—whispered to one another in warm, gusting breezes that swept across the valley. The hot salt crust of the valley floor, under the cold light of swirling stars, emitted a faint glow, like the radium dial of a watch. The world was an ocean of salt and sand, so flat the eye saw a ridge, nuclear white, that rose on all sides.

I stood stock-still, wondering if I was sinking any deeper. The hot mud had been knee-deep on me, or so I thought but now my legs felt braised to midthigh. It seemed hard to breathe out there, alone, in the middle of the night. I felt slightly faint and realized that in this condition I could very well commit philosophy.

"DEATH DESTROYS A MAN, BUT THE IDEA OF DEATH SAVES HIM" —WILLIAM FORESTER

Every life offers certain challenges that require grit, intelligence, spirit, spunk, careful planning, and nifty interpersonal skills. Try cashing an out-of-state check in New York City on a Sunday. Buy a used car from a friend of your brother-in-law. Ask for a promotion. Or a divorce.

Few of the challenges we face every day, as a matter of course, are physical, however, and a growing number of people seem to feel that lack keenly. Some have taken up individual sports as a kind of antidote to physical stagnation. People run marathons, they compete in triathlons, cycle the breadth of the country nonstop, or attempt to get their names in the *Guinness Book of World Records* by doing cartwheels across the state of Nebraska.

My problem with most athletic challenges is training. I am lazy and find that workouts cut into my drinking time. The thought of a new personal best no longer fills me with ambition or a burning desire to win. I need incentives.

Consequently, every once in a while, I like to flirt with some physical challenge in which the price of failure is death. Amazing how easy training becomes in such a situation, how carefully one plans, how intently the mind focuses.

Over the past decade, I've jumped out of quite a few perfectly good airplanes—"When the people look like ants," my first instructor said, "pull the chute; when the ants look like people, pray." I've been diving with tiger sharks on the Great Barrier Reef and have crawled creepy damp through caverns half a mile below the surface of the earth; I've run some nasty rapids, climbed a few mountains, traded bolts of red cloth for food in the Amazon basin, and surfed my kayak through ice floes on waves thrown up by calving tidewater glaciers.

A number of magazines have actually paid me to do these things, to realize a lot of adolescent fantasies. These stories want to write themselves, and the work seems effortless. Research, for instance, in a situation where shoddy research can be deadly is scintillating. On location, and at risk, the senses are bombarded, and the world seems to vibrate with color and sound and life. Impressions are hard edged, settling permanently inside the brain, where at odd and frequent intervals they combine to explode in star showers of adrenaline-charged images and ideas. Writing the subsequent article requires a good deal less head banging than, say, analyzing the national deficit. We are talking about adventure here, about fear recollected in tranquility. And fear—we've all felt it—is unforgettable.

One day last year I spoke with my editor about self-imposed physical risk in the natural world. In a nation where signs in the national parks warn visitors not to fall over the lip of a cliff, there is a sense of something sadly lost. Survivors of people who step out into space and go hurtling tragically off the tops of cliffs can file, and win, lawsuits. The Park Service feels obliged to install guardrails and erect signs explaining the concept of gravity.

Nothing is safe in a world where lawyers define what is dangerous. As it happens, a growing number of people have discovered that they enjoy a view unencumbered by guard-rails and warning signs. These folks feel that they have enough sense not to fall off the nearest cliff. In point of fact, many of them search out spectacular cliffs for the sole and specific purpose of seeing them without plunging to their doom.

There is a whole industry that caters to the impulse. What I've been doing out in the jungles or in the Arctic is called "adventure travel" these days. My colleague Dave Roberts recently described the growth of the adventure-travel industry:

> A 1979 estimate . . . postulated some 2,000 tour operators worldwide [outfitters who supply river rafts, climbing gear, et cetera] . . . a figure which is certainly already obsolete. Another estimate, from 1980, suggests that 2 million Americans have participated in an adventure-travel trip—with rafting, backpacking, and skiing leading the way. Yet public perception of this incredible growth has been lacking, and travel agents themselves have lagged behind their own clientele. The ASTA *Travel News* . . . issued this caveat: "Travel agents had better be prepared to book adventure whether they like it or not."

What I proposed to *Rolling Stone* was a series of articles about various natural challenges, taken to extremes, and written for the edification of adventure travelers and arm-chair adventure travelers. In one part of the series, I'd personally walk from the lowest point in the continental United States to the highest point. It was not an original idea—lots of peoples have maps and can see that these two extremes are separated by a mere hundred or so miles. It is certainly not impossible. Perhaps a dozen men and women of my acquaintance could make this trek at a dead run. But adventure is relative. I knew next to nothing about the desert, and unless I got into some kind of shape and learned a lot, real quick, death was one of the more extreme consequences.

THE INDIANA JONES OF PHOTOGRAPHY

"I know what you did," Nick Nichols raged. We were kneeling on the floor of a hotel room in San Francisco, and topographical maps of Death Valley were spread out on the floor. The bed was littered with desert-survival books, wide-brimmed hats, long-sleeved shirts, heavy cotton jeans, homemade turbans, backpacks, cookstoves, canteens, and cameras. "You were sitting in some editor's office in New York and came up with this, this . . . ."

"Idea?"

"This insanity. You can't propose a story on the girls of Tahiti. Oh, no. Or the four-star restaurants of France. Or the grand hotels of Europe. You come up with this, this . . . ."

I could tell Nick loved the idea. Together we have made something of a living working the adventure-travel beat. We've trekked through sections of the Amazon and Congo basins; studied and lived with mountain gorillas in the Virunga volcanoes of central Africa; swum in the pool under Angel Falls; flown with the air force into the eye of a hurricane, and made the first rainy-season ascent of Arthur Conan Doyle's "Lost World," Mount Roraima, in Venezuela.

Nick has made rafting descents down white-water rivers in Pakistan, cycled partway across China, rappelled nearly a mile down the face of a cliff in the Arctic—a world free-fall record that maybe ten people in the world know about—photographed fire dances in Suriname and been arrested in Zaire for reasons that have yet to be explained. *Photo* magazine calls Nick Nichols "the Indiana Jones of photography."

We studied the maps for a while. The Park Service officials we contacted had actively discouraged the hike. We would be trekking through a blast furnace, they said. Late June was deadly in the valley. Why didn't we hike from Mount Whitney down to the ocean?

Trekking from the highest point to a pretty low point, we said, seemed to lack the proper emotional resonance. Chief Ranger Dick Rayner sighed and said that we'd have to file an itinerary with him and that we'd have to have a support vehicle, a four-wheel-drive rig driven by a third party. If we failed to reach the vehicle at the proper points and times, the driver would report to the rangers and a search party would be dispatched.

Frank Frost, a Northern California photographer, agreed to drive the support vehicle, and we were in business. Dick Rayner had said that natural springs in the valley were undependable and fouled with diarrhea-producing giardia. It was best to stash water, to bury it in plastic jugs. We should plan on two gallons a day, minimum.

Nick and I formulated a set of rules. It was okay to make our packs as light as possible by burying food along with the water. We'd avoid roads and bushwhack cross-country as much as possible. Since the Park Service required a support vehicle anyway, it was okay to pack it full of cold water and iced beer.

"This one ranger I talked to," Nick said, "he told me that a couple of groups a year try this trek, and they've had to rescue a few of them. The guy said he had been in Death Valley for several years, and he never knew anyone who made it. Ninety percent of them, he said, quit the first day. He said it was psychological. Either that or poor planning."

"We're pretty psychological," I pointed out.

Nick didn't reply. He was studying a copy of *Death Valley's Victims,* looking at the photos of desiccated corpses baking out on the valley floor.

"We're going to die," the Indiana Jones of photography said.

HOT DAMN

We managed to slog through the crusted, steaming bog before dawn, according to plan, and found the water and food Nick had buried. Our campsite was Tule Spring, on the valley floor, at the foot of the Panamint Mountains. We pitched the rain flies from the one-man tents we carried and settled down for a long, mindless sleep. The tents would provide shelter from the sun: we had tested them out on the grass at the Death Valley Visitors Center, in the village of Furnace Creek. The temperature inside the tents had been twelve degrees cooler than the outside air.

By eleven that morning I felt like a side of beef, and my skin was the color of medium-rare prime rib. The pores on the back of my hand were the size of a quarter, or so it seemed, and dozens of tiny but cruel dwarfs were building a condominium inside my skull. The thermometer registered 128 inside the tent. The record high temperature in the United States is 134, recorded on July tenth, 1913, in Death Valley. If it was 12 degrees hotter outside of the tent— 140 degrees—I was dying through the hottest American day on record.

But I found the temperature outside was only 113. It didn't make any sense. I laid the thermometer down on the ground, next to my boots, and the mercury pegged at 150 degrees. The thermometer wasn't made to measure temperatures any higher. What we'd failed to consider, Nick and I, is the fact that gravel and sand—like white cement highways under the summer sun—get hot. Real hot. A lot hotter than grass or even the air itself. The Indian name for Death Valley, Tomesha, means "ground on fire." In 1972, a record ground temperature of 201 degrees was recorded on the valley floor.

Instead of protecting us from the 113-degree outside air, the tents were concentrating the 150-plus-degree ground temperature and literally baking us. Nick and I moved outside. We sat on foam pads, under lean-tos we had made with space blankets. With the noonday sun directly overhead, the blankets provided perhaps two square feet of shade. It was now 121 degrees. Hot air, rising off the superheated sand and salt, scalded our lungs. Sleep was impossible.

"The tents were a dumb idea," Nick said.

"Poor planning," I muttered.

Then we didn't say anything for nine hours.

I could feel the hot air rising all around us. The laws of physics demand that heavier, cooler air should fall from the heavens, and that is what happens in Death Valley. It falls, comes into contact with the ground, becomes superheated, and rises. The mountains surrounding the valley allow no air to escape, so that as the day wears on, the upper levels of air—which have made several passes over the ground—are not really cool anymore, only less hot than air at ground level. In effect, the valley is a giant convection oven.

All this rising and falling air whistles across the valley floor in gusting waves of arid wind that suck the moisture out of a man's body the way a hand wrings water from a sponge. You sweat, of course, but you do not feel sweat on your body in Death Valley, even at 121 degrees. The killing convective wind will allow no moisture to form, but all that rapid evaporation is a cooling process, so the wind feels good, almost pleasant, as it desiccates the body. And that is why people who die in the desert are often found naked lying face down on the skillet of the valley floor.

Some victims have been found with a quart or two of water in their possession. Apparently, they intended to save the water until they felt they really needed it. Staggering, suffering from dehydration or heat exhaustion or heat stroke, they fell unconscious, and the ground on fire killed them in a matter of hours. Other victims, too weak to walk simply fell to the sand and couldn't rise to their feet. In 1973, Death Valley killed three people this way.

Unpleasant thoughts. Huddled there in my small square of shade with the circumstances of various tragedies stumbling slowly through my mind like terminal winos, I began dreaming of the Man in the Freon Suit. What a guy! In Death Valley, certain legends exist and have the ring of truth about them because everyone knows them, everyone repeats them, and they are so poetically morbid as to live in memory, whether one wills them to or not.

Such is the tale of the Man in the Freon Suit. I heard it my first day in the valley, at Furnace Creek, a tale eagerly told concerning an unfortunate inventor who constructed a kind of space suit, using the tubing and coils of an old Frigidaire. The man ventured out into Death Valley, wearing his air-conditioned suit and pulling a battery behind him in a small wagon. The suit malfunctioned, however: it apparently began pumping out great blasts of freezing air. According to legend, the man couldn't remove the suit. Perhaps his fingers had frozen beyond the point of movement. At any rate, he was found lying on the floor of Death Valley, lying there on the baking 180-degree plain, frozen solid.

Many people knew the story, though no one was quite sure when it happened. Most folks put the tragedy sometime around 1950. One person thought he remembered the man's name: John Newbury or Newhouse or Newton. Something like. Poor son of a bitch.

In point of fact, the story of the Man in the Freon Suit was first reported by Dan De Quille in the July second, 1874, issue of Virginia City's *Territorial Enterprise* under the headline SAD FATE OF AN INVENTOR. De Quille's story concerned a "man of considerable inventive genius" named Jonathan Newhouse who had constructed a "'solar armor," which consisted of a hood, jacket, and pants, of "common sponge," all about an inch thick. "Under the right arm," De Quille reported,

> was suspended an India-rubber sack filled with water and having a small gutta-percha tube leading to the top of the hood. In order to keep the armor moist, all that was necessary to be done by the traveler as he progressed over the burning sands, was to press the sack occasionally, when a small quantity of water would be forced up and thoroughly saturate the hood and the jacket below it. Thus, by the evaporation . . . it was calculated might be produced any degree of cold. Mr. Newhouse went down to Death Valley, determined to try the experiment of crossing that terrible place in his armor.

According to De Quille, Newhouse was found the next day, about twenty miles into the desert,

> a human figure seated against a rock. . . . His beard was covered with frost, and—though the noonday sun poured down its fiercest rays—an icicle over a foot in length hung from his nose. There he had perished miserably, because his armor had worked but too well, and because it was laced up behind where he could not reach the fastenings.

The story was reported as news worldwide, and the boys in the newsroom at the *Terminal Enterprise* must have had a good laugh over that one. De Quille, like his contemporary Mark Twain, could tell a story so patently false that truth smirked out from around the edges.

In my mind's eye, I could see the foot-long icicle, blue white under a molten sun. Slowly, the thing began to grow, and it floated dumbly out into the shimmering salt pan of the valley floor, where it stood like a massive religious icon, a monolithic icicle plunged into the heart of Death Valley.

By nine that night it had cooled off enough to walk. Neither of us had slept for over forty hours. Worse, we had lain our boots on the ground to dry. The boots had been wet and caked with muddy salt. The ground on fire had baked them into weird, unfootlike shapes. Mine seemed to weigh fifteen or twenty pounds apiece. The canvas and leather felt like cement.

We hammered on the boots with rocks, cracking away the caked adobe.

"Leaving the boots on the ground," Nick pointed out, "was dumb."

There was poor planning involved, all right. The next water stash was only five miles away, at a place where a scrubby bush grew beside a rocky four-wheel-drive road. Unfortunately, in that area there had been a number of springtime flash floods. Water had thundered down the mountainsides in several temporary rivers, and each wash, in the light of our headlamps, appeared to be a four-wheel-drive road. We couldn't find the water. Poor planning.

The evaporative wind had cranked up to about forty miles an hour. This was serious. We retraced our steps searching for the stash, walking like a pair of Frankenstein monsters in our adobe boots. We both were developing severe blisters, but there was no stopping now. Finding the water was more important than some little excruciatingly crippling pain.

About 2:30 that morning we stumbled over the water and food. We had been out on the Valley floor for twenty-six hours, in temperatures sometimes exceeding 120 degrees. My feet looked and felt like I'd been walking across hot coals. We both carried extra boots, but walking over ground on fire makes feet expand. Mine looked sort of like big red blistered floppy clown feet. My second pair of boots simply didn't fit, not even a little bit. Another bit of poor planning that meant I'd have to walk forever in cruel shoes, limping pathetically.

We'd made too many mistakes, Nick and I, and the errors had compounded themselves exponentially, so that we had completely lost the will to push on. In the distance, seventeen miles away, we could see the lights of Furnace Creek. We doctored our feet—break the blister, apply the antiseptic, coat with Spenco Second Skin tape—and discussed complete capitulation. In our condition, with blisters and thirty-pound packs, we could probably make two miles in an hour. It would take six and a half hours to walk to Furnace Creek just to surrender.

On the other hand, the next stash was three miles away, in the Panamint Mountains, at an elevation of 2,300 feet. Say, four hours to cover eight miles, and give it another hour for each 1,000 feet of elevation. The stash was about six and a half hours away. It would hurt just as bad to give up as to push on, and it would be ten or fifteen degrees cooler at 2,300 feet.

Still, if blisters and exhaustion kept us from reaching the rocks before noon, that could be fatal. We decided to gamble and headed for the high country, hoping the idea of death could save us.

INTO THE FIRE

We were perhaps 1,500 feet up into the Panamints, walking up a long, bare slope littered with sage. There was no shade anywhere on the slope. We had miles to walk before the rock would rise above us and provide some protection from the sun. Quite clearly, neither of us could survive another day crouched under a space blanket.

Nick was wearing shorts, and I could see the muscles in his thighs twitching spasmodically. It was only two hours until sunup. There was a full moon that night, and in its light I suddenly saw, sloping off to my right, a long, narrow valley. In that valley, almost glittering in the moonlight, was a town full of large frame houses, all of them inexplicably painted white. The houses seemed well maintained but were clearly abandoned. There was nothing on the map that indicated a ghost town here in Trail Canyon.

"Jesus, Nick, look." I pointed to the ghost town, perhaps 250 feet below us.

"What?"

"We can hole up down there."

"Where?"

"Down there."

Nick stared down into the valley for a full thirty seconds. "You're pointing to a ditch," he said finally. "You want to hole up in a ditch?"

I squinted down at the ghost town. Slowly, it began to rise toward me. The neatly painted white houses became strands of moon-dappled sage in a ditch perhaps five feet deep.

"I been having 'em, too," Nick said.

Nick wouldn't say what his hallucinations were like. I had to coax it out of him.

"Graveyards," Nick said finally. "I been seeing graveyards."

An hour and a half later we sat to rest. To the east, over the Funeral Mountains, on the other side of Death Valley, the pale light of false dawn had given way to a faint pastel pink. The sky suddenly burst into flame, filling the high canyons with a crimson that flowed down the ridges and flooded the valley floor with blood. Then the sun rose over the Funeral Mountains, fierce and blindingly hot, like molten silver, and its white heat scattered the crimson, so that for a moment the full weight of the sun lay glittering and triumphant on the great lifeless salt pan below. It was still cool—perhaps 85 degrees—but, within a matter of hours, the temperature could rise to 120 or more.

It was the first time in my life I'd ever found a beautiful sunrise terrifying. It was like seeing a huge mushroom cloud rise in the distance, that sunrise.

THE POST CARD OF THE RESCUE

In the rocks above the bare sage slope, we found a narrow S-shaped canyon, where we lay down to sleep. Throughout the day, the sun chased us around the bends of the S, but there was always shade somewhere. We shared the canyon with a small, drab, gray sparrowlike bird that seemed to be feeding on some thorny red flowers that grew in the shade. I loved Death Valley. It was, as the ranger said, psychological, this place. It slammed you from one extreme to another. My heart seemed to expand inside my chest, and I could feel tears welling up in my eyes. I turned away from Nick, and we sat like that for a time, back to back.

"Nice here," I said finally. "Comfortable."

"Birds and shit," Nick agreed. His voice was shaky.

We slept for twelve hours, ate at nine that night, then slept until six the next morning. The swelling in our feet had gone down after twenty-one hours of sleep, and we could wear our extra boots. I felt like skipping. By noon that

day we had reached an abandoned miner's cabin where we had stashed six gallons of water.

"How do you feel?" I asked Nick.

"Real good."

"Me too."

"We're going to make it," he said.

"I know."

It was cool enough to cook inside the cabin, and Nick was whipping up
one of his modified freeze-dried Creole shrimp dinners when we heard the
plane.

It was moving up the slope, circling over the route we had given Dick
Rayner, and we couldn't believe they were looking for us now that we felt like
gods of the desert. It was still twenty-four hours to the first checkpoint. Why
were they searching for us? I laid out a yellow poncho so the rangers could spot
us. Beside the poncho I arranged several dozen rocks to read "OK" in letters ten
feet high. The plane came in close and dipped a wing. The pilot looked like
Rayner. He circled twice more, then flew back down Furnace Creek.

It was an odd sensation, having them out spending taxpayers' money
searching for us. I felt like some boy scout had just offered to help me across
the street.

APOTHEOSIS

We had, it seemed, acclimated to the desert. It was easier, now, to walk during
the day and sleep in the cool of the evening. We took the Panamint Valley at
midday in temperatures that rose to 115 degrees. The next day, climbing
another range of mountains, we came upon a series of enclosing rock walls that
reminded us of our good friend the S-shaped canyon. It rose up into the mountains, and there was a small, clear creek running down the middle of the canyon
where green grass and bulrushes and coyote melons and trees—actual willow
trees—grew. Ahead, water cascaded over some boulders that had formed a natural dam. The pool beyond the boulders was clear green with a golden sandy
bottom. It was deep enough to dive into, and the water was so cold it drove the
air from my lungs like a punch to the chest. Above, several waterfalls fell down
a series of ledges that rose like steps toward the summit of the mountains.

The same sun that had tried to kill us in Death Valley offered its apologies,
and we lay out on the rocks, watching golden-blue dragonflies flit over the
pool. It was 111 degrees, and we were sunbathing.

The next day we made twenty miles overland. The day after that, almost
thirty. We crossed the salt flats of the Owens Valley in the middle of the day,
roared into the town of Lone Pine, registered with the rangers to climb Mount
Whitney, and reached the summit in a day and a half. It is, perhaps, the easiest
pretty high mountain in the world to climb: a walk up.

About forty people made the summit that day, but only half a dozen of us camped there. I was using the stove to melt snow for drinking water and shivering slightly because my summer sleeping bag wasn't keep me entirely warm. Nick was shooting the sunset over the headwaters of the Kern River far below.

"My fingers are numb," he said. "It's hard to focus." The thermometer read sixteen degrees.

"Yeah, well, you know what they say."

"What do they say?"

"If it's not one thing, it's another."

"Nice view, though," the Indiana Jones of photography said.

THE KINDNESS OF RANGERS REVISITED

When we walked into Dick Rayner's office, I had a copy of the *The Death Valley Gazette* under my arm. The chief ranger agreed that, yes, according to the plan we'd filed, we hadn't been late. What had happened, he said, was that Frank Frost, in the support vehicle, had climbed to the top of a mountain with a commanding view of our route and had spent a day scanning the trails with high-powered binoculars. It was the day we had spent sleeping in the S-shaped canyon. Frank couldn't find us anywhere.

He reported to the rangers, who had immediately set out to save our lives. The foul-up hadn't been anyone's fault really, and I suppose I was glad that the Park Service employs men like Dick Rayner who are willing to leave an air-conditioned office to save a couple of nincompoops like us.

Still, I couldn't help zapping him a little. "The article says we were more than twenty-four hours overdue. I mean, look at our trip plan. We still had twenty-four hours to the first checkpoint."

Rayner said, "I didn't write the article."

"They quote you directly, though. You say we were in 'good but fatigued condition.'"

"Well, we saw your footprints across the valley," Rayner said. "That's a tiring walk. And we could see you were in good condition when we flew over. So: 'good but fatigued condition.'"

The chief ranger seemed a little embarrassed. He recounted some of the rescues he'd participated in, and one of the deaths he knew about. Rayner seemed to be saying that he'd just as soon nobody walked across Death Valley in the summertime. It was his job to discourage such treks—to put guardrails along the cliffs—and he apparently felt that newspaper articles about half-dead dumbshits in the desert were something of a public service. He was a good man who just purely hated the idea of people getting hurt in his park.

"Would you do it again?" Rayner asked.

I glanced over at the Indiana Jones of photography, who was smiling in a manner that made him look somewhat psychological. "We could change the rules," he said. "No stashes. We walk from spring to spring and carry portable water purifiers. Badwater to Tule Spring to Trail Canyon . . ."

Dick Rayner seemed intrigued. Certainly against his better judgment, he pointed to the map on his desk and said, "There's a spring here that would get you into the canyon in better shape."

RESPONDING JOURNAL

Do you find any evidence of heroism in the essay? How would you characterize the behavior of the two men? What would you have done in their place? Would you have gone on with the expedition after the warning from the rangers? Explore your reactions in an entry of several paragraphs.

QUESTIONS FOR CRITICAL THINKING

STRATEGY

1. How does Cahill develop his personality as a writer? List the devices that he uses to create a sense of identity for his reader; pay particular attention to the early part of the essay.
2. Did you find that the old stories of death in Death Valley added interest to the essay? How do such tall tales persist over time? Would the essay have been as effective without the stories of buried animals or an icicle on a corpse's nose?

ISSUES

1. The idea of "poor planning" plagues the two travelers throughout the essay. Could they have foreseen the difficulties they encountered? How might every adventurer in the wild prepare to face such difficulties?
2. Do you feel that the readership of *Rolling Stone* would expect the sort of essay that Cahill presents here? What distinguishes this essay from others you have read? Do you find this sort of writing—which features unusual, even bizarre descriptions and situations—successful? Does the attempt to derive a humorous perspective make the adventure more or less successful? Why?

COLLABORATIVE WRITING ACTIVITY

Think of a trip—or a picnic, or another sort of outdoors exploration—that you took and on which you encountered difficulties you had not anticipated. How did you deal with these problems? How did you solve your dilemma? Think of the trip on which you had the most considerable difficulties to overcome and describe that trip for your group. After you have presented your trip to the group, you can decide on the "most difficult," "most dangerous," "most humorous," or other "most" that you want to present to the rest of the class when you report. You may want to delay deciding which category you want to use until after you have heard all of the individual reports.

"Destroying the Cormorant Eggs"
Judith Minty

Judith Minty, now living again in her native Michigan, is a poet and fiction writer; she has also served as the Director of the Creative Writing Program at Humboldt State University in Arcata, California. Having won numerous awards for her writing, she is very much at home in the natural world. Her work has appeared in The New Yorker, Atlantic Monthly, Ladies' Home Journal, Missouri Review, *and* Black Warrior Review. *Among her books are the collections* Yellow Dog Journal, In the Presence of Mothers, *and* Dancing the Fault. *Her first collection of poetry,* Lake Songs and Other Fears, *won the United States Award of the International Poetry Forum.*

"Destroying the Cormorant Eggs" presents a picture of life at the edge of Lake Michigan, where Minty grew up. The poem's title describes a process in the natural world being thwarted by a human process.

EXPLORING JOURNAL

Compose an entry on the idea of humans destroying the eggs of birds. Have you ever read about or been aware of such an activity? How about when you were a child? Why might people destroy birds' eggs?

———⟶•✦•⟵———

Black, black as the plumage
of the Double-Crested Cormorant, all black
except for the orange chin pouch below its slender,
curved bill, who nests by the shore in shadow and crack
of rock along with the lighter, tan or gray or white
gulls and terns on Little Gull Island and Gravelly Island
in the middle of Lake Michigan— Black as the long shadow
of this fisherman, or madman, slipping

over these rocks, these nests, an eclipse or is it God,
some thing without conscience between sun and earth/water,
his staff much like a shepherd's crook, but
this time carried for balance and for the rest of it: the
choosing, knowing which eggs, only lovely pale blue,
not the gulls' and terns' brown or buff, then
to lift out, to hurl against the granite,
to punish them for fishing these waters,
to crush under boot or beat with his stick,
2000 eggs, the silent cormorants now emitting faint squawks,
flapping their wings over this darkness,
the albumin and yolk, the embryos shining on dull rock,
the small pieces of sky fallen down— Black
as the night waters of a man's dream where he gropes
below the surface, groaning with the old hungers,
the luminescence of his skin now covered by something
so thick his arms stroke heavy with it, the water
without end, and no island, no island in sight.

RESPONDING JOURNAL

Explain in an entry what process analysis offers by way of explanation to the reader of this poem. Does looking at the poem's action as a process help you to understand motivation? Explain your response.

QUESTIONS FOR CRITICAL THINKING

STRATEGY

1. How does Minty use the color black as a means of beginning her poem? How does that color help to set the mood?
2. The conclusion also features the color black prominently; how has the poem moved from the first use of black to the last? What new ideas appear at the end?

ISSUES

1. Explain the idea of point of view in the image of the "fisherman, or madman, slipping / over these rocks, these nests, an eclipse or is it God"; what perspective does the poet seem to be taking on these events?
2. Characterize the response of the cormorants to the action being depicted. What does their reaction seem to suggest about the natural world and its response to humans?

COLLABORATIVE WRITING ACTIVITY

Describe a parallel situation you have either read about or experienced in which some animal was destroyed as a result of a decision made to "protect" another species. After reading your description to the group, work together to produce a statement on how humans can find ways to explain acts of destruction in the natural world.

SUGGESTIONS FOR FORMAL WRITING ASSIGNMENTS

1. In a substantial essay, explore the validity of the idea "business as usual" with reference to the natural world . Explain your process of thought that led you to question or to doubt the wisdom of the saying "peace in our time" with regard to our actions toward the natural landscape and the organisms that inhabit it. Be sure to emphasize the key points in the course of your thinking.

2. What is the process through which we can stop being "Wasichus"? Is recycling and composting enough? Write a well-considered essay in which you provide a plan of attitude adjustment that will allow us to be more considerate residents of Turtle Island.

3. Does Alice Walker "think like a mountain"? How is thinking like a tree similar to thinking like a mountain? Write an essay in which you describe how the same sort of process would go on in thinking like either one.

4. Rewrite the events of Cahill's essay from a more serious point of view. Decide what he has told that can be omitted, and concentrate on the essentials of the process of hiking that suggest high adventure and not humor. Alternately, you may describe the process of an unusual hike you experienced. In your essay, determine the tone you want to strike for your reader and pay careful attention to how you will achieve that effect.

Chapter 9

COMPARATIVE STRATEGIES I: LITERAL COMPARISON AND CONTRAST

Samuel Colman, *Storm King on the Hudson,* 1866

Thomas Worthington Whittredge, *Crossing the Ford, Platte River, Colorado,*
1868–70

1. Explain the basic parallel between these two paintings. On what
   similar visual characteristics do both seem to depend?

2. Describe the most important points of difference you find
   between these two paintings. Defend your assertions by referring
   to both paintings.

3. Develop in detail a statement of what you have learned about the
   nature of visual reproduction by comparing and contrasting different
   aspects of these two paintings.

Chapter 9

COMPARATIVE STRATEGIES I: LITERAL COMPARISON AND CONTRAST

Comparison and contrast is a strategy of writing best suited to making comments about two parallel entities. The method employs two of our most basic learning processes: the comparison, which studies likenesses, and contrast, which explores differences. By examining likeness and difference in carefully organized ways, we learn not only much about both of the two areas of our study, but in a good comparison and contrast situation, we learn much in a third area from seeing what the two together reveal.

For example, in the last chapter, when you examined Abbey's description of the packs used by De Puy and Peacock, you saw two extremes in the philosophy of what to take on a desert hike. What you were probably doing, at least subconsciously, was thinking about what *you* would have taken from each hiker's philosophy and methodology and included in or excluded from the imaginary pack you were assembling as you read the excerpt. We can set this up in another way by visualizing the relations:

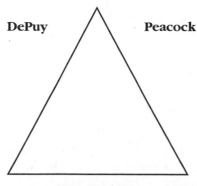

DePuy   **Peacock**

Your Version
(Result of Comparison and Contrast)

This diagram, with the product of the comparison/contrast on the bottom, suggests the ultimate result of this process.

In this way, you see the importance of the third area. In other words, the process of applying comparison (mentally comparing the two packs to what you would like to have, including what you find helpful) and contrast (mentally seeing what was unnecessary to you and deciding that it would not be in your ideal pack) shows you how this third area could be even more beneficial than the actual comparisons and contrasts would be.

So it will be with your writing. You will have a purpose in mind, one that you will reflect in your thesis, which will work as a long-range goal when you set up your comparison and contrast writing. If the brief section on the packs were an essay, Abbey's thesis would include his words "you may wish to avoid such extremes" as part of his argued statement to prospective hikers. The *point* of the comparison and contrast, then, should always be foremost and be clear to your reader.

In making your point clear to your reader, you will have to observe carefully some key elements in the planning stage. When you set up your comparison and contrast, you will have to determine with genuine clarity what points you wish to explore. This is a great place to use the mapping techniques we mentioned earlier, for you can set up two columns, one for each of the two entities to be compared and contrasted, and list the qualities of each; for example, with the packs:

De Puy	Peacock
like camel	ex-Green Beret
J.C. Penny hightops	like buffalo
plastic pack	90-lb. pannier
sketchbook	guns, ammo, bayonet
jerky	pitons, carabineers
green chilies	cameras
	field book
	rope
	sledge, rock samples, kit
	field glasses
	canteens
	jungle boots
	C-rations
	etc.

In the process of listing all these items, you may well become aware—if you were not already—that Abbey starts with a metaphorical animal to describe each man, gives some idea of how each will move across the desert (the hightops, loading pack at trailhead), describes the pack itself, and then proceeds to contents. The connotations of camel and buffalo might give you some additional help as you think about how to pack, at least imaginatively, your own pack for a jaunt in the desert. You would have a better idea of the specific points on which the two differed, and you would have much material for the contrast aspect of your presentation; since these two individuals represent the extremes, Abbey would not be giving you as a reader as much to compare.

To shift gears for a moment, imagine yourself in Abbey's position, writing a comparison and contrast essay on planning your pack for a desert hike. In setting up your essay, you will want to acknowledge that even though you will be comparing and contrasting, you will be emphasizing the *differences* between two extremes, and so the contrast will be the more crucial of the two parallel philosophies. A list or chart like the one above is invaluable as you decide the order in which to present your material, examples, and evidence, and to develop your essay.

There are several questions that you will want to answer as you move through your plan for comparing and contrasting. In order to sketch out the possibilities, let's label our first item to be compared/contrasted as "A," and the second as "B"; in our continuing example, De Puy would be A and Peacock B. First, you must decide whether to present all of your A information first, and then all of your B information, after which you would make your comments regarding the likenesses and differences. Or, you could decide on certain important points—such as the animal imagery used for each, the nature of the pack itself, the contents of each pack, and the conclusions we can draw about each packer—and then examine A and B for point #1 (animal image), A and B for point #2 (pack itself), and so on. This we call the point-by-point method.

With comparison and contrast, deciding on which of these two organizing principles to use is the most crucial decision you have to make in planning. The "all A, then B" method is good when you want to preserve the integrity of each item being examined, and you feel that such a focus is better than emphasizing points of similarity and difference. The point-by-point method is useful for drawing fine distinctions between two entities so you can match up categories and draw your conclusions immediately. You can see how Abbey creates an interesting dramatic and comic effect by describing each man and his pack in a single section, but for a projected essay evaluating the two extremes, you can also see that you would probably want to look carefully at each category to see how you would decide to set up your own pack.

Let's take a look at a slightly different situation, this time involving two works of visual art by two highly regarded nineteenth-century painters. The first, *Storm King on the Hudson,* was painted by Samuel Colman in 1866; the second, *Crossing the Ford, Platte River, Colorado,* was painted by Thomas Whittredge between 1868 and 1870. We might compare these two paintings for several reasons. First, they were both painted by American artists; second, they were both painted within a four-year time span, a relatively brief period for a painting; third, they both portray scenes on major American rivers; fourth, they depict similar structures of human figures on rivers in middle distances against backdrops of distant mountains.

We might contrast them for several other reasons. First, one is of an eastern, the other a western scene; second, one has an identifiable foreground (the river-bank closest to the painter's position) while the other does not; third, one clearly portrays a mechanical intrusion on the natural scene, while the other empha-sizes the peaceful relationship between an indigenous people and their land; fourth, the eastern scene explores the effects of clouds in the atmosphere, while the western scene displays a clear, crystalline sky. The distinctions offer a means of exploring some differences between visions of the land in the east and west after the Civil War, most notably in the contrast between the coming of indus-trialization in the east and the calm perfection of an unspoiled western scene.

If you were to write an essay comparing and contrasting these paintings as examples of post-Civil War attitudes toward change in the landscape as reflected in art, you might use for a thesis the idea that the Colman painting presents the Hudson as a fertile field for the industrialization represented by the steamboat harmoniously moving into the landscape, where the western view emphasizes a clear relation between a calm land and a native village. More simply, the east recognizes and accepts the promise of industrialization of the landscape while the west emphasizes the unchanging nature of the land. Here, then, we have the third part of the comparison-contrast triangle: the reason for the compari-son being made in the first place.

In such a case as this one involving the two paintings, your thesis deter-mines whether you take the entire A, entire B approach, or the point-by-point approach. Whenever you find it important to give the full effect of each entity, as Abbey does in describing his two packers, you will want to use the entire approach; whenever you want to build a carefully detailed argument involving the grounds on which you draw careful comparisons and minute distinctions, you will find the point-by-point method more useful. It is possible to combine the two modes, most often in a situation where you expect your audience may not be familiar with a topic and you therefore need to provide more information than usual. In this case, you might present almost all of A, then almost all of B, including several points for each, before you move to a point-by-point analysis of several commonalities you wish to emphasize. In these situations, you will want to use an overall organizational pattern and clear transitions that will let your reader know exactly where you are in terms of your overall argument.

Perhaps more than any other rhetorical strategy, comparison and contrast will help you to see how much you can help yourself as a writer by carefully organizing your data, by seeing exactly how you will present your material for a reader. Since our brains, scientists tell us, operate so well on such binary structures as up/down, high/low, easy/hard, near/far, etc., we seem to learn particularly quickly when presented with such information. Comparison/contrast then becomes an excellent way for us to teach others about new information; we can even do such teaching in ways that preserve the complex integrity of the material we are explaining. We may not realize it when we are doing it, but surfing through the television channels with our remotes is a form of compar-ing and contrasting the choices available to us and making a decision based on criteria we have supposedly weighed and evaluated.

The comparison-contrast method works in various ways for the authors represented in this chapter. Freeman suggests parallels between a human and a tree in a surprising manner. The Watermans provide an example of classic comparison and contrast when they compare and contrast two theories of backpacking by using two very different female models. William Least Heat Moon contrasts different houses along his routes in order to consider the people who live in them. Ella Cara Deloria shows how Waterlily's experience of marriage differs from what she had expected. In these selections, comparison and contrast are subtle devices used by the authors to alter our usual responses.

"The Lombardy Poplar"
Mary E. Wilkins Freeman

Born in Massachusetts in 1852, Mary E. Wilkins Freeman did not get her first story published until 1881, when she was 29. Her family suffered financial reverses despite a move to Brattleboro, Vermont, to improve their fortunes, and Freeman worked diligently at supplementing the meager family income as a teacher, maid, and ultimately, professional writer. Her writing often reflects the sparse, sober routine of the New England family life she knew well. Fortunately for Freeman, her short stories became increasingly popular during the 1880s and 1890s, and along with Sarah Orne Jewett she came to be regarded as a master storyteller of New England life.

Her best-known stories, "A New England Nun" and "The Revolt of Mother," display a fine ear for the inflections of New England speech and a keen feel for the rhythm of life on that rocky soil. She conveys a strong sense of the varieties of loneliness that plague people, particularly women, in a climate and region where people often spoke little to one another. In "The Lombardy Poplar," she examines the quality of the daily life of elderly women for whom the last of a stand of poplar trees serves as a point from which the ladies can draw comparisons about life and human relationships.

EXPLORING JOURNAL

Freewrite for ten minutes on your sense of New England or on the difficulties of earning a living from rocky soil in a region often battered by harsh winters.

———————

There had been five in the family of the Lombardy poplar. Formerly he had stood before the Dunn house in a lusty row of three brothers and a mighty father, from whose strong roots, extending far under the soil, they had all sprung.

Now they were all gone, except this one, the last of the sons of the tree. He alone remained, faithful as a sentinel before the onslaught of winter storms and summer suns; he yielded to neither. He was head and shoulders above the other trees—the cherry and horse-chestnuts in the square front yard behind him. Higher than the house, piercing the blue with his broad truncate of green, he stood silent, stiff, and immovable. He seldom made any sound with his closely massed foliage, and it required a mighty and concentrated gust of wind to sway him ever so little from his straight perpendicular.

As the tree was the last of his immediate family, so the woman who lived in the house was the last of hers. Sarah Dunn was the only survivor of a large family. No fewer than nine children had been born to her parents; now father, mother, and eight children were all dead, and this elderly woman was left alone in the old house. Consumption had been in the Dunn family. The last who had succumbed to it was Sarah's twin-sister Marah, and she had lived until both had gray hair.

After that last funeral, where she was the solitary real mourner, there being only distant relatives of the Dunn name, Sarah closed all the house except a few rooms, and resigned herself to living out her colorless life alone. She seldom went into any other house; she had few visitors, with the exception of one woman. She was a second cousin, of the same name, being also Sarah Dunn. She came regularly on Thursday afternoons, stayed to tea, and went to the evening prayer-meeting. Besides the sameness of name, there was a remarkable resemblance in personal appearance between the two women. They were of about the same age; they both had gray-blond hair, which was very thin, and strained painfully back from their ears and necks into tiny rosettes at the backs of their heads, below little, black lace caps trimmed with bows of purple ribbon. The cousin Sarah had not worn the black lace cap until the other Sarah's twin-sister Marah had died. Then all the dead woman's wardrobe had been given to her, since she was needy. Sarah and her twin had always dressed alike, and there were many in the village who never until the day of her death had been able to distinguish Marah from Sarah. They were alike not only in appearance, but in character. The resemblance was so absolute as to produce a feeling of something at fault in the beholder. It was difficult, when looking from one to the other, to believe that the second was a vital fact: it was like seeing double. After Marah was dead it was the same with the cousin, Sarah Dunn. The clothes of the deceased twin completed all that had been necessary to make the resemblance perfect. There was in the whole Dunn family a curious endurance of characteristics. It was said in the village that you could tell a Dunn if you met him at the ends of the earth. They were all described as little, and sloping-shouldered, and peak-chinned, and sharp-nosed, and light-livered. Sarah and Cousin Sarah were all these. The family tricks of color and form and feature were represented to their fullest extent in both. People said that they were Dunns from the soles of their feet to the crowns of their heads. They did not even use plurals in dealing with them. When they set out together for evening meeting in the summer twilight, both moving with the same gentle, mincing step, the same slight sway of shoulders, draped precisely alike with little, knitted, white wool shawls, the same deprecating cant of heads, identically bonneted, as if they were perpetually avoiding some low-hanging bough of life in their way of progress, the neighbors said, "There's Sarah Dunn goin' to meetin'."

When the twin was alive it was, "There's Sarah and Marah goin' to meetin'." Even the very similar names had served as a slight distinction, as formerly the different dress of the cousins had made it easier to distinguish between them. Now there was no difference between the outward charactertistics of the two Sarah Dunns, even to a close observer. Name, appearance, dress, all were identical. And the minds of the two seemed to partake of this similarity. Their conversation consisted mainly of a peaceful monotony of agreement. "For the Lord's sake, Sarah Dunn, 'ain't you got any mind of your own?" cried a neighbor of an energetic and independent turn, once when she had run in of a Thursday afternoon when the cousin was there. Sarah looked at the cousin before replying, and the two minds seemed to cogitate the problem through the medium of mild, pale eyes, set alike under faint levels of eyebrow. "For the Lord's sake, if you ain't lookin' at each other to find out!" cried the neighbor, with a high sniff, while the two other women stared at each other in a vain effort to understand.

The twin had been dead five years, and the cousin had come every Thursday afternoon to see Sarah before any point of difference in their mental attitudes was evident. They regarded the weather with identical emotions, they relished the same food, they felt the same degree of heat or cold, they had the same likes and dislikes for other people, but at last there came a disagreement. It was on a Thursday in summer, when the heat was intense. The cousin had come along the dusty road between the white-powdered weeds and flowers, holding above her head an umbrella small and ancient, covered with faded green silk, which had belonged to Marah, wearing an old purple muslin of the dead woman's, and her black lace mitts. Sarah was at home, rocking in the south parlor window, dressed in the mate to the purple muslin, fanning herself with a small black fan edged with feathers which gave out a curious odor of mouldy roses.

When the cousin entered, she laid aside her bonnet and mitts, and seated herself opposite Sarah, and fanned herself with the mate to the fan.

"It is dreadful warm," said the cousin.

"Dreadful!" said Sarah.

"Seems to me it 'ain't been so warm since that hot Sabbath the summer after Marah died," said the cousin, with gentle reminiscence.

"Just what I was thinking," said Sarah.

"An' it's dusty too, just as it was then."

"Yes, it was dreadful dusty then. I got my black silk so full of dust it was just about ruined, goin' to meetin' that Sabbath," said Sarah.

"An' I was dreadful afraid I had sp'ilt Marah's, an' she always kept it so nice."

"Yes, she had always kept it dreadful nice," assented Sarah.

"Yes, she had. I 'most wished, when I got home that afternoon, and saw how dusty it was, that she'd kept it and been laid away in it, instead of my havin' it, but I knew she'd said to wear it, and get the good of it, and never mind."

"Yes, she would."

"And I got the dust all off it with a piece of her old black velvet bunnit," said the cousin, with mild deprecation.

"That's the way I got the dust off mine, with a piece of my old black velvet bunnit," said Sarah.

"It's better than anything else to take the dust off black silk."

"Yes, 'tis."

"I saw Mis' Andrew Dunn as I was comin' past," said the cousin.

"I saw her this mornin' down to the store," said Sarah.

"I thought she looked kind of pindlin', and she coughed some."

"She did when I saw her. I thought she looked real miserable. Shouldn't wonder if she was goin in the same way as the others."

"Just what I think."

"It was funny we didn't get the consumption, ain't it, when all our folks died with it?"

"Yes, it is funny."

"I s'pose we wa'n't the kind to."

"Yes, I s'pose so."

Then the two women swayed peacefully back and forth in their rocking-chairs, and fluttered their fans gently before their calm faces.

"It is too hot to sew to-day," remarked Sarah Dunn.

"Yes, it is," assented her cousin.

"I thought I wouldn't bake biscuit for supper, long as it was so dreadful hot."

"I was hopin' you wouldn't. It's too hot for hot biscuit. They kind of go against you."

"That's what I said. Says I, now I ain't goin' to heat up the house bakin' hot bread to-night. I know she won't want me to."

"No, you was just right. I don't."

"Say I, I've got some good cold bread and butter, and blackberries that I bought of the little Whitcomb boy this mornin', and a nice custard-pie, and two kinds of cake besides cookies, and I guess that'll do."

"That's just what I should have picked out for supper."

"And I thought we'd have it early, so as to get it cleared away, and take our time walkin' to meetin', it's so dreadful hot."

"Yes, it's a good idea."

"I s'pose there won't be so many to meetin', it's so hot," said Sarah.

"Yes, I s'pose so."

"It's queer folks can stay away from meetin' on account of the weather."

"It don't mean much to them that do," said the cousin, with pious rancor.

"That's so," said Sarah. "I guess it don't. I guess it ain't the comfort to them that it is to me. I guess if some of them had lost as many folks as I have they'd go whether 'twas hot or cold."

"I guess they would. They don't know much about it."

Sarah gazed sadly and reflectively out of the window at the deep yard, with its front gravel walk bordered with wilting pinks and sprawling peonies, its horse-chestnut and cherry trees, and its solitary Lombardy poplar set in advance, straight and stiff as a sentinel of summer. "Speakin' of losin' folks," she said, "you 'ain't any idea what a blessin' that popple-tree out there has been to me, especially since Marah died."

Then, for the first time, the cousin stopped waving her fan in unison, and the shadow of a different opinion darkened her face. "That popple-tree?" she said, with harsh inquiry.

"Yes, that popple-tree." Sarah continued gazing at the tree, standing in majestic isolation, with its long streak of shadow athwart the grass.

The cousin looked, too; then she turned towards Sarah with a frown of puzzled dissent verging on irritability and scorn. "That popple-tree! Land! How you do talk!" said she. "What sort of a blessin' can an old tree be when your folks are gone, Sarah Dunn?"

Sarah faced her with stout affirmation: "I've seen that popple there ever since I can remember, and it's all I've got left that's anyways alive, and it seems like my own folks, and I can't help it."

The cousin sniffed audibly. She resumed fanning herself, with violent jerks. "Well," said she, "if you can feel as if an old popple-tree made up to you, in any fashion, for the loss of your own folks, and if you can feel as if it was them, all I've got to say is, I can't."

"I'm thankful I can," said Sarah Dunn.

"Well, I can't. It seems to me as if it was almost sacrilegious."

"I can't help how it seems to you." There was a flush of nervous indignation on Sarah Dunn's pale, flaccid cheeks; her voice rang sharp. The resemblance between the two faces, which had in reality been more marked in expression, as evincing a perfect accord of mental action, than in feature, had almost disappeared.

"An old popple-tree!" said the cousin, with a fury of sarcasm. "If it had been any other tree than a popple, it wouldn't strike anybody as quite so bad. I've always thought a popple was about the homeliest tree that grows. Much as ever as it does grow. It just stays, stiff and pointed, as if it was goin' to make a hole in the sky; don't give no shade worth anything, don't seem to have much to do with the earth and folks, anyhow. I was thankful when I got mine cut down. Them three that was in front of our house were always an eyesore to me, and I talked till I got father to cut them down. I always wondered why you hung on to this one so."

"I wouldn't have that popple-tree cut down for a hundred dollars," declared Sarah Dunn. She had closed her fan, and she held it up straight like a weapon.

"My land! Well, if I was goin' to make such a fuss over a tree I'd have taken something different from a popple. I'd have taken a pretty elm or a maple. They look something like trees. This don't look like anything on earth besides itself. It ain't a tree. It's a stick tryin' to look like one."

"That's why I like it," replied Sarah Dunn, with a high lift of her head. She gave a look of sharp resentment at her cousin. Then she gazed at the tree again, and her whole face changed indescribably. She seemed like another person. The tree seemed to cast a shadow of likeness over her. She appeared straighter, taller; all her lines of meek yielding, or scarcely even anything so strong as yielding, of utter passiveness, vanished. She looked stiff and uncompromising. Her mouth was firm, her chin high, her eyes steady, and, more than all, there was over her an expression of individuality which had not been there before. "That's why I like the popple," said she, in an incisive voice. "That's just why. I'm sick of things and folks that are just like everything and everybody else. I'm sick of trees that are just trees. I like one that ain't."

"My land!" ejaculated the cousin, in a tone of contempt not unmixed with timidity. She stared at the other woman with shrinking and aversion in her pale-blue eyes. "What has come over you, Sarah Dunn?" said she, at last, with a feeble attempt to assert herself.

"Nothin' has come over me. I always felt that way about that popple."

"Marah wa'n't such a fool about that old popple."

"No, she wa'n't, but maybe she would have been if I had been taken first instead of her. Everybody has got to have something to lean on."

"Well, I 'ain't got anything any more than you have, but I can stand up straight without an old popple."

"You 'ain't no call to talk that way," said Sarah.

"I hate to hear folks that I've always thought had common-sense talk like fools," said the cousin, with growing courage.

"If you don't like to hear me talk, it's always easy enough to get out of hearin' distance."

"I'd like to know what you mean by that, Sarah Dunn."

"I mean it just as you want to take it."

"Maybe you mean that my room is better than my company."

"Just as you are a mind to take it."

The cousin sat indeterminately for a few minutes. She thought of the bread and the blackberries, the pie and the two kinds of cake.

"What on earth do you mean goin' on so queer?" said she, in a hesitating and somewhat conciliatory voice.

"I mean just what I said. That tree is a blessin' to me, it's company, and I think it's the handsomest tree anywheres around. That's what I meant, and if you want to take me up for it, you can."

The cousin hesitated. She further reflected that she had in her solitary house no bread at all; she had not baked for two days. She would have to make a fire and bake biscuits in all that burning heat, and she had no cake nor berries. In fact, there was nothing whatever in her larder, except two cold potatoes, and a summer-squash pie, which she suspected was sour. She wanted to bury the hatchet, she wanted to stay, but her slow blood was up. All her strength of character lay in inertia. One inertia of acquiescence was over, the other of dissent was triumphant. She could scarcely yield for all the bread and blackberries and cake. She shut up her fan with a clap.

"That fan was Marah's," said Sarah, meaningly, with a glance of reproach and indignation.

"I know it was Marah's," returned the cousin, rising with a jerk. "I know it was Marah's. 'Most everything I've got was hers, and I know that too. I ought to know it; I've been twitted about it times enough. If you think I ain't careful enough with her things, you can take them back again. If presents ain't mine after they've been give me, I don't want 'em."

The cousin went out of the room with a flounce of her purple muslin skirts. She passed into Sarah's little room where her cape and bonnet lay carefully placed on the snowy hill of the feather bed. She put them on, snatched up her green silk parasol, and passed through the sitting-room to the front entry.

"If you are a mind to go off mad, for such a thing as that, you can," said Sarah, rocking violently.

"You can feel just the way you want to," returned the cousin, with a sniff, "but you can't expect anybody with a mite of common-sense to fall in with such crazy ideas." She was out of the room and the house then with a switch, and speeding down the road with the green parasol bobbing over-head.

Sarah gave a sigh; she stared after her cousin's retreating form, then at the poplar-tree, and nodded as in confirmation of some resolution within her own mind. Presently she got up, looked on the table, then on the bed and bureau in the bedroom. The cousin had taken the fan.

Sarah returned to her chair, and sat fanning herself absent-mindedly. She gazed out at the yard and the poplar-tree. She had not resumed her wonted expression; the shadow of the stately, concentrated tree seemed still over her. She held her faded blond head stiff and high, her pale-blue eyes were steady, her chin firm above the lace ruffle at her throat. But there was sorrow in her heart. She was a creature of as strong race-ties as the tree. All her kin were dear to her, and the cousin had been the dearest after the death of her sister. She felt as if part of herself had been cut away, leaving a bitter ache of vacancy, and yet a proud self-sufficiency was over her. She could exist and hold her head high in the world without her kindred, as well as the poplar. When it was tea-time she did not stir. She forgot. She did not rouse herself until the meeting-bell began to ring. Then she rose hurriedly, put on her bonnet and cape, and hastened down the road. When she came in sight of the church, with its open vestry windows, whence floated already singing voices, for she was somewhat late, she saw the cousin coming from the opposite direction. The two met at the vestry door, but neither spoke. They entered side by side; Sarah seated herself, and the cousin passed to the seat in front of her. The congregation, who were singing "Sweet Hour of Prayer," stared. There was quite a general turning of heads. Everybody seemed to notice that Sarah Dunn and her cousin Sarah Dunn were sitting in separate settees. Sarah opened her hymn-book and held it before her face. The cousin sang in a shrill tremolo. Sarah hesitated a moment, then she struck in and sang louder. Her voice was truer and better. Both had sung in the choir when young.

The singing ceased. The minister, who was old, offered prayer, and then requested a brother to make remarks, then another to offer prayer. Prayer and remarks alike were made in a low, inarticulate drone. Above it sounded the rustle of the trees outside in a rising wind, and the shrill reiteration of the locusts invisible in their tumult of sound. Sarah Dunn, sitting fanning, listening, yet scarcely comprehending the human speech any more than she comprehended the voices of the summer night outside, kept her eyes fastened on the straining surface of gray hair surmounted by the tiny black triangle of her cousin's bonnet. Now and then she gazed instead at the narrow black shoulders beneath. There was something rather pitiful as well as uncompromising about those narrow shoulders, suggesting as they did the narrowness of the life-path through which they moved, and also the stiff-neckedness in petty ends, if any, of their owner; but Sarah did not comprehend that. They were for her simply her cousin's shoulders, the cousin who had taken exception to her small assertion of her own individuality, and they bore for her an expression of arbitrary criticism as marked as if they had been the cousin's face. She felt an animosity distinctly vindictive towards the shoulders; she had an impulse to push and crowd in her own. The cousin sat fanning herself quite violently. Presently a short lock of hair on Sarah's forehead became disengaged from the rest, and blew wildly in the wind from the fan. Sarah put it back with an impatient motion, but it flew out again. Then Sarah shut up her own fan, and sat in stern resignation, holding to the recreant lock of hair to keep it in place, while the wind from the cousin's fan continued to smite her in the face. Sarah did not fan herself until the cousin laid down her fan for a moment, then she resumed hers with an angry sigh. When the cousin opened her fan again, Sarah dropped hers in her lap, and sat with one hand pressed against her hair, with an expression of bitter long-suffering drawing down the corners of her mouth.

After the service was over Sarah rose promptly and went out, almost crowding before the others in her effort to gain the door before her cousin. The cousin did the same; thus each defeated her own ends, and the two passed through the door shoulder to shoulder. Once out in the night air, they separated speedily, and each went her way to her solitary home.

Sarah, when she reached her house, stopped beside the poplar-tree and stood gazing up at its shaft of solitary vernal majesty. Its outlines were softened in the dim light. Sarah thought of the "pillar of cloud" in the Old Testament. As she gazed the feeling of righteous and justified indignation against the other Sarah Dunn grew and strengthened. She looked at the Lombardy poplar, one of a large race of trees, all with similar characteristics which determined kinship, yet there was this tree as separate and marked among its kind as if of another name and family. She could see from where she stood the pale tremulousness of a silver poplar in the corner of the next yard. "Them trees is both poplars," she reflected, "but each of 'em is its own tree." Then she reasoned by analogy. "There ain't any reason why if Sarah Dunn and I are both Dunns, and look alike, we should be just alike." She shook her head fiercely. "I ain't goin' to be Sarah Dunn, and she needn't try to make me," said she, quite aloud. Then she went into the house, and left the Lombardy poplar alone in the dark summer night.

It was not long before people began to talk about the quarrel between the two Sarah Dunns. Sarah Dunn proper said nothing, but the cousin told her story right and left: how Sarah had talked as if she didn't have common-sense, putting an old, stiff popple-tree on a par with the folks she'd lost, and she, the cousin, had told her she didn't have common-sense, and then Sarah had ordered her out of her house, and would not speak to her comin' out of meetin'. People began to look askance at Sarah Dunn, but she was quite unaware of it. She had formed her own plan of action, and was engaged in carrying it out. The day succeeding that of the dispute with the cousin was the hottest of a hot trio, memorable long after in that vicinity, but Sarah dressed herself in one of her cool old muslins, took her parasol and fan, and started to walk to Atkins, five miles distant, where all the stores were. She had to pass the cousin's house. The cousin, peering between the slats of a blind in the sitting-room, watched her pass, and wondered with angry curiosity where she could be going. She watched all the forenoon for her to return, but it was high noon before Sarah came in sight. She was walking at a good pace, her face was composed and unflushed. She held her head high, and walked past, her starched white petticoat ratting and her purple muslin held up daintily in front, but trailing in the back in a cloud of dust. Her white-stockinged ankles and black cloth shoes were quite visible as she advanced, stepping swiftly and precisely. She had a number of large parcels pressed closely to her sides under her arms and dangling by the strings from her hands. The cousin wondered unhappily what she had bought in Atkins. Sarah, passing, knew that she wondered, and was filled with childish triumph and delight. "I'd like to know what she'd say if she knew what I'd got," she said to herself.

The next morning the neighbors saw Annie Doane, who went out dress-making by the day, enter Sarah Dunn's yard with her bag of patterns. It was the first time for years that she had been seen to enter there for Sarah and Marah had worn their clothes with delicate care, and they had seldom needed replenishing, since the fashions had been ignored by them.

The neighbors wondered. They lay in wait for Annie Doane on her way home that night, but she was very close. They discovered nothing, and could not even guess with the wildest imagination what Sarah Dunn was having made. But the next Sunday a shimmer of red silk and a toss of pink flowers were seen at the Dunn gate, and Sarah Dunn, clad in a gown of dark-red silk and a bonnet tufted with pink roses, holding aloft a red parasol, passed down the street to meeting. No Dunn had ever worn, within the memory of man, any colors save purple and black and faded green or drab, never any but purple or white or black flowers in her bonnet. No woman of half her years, and seldom a young girl was ever seen in the village clad in red. Even the old minister hesitated a second in his discourse, and recovered himself with a hem of embarrassment when Sarah entered the meeting-house. She had waited until the sermon was begun before she sailed up the aisle. There were many of her name in the church. The pale, small, delicate faces in the neutral-colored bonnets stared at her as if a bird of another feather had gotten into their nest; but the cousin, who sat across the aisle from Sarah, caught her breath with an audible gasp.

After the service Sarah Dunn walked with her down the aisle, pressing close to her side. "Good-mornin'," said she, affably. The cousin in Marah's old black silk, which was matched by the one which Sarah would naturally have worn that Sunday, looked at her, and said, feebly, "Good-mornin'." There seemed no likeness whatever between the two women as they went down the aisle. Sarah was a Dunn apart. She held up her dress as she had seen young girls, drawing it tightly over her back and hips, elevating it on one side.

When they emerged from the meeting-house, Sarah spoke. "I should be happy to have you come over and spend the day to-morrow," said she, "and have a chicken dinner. I'm goin' to have the Plymouth Rock crower killed. I've got too many crowers. He'll weigh near five pounds, and I'm goin' to roast him."

"I'll be happy to come," replied the cousin, feebly. She was vanquished.

"And I'm goin' to give you my clothes like Marah's," said Sarah, calmly. "I'm goin' to dress different."

"Thank you," said the cousin.

"I'll have dinner ready about twelve. I want it early, so as to get it out of the way," said Sarah.

"I'll be there in time," said the cousin.

Then they went their ways. Sarah, when she reached home, paused at the front gate, and stood gazing up at the poplar. Then she nodded affirmatively and entered the house, and the door closed after her in her red silk dress. And the Lombardy poplar-tree stood in its green majesty before the house, and its shadow lengthened athwart the yard to the very walls.

—————⇒•⇐—————

RESPONDING JOURNAL

Explain in a journal entry of at least a page how accurate you feel the events of the story are today. Could we find a similar situation occurring today? What sorts of detail would you have to change in the story in order to update it ?

QUESTIONS FOR CRITICAL THINKING

STRATEGY

1. Look at the opening sentences of the first and third paragraphs. How do you account for the difference in sentence structure in the two sentences? How similar are the functions the two sentences serve?
2. Consider Freeman's extended fourth paragraph. Point out all the devices you see that help her to make the comparisons between the two Sarah Dunns. Which of these would you be most likely to use?

ISSUES

1. Consider the extent to which you find landscape a factor in this 100+-year-old story. Could people find the same importance in local landmarks?

Why or why not? Try to explain for the class some similar situations involving trees or other natural landmarks that seem to anchor human thought or action.

2. Think about the importance of the prayer service in the action of the story. What generalizations can you make about the relation between religion and the importance of objects in the natural world, such as a tree?

COLLABORATIVE WRITING ACTIVITY

Sarah Dunn's red silk dress seems to mean much to her, and its close association in the story's final lines with the Lombardy poplar links both. Write about an item of clothing you feel has an important connection of some sort with the physical world. Do you have a particular pair of boots or shoes that have strong associations for you? Is there an item of clothing that brings up memories of a campus landmark, or perhaps a certain downtown spot? Explain your response by comparing and contrasting qualities or emotions you feel and that you believe are either inherent in the natural or urban site or object, or that you can legitimately ascribe to the object or place. Present your ideas to the class, and work together to draft a statement of the common qualities you detect in your group responses.

"That Pack on the Back: Mae West vs. Twiggy," from *Backwoods Ethics*
Laura and Guy Waterman

Laura and Guy Waterman have made careers of public service to campers, backpackers, and climbing enthusiasts. Long-term volunteers for the United States Forest Service and numerous conservation groups, they have written on varied topics. Their books give a good indication of their specific interests, which include: Wilderness Ethics: Preserving the Spirit of Wildness; Forest and Crag: A History of Hiking, Trail Blazing, and Adventure in the Northeast Mountains; *and* Yankee Rock and Ice: A History of Climbing in the Northeast U.S. *They live a nearly self-sufficient existence on their 27-acre home in Vermont.*

The Watermans begin Backwoods Ethics *with their observation that Thoreau's injunction to us to remember that "In wildness is the preservation of the world" today must caution us to keep in mind that preservation is threatened from many sides and that we must be ever vigilant. In many ways, we can unconsciously allow our wilds to be ruined if we do not maintain our vigilance. In comparing and contrasting two similar yet differing attitudes toward backpacking, the Watermans introduce their readers to the crucial problems of how unobtrusively we move into wilderness areas.*

EXPLORING JOURNAL

Think about the concept of individual responsibility outdoors. Freewrite for ten minutes on what precautions you should observe when you hike, whether on a clearly marked path in a specially designated natural area or in an undeveloped area where you are striking out on your own. If you need additional stimuli, think about the differences in responsibility between the two situations.

> *And seeing the snail, which everywhere doth roam*
> *Carrying his own house still, still is at home,*
> *Follow (for he is easy pac'd) this snail,*
> *Be thine own palace, or the world's thy jail.*
>
> —JOHN DONNE

We meet two distinct types of hikers in the woods: the kitchen sink and the hair shirt. Maybe you know them too. Maybe you're one or the other. Some people carry more than you can possibly use on a two- or three-day weekend; others less than you need to enjoy a reasonably good time.

The kitchen sink believes in carrying it all, everything to make himself comfortable in the woods, along with every precaution against a wide range of possible emergencies, from hypothermia to hangnail. With his pack bulging and stuff sacks strapped on top or hanging below, he grunts along, sweat pouring from his overworked body. Maybe you can't take it with you, but this fellow will obviously try.

The hair shirt subscribes to the go-light school, toothpick-and-a-match, survive on as little as possible. He drills holes in the handle of his toothbrush and bivouacs in a rain poncho.

These two traditions have deep roots. It's Atlas versus Mercury. He who bears the weight of the world on his shoulders versus he of the winged foot. It's Paul Bunyan versus John Muir. Goliath versus David. Connie versus Big Stoop (remember, you "Terry and the Pirates" fans?). Late Gothic versus Rococo. Mae West versus Twiggy. (As we said, some people carry more than you can possibly use on a two- or three-day weekend; others less than you need to enjoy a reasonably good time.) The difference is more than a matter of pack weight; it's a contrast in approach to the outdoors, a divergence of personal style, almost a split in philosophy.

On his pioneering explorations of Yosemite Valley and the Sierra Nevada, John Muir carried incredibly little—sometimes just a couple of blankets and a food box small enough to strap to his belt, containing only grain meal, sugar, and tea. Muir is not only one of America's greatest champions of conservation, he is also the patron saint of the go-light school.

Muir's modern-day successor as lord of the Sierra backcountry was Norman Clyde, who spent 40 years exploring hidden valleys and making solo first ascents of difficult routes almost right up to his death at age 87, in 1972. But Clyde was the opposite of Muir as a backpacker. His typical 100-pound pack might include elaborate fishing gear, two pistols, camera equipment, extra pairs of boots and clothes, several large kettles, a wide assortment of dishes, bowls, and cups, canned food, and perhaps a few chunks of firewood if he were going above timberline; plus the famous Clyde library of books. "The pack that walked like a man," he was called. Surely Clyde was the patron saint of the kitchen sink breed.

The New England inheritors of the Clyde legacy have traditionally been the hut men of the Appalachian Mountain Club's chain of huts in the White Mountains. These college-age lads used to stock all the food and equipment for huts that were as far as six miles from the road over mountain trails, with as much as 3,600 feet of elevation to be gained. Their packs of well over 100 pounds humbled many a tired vacationer whom they steamed past on the trail. Back in the 1950s, the hut boys at the 4,900-foot Madison Springs Hut began packing all the parts of a Model T up to that remote and rocky windswept col, intending to assemble it among the boulders. Sober authorities intervened, and a helicopter flew out the parts as part of a cleanup campaign in the 1960s. These days the AMC is deemphasizing the load-carrying legend of hut life; modern hut men are hired more for their hospitality or their ability to interpret the ecology around the hut, and there are now female "hut persons." Loads are still respectable, but packs over 100 pounds are frowned upon. An era has passed.

The modern backpacking ideal is to cut weight ruthlessly. The western outdoors painter Roy Kerswill says he gets along fine for five days on a 16 1/2 pound pack, including camera, sketch pad, and brushes. He carries no cooking gear because he eats cold food only, and has worked it out so that he survives comfortably on a half pound of food per day.

Many years ago a New Englander of outsized legends, a fiery eccentric by the name of Arthur Comey, boasted of multiday trips with his famous "Ten Pound pack." Incredibly, Comey's 10 pounds found room for many items we would scarcely consider necessary today—razor and shaving soap, washcloth, bathing trunks, moccasins, and axe, with sheath and whetstone.

CONFESSIONS OF A WEIGHT LIFTER

Considering the obvious advantages of the go-light approach, we wish we could report that our normal packs are models of how to go superlight. Not so. Part of our problem is that we do a good deal of winter camping, when some heavy gear is unavoidable. Somehow we can't bring ourselves to part with it in summer.

For example, for years we were accustomed to a roomy, stormproof tent. It's very useful in winter, especially in exposed campsites or spots where you may have to sit out a storm for a full day or more. Being accustomed to this luxury in winter, we went on enjoying it year-round. We carried the Bauer Expedition model, with the front *and* back vestibules, snow flaps (in July?), and a weight of 12 pounds (grunt!).

It took the "new ethic" of clean camping to wean us away from our beloved Bauer for summer camping. Now we swing along the trail with somewhat lightened loads (and fewer backaches), since we made the switch to hammocks. . . .

For winter camping, we have discovered the delicious advantage of the sled over the backpack. As long as you aren't trying to climb too steep a slope, it is much, much easier to drag your overnight gear on a simple plastic sled. When you stop walking with a pack on, that 50 pounds is still on your back unless you struggle to set it down and heft it back on again. With a sled, when you stop pulling, that 50 (or even 100) pounds sits there on the snow, not on your back.

Otherwise, we have learned the lessons of weight too slowly. For many years we were hung up on a model of headlamp that gives strong light while leaving both hands free; but it ran on four D batteries, while the handy miniature flashlights run on two tiny AA batteries. Today we're content with the little flashlights and a jerry-built head rig for converting them to headlamps if we need to.

We were very reluctant to give up the marvelous Optimus 111B camping stove. We knew it was great for melting large amounts of snow (again . . . in July?), but it was also virtually the heaviest of the myriad models of backpacking stoves on the market. . . .

Last fall we noticed that in an obscure corner of our pack we'd been carrying—all summer—a file for sharpening crampons on ice-climbing trips . . .

An anthropologist named Woody Allen claims that there are tribes in Borneo that do not have a word for "no" in their language and turn down requests by nodding and saying "I'll get back to you." Well, our ability to reject articles from backpacking trips sometimes seems about as effective. So, for years in went the extra batteries, the extra sweater, the extra fuel.

WHAT YOU *DON'T* NEED

If you want to avoid getting into this bind, we can suggest a number of things that we've seen in the backcountry that you don't need:

1. A folding foxhole shovel, an old Boy Scout favorite; it might have been great in World War I, but who needs the weight in today's backwoods?
2. A mallet for driving tent pegs; you will need it to set up a circus tent, but not for the typical weekend camping setup.

3. A tool kit, including a wrench, needle nose pliers, screw drivers (regular and Phillips), and scissors; we find that a four-ounce Swiss Army knife, plus a little parachute cord, will suffice for emergency repairs.
4. Various camp stools and folding chairs—even folding toilets, as if these matters can't be taken care of without specialized equipment.
5. Cosmetics for the ladies. One handbook for women in the woods advises that "using a deodorant daily in the backwoods is a must." And we have seen recommended equipment lists that include a nail file and clippers (again, these are handy on a Swiss Army knife).
6. Carbon monoxide detection kits. No kidding—one widely quoted authority on camping says that you should never use a portable cookstove inside a tent without a carbon monoxide detection kit. Who wants to lug a kit around when thoroughly ventilating your tent will solve the problem?

TEN TINY TIPS FOR WEIGHT CARRYING

If you eliminate all such non-necessities, but you still find that your pack outweighs a Notre Dame linebacker, we could suggest 10 tiny tips on how to coexist with the enemy. Alas, these ideas grow from long personal experience. If your pack is heavy:

1. Keep the weight high; load the heavier items near the top of your pack.
2. Keep it all close to your back; avoid bulky items strapped on the outside in such a way as to pull you over backward.
3. Get everything into or onto the pack; don't try to carry anything in your hands, unless you go in for a walking staff (or in winter, an ice ax or ski pole).
4. Use a waist strap to transfer most of the load from your shoulders to your hips: on most modern packs this is standard.
5. Once you put the pack on, plan to walk steadily for long periods and not to stop for "rests" very often; frequent rest stops, taking the pack off and wrestling it back on, will delay your progress interminably and use up more energy than the rests restore.
6. Adopt a slow, sustainable pace, with a steady rhythm of regular steps; if you can sort of roll your weight from one foot to the other you can get a momentum that eases the strain. A stop-and-go, herky-jerky motion continually makes the full presence of the pack felt.
7. In winter, try using a sled for most of the weight, although this approach may be appropriate for some itineraries and not others.
8. Winter or summer, plan itineraries that fan out from base camps, so that you aren't always carrying all the weight. Enjoy a light day pack for most of your upper-elevation walking, and camp low—which eases the impact on fragile alpine ecologies as well as on your back.

9. Cultivate strong young companions who like to show that they can carry enormous weights. Then, as you walk uphill, ask them short questions that require long answers, so they have to do the talking while you gasp for breath as unobtrusively as possible. ("I didn't quite follow the theory of relativity. Would you go through it again?")

10. If all else fails, and you must carry weight, grin as you bear it. Think positive; like so many activities that seem purely physical, packing a heavy load is 75 percent mental. If you can pick up a pack, you can walk all day with it—if your frame of mind is right.

We know a marvelous fellow, Win Thratchett, with whom we have been on several winter camping trips in the Adirondacks. Thratchett's the kitchen sink type, par excellence. He carries an enormous pack, but he's ready for anything. In fact, he's never so happy as when some unusual emergency requires some obscure item that only a pack of his size could possibly provide. When a trip goes smoothly, Thratchett's unhappy—all that extra weight for nothing.

On one trip a young friend broke a snowshoe and was bemoaning the inadequacy of his planned patch job, which made use of a stick, a strip of rawhide, and tape. Along came Thratchett and asked (somewhat eagerly, we thought) if he could help. Our young friend allowed as how what he really needed was a pair of wood screws just the right size.

Thratchett looked delighted. "What size?" he asked as he swung off his enormous pack and started into it.

The other man felt this was just too much and remarked somewhat acidly: "Five-eighths inch, and only flat heads will do."

Thratchett looked momentarily nonplussed, but buried deeper into the dark recesses of the pack. When he came up, triumphantly clutching his tool kit, you could sense his satisfaction as he said:

"Brass or steel?"

RESPONDING JOURNAL

Explore in a substantial journal entry what you find to be the special strengths of the comparison and contrast strategy in the Watermans' essay. Why does this strategy work well for their purposes here? How else might they have set up their presentation? Would another way have been as effective?

QUESTIONS FOR CRITICAL THINKING

STRATEGY

1. What is the main device the Watermans use to make their presentation effective? Do the names Mae West (a voluptuous film star of the 1930s, often cast as a predatory female with comedian W.C. Fields) or Twiggy

(a notoriously thin fashion model of the late 1960s) mean anything to you? How much does the essay depend on this knowledge?

2. This essay is heavily allusive, that is, it refers to the Bible, comics, famous environmentalists, and movie stars in order to make its points. Explain how effective you find this strategy. Are you able to figure out most of the allusions from their contexts?

ISSUES

1. What are the spiritual advantages and drawbacks of the "Mae West" pack? Of the "Twiggy" pack? Which do you think Thoreau would recommend? Why?

2. Explain how your thinking has changed as a result of reading this essay. Has it focused your attention on the subtleties of hiking in new ways? Does it give you some new ideas for organizing your thoughts on a subject? If so, explain.

COLLABORATIVE WRITING ACTIVITY

Explain what your ideal pack would be if you had to limit it to the ten most essential items. Briefly explain why you would include each item. When you bring your list to your group, be prepared to defend your choices. After hearing all choices, vote on the ten most sensible and draft a joint statement explaining why this group of ten would be effective.

from *Blue Highways*
William Least Heat Moon

In order to write his first book, Blue Highways, *published in 1982, Moon set out with a copy of Whitman's* Leaves of Grass *and* Black Elk Speaks! *in his Ford van, which he named "Ghost Dancing." His more recent book,* PrairyErth: A Deep Map, *which was published in 1992, explores one significant place in America: its geographic center, a stand of tallgrass prairie that has escaped the devastation almost all of the great plains tallgrass suffered. Whether exploring across thousands of miles of back roads or concentrating on the people of one small Kansas town, Moon brings aspects of rural life to our attention and suggests that they carry an importance we all share.*

Of Osage ancestry, Moon demonstrates a keen sense of history of all American peoples. He realizes that at the start of his journey, he crosses the Missouri not far from where Lewis and Clark set out to map the unknown terrain for their boss, President Thomas Jefferson. His boss is less certain, but in traveling the blue lines on roadmaps that designate secondary roads, he is discovering an unknown America that will provide as many insights, as much new knowledge about the country, as the famous explorers of the Corps of Exploration did. Though his comparisons are often subtle, he continually suggests those who traveled back roads before him.

EXPLORING JOURNAL

Think about the region where you live or go to school. What would someone who traveled the backroads of that region, and who had not seen it before, experience? What would first impress the new traveler? Would it be a positive or negative experience? What features of the landscape would determine this reaction? Explore these ideas in a substantial entry.

—————

6

Uniontown, Demopolis. The Tombigbee River and blue highway 28. I missed the turnoff to Sucarnochee, Mississippi, and had to enter the state by way of Scooba on route 16, a road of trees and farmhouses. The farmhouses weren't the kind with large, encircling porches and steeply pitched roofs and long windows you used to see, but rather new houses indistinguishable from wet-bar, walk-out basement, Turfbuilder-Plus surburban models.

Then Philadelphia, Mississippi. Here, too, the old, sad history. The town, like others in the area, was built over the site of a Choctaw village. The Choctaw, whose land once covered most of Mississippi, earned a name from their skill in horticulture and diplomacy; they were a sensible people whose chieftains attained position through merit. In the early nineteenth century, they learned from white men and began building schools and adding livestock to their farms. Later, whites would refer to them as one of the "five civilized tribes." Nevertheless, as pressure from white settlement increased, the Choctaw had to cede to the government one piece of land (in million-acre increments) after another. Federal agents pressured tribes to sign treaties through mixed-bloods bribed with whiskey and trinkets; they promised Indians annuities, land grants, and reparations, almost none of which the Congress ever paid. To President Andrew Jackson, it made no difference that Choctaw officers like Ofahoma had fought with him against rebellious Creeks; Jackson pushed on with land-gobbling compacts. With the Treaty of Dancing Rabbit Creek, held in the woods northeast of Philadelphia, the Choctaw gave up the last of their land and reluctantly agreed to leave Mississippi forever. They walked to the arid Indian Territory where they set up their own republic modeled after the government that had just dispossessed them.

It's a sad history not because of the influx of settlers—after all, Indians had encroached upon each other for thousands of years. It's a sad history because of the shabby way the new people dealt with tribal Americans: not just the lies, but the utter unwillingness to share an enormous land.

Yet, a thousand or so Choctaw secretly stayed in Mississippi to claim land promised, although few ever saw a single acre returned. That afternoon their descendants were shopping along the square in Philadelphia, eating a hotdog at the Pow Wow drive-in, taking a few hours away from the reservation west of town. Holding to the token parcel now theirs, they could watch towns white men had built wither: Improve, Enterprise, Increase, Energy, Progress. As for what the land around the towns produced, they could watch that too.

Highway 16 passed through green fields, blue ponds, clumps of pine; it crossed the earthy Yokahockana River, a name that stands with other rivers of strong name in Mississippi: the Yazoo, Yalobusha, Little Flower, Noxubee, Homochitto, Bogue Chitto, Chickasawhay, Skuna, the Singing River.

At Ofahoma, I drove onto the Natchez Trace Parkway, a two-lane running from Natchez to near Nashville, which follows a five-hundred-mile trail first opened by buffalo and Indians. Chickasaws called it the "Path of Peace." In 1810, the Trace was the main return route for Ohio Valley traders who, rather than fight the Mississippi currents, sold their flatboats for scrap in Natchez and walked home on the Trace. The poor sometimes traveled by a method called "ride and tie": two men would buy a mule; one would ride until noon, then tie the animal to a tree and walk until his partner behind caught up on the jack that evening. By mid-century steamboats made the arduous and dangerous trek unnecessary, and the Trace disappeared in the trees.

Now new road, opening the woods again, went in among redbuds and white blossoms of dogwood, curving about under a cool evergreen cover. For miles no powerlines or billboards. Just tree, rock, water, bush, and road. The new Trace, like a river, followed natural contours and gave focus to the land; it so brought out the beauty that every road commissioner in the nation should drive the Trace to see that highway does not have to outrage landscape.

Northeast of Tougaloo, I stopped to hike a trail into a black-water swamp of tupelo and bald cypress. The sun couldn't cut through the canopy of buds and branches, and the slow water moved darkly. In the muck pollywogs were starting to squirm. It was spring here, and juices were getting up in the stalks; leaves, terribly folded in husks, had begun to let loose and open to the light; stuff was stirring in the rot, water bubbled with the froth of sperm and ova, and the whole bog lay rank and eggy, vaporous and thick with the scent of procreation. Things once squeezed close, pinched shut, things waiting to become something else, something greater, were about ready.

I had a powerful sense of life going about the business of getting on with itself. Pointed phallic sprouts pressed up out of the ooze, green vegetable heads came up from the mire to sniff for vegetation of kin. Staminate and pistillate, they rose to the thrall of the oldest rhythms. Things were growing so fast I could almost feel the heat from their generation: the slow friction of leaf against bud case, petal against petal. For some time I stood among the high mysteries of being as they consumed the decay of old life.

Then I went back to the Trace and followed dusk around the spread of Jackson highways that had broken open like aneurisms and leaked out strawberry-syrup pancakes, magic-finger motel beds, and double-cheese pizzas. Across the Pearl River and into Clinton, a hamlet that Sherman pillaged but decided not to burn. The place was shut down. Near the campus of old Mississippi College, I parked for the night and ate a tin of tuna and three soft carrots. Rejected the chopped liver. I ate only because I didn't know what else to do. I'd got uppity about multilane America and was paying the price. Secretly, I hungered for a texturized patty of genetically engineered cow.

7

A century and a half ago, the founders of Mississippi College hoped the school would become the state university. But that didn't work out, so they gave it to the Presbyterians; that didn't work out either, and the Presbyterians gave it back. The Baptists had a go at it, and the college got on in its own quiet way, eventually turning out three governors. Actually, all the changing around may have made little difference. A student told me that everyone in town was a Baptist anyway, even the Presbyterians.

I was eating breakfast in the cafeteria. A crewcut student wearing mesh step-in casuals sat down to a tall stack of pancakes. He was a methodical fellow. After a prayer running almost a minute, he pulled from his briefcase a Bible, reading stand, clips to hold the book open, a green felt-tip, a pink and yellow; next came a squeeze-bottle of liquid margarine, a bottle of Log Cabin syrup wrapped in plastic, a linen napkin, and one of those little lemony wet-wipes. The whole business looked like the old circus act where twelve men get out of a car the size of a trashcan.

A woman with a butter-almond smile sat down across from me. Her hair, fresh from the curling wand, dropped in loose coils the color of polished pecan, and her breasts, casting shadows to her waist, pressed full against a glossy dress that looked wet. A golden cross swung gently between, and high on her long throat was a small PISCES amulet. Her dark, musky scent brought to mind the swamp. We nodded and she said in soft Mississippian, "You were very interested in Jerry's pancakes."

"It was the briefcase. I thought he was going to pull out a Water-Pik and the Ark of the Covenant next."

"He's a nice boy. His parameters just aren't yours." She couldn't have surprised me more had she said floccinaucinihilipilification. "The bottom line is always parameters no matter what the input."

"Let me make a crazy guess. You're in computer programming."

"I'm in business, but my brother is a computer programmer in Jackson. He's got me interested in it. He plays with the computer after hours. Made up his Christmas cards on an IBM three-sixty-one-fifty-eight last year and did his own wedding invitations two years ago. But we're channelized different. I want to use the computer to enrich spiritual life. Maybe put prayers on a computer like that company in California that programs them. For two dollars, they run your prayer through twice a day for a week. They send up ten thousand a month."

"What if God doesn't know Fortran?"

"Come on, you! People are critical, but they don't ridicule prayer wheels or rosaries and those are just prayer machines."

"Does God get a printout?"

"Quit it! You get the printout. Suitable for framing. Quit smiling!"

"Sorry, but you said they send the prayers 'up,' and I just wondered what kind of hard copy we're dealing with here."

"You're a fuddydud! It's all just modalities. The prayer still has to come from a heart. Japanese write prayers on slips of paper and tie them to branches so the wind sort of distributes them. Same thing—people just trying to maximize the prayer function."

"You're a Pisces?"

"Would a Sagittarius wear a Pisces necklace?"

"How can you believe in astrology and wear a cross?"

"What a fuddydud! Who made the stars? Astrology's just another modality too." She took a computer card from her notebook. "I've got to get to class, but here's one more modality. In India, people pray when they eat—like each chew is a prayer. Try it sometime. Even grumpy fuddyduds like it."

She handed me the card and hurried off. Here it is, word for word:

SCRIPTURE CAKE

2 cups Proverbs 30:33	1 cup Genesis 43:11
3 1/2 cups Exodus 29:2	6 Isaiah 10:14
3 cups Jeremiah 6:20	2 tbsp I Corinthians 5:6
2 cups I Samuel 30:12	1 tbsp I Samuel 14:25
2 cups Nahum 3:12	Season with I Kings 10:10
1/2 cup Judges 4:19	Follow Leviticus 24:5

SERVE WITH LOVE . . . SALLY

8

I went to the Trace again, following it through pastures and pecan groves and tilled fields; wildflowers and clover pressed in close, and from trees, long purple drupes of wisteria hung like grape clusters; in one pond a colony of muskrats. I turned off near Learned and drove northwest to cross the Mississippi at Vicksburg. South of town, I ate a sandwich where Civil War earthworks stuck out on a bluff high above the river. From these aeries, cannoneers had lobbed shells onto Union gunboats running the river. Anything—a rock, a stick—falling from that height must have hit with a terrible impact.

The western side of the river was Louisiana, and the hills of Mississippi gave way to low and level cotton fields where humid heat waves boiled up, turning dusty tractors into shimmering distortions. The temperature climbed to eighty-six. Once, a big oak or gum grew in the middle of each of these fields, and under them, the farmer ate dinner, cooled the team, took an afternoon nap. Now, because they interfered with air-conditioned powerhouse tractors plowing the acres, few of the tall trees remained.

The traffic on U.S. 80 had gone to I-20, and the two-lane carried only farm trucks and tractors pulling big cannisters of liquid fertility. The federal highway, like most I'd driven, was much rougher than state or county highways, so we all went slowly, just trundling along in the heat.

A traveler who leaves the journey open to the road finds unforeseen things come to shape it. "The fecundity of the unexpected," Proudhon called it. The Cajun Fried Chicken stand in Monroe (accent the first syllable), where I'd stopped for gas, determined the direction of the next several days. I wasn't interested in franchise chicken, but the word *Cajun* brought up the scent of gumbo, hot boudin, and dirty rice. Monroe is a long way from Cajunland, but while the tank filled, I decided to head south for some genuine Cajun cooking.

On the other side of the pump, a man with arms the size of my thighs waited for the nozzle. He said, "You driving through or what?"

"On my way south."

"You want some meat?" It sounded aggressive, like, "Want a knuckle sandwich?"

"Pardon me?"

"You want meat? I'm flying out of Shreveport this afternoon. Can't carry the steaks with me. Just got called to Memphis. If you're cooking out, might as well take them. It's you or the garbage can."

He had a way with words.

"Get him the steaks, Roger." A boy, about ten, came around and handed me four nice flank cuts still frozen. I thanked the man.

"What'd you pay for your Ford?" the boy said.

"Three thousand in round numbers."

"How much to build the insides?"

"Couple hundred dollars."

"How about that homemade bed? Could I try it?" I opened the door, he jumped on the bunk, stretched out, and made a loud snoring noise. Dreaming of far places. His eyes popped open. "Inflation's added about twelve percent. These models run higher now too. How's the gas mileage?"

"Around twenty-five to the gallon."

"Can't be."

"Can be and is. Straight shift, no factory options except highback seats, lightweight, and I drive around fifty." That short man of a boy depressed me. Ten years old and figuring the rate of interest and depreciation instead of the cost of adventure. His father handed me a loaf of bread.

"Thanks very kindly," I said, "but I'm not much for white bread."

"Just have to leave it along the interstate for possums and niggers." He did it again.

With the steaks and white bread (would go well with chopped liver) I drove south toward the flat, wet triangle of gulf-central Louisiana that is Cajunland. The highway clattered Ghost Dancing and shook me so that my head bounced like one of those plastic dogs in car rear windows. The heat made me groggy, and I couldn't shake it, and I didn't want to stop. After a while, the road seemed a continuum of yellow-lined concrete, a Möbius strip where I moved, going neither in nor out, but around and up and down to all points of the compass, yet always rolling along on the same plane.

My eyes were nearly closed. Then a dark face staring in. My head snapped back, and I pulled the truck out of the left lane. A hitchhiker. I stopped. His skin shone like wet delta mud, and his smile glittered like a handful of new dimes. He was heading home to Coushatta after spending two days thumbing along I-20 from Birmingham, where he'd looked for work as a machinist. He'd found nothing. Usually he got long rides on freeways if he could manage one, but it was easier for a black man to get a lift on the small roads where there were more Negro drivers. Sometimes the ride included a meal and bed, but last night he'd slept in a concrete culvert. I asked where he learned his trade. "In the Army. I was a Spec Four. "

"Were the jobs filled in Birmingham?"

"They said they were. I don't know."

"Was it a racial question, do you think?"

He moved warily in his seat. "Can't always tell. It's easy to say that."

"What will you do now?"

"Go home and wait for something to open up." We rode quietly, the even land green and still. He was a shy man and appeared uncertain about what to say. I filled some silence, and then he said, "Seems things I wait for don't come along, and the ones I want to see pass on by, stop and settle in."

"I'm between jobs myself. Waiting for something to open up too."

"I hope I'm just between jobs. I went in the Army to learn a trade. Figured I'd found a good one for civvy life. Now I'm looking like my uncle. He only had one good job in his life. Good for his time anyway. Ran an elevator at the Roosevelt Hotel in New Orleans. Then they put in push-button elevators. He said he drove his old elevator a hundred thousand miles. He came back to Coushatta and did a little field work, then went hunting a better job in Dallas and got shot dead. I used to think he musta been a bum. Don't see it like that now."

The rest of the way was mostly quiet. "I'll get out here," he said at last.

"A man gave me some steaks. My cooler won't keep them in this heat. Why don't you take a couple?" I pulled out a steak and handed him the rest. "Gave me this bread too. Take it if you like."

He put the steaks in his plaid suitcase but had to carry the bread in his hand. "Can I ask you a question? Why did you give me a ride?"

"I was dozing off. Owed you for waking me up."

He shook his head. "Maybe. It'll be a good night at home. Mama loves steak."

Up the road he went, thumb out, smiling into the tinted windshields. Home is the hunter, home from the hill; home the sailor, home from the sea. And what about the Specialist Four home from Birmingham?

9

All the way to Opelousas, I thought of the machinist whose name I never learned. He had gone out and come back only to find a single change: he was older. Sometimes a man's experience is like the sweep second hand on a clock, touching each point in its circuit but always the arcs of movement repeating.

Near Ville Platte a scene of three colors: beside a Black Angus, in a green pasture, a white cattle egret waited for grubbings the cow stirred up. The improbable pair seemed to know each other well, standing close yet looking opposite directions. I don't know what the egret did before it flew into the New World; I suppose it took its long, reedy legs to shallow water and picked in the bottoms for a couple of million years, each bird repeating until the new way of life came to it.

I switched on the radio and turned the dial. Somewhere between a shill for a drive-up savings and loan and one for salvation, I found a raucous music, part bluegrass fiddle, part Texas guitar, part Highland concertina. Cajun voices sang an old, flattened French, part English, part undecipherable.

Looking for live Cajun music, I stopped in Opelousas at the Plantation Lounge. Somebody sat on every barstool; but a small man, seeing a stranger, jumped down, shook my hand, and insisted I take his seat. In the fast roll of Cajun English, he said it was the guest stool and by right belonged to me. The barmaid, a woman with coiled eyes, brought a Jax. "Is there Cajun music here tonight?" I asked.

"Jukebox is our music tonight," she snapped.

A man called Walt, with dark hair oiled and slicked back in the style of an older time, squeezed in beside me. "If you're lookin' for French music, you need to get yourself to laugh yet."

"What's that mean?"

"Means haul your butt to laugh yet. Biggest Coonass city in the world."

"Lafayette?" I made it three syllables.

"You got it, junior, but we don't say Lah-fay-et."

"Where should I go in Laughyet?"

He drew a map so detailed I could almost see chuckholes in the streets. "Called Eric's. That's one place. In Laughyet they got whatever you want: music, hooch, girls, fights, everything." He passed the bar peanuts. "By the way, junior," he asked casually, "ever had yourself a Cajun woman?"

His question silenced the bar. "Don't think I have."

"Got some advice for you then—if you find you ever need it."

It was the quietest bar I'd ever been in. I answered so softly no sound came out, and I had to repeat. "What advice?"

"Take off your belt before you climb on so you can strap your Yankee ass down because you will get taken for a ride. Up the walls and around."

Now the whole bar was staring, I guess to surmise whether my Yankee ass was worth strapping down. One rusty geezer said, "Junior ain't got no belt."

Walt looked at my suspenders and pulled one, letting it snap back. "My man," he said, "tie on with these and you will get zanged out the window like in a slingshot."

The men pounded the bar and choked on their Dixie beer. One began coughing and had to be slapped on the back. Two repeated the joke.

Walt shouted to the barmaid, "Let's get junior another Jax." To me he said, "Don't never take no offense at a Coonass. We're all fools in God's garden. Except for bettin'. Now that's serious. These boys'll bet on anything that moves or scores points and even some things that don't do neither. Charles, here, for example, will bet he can guess to within four how many spots on any Dalmatian dog. I bet on movement because I don't know dogs and not too many things score points. But everything moves—sooner or later. Even hills. Old Chicksaw taught me that."

10

If you've read Longfellow, you can't miss Cajunland once you get to the heart of it: Evangeline Downs (horses), Evangeline Speedway (autos), Evangeline Thruway (trucks), Evangeline Drive-in, and, someone had just said, the Sweet Evangeline Whorehouse.

I found my way among the Evangelines into an industrial area of Lafayette, a supply depot for bayou and offshore drilling operations. Along the streets were oil-rig outfitters where everything was sections of steel: pipes, frames, ladders, derricks, piles, cables, buoys, tanks. Crude oil opened Acadian Louisiana as nothing in the past three centuries had, and it seemed as if little could be left unfound in Cajun hamlets once quite literally backwaters.

Eric's, on the edge of the outfitters' district, was a windowless concrete-block box with a steel door and broken neon and a parking lot full of pickups, Cadillacs, and El Caminos ("cowboy Cadillacs"). But no French music.

I drank a Dixie and ate bar peanuts and asked the bartender where I could hear "chanky-chank," as Cajuns call their music. She, too, drew a map, but her knowledge gave out before she got to the destination. It's called Tee's. "It's down one of these roads, but they all look alike to me out there."

"Out there?"

"It's in the country. Follow my map and you will be within a couple miles."

When I left she said good luck. The traveler should stand warned when he gets wished luck. I followed her map until the lights of Lafayette were just a glowing sky and the land was black. I wound about, crossing three identical bridges or crossing one bridge three times. I gave up and tried to find my way back to town and couldn't do that either.

Then a red glow like a campfire. A beer sign. Hearty music rolled out the open door of a small tavern, and a scent of simmering hot peppers steamed from the stovepipe chimney. I'd found Tee's. Inside, under dim halos of yellow bug lights, an accordion (the heart of a Cajun band), a fiddle, guitar, and ting-a-ling (triangle) cranked out chanky-chank. The accordionist introduced the numbers as songs of *amour* or *joie* and the patrons cheered; but when he announced *"un chanson de marriage,"* they booed him. Many times he cried out the Cajun motto, *"Laissez les bons temps rouler!"*

While the good times rolled, I sat at the bar next to a man dying to talk. My Yankee ass and his were the only ones in the place. His name was Joe Seipel and his speech Great Lakes. I asked, "You from Wisconsin?"

"Minnesota. But I been here seven years working for P.H.I."

"What's P.H.I.?"

He put down his bottle and gave me an exaggerated, wide-eyed, open-mouthed look to indicate my shocking ignorance. "You gotta be kidding!"

"About what?"

"Petroleum Helicopters Incorporated!" He shook his head. "Jees!"

"Oh, that's right. What kind of helicoptering do you do?" I tried to talk between numbers, but he talked through it all.

"I don't fly. I'm a mechanic. But Stoney here flies out to the offshore rigs. Delivers materials, crews. You know."

The pilot, in his fifties, wore cowboy boots and a jaunty avocado jump-suit. He was applying a practiced *Bridges-at-Toko-Ri* machismo to a hugely mam-married woman who had painted on a pair of arched, red lips the likes of which the true face of womankind has never known.

Seipel said, "I was just like you when I came here—dumb as hell. But I've read about Louisiana. Learned about Coonasses from that yellow book."

"What yellow book is that?"

"That one comes out every month."

"National Geographic?"

"That's it. They had a story on Coonasses."

"Did they explain the name 'Coonass'?"

"I think they missed that."

A small, slue-footed Gallic man wearing a silky shirt with a pelican on it dragged an upturned metal washtub next to the band and climbed on. I think he'd taken out his dentures. A mop handle with baling twine tied to it projected from the tub, and he thrust the stick about in rhythm with the music, plucking out the sound of a double bass.

"That's DeePaul on the gut bucket," Seipel said. "He's not with the band."

After a couple of numbers on the tub, the small man hopped down and waltzed around the floor, quite alone, snapping his wrists, making sharp rapid clacks with four things that looked like big ivory dominoes.

"Those are the bones," Seipel said. "Sort of Cajun castanets."

When the band folded for the night, the little fellow sashayed to the lighted jukebox, drawn to it like a moth, and clacked the bones in fine syncopation, his red tongue flicking out the better to help him syncopate, his cropped orb of a head glowing darkly. Seipel hollered him over.

He showed how to hold the bones one on each side of the middle fingers, then flung out his wrist as if throwing off water and let loose a report like the crack of a bullwhip. "Try dem in you hands."

The bones were smooth like old jade. I laboriously inserted the four-inch counters between my fingers and snapped my wrist. *Cluk-cluk.* "Lousy," Seipel said. I tried again. *Cluk-cluk.* Wet sponges had more resonance. Seipel shook his head, so I handed them to him. He got them mounted, lashed out an arm, and a bone sailed across the room.

"You boys don't got it," DeePaul said, his words looping in the old Cajun way. DeePaul's name was in fact Paul Duhon. He had cut the clappers from a certain leg bone in a steer and carved them down to proper shape and a precise thickness. "You got to have da right bone, or da sound she muffle. And da steer got to be big for da good ringin' bones."

I tried again. *Cluk-cluk.* "I work at dis forty years," Duhon said, "and just now do I start gettin' it right. Look at me, gettin' ole and just now gettin' good. Dat's why only ole, ole men play da good bones."

"Where'd you learn to make them?"

"Ole color man, he work on da rayroad. He got nuttin' but he love music so he play da bones. He play dem in da ole minstrel shows. He da one day call 'Mister Bones,' and it Mister Bones hisself he show me carvin'. Now people say, 'Come play us da bones in Shrevepoat.' But da bones just for fun."

"DeePaul flies kites," Seipel said. "Wants in the *Guinness Book.*"

"My kites day fly for time in da air, not how high. Someday I want people to be rememberin' Duhon. I want 'Duhon' wrote down."

"I can play the musical saw," Seipel said and called to the barmaid, "Got a saw here?" She pushed him a saltshaker. "What's this?"

"That's the salt you're yellin' for." Seipel and I laughed, holding on to the bar. Duhon went home. Everybody went home. The barmaid watched us wearily. "Okay," she said, "come on back for some hot stuff."

"Is this where we find out why they call themselves 'Coonasses'?" I said, and we laughed again, holding on to each other.

"All right, boys. Settle down." She led us not to a bedroom but to a large concrete-floor kitchen with an old picnic table under a yellow fluorescent tube. We sat and a young Cajun named Michael passed a long loaf of French bread. The woman put two bowls on the oil cloth and ladled up gumbo. Now, I've eaten my share of gumbo, but never had I tasted anything like that gumbo: the oysters were fresh and fat, the shrimp succulent, the spiced sausage meaty, okra sweet, rice soft, and the roux—the essence—the roux was right. We could almost stand our spoons on end in it.

The roots of Cajun cookery come from Brittany and bear no resemblance to Parisian cuisine and not even much to the Creole cooking of New Orleans. Those are *haute cuisines* of the city, and Cajun food belongs to the country where things got mixed up over the generations. No one even knows the source of the word *gumbo*. Some say it derives from an African word for okra, *chinggombo,* while others believe it a corruption of a Choctaw word for sassafras, *kombo,* the key seasoning.

The woman disappeared, so we ate gumbo and dipped bread and no one talked. A gray cat hopped on the bench between Seipel and me to watch each bite of both bowls we ate. Across the room, a fat, buffy mouse moved over the stove top and browsed for drippings from the big pot. The cat eyed it every so often but made no move away from our bowls. Seipel said, "I've enjoyed the hell out of tonight," and he laid out a small shrimp for the cat. Nothing more got spoken. We all went at the gumbo, each of us, Minnesotan, Cajun, cat, mouse, Missourian.

RESPONDING JOURNAL

The subtitle of *Blue Highways* is *A Journey into America*. If we were to emphasize the *into* in the subtitle, what would you say you have learned about the interior of America? Does anything that Moon tells you surprise you?

QUESTIONS FOR CRITICAL THINKING

STRATEGY

1. Explain the benefits a book like Moon's has. What does he derive from following a road map? From following the secondary roads on that map?
2. Does the Choctaw tribe's situation parallel that of any other group Moon mentions? Explain the reasons for your response.

ISSUES

1. What justification does Moon have to ask the questions about racial relations that he puts to the people with whom he comes in contact? Should such questions be a part of the experience of traveling along America's back roads?
2. Do you think Moon would find a similar situation if he were to repeat in 1995 his earlier visit? Would he find similar changes if he were to repeat his entire trip throughout the United States?

COLLABORATIVE WRITING ACTIVITY

Think of a trip that you took across the landscape which you feel taught you something. Did you learn about relations between humans and the natural world? Do some details from the natural world stay with you when you think back to the trip and the lessons you learned from it? Write about your experiences in the form of an essay, and read it to your group. After your group has heard the other essays, plan together a comparison and contrast essay that you could write as a group to show the similarities and differences in the sort of learning that took place in the group members' examples.

"Blue Bird's Offering" from *Waterlily*
Ella Cara Deloria

Long more widely known for her anthropological work than for her fiction, Ella Cara Deloria worked extensively with the noted linguist-ethnographer Franz Boas during much of her career. She published linguistic studies of her native Dakota language and stories in the 1930s and 1940s, all of which were influential in determining the growth of Native American studies. At the same time, she was writing the manuscript that was to become Waterlily, *though this would not see publication until 1988.*

In this classic story of life in pre-immigrant America, Deloria tells the story of Blue Bird and her husband Star Elk. In terms that seem more modern and soap-operaish than traditional Dakota, Deloria examines the difficulty of a strong woman attempting to maintain a relationship with a weak and vain man. The following excerpt comprises the second and third chapters of the novel.

EXPLORING JOURNAL

Think about the nature of romantic involvement in Native American communities. What do you know about traditional customs and practices in romantic relationships? What might be different from contemporary practices in such situations? What are the sources of your knowledge? Try to move beyond stereotypes provided by the media as you develop your response.

Blue Bird had never been entirely happy either in her marriage or in her life in a camp circle that was not her own. It was not that the people were unkind; quite the contrary. But she could not feel satisfied there. She never ceased to yearn for her own people. It was almost four years now she and her grandmother had been staying there. Sometimes she wished they had risked everything and struck out alone in search of their own camp circle. Even if they had perished on the way, it would have been worth trying.

Her childhood among her own many loving kinsmen was a happy one, but that time was like a dream vanished. Try as she would, she could never recapture the feel of that carefree life, so cruelly ended in a day. Tonight for the first time, with her infant, Waterlily, asleep beside her, she was again completely happy. This was a different kind of happiness, satisfying, if subdued. But it was good. She lay idly reminiscing in the dark tipi of her cousin, who was out somewhere. With singular detachment she was able for the first time to recall in detail the events of that tragic day that had robbed her of her family. Tonight it seemed remote, like something that had happened to someone else long ago in a far-off place.

She had been fourteen years old at the time. Her father had decided to leave the camp circle for a few days of deer hunting because their supply of meat was dwindling fast and no buffalo had been sighted in a long time. He took his wife and his mother and his three children, Blue Bird and her brothers, ten and six years of age.

The family made their temporary camp near a wood and immediately went out to cut the poles and boughs needed to set up their working arrangements, drying racks for the meat, and an arbor of leaves. Soon the father was bringing in deer and other game at such a rate that the two women had to work steadily from dawn to dusk caring for the meat, for they were a frugal family and saved every bit that could be used.

But busy as they were, the old grandmother took time out one afternoon for a walk under the tall cottonwoods nearby. The next day she announced, as they sat eating, "There is a large cache of earth beans over yonder, where many little paths under the matted grass come together from all directions. The field mice, too, have been busy preparing for winter."

"Well, that is indeed good news," her son's wife said. "Now maybe we can take back earth beans as well as meat. I hope you can remember where the cache is."

"Oh, yes. And anyway I set up a stick to mark the spot," the old woman assured her, adding, "I could easily have thumped in the dirt roof right then with my club and brought the beans home. But of course I waited."

The younger boy, who dearly relished them, pouted, "Oh, Grandmother, you should have! Then I could be having some beans now."

Blue Bird cut in, "You can't do that, silly! Don't you know that you have to leave a return gift for the mice when you take away their food? They have to have something to live on, too."

That was no way for a girl to speak to a brother, and few adults would fail to correct such a slip. Blue Bird's mother said gently but firmly, "Daughter, one does not call one's brother 'silly.'" And then she turned to her mother-in-law, saying, "I have some dried corn in that rawhide box. Will it answer?"

The old women was delighted. "Of course. It will be just the thing. Too good, I should say. For who are they, to have green corn dried for them? They should be too happy with it to think of bewitching me—I hope." She said this laughingly, but it was plain she half feared the common belief about the powers of resentful mice.

Blue Bird went with her grandmother to open the cache and they found an abundance of beans, unusually large and meaty. They would cook up rich and sweet, the old woman said. She found more caches and went to work at once, happily drawing out handfuls of the black, earth-caked store and piling it on her blanket, spread out to receive it. For each handful she religiously returned a handful of green corn that had been parboiled and then sundried, a treat for the mice indeed. When she was through, she and Blue Bird grasped the corners of the blanket and tossed the beans high to winnow out the dirt. The loose, fine dust was carried off on the breeze. The clumps settled to the bottom and could easily be removed later.

The old woman gathered the blanket to form a bag holding the beans and tied a thong around it tightly, bending over. When she straightened up, she groaned a bit over a kink in her back, as was her habit. Then, shielding her eyes with one hand, she studied the sun. "Come, child, we must be going now. It is getting on," she said. But she could not resist hurriedly picking up a few dried sticks. "We can always do with more firewood." She made a bundle of them with the pack strap she always wore for a belt. From a lifetime of practice she flung the bundle expertly onto her back. At last she and Blue Bird started home, carrying the beans suspended between them.

Out of the woods and into the clear they came, but when they looked toward their camp, it was not there. Everything was in ruins. How could it have happened so quickly, so quietly? It was unbelievable that in the short time since Blue Bird and her grandmother had left it, an enemy war party had raided their camp. Yet that was the case.

The destruction could hardly have been more thorough. The skin tent was slashed beyond mending, the poles were all broken or askew, and the drying racks filled with jerked meat were completely dismantled. The two boys lay dead, flat on their faces, not far from the tipi. They had been shot while running away, impaled by arrows in their backs. The parents had vanished without a trace. Whether they too had been killed or were taken captive the two survivors were never to know.

They hurriedly covered the bodies of the boy, but this was no time to mourn. Shocked as they were, they could not entertain both grief and fear at the same time. One emotion must wait, and fear took precedence. Lest the marauders return and find them, the grandmother decided they must hasten into hiding. She and Blue Bird did so, lying concealed under bushes and behind rocks by day and traveling by night. Under an overcast sky they lost their way and wandered blindly, with no idea of their destination.

The second day at dawn they happened upon a large camp circle, though it was not theirs. But the people were their kind and spoke their dialect, so they knew they had found refuge. On learning of their plight and their recent tragedy, the magistrates sent the crier out from the council tipi to announce their arrival and rally the people to their aid.

The response was quick. Someone gave the newcomers a tipi to live in, while public-spirited collectors carried around the circle a great bull hide into which contributions were placed. Women came running out of their tipis to add their gifts, such items as clothing and food. And thus all in a day Blue Bird and her grandmother were equipped to start life anew. From time to time the wives and mothers of hunters brought them meat, and at the next several feasts they were invited as special guests. In such ways did the people help them establish themselves in their new camp circle. And when at last, in the privacy of their newly acquired home, the two could give way to their grief and wail at leisure, their women neighbors came in to wail with them in sympathy, as was the custom. And the members of the camp circle adopted the newcomers as relatives.

It was true enough that here Blue Bird and her grandmother fell into the category of the humbler folk of the community. Without any male relatives to give them backing, they made no pretensions to importance in the social life of the camp circle. Nor were they expected to. Nevertheless, their lowly station in no way degraded them in the popular esteem. The Tetons did not have to put on airs in that way. If one's circumstances did not allow it, one did not need to give feasts or take part in the conspicuous give-away ceremonies. The grandmother and grandchild, accepting their situation, were content to remain quietly in the background. Since they could not return to the camp circle where they did have position, and with it certain social obligations, it was enough that they had fallen in with their own kind of people and that they had been taken in as relatives in social kinship.

As the seasons passed, the young men of the village could not fail to see that Blue Bird was maturing and that her growing beauty was remarkable. But this fact troubled her grandmother greatly, and she felt the need of someone with whom to share the responsibility for the girl until she should be safely married. Knowing well how some reckless young men played at courtship, she feared for Blue Bird. She must be warned at once that many a girl had come to ruin by taking their smooth wooing seriously, and the grandmother was the only one to tell her. "I shall have a talk with her tomorrow." But each day she put it off, dreading the ordeal. "I am too old for this; would that her mother were here," she said to herself. "Or perhaps I should simply give the girl away in marriage now, to some kind and able householder, to be a co-wife. Then she can be honorably married before any trouble can befall her. Yes, that would be best."

But just whom to give her to was a puzzle. And what wife would want her? Being co-wife was not necessarily bad, provided the man was kind. She had been a co-wife herself. But then, she was the wife's sister and therefore was well received. In fact, as she remembered now, it was that elder sister who had offered to take her into the family. Ah, but Blue Bird had no sister in this camp circle. A head wife might resent her. That too must be considered. Slowly and timidly the old woman turned the problem over in her mind many times. But she had not yet acted when Blue Bird said to her one day, "Grandmother, one of the young men at the courting place has been urging me to marry him. His name is Star Elk."

The old woman shook her head emphatically, "No! No! Not that one. It would be good for you to marry, grandchild. We are so alone and helpless without a man to provide for our home. But not that one. Only last night the women around the campfire were talking about him. 'He is no hunter,' they said. 'He takes no interest in anything. Always he has been headstrong and unfriendly, even as a boy,' they said. That is not the kind of man for you, grandchild."

"But I have told him I would marry him, Grandmother."

The silence that followed was ominous. When the old woman again found her voice, she said, "Ah, if only you had told me he was courting you so I could have warned you, grandchild. Since you have promised already, there is nothing I can do. Once she gives it, an honorable Dakota woman does not break her word to a man. Those who make false promises are ever after derided. To give your word is to give yourself." With that she stumbled out of the little tipi and began to wail in a quavering voice the following lament: "Ah, my son! Ah, my daughter-in-law! You have left me alone to struggle on. What can I do, frail and full of years as I am?" Far into the night she wailed.

Blue Bird's marriage was inevitable now. But even after resigning herself to it, the grandmother went about with a heavy heart. "If only I had had someone to help me arrange a suitable marriage for her," she muttered to herself from time to time. That the girl might run off with Star Elk was a dreaded possibility, even while there was a feeble hope that perhaps the young man had been misrepresented as altogether undesirable. Perhaps he was not that bad after all, and perhaps he would soon do the honorable thing—marry the girl openly, with tribal approval.

The most glamorous kind of marriage was by purchase. A woman who married in that way was much respected, for it meant that she had kept herself so unattainable that the man, who wanted her at all costs, thought nothing of giving horses for her, even at the risk of her rejecting him publicly. "I do not aspire to that for my poor orphaned grandchild," the grandmother said. "All I ask is a valid marriage, and then I should die happy." He might come to live with them or take her to his people openly. Whichever way, it should be planned and aboveboard, and then Blue Bird would be respected.

But Star Elk lived up to his reputation. He lured the girl away, the very thing her grandmother had feared. What a gamble that was! A Teton girl who accepted marriage on such shabby terms took the supreme risk with her honor. How did she know that the man was not just trifling with her? Too often an elopement ended disastrously for the girl, while the man always went free. Momentarily the old woman looked for the girl's return, after the man tired of her. She would of course receive her back; was she not her grandchild? But the disgrace would be lasting.

As it happened, though, Blue Bird was lucky. Star Elk did not dally on the way but took her straight home to his people. And for her grandmother that was the one mitigating fact. Soon the new relatives-in-law came for her and placed her tipi in among their family group that they might care for her. Thus, even though Blue Bird had married in the least honorable way, the material condition of her grandmother was bettered, and that was something.

Because Star Elk had taken Blue Bird home to his people, the marriage was accepted as tolerably valid, and in due time the foolish step was forgiven the girl. The usual sharp censure of the eloping woman was toned down considerably by the circumstances. Condoning it, women began saying, "Well, what could you expect since the girl is very young and pretty and lacks a mother to guide her? What could a tottering grandmother do, anyway? It is good that the girl did not get into real trouble and bear a fatherless child." Thus Blue Bird, by not being "discarded in the wilds" (abandoned far from the camp circle) or bearing a "fatherless child," came in time to be counted among the blameless women of the camp circle. By the slimmest of margins—but she was in.

Even so, things were far from ideal. Star Elk was lazy and petulant and given to jealous fits. It was his relatives who received her kindly and treated her well, partly for their own reputation as correct in-laws, and partly to compensate for his failures. This made up in a measure for the poor bargain Blue Bird had made. But, kind as they were, it might have been better if they had been remiss in some details, or even outright hostile to her, if only her husband had been more satisfactory. The marriage had been unfortunate from the beginning.

The tipis of skin were opaque at night. Unless the fire was flaming, occupants must feel their way about. The cousin entered the tipi at a late hour and moved with caution toward her place. Perhaps Blue Bird was asleep. Nevertheless she spoke to her softly, "Cousin, are you resting well? We shall be by ourselves. I have sent the children and their father to sleep in his mother's tipi. Wake me at once if you need anything."

"I am quite all right, cousin, and I think I can sleep now. I was just thinking of the past; that is all."

The two women said nothing further. There was in the language no formality equivalent to "Good night." One quite well indicated one's goodwill and good wishes by tonal quality. After a little, Blue Bird reached out drowsily and touched her baby, smiling in the dark as she did so. "This is all I want," she murmured. "Let him do his worst!" And presently she slept. How could she guess to what lengths he would go to spite her for staying away?

Star Elk had grown more and more ill-tempered each day of his married life. The truth was that he was tormented with jealousy over his wife. He continually imagined that other men looked upon her with desire, and accused her of encouraging them furtively. She used to enjoy looking on at the celebrations and dances, but after a time he even forbade her to do that. And when she was with child, he once declared in a rage that it was not his. That was pure nonsense, as was nearly everything he said of or to his wife.

And now that the baby was born, he behaved still more outrageously. When he learned that Blue Bird was staying away for a while, he pouted and refused to see her and the baby. It was reported that all day and late into the night he lay out in the tall grass, among the horses that were picketed behind the cousin's tipi, and watched to see who went to call on Blue Bird. When this was rumored around, people laughed at him. "How shameful, to pout like a woman," the men said. "He has always been a queer fellow . . . without close friends . . . even as a boy he had none . . ." "His kind do not take to fasting or warfare—but if at least he were a tolerable hunter."

But neither the disapproval of the older men nor the ridicule of his contemporaries spurred Star Elk to be more dutiful as a husband. If anything, the effect was to make him more and more active in finding ways to hurt his wife. He reached the limit when he decided to "throw her away publicly."

During a great victory dance, attended by many visitors from neighboring camp circles, he waited until the crowd was biggest, and then, at an intermission, he forced his way in, snatched one of the highly decorated sticks used by the drummers, and stood motionless, holding it high. That was the way to gain an audience's full attention when one wanted to take part in the ceremonial give-away. Only persons of social standing were qualified to do this, because of their past record of hospitality, generosity, war achievements, or the like. But here was Star Elk, who had no such record. The crowd waited in silence. What was this fellow going to do?

Arrogantly he held the stick high for one dramatic moment and then cried out, "This is that woman! Whoever needs a woman to fetch his fuel and water can have her!" He flung the heavy drumstick into the crowd. Fearful of being struck, the people pushed back in waves. Nobody scrambled for the stick. He knew nobody would; all he wanted was to hurt and shame Blue Bird—and that he did.

It was a foolish and uncalled-for act but wholly characteristic. Instead of enhancing the man, as it might if he had cried, "This is a horse for the needy!" what he said only lowered his already low standing. Moreover, Star Elk had insulted a victory dance. "Throwing away a wife" was a custom, to be sure, but this was not the place. If it must be done at all, it should be at some social dance where the mood was properly light and reckless. Even then it was a custom shunned by men of standing, who considered it beneath them to air their emotions publicly. The way to leave an unfaithful wife was to send her away or to walk out of her life without so much as a backward glance. Only vain and weak men gave vent to their temper in public as Star Elk had done. It was also wrong because Blue Bird had not been unfaithful, and this was generally known. Star Elk not only succeeded in losing a good wife and making a fool of himself; he earned such public disfavor that he could not remain in the camp circle. He left immediately, his destination unknown.

Naturally, Blue Bird was hurt by that undeserved public insult. But she was not nearly so crushed as Star Elk had intended her to be. The fact was that at the time of the festivities in the center of the camp circle she sat in her cousin's tipi, sick at heart over something infinitely more vital to her even than her honor, for her baby was dying.

Do what she would, Blue Bird could see that the little Waterlily was growing steadily weaker. All that day she had lain motionless and refused to eat. Nor did any of the medicinal roots the cousin brought from neighbors help her. When a mother from the opposite side of the circle heard of the symptoms, she came hurrying across with a powdery substance obtained in the badlands and known as "earth-smoke," saying, "Try this next. It always cured my children—a pinch of it in a little water." But Waterlily could not swallow it.

All hope was dwindling fast. In a few hours the child would die unless something was done. Blue Bird must pray. Inexperienced in such holy matters, she nevertheless determined to make some kind of appeal for divine aid. But how to proceed?

Mechanically she thrust her hand into the flat rawhide bag which was her purse and which contained her personal effects and a small otterskin very evenly painted a brilliant red. She had hurriedly salvaged this one object from her father's belongings that fatal day and had prized it ever since. She was sure it must be potent with supernatural power, although her father had never said so. She only knew that he had venerated it above all else. Perhaps, possibly, it would save her baby.

Then she took some smoking mixture of tobacco and red willow bark and tied a bit of it into squares of deerskin, making ten tiny bundles no bigger than her thumb. She knew she must make some sacrificial offerings. Fumbling in her haste, she muttered to herself, "Is that right? Alas, what do I know about it? Those who know tell of the Something Holy—*Taku Wakan*—that has supreme power, but I never understood. It is so remote. What right have I?" All the while she worked in a desperate race with death.

Throughout the Great Plains and the wooded country near the Rockies, wherever the people moved, she had seen here a rock or there a tree with red paint on it, and sometimes a once beautiful blanket or other gift rotting there. She knew what that meant. Everyone knew. Those rocks and those trees had been set apart and consecrated; they were individual altars where people had prayed in times of stress. She was going to make her own altar now.

Taking up the baby and clutching in her free hand the bag with its special contents, Blue Bird left the tipi and walked away unnoticed. On and on she walked, until the noise of the festivities died away and the camp circle was no longer in sight. She stopped in utter stillness beside a great rock in the midst of an empty plain. She studied the rock and saw that though it was well embedded in the earth, the exposed part was nearly as tall as she. It sat apart from any other object as though already reserved for her. It sat on virgin ground. Surely no human had ever stood where she now stood.

She carefully laid the all but lifeless infant on her wrap spread on the ground and set immediately to work, covering one side of the rock with red ocher face paint. Then she carefully spread her father's hallowed otterskin on the top of the rock, like a covering for its head. Next she planted a stick upright in front of it. (The painted side had become the rock's face. She had personalized it.) To the top of the stick, which stood waist high, she tied the ten little tobacco bundles. They were on short strings of uneven lengths and dangled, now clustering and now separating, in the faint breeze that came and went. In descriptions of sacrificial altars where men fasted and prayed in some lonely spot, the essential property was unvarying: one hundred bundles of tobacco. But ten was all she could manage; it would have to do.

The hot summer sun beat down in all its fury. The earth danced in continuous ripples around the rim where it met the sky. Once the baby whimpered weakly. Everything was as ready as Blue Bird could make it. Now for the prayer, which was of utmost importance. "Prayer should be audibly released into the infinite" she had heard somewhere. She began speaking to the Great Spirit (*Wakan Tanka*) in the rock. Aloud but haltingly, fearful lest she not pray correctly, she said: "O, Grandfather, hear me! Since the very beginning you have been here. Before there were any men you were here. And it is certain that long after we are all gone you will remain. Hear me, Grandfather, and pity me. I want my baby to live."

Right or wrong, that was her prayer. Overwhelmed by her daring, she stood motionless, waiting—for what, she did not know. Presently someone said in her ear quite clearly, *"Hao!"* It was the Dakota word of approval and consent.

Quickly Blue Bird covered her face with both hands and bowed her head for whatever was to follow. A man must have stolen up from behind while she was praying, maybe an enemy, ready to kill her. So be it. Inert and without clear thought she waited a long time, but nothing happened. Very slowly she raised her head and uncovered her face to look about her—behind her and then farther off and finally to the whole round horizon. In that vast emptiness she stood alone.

"I have prayed aright and my prayer is heard! My baby, my baby will live!" A strange lightness, an unearthly joy, seized her, and a new boldness born of confidence swept away all her girlhood diffidence. Holding her hands to her mouth like a trumpet, she first called out, then shouted, and finally screamed her enraptured thanks: "Grandfather, you have made me thankful!" Screaming this over and over until her throat ached, she whirled about, throwing her voice in all directions in a frantic aim to reach the whole of space with her thanks.

But that sort of elation could not last. Soon enough she was back to reality and began to contemplate her unhappy lot. Through no fault of her own she had just been cruelly shamed in public. It was so merciless, so unfair. Feeling sorry for herself, she wailed aloud. In the customary way she addressed all her dead relatives in turn, her parents, brothers, aunts, and uncles, ending with the forlorn question no one can answer. "Where have you gone? Where have you gone?"

It was always good to let sorrow out and bad to hold it in. She felt much better when she had had her fill of unrestrained weeping. She dried her eyes, fitting the base of her palm into her eye sockets as all women did. She picked up her child, but its limp body no longer distressed her as she departed, leaving the rare otterskin to be the Great Spirit's forever. To no one less would she ever have relinquished what her father once venerated.

Everyone was at the dancing when she reentered her cousin's tipi. She was glad it was vacant, that she would not have to account for herself. What she had been through was not for common telling. Once more she offered the earth-smoke and Waterlily was able to swallow it this time. Gradually the infant seemed not to be in such pain as before. She fell asleep and after a while was recovered. Blue Bird was not surprised. She had prayed for that and had her answer already.

Elated about her child, she heard almost without emotion that visitors from another camp circle had recognized her old grandmother as one of the family who had never come back, and would take the news of her and her grand-daughter back home. "Your grandson Black Eagle has long mourned for you," the grandmother was told. "As soon as we get home he will be coming for you." And so he did, and at last Blue Bird, her child, and her grandmother were back where they belonged, where their many relatives welcomed them with tears of grief for the dead and of joy for the living.

—➤◦◦◄—

Responding Journal

What sorts of contemporary customs might parallel "throwing away a wife"? Under what circumstances might such an extreme measure be warranted? Would such an action ever be condoned by the social group at large? Respond to this action by Star Elk in a journal entry.

Questions for Critical Thinking

Strategy

1. Suspense has long been valued as a component of effective story-telling. Explain how Deloria uses suspense to create an absorbing narrative here.
2. Detail is another key element of effective narration. Point out those places where the level of detail moves the storyline effectively. Can you find two or three points where the detail makes the story noticeably stronger?

Issues

1. Think about the sacrifices Blue Bird sees in the woods and which she assumes other people in desperate situations have made in hopes of having their prayers answered. What parallels do such actions have in our contemporary society?
2. As Blue Bird thinks of the way in which she has been shamed, she calls on her deceased relatives and asks, "Where have you gone?" Relate this experience to the concepts you have been considering in relation to "then and now," "here and there." What conclusions can you draw?

Collaborative Writing Activity

Consider how the human relationships in this story relate to the relationships between humans and the land. Even though Blue Bird mentions little about the natural world, her story transpires on the land the Indians regarded as holy and intimately related to their daily lives. Based on what you already know, speculate on how Blue Bird's experiences can be related to more general relations between humans and the natural landscape. How would you describe those elements which she compares and contrasts most effectively?

"Bluegrass"
Jean Nordhaus

A poet, Jean Nordhaus has composed verse on a variety of subjects. The range of periodicals that have published her poetry indicates the different audiences to which she appeals: Poetry, Ploughshares, The Gettysburg Review, American Poetry Review. *After completing her undergraduate degree at Barnard, she earned a Ph.D. in German at Yale. In 1982* A Language of Hands, *her first book of poetry, was published. Her other collections are* A Bracelet of Lies *and* My Life in Hiding; *the latter was published in the* Quarterly Review of Literature *in 1991.*

"Bluegrass" appeared in the collection Poems for a Small Planet. *Like many of the other poems in that book, "Bluegrass" suggests the importance of outdoor activities keenly observed. Nordhaus displays her characteristic gifts of close examination and sharp analysis of important detail in this account of time spent on hillsides. Whether she describes New England or Europe, Nordhaus' poetry compels us to see the significance of everyday life on our small planet.*

EXPLORING JOURNAL

What do you know about bluegrass music? What possible connection could it have to the natural world? Explore your ideas in a freewriting entry.

We drive to water
Sunday afternoons through second growth,
rivers of bluegrass tumbling from the speaker.
Trees thwang past like banjo strings,
the crickets frail.

Climbing, with a camera,
like carrying a child or trying out
a new, vulnerable limb,
we relearn the perils of walking,
cautious over rock.

The trail threads downstream
like a melody, gropes for water,
misses, runs ahead
down blind alleys of rock
toward a promise of green,

plunges back into the woods and
climbs to reach another outcrop.
Along the bank, the juts of rock
lean out and point upstream like cannon
single-sighted, while imagination

edging toward the rim
creeps forward hand by hand, then falters
where the heart drops away like a cliff
to a rope of silt-green river
twisting in the gorge.

Hiking home through spangled woods,
we pass young couples starting out with ropes.
They will lower themselves like grasshoppers
just for sport over the sheerest cliffs,
the ones we couldn't contemplate,

run lightly up and down the strings.

RESPONDING JOURNAL

What other sorts of connections might you make between kinds of music and phenomena in the natural world? Explore your ideas in an entry; use whatever suggestions you see in Nordhaus' poem.

QUESTIONS FOR CRITICAL THINKING

STRATEGY

1. Explain the images in the opening stanza. What is the situation the poet describes? To what other experiences in the outdoors might you compare Nordhaus' description?
2. How appropriate do you find the poem's final image? On what principles does its strength depend?

ISSUES

1. Explain the point of view the speaker in the poem takes. Does the age of the speaker seem to determine any of the feeling in the poem? Justify your response.
2. "The trail threads downstream / like a melody," says Nordhaus. What other images can you suggest to develop the idea the poet conveys in the poem?

COLLABORATIVE WRITING ACTIVITY

Compose a journal entry on the most appropriate music for an outing such as a hike or a climb. What points would you single out as most appropriate for the music you have chosen? Think of the concepts of comparison and contrast. Bring your work to class and, if possible, bring a tape of the music you would like to use to accompany your writing. After you share your work with your group, decide together how you would like to proceed with your projects; perhaps you would like to present a symphony of writing and music.

SUGGESTIONS FOR FORMAL WRITING ASSIGNMENTS

1. Think about the ways in which many Native Americans describe the natural world. Do they use means different from the ones you would use? What characterizes these differences? What could account for such differences? Write an essay of at least 500 words in which you compare and contrast your ideas of the natural world with those you have found in your reading of texts by Native Americans.

2. Write a "then and now" essay describing what you have learned about the experiences of Native Americans. For the "then," use the accounts of their lives before they were condemned to reservations or while their family units were still intact. For "now," use either contemporary ideas or situations that occurred after significant interaction with Whites.

3. Compose an essay in which you compare and contrast joggers and walkers. Do they share equally in the enjoyment of the natural world? Do not forget about people like "mall walkers" who do their walking (or running) in specially designated areas.

4. Write a substantial essay in which you compare two different people's attitudes, clothing, and equipment in some sort of outdoors walking activity, from city shopping to mountain climbing. Think about the treatments you have seen in this chapter's readings as you do your planning.

Chapter 10

COMPARATIVE STRATEGIES II: FIGURATIVE EXTENSIONS

Jack Delano, *Puxnawatsney, PA* 1940

1. How does the spectator's elevated position cause you to respond to this photograph?

2. Explain the horizontal lines in this photo. How do they help to hold the different areas of the photograph in unity?

3. Compare and contrast in detail the differences in point of view and medium (painting vs. photograph) you find between Delano's photo and either or both of the paintings by Colman and Whittredge.

Chapter 10

COMPARATIVE STRATEGIES II: FIGURATIVE EXTENSIONS

The ways in which we conceive our world determine not only our relationship to it, but to other people as well. Think of the difference in meaning between *home* and *environment:* both mean the physical space that surrounds us, but the connotations of the two are quite different. Why does the first carry such a weight of meaning, yet the second seem so clinical—perhaps even detached— when they both signify the situation of a person or organism's immediate surroundings?

Figurative language encompasses numerous uses of language: whenever we create meaning through terms other than saying directly what we mean, we are using figurative language. We also use figurative language when we describe something with the aid of something else. For example, when we use a figure of speech like "higher than the tallest mountain," we attempt to explain a concept of great height by calling up an image of the tallest mountain the listener can imagine; in this way, the sense of the greatest height the listener knows adds to the meaning the speaker attempts to evoke.

Figurative language also works in effective writing in more sustained ways; when Rachel Carson writes about the nature of the ocean floor, she uses the image of the mountain to explain what the bottom of the sea is like. In other words, she employs more familiar information in order better to explain far less familiar information. Such writing is often extremely effective because it can cause us to look at ideas in fresh and exciting new ways.

Such a situation occurs in the book *Lame Deer Seeker of Visions* when John Fire (Lame Deer) describes the concepts held in a simple pot of soup cooking on the stove:

> What do you see here, my friend? Just an ordinary old cooking pot, black with soot and full of dents.
>
> It is standing on the fire on top of that old wood stove, and the water bubbles and moves the lid as the white steam rises to the ceiling. Inside the pot is boiling water, chunks of meat with bone and fat, plenty of potatoes.

It doesn't seem to have a message, that old pot, and I guess you don't give it a thought. Except the soup smells good and reminds you that you are hungry.

Lame Deer then goes on to tell his readers that even such an everyday item has special significance for Indians, and everyone should look more closely:

But I'm an Indian. I think about ordinary, common things like this pot. The bubbling water comes from the rain cloud. It represents the sky. The fire comes from the sun which warms us all—men, animals, trees. The meat stands for the four-legged creatures, our animal brothers, who gave of themselves so that we should live. The steam is living breath. It was water; now it goes up to the sky, becomes a cloud again. These things are sacred.

In order to explain how the soup carries special meaning for him, Lame Deer indicates as well how even the most common of items—our daily food—have significance that we must train ourselves to recognize. He explains how his people accomplish this:

We Sioux spend a lot of time thinking about everyday things, which in our mind are mixed up with the spiritual. We see in the world around us many symbols that teach us the meaning of life.

Lame Deer explains how the Indian eye, accustomed to looking at the natural world slowly and with insight, differs from the point of view of White Americans:

We Indians live in a world of symbols and images where the spiritual and the commonplace are one. To you symbols are just words, spoken or written in a book. To us they are part of nature, part of ourselves—the earth, the sun, the wind and the rain, stones, trees, animals, even little insects like ants and grasshoppers. We try to understand them not with the head but with the heart, and we need no more than a hint to give us the meaning.

When we reconfigure our world, we gain insights. When we adjust our perspective, we learn things we had not known previously. All exploration is an adjustment of perspective, a filling-in of what was suspected but not known for certain. Looking through the eyes of another person, or "walking in another person's shoes," helps us to understand multiple perspectives. Figurative language works in the same way: different figures of speech or thought cause us to rethink information in new ways, to adjust our way of looking at the world. If we could see our common nourishment in the same way that Lame Deer does, perhaps we could better understand how we affect our identities and our bodies by the way we put food into ourselves.

Think of other rituals involved with food: many people say grace or ask a blessing before eating, and in the course of such an activity the speakers give thanks for what they are about to eat. They often attribute the food to the grace or goodness of another power, much as Lame Deer thinks of his "animal brothers" when he smells the meat cooking in the stew. You may also consider what happens when we are further removed from the sources of our food, when we buy fast food or when we open a can; we are removed from not only the securing of the food, but also of the preparation process. The styrofoam containers that take so long to decompose provide only one example of what comes between us and our food, and we know our landfills cannot continue to absorb our unending waste. What would Lame Deer say about such eating habits? Even if we are not vegetarians, we can see that Lame Deer's symbolic sense of meat differs greatly from the bargain burgers hawked ceaselessly in our world.

The adjustment of focus, then, brought about by more careful thought patterns comes about more quickly when we attune ourselves to the deeper meanings behind appearances. Steinbeck's *Log from the Sea of Cortez* shows him looking beyond the scientist's position of merely gathering data to seeing how profoundly scientific knowledge can change the world when it is coupled with a strong emotional sense of the interconnections among all living things on the planet.

Many of the emotional effects created in the narratives that follow in this section derive their strength from the unique perspective they bring to their subject matter. When Steinbeck in *The Grapes of Wrath* tries to describe the feelings of sharecroppers at seeing the mechanized hordes sweeping over the landscapes, he pictures them as motorized locusts, goggle-eyed and oblivious to their former neighbors, crawling across the fields and destroying what formerly had been homes.

Other examples of the power of figurative language include providing the means to sway one side or the other in a debate. For example, James Watt used his position as Secretary of the Interior during the Reagan administration to portray environmentalists as radicals whose ideas ran counter to the thinking of middle-class America. But Watt's effort at mobilizing support to extend timbering and mining activities in the American west backfired when it turned out that the American public had moved further in the direction of environmentalism than he thought, because far more Americans than he realized labeled themselves as environmentalists! Successful arguing requires careful consideration of the intended audience and how members of that audience think of themselves.

In *Wilderness and the American Mind,* his landmark study of the environmental movement and the ideas associated with it, Roderick Nash explores the concept of irony in relation to wilderness. He points out that from Thoreau to Muir to legendary Sierra Club Director David Brower, people have worked mightily to preserve the wilderness area from development. But they have, ironically, succeeded too well. People are visiting the better known natural areas such as Yosemite and Grand Canyon in such numbers that the sites are being endangered by their supporters! Such irony is one of the modes of figurative language that writers can use to make their points; exaggeration and paradox are two others that you might find writers employing as they respond to our society's seeming inability to protect our lands.

from *Refuge*
Terry Tempest Williams

Terry Tempest Williams is Naturalist-in-Residence at the Utah Museum of Natural History in Salt Lake City. Also a writer, she published Pieces of White Shell: A Journey to Navajoland *in 1984. A native of Utah, she grew up in a Mormon household with a strong sense of commitment to family and to home; the Tempests trace their Mormon Utah roots back to 1847.* Refuge: An Unnatural History of Family and Place *narrates her mother's death from cancer and Williams' recognition that she is probably genetically predisposed to the disease—as evidenced by her two breast cancer biopsies and a "borderline malignancy" between her ribs. Williams' essay "The Clan of the One-Breasted Women" in* Refuge *traces this family trait and links it to nuclear testing conducted in Utah in the early 1960s. The book is also about how death affects a family and how her profession as a naturalist parallels the experience of grieving.*

Heavy rains in 1982 and 1983 raised the level of the Great Salt Lake to the point of endangering the Bear River Migratory Bird Refuge, where Williams spent much of her time working as a naturalist. The refuge in the title of her book refers to more than one safe haven, though the threats to the bird refuge constitute the obvious reason for the title. She has organized her book according to various species of birds she encounters. The chapter excerpted here treats "Birds of Paradise," both the actual creatures and a more figurative one: Williams describes the moment of her mother's death at the end of the preceding chapter.

EXPLORING JOURNAL

What is the place of mourning and funeral services in American culture? Does your family or community have any particular rituals that you feel help people through the mourning process? You may want to move through this assignment more slowly than you do your normal "Exploring" freewrites.

PROLOGUE

Everything about Great Salt Lake is exaggerated—the heat, the cold, the salt, and the brine. It is a landscape so surreal one can never know what it is for certain.

In the past seven years, Great Salt Lake has advanced and retreated. The Bear River Migratory Bird Refuge, devastated by the flood, now begins to heal.

Volunteers are beginning to reconstruct the marshes just as I am trying to reconstruct my life. I sit on the floor of my study with journals all around me. I open them and feathers fall from their pages, sand cracks their spines, and sprigs of sage pressed between passages of pain heighten my sense of smell— and I remember the country I come from and how it informs my life.

Most of the women in my family are dead. Cancer. At thirty-four, I became the matriarch of my family. The losses I encountered at the Bear River Migratory Bird Refuge as Great Salt Lake was rising helped me to face the losses within my family. When most people had given up on the Refuge, saying the birds were gone, I was drawn further into its essence. In the same way that when some-one is dying many retreat, I chose to stay.

Last night, I dreamed I was walking along the shores of Great Salt Lake. I noticed a purple bird floating in the waters, the waves rocking it gently. I entered the lake and, with cupped hands, picked up the bird and returned it to shore. The purple bird turned gold, dropped its tail, and began digging a bur-row in the white sand, where it retreated and sealed itself inside with salt. I walked away. It was dusk. The next day, I returned to the lake shore. A wooden door frame, freestanding, became an arch I had to walk through. Suddenly, it was transformed into Athene's Temple. The bird was gone. I was left standing with my own memory.

In the next segment of the dream, I was in a doctor's office. He said, "You have cancer in your blood and you have nine months to heal yourself." I awoke puzzled and frightened.

Perhaps, I am telling this story in an attempt to heal myself, to confront what I do not know, to create a path for myself with the idea that "memory is the only way home."

I have been in retreat. This story is my return.

TTW
July 4, 1990

BIRDS-OF-PARADISE

lake level: 4211.65'

Mother was buried yesterday.

These days at home have been a meditation as I have scoured sinks and tubs, picked up week-worn clothes, and vacuumed.

I have washed and wiped each dish by hand, dusted tables, even under the feet of figurines.

I notice my mother's hairbrush resting on the counter. Pulling out the nest of short, black hairs, I suddenly remember the birds.

I quietly open the glass doors, walk across the snow and spread the mesh of my mother's hair over the tips of young cottonwood trees—

For the birds—
For their nests—
In the spring.

"Wait here, I want to show you something . . ." My friend, who runs a trading post in Salt Lake City, disappeared into the back room and returned with a pair of moccasins.

They took my breath away. The moccasins were ankle-high and fully beaded, including the soles, which were an intricate design of snakes. Cut glass beads: red, blue, and green, hand-sewn on white deerskin. As I carefully turned them, I wondered how anyone on earth could wear these. To walk in these moccasins would destroy the exquisite handwork.

An Indian woman who had been browsing, smelling the baskets of sweet grass, quietly walked over to the counter.

"Those are burial moccasins," she said. I handed one to her, but she would not touch it. "You won't see many of these."

My friend looked at the woman and then at me. "She's right. A Shoshone woman from Grantsville, ten miles south of Great Salt Lake, brought them in yesterday. They had just buried her grandmother in Skull Valley with the best they had: a buffalo robe, pendleton blankets, jewelry, a beaded dress of buckskin, and the moccasins. The granddaughter made two pairs."

The Indian woman in the trading post identified herself as Cherokee. She explained how, among her people, they sew only one bead on the soles of their burial moccasins.

I thought of the Mormon rituals that surround our dead: the care Mimi and I took in preparing Mother's body with essential oils and perfumes, the way we dressed her in the burial dress Ann had made of white French cotton; the high collar that disguised her weight loss, the delicate tucks from the neck down, the simple elegance of its lines. I recalled the silk stockings; the satin slippers; and the green satin apron, embroidered with leaves, symbolic of Eve and associated with sacred covenants made in the Mormon temple, that we tied around her waist—how it had been hand-sewn by my great-grandmother's sister at the turn of the century. A gift from Mimi. And then I remembered the white veil which framed Mother's face.

I tried to forget my encounter with the mortician in the hallway of the mortuary prior to the dressing, the way he led me down two flights of stairs, through the maze of coffins, and then abruptly drew the maroon velvet curtains that revealed Mother's body, now a carapace, naked, cold, and stiff, on a stainless steel table. Her face had been painted orange. I asked him to remove the make-up. He told me it was not possible, that it would bruise the skin tissues. I told him I wanted it off if I had to remove it myself. The mortician left in disgust and returned with a rag drenched in turpentine. He reluctantly handed me the cloth and for one hour, I wiped my mother's face clean.

I remember arriving at the chapel early, so I could check on the flowers and have some meditative time with Mother's body before the funeral. The face paint was back on. I stood at the side of my mother's casket, enraged at our inability to let the dead be dead. And I wept over the hollowness of our rituals.

The same funeral director put his hand on my shoulder. I turned.

"I'm sorry, Mrs. Williams, she did not pass our inspection. We felt she had to have some color."

"Won't you sit down." he said. "Death is most difficult on the living."

"I'll stand, thank you." I said taking my handkerchief to Mother's face once again.

One by one, family members entered the room, walked to the open coffin and paid their respects. This was the first time my grandmother Lettie had seen her daughter since Christmas Eve. Confined to a wheelchair in a nursing home, her only contact had been by phone. My grandfather Sanky stood behind her with his hands on her shoulders. She mourned like no other.

As is customary in Mormon tradition, Steve and I brought the white veil down over Mother's face and tied the bow beneath her chin. I had hidden sprigs of forsythia down by her feet. The casket was closed. Dan and Hank placed the large bouquet of tulips, lilacs, roses, and lilies, across the top. Dad stood back, frozen with protocol.

Friends came to call. The line grew longer and longer. We became public greeters, entertaining their sorrow as we put aside our own.

I cannot escape these flashbacks. Some haunt. Some heal.

Today is Mother's birthday. March 7, 1987. She would be fifty-five. I lay one bird-of-paradise across her grave.

In a dugout canoe, Brooke and I paddle through a narrow channel of mangroves. A four-foot tiger heron peers out with golden eyes, more mysterious, perhaps, than any bird I have ever seen. The canal widens and we find ourselves in a salt water bay reminiscent of home.

We are in Rio Lagartos, Mexico.

Row upon row of flamingos are dancing with the current. It is a ballet. The flamingos closest to shore step confidently, heads down as they filter small molluscs, crustaceans, and algae through their bills before the water is expelled through either side. These are not quiet birds.

Behind the feeders, a corp de ballet tiptoes in line, flowing in the opposite direction like a feathered river. They too are nodding their heads, twittering, gliding with the black portion of their bills pointing upward. They move with remarkable syncopation.

American flamingos. Gray. White. Fuchsia and pink. They span the red spectrum. Feathers float in the water. Delicately. Brooke leans over the gunnels of the canoe and retrieves one. It contracts out of water. He blows it dry.

The birds are a pink brushstroke against the dark green mangroves. A flock flies over us, their necks extended with their long legs trailing behind them. Pure exotica. In the afternoon light, they become flames against a cloudless blue sky. Early taxonomists must have had the same impression: the Latin family name assigned to flamingos is *Phoenicop-teridae,* derived from the phoenix, which rose from its ashes to live again.

There is a holy place in the salt desert, where egrets hover like angels. It is a cave near the lake where water bubbles up from inside the earth. I am hidden and saved from the outside world. Leaning against the back wall of the cave, the curve of the rock supports the curve of my spine. I listen:

Drip. Drip-drip. Drip. Drip. Drip-drip.

My skin draws moisture from the rocks as my eyes adjust to the darkness.

Ancient murals of ceremonial art bleed from the cavern walls. Pictographs of waterbirds decorate the interior of the cave. Herons, egrets, and cranes. Tadpoles and serpents stain the walls red. Human figures dance wildly, backs arched, hips thrust forward. A spear-thrower lunges toward fish. Beyond him stands a water-jug maiden faintly painted above ferns. So lucent are these forms on the weeping rocks, they could be smeared without thought.

I kneel at the spring and drink.

This is the secret den of my healing, where I come to whittle down my losses. I carve chevrons, the simple image of birds, on rabbit bones cleaned by eagles. And I sing without the embarrassment of being heard.

The men in my family have migrated south for one year to lay pipe in southern Utah.

My keening is for my family, fractured and displaced.

―――――→•◦•←―――――

RESPONDING JOURNAL

Williams takes some strands of her mother's hair from a hairbrush and lays them on bushes and small trees outdoors. Explain in a journal entry why you would or would not feel comfortable performing a similar action after the death of someone close to you.

QUESTIONS FOR CRITICAL THINKING

STRATEGY

1. A bird of paradise is a plant rather than an actual bird. Explain why Williams might have used this flower as a chapter title when all the other chapters have a bird as the heading title.
2. Explain the organization of this chapter. How does Williams proceed? What effect does this organizational pattern have on you as a reader?

ISSUES

1. Williams dedicates *Refuge* to her mother, "Who understood landscape as refuge." Explain the sense of refuge you have gained from reading this selection.

2. How do you see Mormon traditions of mourning fitting in with the other traditions of mourning that Williams mentions? Explain your answers by referring to the examples Williams uses.

COLLABORATIVE WRITING ACTIVITY

How can writing be healing? More particularly, how can writing about landscape be healing? You have had sufficient experience by now to be able to address this idea. Begin writing your response to these questions, and bring a draft of your best ideas to your collaborative group meeting. You may want to look particularly at Williams' final comment in this chapter: "My keening is for my family, fractured and displaced." Keening is a form of mourning; what healing properties might it have?

When you work in your group, read one another's writing and in your followup discussion be sure you have answered these questions: What is the specific healing process? What qualities of what landscape have aided or slowed this process? Does the writer explicitly link this healing process with writing? Use questions and suggestions in your discussions to help one another with this assignment. Remember that there can be many different forms of healing.

"The Long Snowfall"
from *The Sea Around Us*
Rachel Carson

Rachel Carson (1907-1964) was trained as a marine biologist, an education that prepared her admirably for her writing about the natural world. During the 1950s, Carson reached an extremely wide audience with her books and essays explaining the mysteries of the maritime world. Her first book, Under the Sea Wind, *was published in 1941. The Sea Around Us, published in 1951, reached the bestseller lists, was translated into numerous languages, and was condensed for greater accessibility. Carson clearly had a gift for making the factual information of science not only comprehensible but exciting as well.*

In 1962, she reached an even wider audience with the publication of Silent Spring, *a grim warning about the dangers potential in the widespread use of DDT and other pesticides. Alarmed by their effect on living things, Carson sent out a distress signal that many people think began the environmental movement worldwide. No other single book has had such a dramatic effect on collective thinking. The excerpt presented here shows her ability to relate things known (lands and snowfall) to things unknown (terrain and composition of the sea floor).*

EXPLORING JOURNAL

How might we find a sense of a "long snowfall" in a book about the sea? What might be a snowfall under the surface of the sea? Freewrite for ten minutes on the sort of activity that might be going on under the surface waters.

———◦◦◦———

A deep and tremulous earth-poetry.

—LLEWELYN POWYS

Every part of earth or air or sea has an atmosphere peculiarly its own, a quality or characteristic that sets it apart from all others. When I think of the floor of the deep sea, the single, overwhelming fact that possesses my imagination is the accumulation of sediments. I see always the steady, unremitting, downward drift of materials from above, flake upon flake, layer upon layer—a drift that has continued for hundreds of millions of years, that will go on as long as there are seas and continents.

For the sediments are the materials of the most stupendous 'snowfall' the earth has ever seen. It began when the first rains fell on the barren rocks and set in motion the forces of erosion. It was accelerated when living creatures developed in the surface waters and the discarded little shells of lime or silica that had encased them in life began to drift downward to the bottom. Silently, endlessly, with the deliberation of earth processes that can afford to be slow because they have so much time for completion, the accumulation of the sediments has proceeded. So little in a year, or in a human lifetime, but so enormous an amount in the life of earth and sea.

The rains, the eroding away of the earth, the rush of sediment-laden waters have continued, with varying pulse and tempo, throughout all of geologic time. In addition to the silt load of every river that finds its way to the sea, there are other materials that compose the sediments. Volcanic dust, blown perhaps half way around the earth in the upper atmosphere, comes eventually to rest on the ocean, drifts in the currents, becomes waterlogged, and sinks. Sands from coastal deserts are carried seaward on off-shore winds, fall to the sea, and sink. Gravel, pebbles, small boulders, and shells are carried by icebergs and drift ice, to be released to the water when the ice melts. Fragments of iron, nickel, and other meteoric debris that enter the earth's atmosphere over the sea—these, too, become flakes of the great snowfall. But most widely distributed of all are the billions upon billions of tiny shells and skeletons, the limy or silicious remains of all the minute creatures that once lived in the upper waters.

The sediments are a sort of epic poem of the earth. When we are wise enough, perhaps we can read in them all of past history. For all is written here. In the nature of the materials that compose them and in the arrangement of their successive layers the sediments reflect all that has happened in the waters above them and on the surrounding lands. The dramatic and the catastrophic in earth history have left their trace in the sediments—the out-pourings of volcanoes, the advance and retreat of the ice, the searing aridity of desert lands, the sweeping destruction of floods.

The book of the sediments has been opened only within the lifetime of the present generation of scientists, with the most exciting progress in collecting and deciphering samples made since 1945. Early oceanographers could scrape up surface layers of sediment from the sea bottom with dredges. But what was needed was an instrument, operated on the principle of an apple corer, that could be driven vertically into the bottom to remove a long sample or 'core' in which the order of the different layers was undisturbed. Such an instrument was invented by Dr. C. S. Piggot in 1935, and with the aid of this 'gun' he obtained a series of cores across the deep Atlantic from Newfoundland to Ireland. These cores averaged about 10 feet long. A piston core sampler, developed by the Swedish oceanographer Kullenberg about 10 years later, now takes undisturbed cores 70 feet long. The rate of sedimentation in the different parts of the ocean is not definitely known, but it is very slow; certainly such a sample represents millions of years of geologic history.

Another ingenious method for studying the sediments has been used by Professor W. Maurice Ewing of Columbia University and the Woods Hole Oceanographic Institution. Professor Ewing found that he could measure the thickness of the carpeting layer of sediments that overlies the rock of the ocean floor by exploding depth charges and recording their echoes; one echo is received from the top of the sediment layer (the apparent bottom of the sea), another from the 'bottom below the bottom' or the true rock floor. The carrying and use of explosives at sea is hazardous and cannot be attempted by all vessels, but this method was used by the Swedish *Albatross* as well as by the *Atlantis* in its exploration of the Atlantic Ridge. Ewing on the *Atlantis* also used a seismic refraction technique by which sound waves are made to travel horizontally through the rock layers of the ocean floor, providing information about the nature of the rock.

Before these techniques were developed, we could only guess at the thickness of the sediment blanket over the floor of the sea. We might have expected the amount to be vast, if we thought back through the ages of gentle, unending fall—one sand grain at a time, one fragile shell after another, here a shark's tooth, there a meteorite fragment—but the whole continuing persistently, relentlessly, endlessly. It is, of course, a process similar to that which has built up the layers of rock that help to make our mountains, for they, too, were once soft sediments under the shallow seas that have overflowed the continents from time to time. The sediments eventually became consolidated and cemented and, as the seas retreated again, gave the continents their thick, covering layers of sedimentary rocks—layers which we can see uplifted, tilted, compressed, and broken by the vast earth movements. And we know that in places the sedimentary rocks are many thousands of feet thick. Yet most people felt a shock of surprise and wonder when Hans Pettersson, leader of the Swedish Deep Sea Expedition, announced that the *Albatross* measurements taken in the open Atlantic basin showed sediment layers as much as 12,000 feet thick.

If more than two miles of sediments have been deposited on the floor of the Atlantic, an interesting question arises: has the rocky floor sagged a corresponding distance under the terrific weight of the sediments? Geologists hold conflicting opinions. The recently discovered Pacific sea mounts may offer one piece of evidence that it has. If they are, as their discoverer called them, 'drowned ancient islands,' then they may have reached their present stand a mile or so below sea level through the sinking of the ocean floor. Hess believed the islands had been formed so long ago that coral animals had not yet evolved; otherwise the corals would presumably have settled on the flat, planed surfaces of the sea mounts and built them up as fast as their bases sank. In any event, it is hard to see how they could have been worn down so far below 'wave base' unless the crust of the earth sagged under its load.

One thing seems probable—the sediments have been unevenly distributed both in place and time. In contrast to the 12,000-foot thickness found in parts of the Atlantic, the Swedish oceanographers never found sediments thicker than 1000 feet in the Pacific or in the Indian Ocean. Perhaps a deep layer of lava, from ancient submarine eruptions on a stupendous scale, underlies the upper layers of the sediments in these places and intercepts the sound waves.

Interesting variations in the thickness of the sediment layer on the Atlantic Ridge and the approaches to the Ridge from the American side were reported by Ewing. As the bottom contours became less even and began to slope up into the foothills of the Ridge, the sediments thickened, as though piling up into mammoth drifts 1000 to 2000 feet deep against the slopes of the hills. Farther up in the mountains of the Ridge, where there are many level terraces from a few to a score of miles wide, the sediments were even deeper, measuring up to 3000 feet. But along the backbone of the Ridge, on the steep slopes and peaks and pinnacles, the bare rock emerged, swept clean of sediments.

Reflecting on these differences in thickness and distribution, our minds return inevitably to the simile of the long snowfall. We may think of the abyssal snowstorm in terms of a bleak and blizzard-ridden arctic tundra. Long days of storm visit this place, when driving snow fills the air; then a lull comes in the blizzard, and the snowfall is light. In the snowfall of the sediments, also, there is an alternation of light and heavy falls. The heavy falls correspond to the periods of mountain building on the continents, when the lands are lifted high and the rain rushes down their slopes, carrying mud and rock fragments to the sea; the light falls mark the lulls between the mountain-building periods, when the continents are flat and erosion is slowed. And again, on our imaginary tundra, the winds blow the snow into deep drifts, filling in all the valleys between the ridges, piling the snow up and up until the contours of the land are obliterated, but scouring the ridges clear. In the drifting sediments on the floor of the ocean we see the work of the 'winds,' which may be the deep ocean currents, distributing the sediments according to laws of their own, not as yet grasped by human minds.

We have known the general pattern of the sediment carpet, however, for a good many years. Around the foundations of the continents, in the deep waters off the borders of the continental slopes, are the muds of terrestrial origin. There are muds of many colors—blue, green, red, black, and white—apparently varying with climatic changes as well as with the dominant soils and rocks of the lands of their origin. Farther at sea are the oozes of predominantly marine origin—the remains of the trillions of tiny sea creatures. Over great areas of the temperate oceans the sea floor is largely covered with the remains of unicellular creatures known as foraminifera, of which the most abundant genus is Globigerina. The shells of Globigerina may be recognized in very ancient sediments as well as in modern ones, but over the ages the species have varied. Knowing this, we can date approximately the deposits in which they occur. But always they have been simple animals, living in an intricately sculptured shell of carbonate of lime, the whole so small you would need a microscope to see its details. After the fashion of unicellular beings, the individual Globigerina normally did not die, but by the division of its substance became two. At each division, the old shell was abandoned, and two new ones were formed. In warm, lime-rich seas these tiny creatures have always multiplied prodigiously, and so, although each is so minute, their innumerable shells blanket millions of square miles of ocean bottom, and to a depth of thousands of feet.

In the great depths of the ocean, however, the immense pressures and the high carbon-dioxide content of deep water dissolve much of the lime long before it reaches the bottom and return it to the great chemical reservoir of the sea. Silica is more resistant to solution. It is one of the curious paradoxes of the ocean that the bulk of the organic remains that reach the great depths intact belong to unicellular creatures seemingly of the most delicate construction. The radiolarians remind us irresistibly of snow flakes, as infinitely varied in pattern, as lacy, and as intricately made. Yet because their shells are fashioned of silica instead of carbonate of lime, they can descend unchanged into the abyssal depths. So there are broad bands of radiolarian ooze in the deep tropical waters of the North Pacific, underlying the surface zones where the living radiolarians occur most numerously.

Two other kinds of organic sediments are named for the creatures whose remains compose them. Diatoms, the microscopic plant life of the sea, flourish most abundantly in cold waters. There is a broad belt of diatom ooze on the floor of the Antarctic Ocean, outside the zone of glacial debris dropped by the ice pack. There is another across the North Pacific, along the chain of great deeps that run from Alaska to Japan. Both are zones where nutrient-laden water wells up from the depths, sustaining a rich growth of plants. The diatoms, like the radiolaria, are encased in silicious coverings—small, boxlike cases of varied shape and meticulously etched design.

Then, in relatively shallow parts of the open Atlantic, there are patches of ooze composed of the remains of delicate swimming snails, called pteropods. These winged mollusks, possessing transparent shells of great beauty, are here and there incredibly abundant. Pteropod ooze is the characteristic bottom deposit in the vicinity of Bermuda, and a large patch occurs in the South Atlantic.

Mysterious and eerie are the immense areas, especially in the North Pacific, carpeted with a soft, red sediment in which there are no organic remains except sharks' teeth and the ear bones of whales. This red clay occurs at great depths. Perhaps all the materials of the other sediments are dissolved before they can reach this zone of immense pressures and glacial cold.

The reading of the story contained in the sediments has only begun. When more cores are collected and examined we shall certainly decipher many exciting chapters. Geologists have pointed out that a series of cores from the Mediterranean might settle several controversial problems concerning the history of the ocean and of the lands around the Mediterranean basin. For example, somewhere in the layers of sediment under this sea there must be evidence, in a sharply defined layer of sand, of the time when the deserts of the Sahara were formed and the hot, dry winds began to skim off the shifting surface layers and carry them seaward. Long cores recently obtained in the western Mediterranean off Algeria have given a record of volcanic activity extending back through thousands of years, and including great prehistoric eruptions of which we know nothing.

The Atlantic cores taken more than a decade ago by Piggot from the cable ship *Lord Kelvin* have been thoroughly studied by geologists. From their analysis it is possible to look back into the past 10,000 years or so and to sense the pulse of the earth's climatic rhythms; for the cores were composed of layers of cold-water globigerina faunas (and hence glacial stage sediments), alternating with globigerina ooze characteristic of warmer waters. From the clues furnished by these cores we can visualize interglacial stages when there were periods of mild climates, with warm water overlying the sea bottom and warmth-loving creatures living in the ocean. Between these periods the sea grew chill. Clouds gathered, the snows fell, and on the North American continent the great ice sheets grew and the ice mountains moved out to the coast. The glaciers reached the sea along a wide front; there they produced icebergs by the thousand. The slow-moving, majestic processions of the bergs passed out to sea, and because of the coldness of much of the earth they penetrated farther south than any but stray bergs do today. When finally they melted, they relinquished their loads of silt and sand and gravel and rock fragments that had become frozen into their under surfaces as they made their grinding way over the land. And so a layer of glacial sediment came to overlie the normal globigerina ooze, and the record of an Ice Age was inscribed.

Then the sea grew warmer again, the glaciers melted and retreated, and once more the warmer-water species of Globigerina lived in the sea—lived and died and drifted down to build another layer of globigerina ooze, this time over the clays and gravels from the glaciers. And the record of warmth and mildness was again written in the sediments. From the Piggot cores it has been possible to reconstruct four different periods of the advance of the ice, separated by periods of warm climate.

It is interesting to think that even now, in our own lifetime, the flakes of a new snow storm are falling, falling, one by one, out there on the ocean floor. The billions of Globigerina are drifting down, writing their unequivocal record that this, our present world, is on the whole a world of mild and temperate climate. Who will read their record, ten thousand years from now?

————————

RESPONDING JOURNAL

Carson finds much to think about in the sediment she sees in the ocean. What have you seen in our waters that could one day tell a scientist about our civilization? Write a substantial entry of several paragraphs; be as specific as possible in detailing the lessons that might be learned.

QUESTIONS FOR CRITICAL THINKING

STRATEGY

1. Does Carson's final question put her remarks in a context that gives "the long snowfall" special meaning? Do you find that your reading of her work—or that of other nature writers—changes when you realize that the processes you are studying right now might still be going on "ten thousand years from now"?

2. How well does this image of a "long snowfall" work to explain the various accumulations of sediment on the ocean floor? Can you think of other possible metaphors or similes?

ISSUES

1. Carson states that "the sediments are a sort of epic poem of the earth." Can you think of other natural phenomena that might as accurately be labeled "the epic poem of the earth"? Justify your response.

2. The uneven distribution of the sediments seems to suggest a further reason for comparing snowfall and sediment building. What other situations outside the sea can you think of where we might productively measure sediment?

COLLABORATIVE WRITING ACTIVITY

In her writing, Carson talks of earlier, supposedly "scientific" ideas that postulated situations or conditions in the natural world that later proved to be false. What other scientific ideas can you think of that were once widely believed but are now discredited? Think of another process of decay in the natural world that leaves behind elements which indicate this process of decomposition. Brainstorm a list and bring it to your group; after you have listened to all the lists and discussed them, state some principles that seem to indicate how such a process can serve as a parallel to illuminate a different situation.

"Jeremy Bentham, the *Pietà,* and a Precious Few Grayling"
David Quammen

David Quammen, the author of Outside *magazine's "Natural Acts" column, writes fiction as well as nonfiction related to natural science issues. His* Blood Line: Stories of Fathers and Sons *(1987) won him wide recognition quite apart from his impressive work in scientific prose. His essays are collected in* Natural Acts *(1985), in which the following essay appeared, and* The Flight of the Iguana: A Sidelong View of Science and Nature *(1988).*

Quammen's undergraduate degree from Yale, followed by an Oxford degree completed as a Rhodes Scholar, indicate his preparation as a writer and humanist, while his training in entomology at the University of Montana gives him his scientific knowledge of the sort displayed in "Jeremy Bentham, the Pietà, *and a Precious Few Grayling." Here he displays his ability to bring together various strands of knowledge to argue for the continued vigilance over a distressed and endangered species in the United States. The essay first appeared in* Audubon *in 1982.*

EXPLORING JOURNAL

Freewrite for ten minutes on the threats posed to fish by forces—such as humans—over which the fish have no control. What are some of the problems, and what seem to be some solutions?

Rumor had it they were gone, or nearly gone, killed off in large numbers by dewatering and high temperatures during the bad drought of 1977. The last sizable population of *Thymallus arcticus*—Arctic grayling—indigenous to a river in the lower forty-eight states: *ppffft.* George Liknes, a graduate student in fisheries at Montana State University, was trying to do his master's degree on these besieged grayling of the upper Big Hole River in western Montana, and word passed that his collecting nets, in late summer of 1978, were coming up empty. The grayling were not where they had been, or if they were, Liknes for some reason wasn't finding them. None at all? "Well," said one worried state wildlife biologist, "precious few."

Grayling are not set up for solitude. Like the late lamented passenger pigeon, grayling are by nature and necessity gregarious, thriving best in rather crowded communities of their own kind. When the size of a population sinks below a certain unpredictable threshold, grayling are liable to disappear altogether, poof, evidently incapable of successful pairing and reproduction without the circumstantial advantage of teeming fellowship. This may have been what happened in Michigan. Native grayling were extinguished there, rather abruptly, during the 1930s.

The Michigan grayling and the Montana strain had been from time beyond memory the unique and isolated representatives of the species in temperate North America. They were glacial relics, meaning they had gradually fled southward into open water during the last great freeze-up of the Pleistocene epoch, then, when the mile-thick flow of ice stopped just this side of the Canadian border and began melting back northward, they were left behind in Michigan and Montana as two separate pockets of grayling. These were trapped, as it turned out, cut off by hundreds of miles from what became the primary range of the species, across northern Canada and Alaska. They were stuck in warmish southern habitats overlapping the future range of dominance of *Homo sapiens;* their own future, consequently, insecure.

The Michigan grayling went first. They had been abundant in the upper part of Michigan's Lower Peninsula and in the Otter River of the Upper Peninsula. One report tells of four people catching 3,000 grayling in fourteen days from the Manistee River and hauling most of that catch off to Chicago. By 1935, not surprisingly, the Manistee was barren of grayling. Before long, so was the rest of the state. Sawlogs had been floated down rivers at spawning time, stream banks had been stripped of vegetation (causing water temperatures to rise), exotic competing fish had been introduced, and greedy pressure like that on the Manistee had continued. By 1940, the people of Michigan had just the grayling they were asking for: none.

In Montana, where things tend to happen more slowly, some remnant of the original grayling has endured—against similar adversities in less intense form—a bit longer. Even while disappearing during the past eighty years from parts of their Montana range, grayling have expanded into other new habitat. More accurately, they have been introduced to new habitat, in the zoological equivalent of forced school-busing: hatchery rearing and planting. As early as

1903, soon after the founding of the Fish Cultural Development Station in Bozeman, the state of Montana got into the business of manufacturing grayling; and for almost sixty years thereafter the planting of hatchery grayling was in great vogue.

The indigenous range of the Montana grayling was in the headwaters of the Missouri River above the Great Falls; they were well established in the Smith River, in the Sun River, and in the Madison, the Gallatin, and the Jefferson and their tributaries—notably, the Big Hole River. They had evolved as mainly a stream-dwelling species and existed in only a very few Montana lakes. However, they happened to be rather tolerant of low dissolved oxygen levels, when those levels occurred in cold winter conditions (but not when the oxygen was driven out of solution by summer warming). This made them suitable for stocking in high lakes, where they could get through the winter on what minimal oxygen remained under the ice. In 1909, 50,000 grayling from the Bozeman hatchery were planted in Georgetown Lake. Just a dozen years later, 28 million grayling eggs were collected from Georgetown, to supply hatchery brood for planting elsewhere. And the planting continued: Ennis Lake, Rogers Lake, Mussigbrod Lake, Grebe Lake in Yellowstone National Park. Between 1928 and 1977, millions more grayling were dumped into Georgetown Lake.

Unfortunately, that wasn't all. Back in 1909, hatchery grayling were also planted in the Bitterroot and Flathead rivers, on the west side of the Continental Divide, in stream waters they had never colonized during their ancestral migration. An innocent experiment, and without large consequences, since the grayling introduced there evidently did not take hold. But then, in what may have seemed a logical extension of all this hatchery rearing and planting, the Big Hole River was planted with grayling. The Big Hole already had a healthy reproducing population of wild grayling, but that was not judged to be reason against adding more. From 1937 until 1962, according to the records of the Montana Department of Fish, Wildlife, and Parks (FWP), more than five million grayling from the Anaconda hatchery were poured into the Big Hole, from the town of Divide upstream to the headwaters: hothouse grayling raining down on wild grayling.

This was before FWP biologists had come upon the belated realization that massive planting of hatchery fish in a habitat where the same species exists as a reproducing population is the best of all ways to make life miserable for the wild fish. Things are done differently these days, but the mistake was irreversible. The ambitious sequence of plantings was very likely the most disastrous single thing that ever happened to the indigenous grayling of the Big Hole.

At best, each planting instantaneously created tenement conditions of habitat and famine conditions of food supply. In each place where the hatchery truck stopped, the river became a grayling ghetto. At worst, if any of the planted fish survived long enough to breed with each other and interbreed with the wild fish, the whole planting program served to degrade the gene pool of the Big Hole grayling, making them less capable of surviving the natural adversities—drought, flood, temperature fluctuation, predation—of their natural habitat.

But here's the good news: Very few of those planted grayling would have survived long enough to breed. The mortality rate on hatchery grayling planted in rivers is close to 100 percent during the first year, and most don't last even three months, whether or not they are caught by a fisherman. These planted grayling come, after all, from a small sample of lake-dwelling parents, with little genetic variety or inherited capacity for coping with moving water. Reared in the Orwellian circumstances of the hatchery, cooped in concrete troughs, without a beaver or a merganser to harry them, eating Purina trout chow from the hand of man, what chance have they finally in the most challenging of habitats, a mountain river? The term "fish planting" itself is a gross misnomer, when applied to dropping grayling or trout into rivers; there is no delusion, even among the hatchery people, that these plants will ever take root. More realistically, it's like providing an Easter egg hunt for tourists with fishing rods.

In 1962 the Big Hole planting ceased, and the remaining wild grayling, those that hadn't died during the famine and tenement periods, were left to get on as best they could. Then came the 1977 drought and, the following year, the George Liknes study. One of Liknes's study sections on the Big Hole was a two-mile stretch downstream from the town of Wisdom to just above the Squaw Creek bridge. On a certain remote part of the stretch a rancher had sunk a string of old car bodies to hold his hayfield in place. From that two-mile stretch, using electroshocking collection equipment that is generally reliable, Liknes did not take a single grayling. This came as worrisome news to me because, on a morning in late summer of 1975, standing waist deep within sight of the same string of car bodies and offering no great demonstration of angling skill, I had caught and released thirty-one grayling in four hours. Now they were either gone or in hiding.

Grayling belong to the salmonid family, as cousins of trout and whitefish. In many ways they seem a form intermediate between those two genera; in other ways, they depart uniquely from the salmonid pattern.

The first thing usually noted about them, their distinguishing character, is the large and beautiful dorsal fin. It sweeps backward twice the length of a trout's, fanning out finally into a trailing lobe, and it is, under certain specific conditions, the most exquisitely colorful bit of living matter to be found in the state of Montana; spackled with rows of bright turquoise spots that blend variously to aquamarine and reddish-orange toward the front of the fin, a deep hazy shading of iridescent mauve overall, and along the upper edge, in some individuals, a streak of shocking rose. That's in the wild, or even stuck on a hook several inches underwater. Lift the fish into air, and it all disappears. The bright spots and iridescence drain away instantaneously, the dorsal folds down to nothing, and you are holding a drab gun-metal creature that looks very much like a whitefish. The grayling magic vanishes, like a dreamed sibyl, when you pull it to you.

Except for this dorsal fin, the grayling does resemble that most maligned and misunderstood of Montana fish, its near relative, the mountain whitefish. Both are upholstered—unlike the trout—with large stiff scales, scales you wouldn't want to eat. Both have dull-colored bodies, grayish-silver in the grayling, brownish-silver in the whitefish—though the grayling is marked along its forward flank with another smattering of spots, these purplish-black, playing dimly off the themes in the dorsal. They are also distinguishable (from each other and from their common salmonid relatives) by the shape of the mouth. A trout has a wide, sweeping, toothy grin; a whitefish's mouth is narrow and toothless—worse, it is set in a snout that is pointed and cartilaginous, like a rat's, probably the main single cause of the whitefish's image problem. The grayling, as you can see if you look closely, has been burdened with a mouth that is an uneasy compromise between the two: The narrow mouth is set with numerous tiny teeth and fendered with large cartilaginous maxillaries, but its shortened nose couldn't fairly be called a snout. The point is this: The grayling is one of America's most beautiful fish, but only a few subtle anatomical strokes distinguish it from one of the most ugly. A lesson in hubris.

But a superfluous lesson, since the grayling by character is anything but overweening. It is dainty and fragile and relatively submissive. With tiny teeth and little moxie, it fails in all territorial competition against trout—and this is another reason for its decline in the Big Hole, where rainbow and brown and brook trout that have been moved into the neighborhood now bully it mercilessly. Like many beautiful creatures that have known fleeting success, it is dumb. It seeks security in gregariousness and these days is liable to find, instead, carnage. When insect food is on the water, and the fish are attuned to that fact, a fisherman can stand in one spot, literally without moving his feet, and catch a dozen grayling. Trout are not so foolish: Drag one from a hole and the word will be out to the others. The grayling cannot take such a hint. In the matter of food it is an unshakable optimist; the distinction between a mayfly on the water's surface and a hook decorated with feathers and floss is lost on it. But this rashness, in the Big Hole for example, might again be partly a consequence (as well as a cause) of its beleaguered circumstances. The exotic trouts, being dominant, seize the choice territorial positions of habitat, and the grayling, pushed off into marginal water where a fish can only with difficulty make a living, may be forced to feed much more recklessly than it otherwise would.

At certain moments the grayling seems even a bit stoic, as though it had seen its own future and made adjustments. This is noticeable from the point of view of the fisherman. A rainbow trout with a hook jerked snug in its mouth will leap as though it were angry, furious—leap maybe five or six times, thrashing the air convulsively each time. If large, it will run upstream, finally to go to the bottom and begin scrabbling its head in the rubble to scrape out the hook. A whitefish, unimaginative and implacable, will usually not jump, will never run, will stay near the bottom and resist with pure loutish muscle. A grayling will jump once if at all and remain limp in the air, leaping the way a Victorian matron would faint into someone's arms—with demure, trusting abdication. Then, possibly after a polite tussle, the grayling will let its head be pulled above the water's surface, turn passively onto its side, and allow itself to be hauled in. Once beaten, a rainbow can be coaxed with certain tricks of handling to give you three seconds of docility while you get the hook out to release it. A whitefish will struggle like a hysterical pig no matter what. A grayling will simply lie in your hand, pliant and fatalistic, beautiful, placing itself at your mercy.

So no one has much use for the grayling, not even fishermen. It grows slowly, never as large as a trout, and gives unsatisfactory battle. It is scaly, bony, and not especially good to eat. Montana's fish and game laws will allow you to kill five of them from the Big Hole River in a day,* and five more every day all summer—but what will you do with them? Last year a Butte man returned from a weekend on the river and offered a friend of mine ten grayling to feed his cat. The man had killed them because he caught them, very simple logic, but then realized he had no use for them. This year my friend's cat is dead, through no fault of the grayling, so even that constituency is gone. A grayling does not cook up well, it does not fight well. It happens to have an extravagant dorsal fin, but no one knows why. If you kill one to hang on your wall, its colors will wilt away heart-breakingly, and the taxidermist will hand you back a whitefish in rouge and eye shadow. The grayling, face it, is useless. Like the auk, like the zebra swallowtail, like Angkor Wat.

In June of 1978, the U.S. Supreme Court ruled that completion of the Tellico Dam on the Little Tennessee River was prohibited by law, namely the 1973 Endangered Species Act, because the dam would destroy the only known habitat of the snail darter, a small species of perch. One argument in support of this prohibition, perhaps the crucial argument, was that the snail darter's genes might at some time in the future prove useful—even invaluable—to the balance of life on Earth, possibly even directly to humanity. If the *Penicillium* fungus had gone extinct when the dodo bird did, according to this argument, many thousands of additional human beings by now would have died of diphtheria and pneumonia. You could never foresee what you might need, what might prove useful in the line of genetic options, so nothing at all should be squandered, nothing relinquished. Thus it was reasoned on behalf of snail darter preservation (and thus I have reasoned elsewhere myself). The logic is as solid as it is dangerous.

The whole argument by utility may be one of the most dangerous, even ominous, strategic errors that the environmental movement has made. The best reason for saving the snail darter was this: precisely because it is flat useless. That's what makes it special. It wasn't put there, in the Little Tennessee River, it has no ironclad reason for being there; it is simply there. A hydroelectric dam, which can be built in a mere ten years for a mere $119 million, will have utility

_____

*At the time this essay appeared in *Audubon;* since then, the regulations have been changed in the grayling's favor.

on its side of the balance against snail darter genes, if not now then at some future time, when the cost of electricity has risen above the cost of recreating the snail darter through genetic engineering. A snail darter arrived at the hard way, the Darwinian way, across millions of years of randomness, reaching its culmination as a small ugly perch roughly resembling an undernourished tadpole, is something far more precious than a net asset in potential utility. What then, exactly? That isn't easy to say, without gibbering in transcendental tones. But something more than a floppy disc storing coded genetic lingo for a rainy day.

Another example: On a Sunday in May, 1972, an addled Hungarian named Laszlo Toth jumped a railing in St. Peter's Basilica and took a hammer to Michelangelo's *Pietà,* knocking the nose off the figure of Mary, and part of her lowered eyelid, and her right arm at the elbow. The world groaned. Italian officials charged Toth with crimes worth a maximum total of nine years' imprisonment. Some people, but no one of liberal disposition, said aloud at the time that capital punishment would be more appropriate. In fact, what probably should have been done was to let Italian police sergeants take Toth out into a Roman alley and smack his nose off, and part of his eyelid, and his arm at the elbow, with a hammer. The *Pietà* was at that time 473 years old, the only signed sculpture by the greatest sculptor in human history. I don't know whether Laszlo Toth served the full nine years, but very likely not. Deoclecio Redig de Campos, from the Vatican art-restoration laboratories, said at the time that restoring the sculpture, with glue and stucco and substitute bits of marble, would be "an awesome task that might take three years," but later he cheered up some and amended that to "a matter of months." You and I know better. The Michelangelo *Pietà* is gone. The Michelangelo/de Campos *Pietà* is the one now back on display. There is a large difference. What, exactly, is the difference? Again hard to say, but it has much to do with the snail darter.

Sage editorialists wrote that Toth's vandalism was viewed by some as an act of leftist political symbolism: "Esthetics must bow to social change, even if in the process the beautiful must be destroyed, as in Paris during *les èvènements,* when students scrawled across paintings 'No More Masterpieces.' So long as human beings do not eat, we must break up ecclesiastical plate and buy bread." The balance of utility had tipped. The only directly useful form of art, after all, is that which we call pornography.

Still another example: In May of 1945 the Target Committee of scientists and ordnance experts from the Manhattan Project met to hash out a list of the best potential Japanese targets for the American atomic bomb. At the top of the list they placed Kyoto, the ancient capital city of Japan, for eleven centuries the source of all that was beautiful in Japanese civilization, the site of many sacred and gorgeous Shinto shrines. When he saw this, Henry L. Stimson, a stubbornly humane old man who had served as Secretary of State under Herbert Hoover and was now Truman's inherited Secretary of War, got his back up: "This is one time I'm going to be the final deciding authority. Nobody's going to tell me what to do on this. On this matter I am the kingpin." And he struck the city of shrines off the list. Truman concurred. Think what you will about the subsequent bombing of Hiroshima—unspeakably barbarous act, most justifiable act in the given circumstances, possibly both—think what you will about that; still the sparing of Kyoto, acknowledged as a superior target in military terms, was very likely the most courageous and imaginative decision anyone ever talked Harry Truman into. In May of 1945, the shrines of Kyoto did not enjoy the balance of utility.

"By utility is meant that property in any object, whereby it tends to produce benefit, advantage, pleasure, good, or happiness (all this in the present case comes to the same thing), or (what comes again to the same thing) to prevent the happening of mischief, pain, evil, or unhappiness to the party whose interest is considered: if that party be the community in general, then the happiness of the community; if a particular individual, then the happiness of that individual." This was written by Jeremy Bentham, the English legal scholar of the eighteenth century who was a founder of that school of philosophy known as utilitarianism. He also wrote, in *Principles of Morals and Legislation,* that "an action then may be said to be conformable to the principle of utility . . . when the tendency it has to augment the happiness of the community is greater than any it has to diminish it." In more familiar words, moral tenets and legislation should always be such as to achieve the greatest good for the greatest number. And *the greatest number* has generally been taken to mean (though Bentham himself might not have agreed: see "Animal Rights and Beyond") the greatest number of *humans.*

This is a nefariously sensible philosophy. If it had been adhered to strictly throughout the world since Bentham enunciated it, there would now be no ecclesiastical plate or jeweled papal chalices, no Peacock Throne (vacated or otherwise) of Iran, no Apollo moon landings, no Kyoto. Had it been retroactive, there would be no Egyptian pyramids, no Taj Mahal, no texts of Plato; nor would there have been any amassing of wealth by Florentine oligarchs and hence no Italian Renaissance; finally, therefore, no *Pietà,* not even a mangled one. And if Bentham's principle of utility—in its economic formulation, or in thermodynamic terms, or even in biomedical ones—is applied today and tomorrow as the ultimate touchstone for matters of legislation, let alone morals, then there will eventually be, as soon as the balance tips, no snail darter and no. . . .

But we were talking about the Big Hole grayling. George Liknes was finding few, and none at all near the string of car bodies, and this worried me. I had some strong personal feelings toward the grayling of the Big Hole—proprietary is not the right word, too presumptuous; rather, feelings somewhere between cherishing and reliance. I had come to count on the fact, for cheer and solace in a very slight way, that they were there, that they existed—beautiful, dumb, and useless—in the upper reaches of that particular river. It happened because I had gone up there each year for a number of years—usually in late August, which is the start of autumn in the upper Big Hole Valley, or in early

September—with two hulking Irishmen, brothers. Each year, stealing two days for this pilgrimage just as the first cottonwoods were taking on patches of yellow, we three visited the grayling.

At that time of year the Big Hole grayling are feeding, mainly in the mornings, on a plague of tiny dark mayflies known as *Tricorythodes* (or, for convenience, trikes). One of these creatures is roughly the size of a caraway seed, black-bodied with pale milky wings; but they appear on the water by the millions, and the grayling line up in certain areas to sip at them. The trike hatch happens every August and September, beginning each morning when the sun begins warming the water, continuing daily for more than a month, and it is one of the reasons thirty-one grayling can be caught in a few hours. The trike hatch was built into my understanding with the Irishmen, an integral part of the yearly ritual. Trike time, time to visit the Big Hole grayling.

Not stalk, not confront, certainly not kill and eat; visit. No great angling thrills attach to catching grayling. You don't fish at them for the satisfaction of fooling a crafty animal on its own terms, or fighting a wild little teakettle battle handicapped across a fine leader, as you do with trout. The whole context of expectations and rewards is different. You catch grayling to visit them: to hold one carefully in the water, hook freed, dorsal flaring, and gape at the colors, and then watch as it dashes away. This is good for a person, though it could never be the greatest good for the greatest number. I had visited them regularly at trike time with the two Irishmen, including the autumn of the younger brother's divorce, and during the days just before the birth of the older brother's first daughter, and through some weather of my own. So I did not want to hear about a Big Hole River that was empty of grayling.

A fair question to the Montana Department of Fish, Wildlife and Parks is this: If these fish constitute a unique and historic population, a wonderful zoological rarity within the lower forty-eight states, why let a person kill five in a day for cat food? FWP biologists have offered three standard answers: (1) Until George Liknes finished his master's thesis, they possessed no reliable data on the Big Hole grayling, and they do not like to make changes in management procedures except on the basis of data; (2) grayling are very fecund—a female will sometimes lay more than 10,000 eggs—and so availability of habitat and infant mortality and competition with trout are the limiting factors, not fishing pressure; and (3) these grayling are glacial relicts, meaning they have been naturally doomed to elimination from this habitat, and mankind is only accelerating that inevitability.

Yet, (1) over a period of twenty-five years, evidently without the basic data that would have showed that it was all counterproductive, the department spent a large pile of money to burden the Big Hole grayling with five million hatchery outsiders; and (2) though fishermen are admittedly not the limiting factor on total number of grayling in the river, they can easily affect the number of large, successful, genetically gifted spawning stock in the population, since those are precisely the individual fish that fishermen, unlike high temperatures or low oxygen or competitive trout, kill in disproportionate number. There might be money for more vigorous pursuit of data, there might be support for protecting the grayling from cats, but the critical constituency involved here is fishermen, and the balance of utility is not on the side of the grayling; as for (3), not only are the rivers of Montana growing warmer with the end of the Ice Age, but the Earth generally is warming; it is in fact falling inexorably into the sun, and the sun itself is meanwhile dying. So all wildlife on the planet is doomed to eventual elimination, and mankind is only et cetera.

The year before last, the Irishmen and I missed our visit: The older brother had a second daughter coming, and the younger brother was in Germany, in the Army, soon to have a second wife. I could have gone alone but I didn't. So all I knew of *Thymallus arcticus* on the upper Big Hole was what I heard from George Liknes: not good. Through the winter I asked FWP biologists for news of the Big Hole grayling: not good.

Then one day in late August last year, I sneaked away and drove up the Big Hole toward the town of Wisdom, specifically for a visit. I stopped when I saw a promising arrangement of water, a spot I had never fished or even noticed before, though it wasn't too far from the string of car bodies. I didn't know what I would find, if anything. On the third cast I made contact with a twelve-inch grayling, largish for the Big Hole within my memory. Between sun-on-the-water and noon, using a small fly resembling a *Tricorythodes,* I caught and released as many grayling as ever. As many as I needed.

I could tell you where to look for them, I could suggest how you might fish for them, but that's not the point here. You can find them yourself if you need to. Likewise, it's tempting to suggest where you might send letters, whom you might hector, what pressures you might apply on behalf of these useless fish; also not exactly the point. I merely wanted to let you know: They are there.

Irishmen, the grayling are still there, yes. Please listen, the rest of you: They are there, the Big Hole grayling. At least for now.

RESPONDING JOURNAL

Write a substantial journal entry of at least two pages on the implications of some form of wildlife you consider important being lost through extinction. Though it most often happens with animals, it also happens to plants. The American Chestnut, for example, is almost extinct because of chestnut blight. If you find this length a challenge, do some brainstorming to see if you can generate additional ideas.

QUESTIONS FOR CRITICAL THINKING

STRATEGY

1. Explain the title of Quammen's essay. What does he gain by uniting the three categories? Can you think of another category of knowledge that he might have used here?

2. Look at the sentence on grayling propagation: "Reared in the Orwellian circumstances of the hatchery, cooped in concrete troughs, without a beaver or a merganser to harry them, eating Purina trout chow from the hand of man, what chance have they finally in the most challenging of habitats, a mountain river?" How has Quammen set up this syntax? What sort of repetition does he use here? To what extent do you think this sort of careful construction contributes to the impression his essay makes on you?

ISSUES

1. The paragraph on page 435 describing the grayling in the water and out of the water might suggest other parallels to you. What other animals or plants undergo similar transformations? What causes such transformations?
2. Quammen's title suggests if not three equal concerns, then at least two worthwhile ideas to go along with the first topic he covers—the grayling. After summarizing the ideas from Bentham, explain what the phrase "No More Masterpieces" means in relation to the grayling.

COLLABORATIVE WRITING ACTIVITY

Explore the idea of "beautiful, dumb, and useless" that Quammen applies to the grayling. How appropriate are these labels? Does he really feel so strongly about the poor fish? Explore Quammen's idea in an essay of at least a page that begins here and moves toward your sense of where this essay should take your thoughts. You may have had, if you are a fisherperson, similar thoughts, or you may have had some parallel responses after seeing images of animals killed by hunters. See how you can take Quammen's ideas and move them into the areas of your own personal response to the ideas he has suggested.

Pair up with another group member and read each other's essays. See what questions your partner's essay suggests to you and discuss those issues with your partner. On the basis of this discussion, revise the essay together into a fuller exploration of the issues in your first draft. This work may well lead to a longer work later on, or it may find its way into your portfolio.

"The Dunes, My Father"
Kathleen Stocking

Kathleen Stocking has written for several newspapers and journals; in her writing she focuses on the Leelanau Peninsula where she grew up. This area of northwestern Michigan, extending into Lake Michigan, is undergoing a fairly typical change in that its rural character is being challenged by its desirability as a resort area. Tourist dollars are, of course, important to the local economy, and the balance between protection of the area and protection of local livelihoods causes concern for many citizens.

"The Dunes, My Father" comes from the collection Letters from the Leelanau, *published by the University of Michigan Press in 1990. In these essays Stocking reflects on the place of her home in the larger scheme not only of her native Michigan and the midwest, but also in the ebb and flow of human relationships. This meeting of rural ways and values with late twentieth-century life in America furnishes much of her material.*

EXPLORING JOURNAL

Try a ten-minute freewrite on the subject of dunes. What do you know about them? What has your personal contact with them been? Where do you expect to see them? What do you know about them geographically? What would they feel like, smell like? Use your imagination to make up for any gaps in your experience.

When you first see the Sleeping Bear Dunes rising massive and golden two hundred feet above a grassy plain, you think you're in the Serengeti or someplace equally strange, not the Midwest. On its eastern slope, this mountain of flesh-colored sand rises like the Taj Mahal to be mirrored in the placid, cattail-bordered Mill Pond below.

I never thought too much about it, growing up there. When you're a child your environment is your environment; it could be the far side of the moon and you wouldn't think about it. The Sleeping Bear Dunes were where we always took our end-of-the-school-year picnics, the red, white, and blue school buses of the 1950s lining up at the base of the dunes from all over Michigan. It got so that by third grade we'd complain about going to the dunes *again*—as in vanilla ice cream, *again*. We didn't know that our vanilla ice cream was everyone else's pistachio nut ripple.

When the U.S. Army Corps of Engineers first began surveying the land in the dunes in 1957, there were rumblings about a national park some day. This was dismissed as preposterous by most people because everyone knew you didn't put national parks where there were people living. Others thought it was more wishful thinking, grandiose notions, like the one about the mining company that was supposed to come in and mine the sand.

Realists believed no one in his right mind would ever want that old pile of sand. My father, a lumberman who already owned large tracts of land in and around the dunes, began quietly buying up more duneland, even buying some from the State of Michigan. Most people were only too glad to sell their land in the dunes to anyone fool enough to buy it.

Did my father know about the park coming in? It's not clear. His secretary with whom he was having an affair, an affair that I was privy to and the unwitting beard for at the ages of four and five, was active in the statewide Democratic party. My child's mind was aware only of endless maps, wall-sized blueprints, and the army cot that mysteriously materialized one day in my father's office and that he said he needed for "naps." How much my father's secretary may have lobbied for the park with Democratic senators on the one side, and tipped my father's hand in his land speculation on the other, is not known. What is known is that in all original and final plans her land along the base of the dunes on M-109 was excised from the park. The park boundaries snaked along, encompassing everything in their path, until coming to her land; then, just as if an invisible snag had been hit, the boundaries blipped up and around her land and then back down to the road again a mile later.

In 1961, Democratic Senators Philip Hart and Patrick McNamara, both now deceased, first introduced a national park bill to Congress. At this time public hearings were held around the area. Public sentiment against the park was running high. I can remember my mingled embarrassment and pride as my father—in work boots and wool plaid work shirt—walked onto the stage of the Traverse City High School and demanded, red-faced and using poor grammar, that Senator Hart and "the big boys" in Washington give him and other landowners fair value for their land. How smooth and gracious Senator Hart was by comparison, how schooled in public speaking.

Through all this, local people who hated the idea of, as they phrased it, "the government coming in here and running our lives," saw my father as a champion of capitalism, synonymous in their minds with democracy. My family was staunchly Republican and so was the rest of the community. There is nothing like having the majority on your side to lull you into an unthinking certainty of being right. I mean, does water run uphill?

Back in the classroom at Glen Lake Community School in Maple City, my English teacher assigned us a "paper" on the dunes controversy. He was what people in the community called "an odd duck." He wore his hair longer than most, had students call him "Pete" if they came by his house on weekends, and was the first English teacher I ever had who assigned anything that wasn't in the textbook. He was fired the next year. I liked him and never dreamed he would disapprove when I wrote a paper against the park. He gave me a C+, the lowest grade I ever got from him, and told me to think about it some more.

I've had three decades to think about it since then. And watch it. Just recently I read in the newspaper, in March of 1989, that a group of Glen Arbor citizens got together and purchased nonpark land in the village to protect it from development—rampaging development just beyond the boundaries of the park, development that was caused, ironically, by the presence of the national park. This private purchase of land for public use would have been unthinkable thirty years ago; it simply would not have occurred to anyone.

I've seen the issue change, but mainly I've seen the terminology change. My father, who was viewed in the fifties as "a rugged individualist" and "champion of free enterprise," became by the mid-1970s, "a developer." The tide of public opinion changed through the years so the predominant public mood of wanting to fight the bureaucracy became one of wanting to protect the environment. My father went from hero to villain in the public's eyes in the same time period that he made the reverse journey in mine.

In 1974 I came home from New York City and worked for my father managing his scenic drive in the dunes. I was twenty-nine. I had come home from Manhattan emotionally devastated and thirty thousand dollars in debt following a protracted custody battle and divorce. I did not die of cancer, like Debra Winger in *Terms of Endearment,* but I wanted to. My father told me if I could stop crying long enough to rake the four acres in my front yard at the dunes farmhouse, he'd then talk to me about managing the park. I did both, and for the first time in nearly twenty years we were friends again.

He would come into my orange kitchen at 6:00 A.M. and have a cup of coffee and ask how the trucks were running. Was the backhoe holding up? We would talk with the big dunes behind us, outside the long, high farmhouse windows, somnolent in the early morning light. A goddess. Unnamed. But she made her presence felt.

My father was dark, tense, thick-bodied, quiet, sensual. Ours was a silent communication. This was the man who'd taught me awe and joy as a child, simply by pointing. Birds' nests with tiny blue-speckled eggs. Birds' nests with pink, naked, open-mawed hatchlings. Snake babies under rocks. Ants carrying ant eggs in their mouths. Gordian horse hairs swimming in pools. Dappled newborn fawns motionless in dappled sunlight. Trout guts with caddis flies. Gray milt. Pinkish-yellow roe. The single magenta blossom in the center of the million-blossomed head of the Queen Anne's lace. Pitcher thistle. Bearberry. Arbutus. Bloodroot. The red roots of the sweetgrass in the troughs of the interdunal ponds. The way geese coming north in the spring sound like seagulls, high and mewling, and geese going south in the fall sound like a rocking chair, WONKA, WONKA; all business.

One morning he came in and said, "Your mother was a good woman. But she was awful critical of me. At one point she belonged to twelve different clubs and was president of eleven of them." This was a gloss. My father abandoned my mother, leaving her to raise five daughters alone.

In those days I wore shorts and halters until my father told me to "put some clothes on for Chrissake." My father was a female-dependent male. I know because I keep marrying him. He desperately needed and loved women, desperately needed to escape them, too. His mother kept him alive in a shoe box on the back of the wood stove. Women were his "source," and he feared being absorbed by them as much as he was drawn to them.

Everything is connected. My father's personal life and his treatment of my mother were not a separate thing from his land speculation. When I think of the dunes I think of my mother; I think of my father.

My first day on the job my father showed me how to drive the three-quarter-ton truck, the one with the holes in the floor. He took me and my two children, his grandchildren, whom he treated as if they were invisible, up into the shady forests of the back dunes to show us huge red trilliums in bloom.

But we had little contact. "Gran'pa never looks at me," my son said. One morning my daughter said she'd dreamed she'd seen Gran'pa walking up the driveway, bringing a horse on a lead to her. But it never happened.

My children played in the dunes by day, as I had. I knew without asking them how the sand felt like corduroy beneath their feet, how it squeaked. I knew without asking how the dunes were glittery in the daytime heat, heat waves rising like spirits from the sand, diaphanous. At night when the heat went out of the dunes they were cold as ice, dry ice.

One day my father showed up earlier than usual, and he didn't come into the kitchen. I felt, rather than saw, him on my porch. When I pulled on a pair of Levis and a T-shirt and went out onto the still-dark porch, he looked ashen, his skin the color of creek marl. It was August. I looked past him to the yard, at the trees and grass, then back. August is an odd time; the light is odd, making everything look like an overexposed color photo, the grass too yellow, the green trees black-green.

"Cindy died," he said.

So taciturn was my father, so closemouthed, that he almost expected to communicate by ESP, and almost did. It took me several minutes to recall that he had a black lab named Cindy.

"She was floating in the trout pond this morning," he said. His whole personality, or person, had a leaden quality, like weather waiting for rain. He said his wife had left the day before to go shopping downstate. He said it appeared Cindy had died of a heart attack.

My father had a second wife. He had not married his secretary after all, a small, dark Southern woman who'd stayed married to her husband. He had miraculously found another big-boned, tall, redheaded woman, almost identical to my mother in physical type, but unlike her in having an affinity for guns, hunting, and training dogs. She had been married to a doctor, an anesthesiologist who'd committed suicide by shooting himself in the head on a deserted road. She had a martial quality.

My father's big black Buick sat shiny on the gravel driveway. I looked from the car shining in the sun to him in the shadow of my porch. He looked odd to me, oddly weakened. At moments like this I forgave him everything. Forgave him leaving my mother, forgave him for walking away from me and my four sisters. Saw us all as intimate strangers sucked into the brutality of materialism, how we sold ourselves into marriage, into myriad forms of oppression for the sake of procreating, of surviving. My redhaired mother, a burning bush. My father, rain. The dog's death. The shadow of the thing, rather than the thing. It is not this; it is not that. It isn't the first thing; it isn't the second thing. It is all everything.

It was Cindy's death and the way my father had looked that had me in the blackberries, picking berries for a pie for his birthday September third. But he died. Suddenly and bizarrely, the day before his sixty-ninth birthday and the day after the United States government had awarded him three-and-a-half million dollars for his land in the dunes. Heart attack. His wife was in town shopping. A young boy with him had called the ambulance, but by the time it came the thirty miles from Kalkaska, he was dead.

"John Law," my father's wife kept saying and putting her arm around the shoulders of my then companion, a Leelanau County sheriff's deputy, "John Law. We have John Law here to take care of us." She rambled, going on about how her first husband had always cleaned his guns so carefully.

My father was buried in the hot, dry Kalkaska ground. Evergreen Cemetery. The undertaker had a pallor like that of his client, as if he had put powder over his five o'clock shadow. The service was Catholic, although my father was not. My mother attended at a distance, dry-eyed and resolute. My sisters flew in from all over the country and left again. It was not our style to sit around and talk and drink, and so we didn't.

I did not see my father again until November when his apparition accompanied me on the way to cut the Christmas tree up in the dunes. He didn't talk, which was normal, but simply walked along beside me, going up and over the furrows of the rows of planted Christmas trees.

I never saw his widow again. She moved to Sun City, Arizona. I never visited his home again. His house burned down six months after he died; burned to the ground before the fire trucks could arrive. I never saw the woman he had an affair with, his secretary, again either. But one day when I was living in Traverse City I called an ad in the classified section of the *Preview,* a shopping weekly. There was an ad for a waffle iron. When I called the woman, she said it was like new and only ten dollars. She asked my name and when I told her she said she was my father's old secretary, that she was disabled and couldn't drive. She said I should come and get the waffle iron, that if she could drive she'd bring it to me. She said she'd give it to me for nothing. That it would be free. If I'd come and get it. She was insistent, sticky, like insect legs caught in wool.

For days the image of this old crone kept coming back to me, this ghost from the past, fluttering her old lady's handkerchief at me across the years, a tiny speck on the horizon behind me, her tiny voice saying, "come and get the waffle iron." I pictured her in her wheelchair, all the raw passion for land and sex gone. My father, too, gone, his life measured more by what his mistresses did him out of than by his daughters, his grandchildren, his wonderful, visionary sense of land. His body, like an outdated piece of machinery, rusting in the ground, the primal urges and the transcendent alike, equally dead.

Only the dunes stay, glittery in the noonday heat and cold under the stars at night. Mythic, monumental, mysterious, the dunes lie like a lovely lady napping, waiting to inspire the next insignificant mortal to take a run at her heights, her steep slopes, her massiveness, her soft yet hard gemlike beauty, her ability to not only outlast, but confound, those who would try to possess her.

RESPONDING JOURNAL

Stocking's title indicates several possible figures of speech. She could be constructing a metaphor: "My father is the dunes." She could be personifying the dunes by attributing to them the human characteristics of her father. Or she could be setting up a simile to say that her father is like the dunes. In a journal entry, decide why Stocking might be making such a comparison and then use that thesis as the main support for your assessment of how features of a landscape can be used to shed light on a human being.

QUESTIONS FOR CRITICAL THINKING

STRATEGY

1. Does Stocking's attitude toward her subject seem to change in the course of the essay? If you believe it does, where would you locate her shifts from one feeling toward another?
2. Think about Stocking's title. Can you suggest (a) a subtitle for the essay that would suggest her main point, or (b) another title for the essay that you think would work at least as well?

ISSUES

1. Do you find it interesting that Stocking discusses her father in relation to dunes? Think about both the family situation and the personalities of the people involved. How far might you take generalizations regarding people which come from thinking about such different landscapes?
2. Stocking's father is a fairly complex individual, and so are her responses to him. Make lists of characteristics that indicate his ability as a father on the one hand and as a businessman on the other. What are his strengths and weaknesses? How do you relate these qualities to the dunes themselves?

COLLABORATIVE WRITING ACTIVITY

Explain what Stocking means in the section beginning "One morning he came in and said, 'Your mother was a good woman.'" Indicate why you feel she has included the details she mentions in these three paragraphs. Discuss your answers with your group; your recorder will compile a full explanation to present to the assembled class.

EXPLORING JOURNAL

"Every kind of writing relates to natural materials," Gary Snyder says in his essay "Tawny Grammar." Snyder's comment is extreme—or is it? Think of some examples of your experience of this idea, and explore them in a substantial journal entry of several paragraphs.

"Smokey the Bear Sutra"
Gary Snyder

Gary Snyder is known for his poetry of the California mountains even more than for his essays. Winner of the Pulitzer Prize for his poetry collection Turtle Island, *Snyder has published several volumes of verse. He spent considerable time in Japan; the influence of the east is apparent in many of his works, where he situates eastern ideas in the landscapes of the western United States.*

The Beat writer Jack Kerouac presents a thinly disguised picture of Snyder as the character Japhy Ryder in his novel The Dharma Bums. *This figure shows Snyder's place in American literature: he moves from the Beat generation of the 1950s through the eastern-influenced 1960s to our contemporary sense of a physical planet that requires human cooperation for its continued well-being.* Turtle Island *is the Native American name for North America; Snyder has sought throughout his career ways to preserve the sanctity of this particular island.*

Snyder originally published this poem as a giveaway pamphlet; see if you can determine the relation between his interest in eastern religions and the "Smokey the Bear" anti-fire campaigns waged by the United States Forestry Service. Smokey celebrated his 50th anniversary in 1993; you have probably seen his cute image on billboards or advertisements. A sutra is generally a Hindu or Buddhist sacred text, and more specifically a narrative associated directly with or related by Buddha. How does Snyder find ways to unite eastern and western ideas in this poem?

Once in the Jurassic, about 150 million years ago,
the Great Sun Buddha in this corner of the Infinite
Void gave a great Discourse to all the assembled elements
and energies: to the standing beings, the walking beings,
the flying beings, and the sitting beings—even grasses,
to the number of thirteen billion, each one born from a
seed, assembled there: a Discourse concerning
Enlightenment on the planet Earth.

"In some future time, there will he a continent called
America. It will have great centers of power called
such as Pyramid Lake, Walden Pond, Mt. Rainier, Big Sur,
Everglades, and so forth; and powerful nerves and channels
such as Columbia River, Mississippi River, and Grand
Canyon. The human race in that era will get into troubles all
over its head, and practically wreck everything in spite of its
own strong intelligent Buddha-nature."

"The twisting strata of the great mountains and the pulsings
of great volcanoes are my love burning deep in the earth. My
obstinate compassion is schist and basalt and granite, to be
mountains, to bring down the rain. In that future American
Era I shall enter a new form: to cure the world of loveless
knowledge that seeks with blind hunger, and mindless rage
eating food that will not fill it."

And he showed himself in his true form of

SMOKEY THE BEAR.

A handsome smokey-colored brown bear standing on his
hind legs, showing that he is aroused and watchful.

Bearing in his right paw the Shovel that digs to the
truth beneath appearances; cuts the roots of useless attachments, and flings damp sand on the fires of greed and war;

His left paw in the Mudra of Comradely Display—
indicating that all creatures have the full right to live to
their limits and that deer, rabbits, chipmunks, snakes,
dandelions, and lizards all grow in the realm of the Dharma;

Wearing the blue work overalls symbolic of slaves and
laborers, the countless people oppressed by a civilization that
claims to save but only destroys;

Wearing the broad-brimmed hat of the West, symbolic of
the forces that guard the Wilderness, which is the Natural
State of the Dharma and the True Path of beings on earth;
all true paths lead through mountains—

With a halo of smoke and flame behind, the forest fires
of the kali-yuga, fires caused by the stupidity of those who
think things can be gained and lost whereas in truth all is
contained vast and free in the Blue Sky and Green Earth
of One Mind;

Round-bellied to show his kind nature and that the great
earth has food enough for everyone who loves her and trusts
her;

Trampling underfoot wasteful freeways and needless
suburbs; smashing the worms of capitalism and totalitarianism;

Indicating the Task: his followers, becoming free of cars,
houses, canned food, universities, and shoes, master the
Three Mysteries of their own Body, Speech, and Mind; and
fearlessly chop down the rotten trees and prune out the
sick limbs of this country America and then burn the leftover
trash.

Wrathful but Calm, Austere but Comic, Smokey the Bear
will illuminate those who would help him; but for those
who would hinder or slander him,

HE WILL PUT THEM OUT.

Thus his great Mantra:

Namah samanta vajranam chanda maharoshana
Sphataya hum traka ham mam
"I DEDICATE MYSELF TO THE UNIVERSAL
    DIAMOND
BE THIS RAGING FURY DESTROYED"

And he will protect those who love woods and rivers, Gods
and animals, hobos and madmen, prisoners and sick people,
musicians, playful women, and hopeful children;

And if anyone is threatened by advertising, air pollution, or
the police, they should chant SMOKEY THE BEAR'S WAR
SPELL:

DROWN THEIR BUTTS
CRUSH THEIR BUTTS
DROWN THEIR BUTTS
CRUSH THEIR BUTTS

And SMOKEY THE BEAR will surely appear to put the
enemy out with his vajra-shovel.

Now those who recite this Sutra and then try to put it in
    practice will accumulate merit as countless as the sands
    of Arizona and Nevada,

Will help save the planet Earth from total oil slick,
Will enter the age of harmony of man and nature,
Will win the tender love and caresses of men, women, and
    beasts
Will always have ripe blackberries to eat and a sunny spot
    under a pine tree to sit at,

AND IN THE END WILL WIN HIGHEST PERFECT
    ENLIGHTENMENT.

                thus have we heard.

                              (may be reproduced free forever)

RESPONDING JOURNAL

Explain your reaction to Snyder's poem. Does the context help you to recognize unfamiliar words such as *Mundra* and *vajra-shovel?* How effective do you find Snyder's humor as a means of engaging your interest? Record your response in a substantial entry.

QUESTIONS FOR CRITICAL THINKING

STRATEGY

1. How well has Snyder integrated his examples from the natural world with more "academic" ideas? For example, consider Smokey and the "Great Sun Buddha."
2. Do you find that Snyder's presentation of "SMOKEY THE BEAR'S WAR SPELL" reminds you of anything else? Explain the effect it has on you; what might Snyder have hoped to accomplish with this spell?

ISSUES

1. What do you think about mingling elements of different philosophies as Snyder does? Do you feel you gain understanding of Eastern philosophy when you see it applied to the American landscape?
2. What does Snyder seem to think constitutes an accurate or logical means for distinguishing human life from other forms of life? Explain your response.

COLLABORATIVE WRITING ACTIVITY

Snyder printed this poem as a free pamphlet with the words "may be reproduced free forever" on it. How does such an instruction reflect the ideas of the poem itself? Explain how the poem's nature and its method of publication relate to each other. Once you and your classmates have written your responses, share them in order to provide a written overview of possible readings of Snyder's poem. Draft together a response that you would send to Snyder as your reply to the ideas embodied in his poem.

SUGGESTIONS FOR FORMAL WRITING ASSIGNMENTS

1. In her comments on Gary Snyder's collection of essays *The Practice of the Wild,* Gretel Ehrlich states that in these essays Snyder is "using the lessons of the planet to teach us how to live." In a substantial essay, explain "the lessons of the planet" you feel are most central to a responsible pattern for living. Indicate both the human and the organic, non-human aspects of your proposal for using these lessons as a guide to our human behavior.

2. Figurative language often adds another perspective to a situation, as you have probably noticed in your reading. Choose an important issue in your region or community, and write an essay in which you parallel your proposed solution to that dilemma in terms of a figurative analogy; for example, if you wish to explore the overconsumption of water in a suburb, you might parallel the residents to a thirsty monster bullying other animals by sucking up all available water. Try to dramatize the most blatant aspects of the problem in your analogy.

3. Writing about science often seems to provide many analogies to larger human concerns. For example, Peter Matthiessen's study *The Snow Leopard,* winner of the National Book Award, explores links between Himalayan sheep and goats (the objective of the expedition) but also seeks a glimpse of the exceedingly rare snow leopard. In writing about observing the sheep and goats, he muses on the hope of seeing the elusive snow leopard until it becomes a metaphor for humanity's quest to test itself by reaching known limits and then exceeding them. Think of some experience you have undergone that tested your limits, perhaps of courage or patience or endurance. Think of a figurative construction—perhaps metaphor, analogy, or simile—that would parallel your experience, and write an essay in which you extend and explore that figurative construction in relation to your own experience.

4. Research the history of Smokey the Bear. What possible uses beyond preventing forest fires might he have? Write out a plan to describe how Smokey should be used as a symbol for the next 50 years of his life. Does Snyder's poem give you any ideas to help you get started?

PART FOUR

Argumentation
and Research

PRESENTING AN ARGUMENT: LANDSCAPE AS POLEMIC

Arthur Rothstein, *Copper Mine and Miners' Homes, Meaderville, Montana,* 1939

1. How is this photograph organized? What similarities do you see between a landscape painting and this photo?

2. What is the effect of the mine on the landscape depicted in this photograph? What specific details that you see here are directly related to the mine and its operation?

3. Imagine yourself living in the town, Meaderville, depicted in this photo. What would your life be like? Explain the extent to which your life would be affected by the land around you. Consider, too, how that surrounding land has been affected by the people living there.

Chapter 11

PRESENTING AN ARGUMENT: LANDSCAPE AS POLEMIC

When Dave Foreman in *Confessions of an Eco-Warrior* advocates unusual methods to proclaim the primary importance of the planet earth and guarantee its safety from polluters of various kinds, he is arguing in order to persuade. Whenever you try to get someone to do something, you are engaging in persuasion. Argumentation and persuasion, though closely related, refer to the two different methods of getting someone to agree with you. *Argumentation* means using a scientific approach to your presentation by basing your work on logic and clarity. *Persuasion* means using the power of emotion you can arouse in your reader through your presentation as a means of making your points. Many presentations designed to sway a reader employ techniques characteristic of both argumentation and persuasion.

In antiquity, rhetoric was one of the most respected of the sciences; rhetoric is the art of shaping not only people's thoughts but their conduct as well. As you might guess, rhetoric was closely allied to oratory because the art of influencing audiences through effective speeches was a more prevalent form of communication than writing. In our own times, debate takes advantage of the strengths of both argumentation and persuasion. Your intellect may be compelled by the logic of a sound proposal, but if the person who makes the presentation offends you in any way, your appraisal of the argument will be affected. This connection is the reason politicians have strategists and image manipulators to make sure not only that their candidates' presentations rely on hard data to support positions, but also that they created the proper sort of emotional reaction to get a voter to take the plunge and vote for their candidate.

When you write anything, you have a point you want to make—even in a brief note. You have a purpose in putting the words down on paper. You want to let your roommate know where you will be, or you want your mother or sister to bring your sweater when she comes to pick you up. You have a definite idea of the result you want to see from your writing. In the same way, you are expectant when a classmate or instructor is reading a draft of your essay: you want to know what that reader has to say about how well you have achieved your purpose, whether you have answered all the questions your reader might ask, if all your points and examples are clear.

One of your jobs as a writer, then, is to see that your presentation is as effective as possible. In addition to thinking about the three main concerns of the writer that we have already discussed—thesis, audience, and stance—you must now think about the quality of the evidence you offer your reader. To succeed as an argumentative writer, you have to make sure that your message is clear, that you have taken into account the special needs of your readers for this particular piece, and that you have offered sufficient proof to be sure your readers are convinced. You might be thinking that all the writing you have done so far has been argumentative writing, and you would be correct. The difference is that at this point, we will analyze more closely the characteristics of effective writing and think about readings that employ principles of effective argument. In this way, we will focus on the means of convincing others of our ideas.

In order to help you see quickly how you might set up an argument in an essay, here are two models. The first is a suggestion of how to conceptualize an argument. You structure your essay with the following parts:

- provide in your introduction a clear statement of the contention you will prove or demonstrate;
- provide at least three reasons for holding your contention;
- provide specific data (examples, statistics, testimony) to back up your reasons;
- conclude your presentation with a strong restatement of the original contention.

The resulting essay should demonstrate this structure clearly; you should be able to take out topic sentences and transitions to show the skeleton of your argument.

The second method involves a less prescriptive structure. Here, you have a structural guide rather than a blueprint:

- begin with an interesting idea or proposition to catch the reader's attention;
- provide a clear statement of the proposition being argued (a "grabber");
- explain your support and how you will demonstrate the accuracy and adequacy of that support (here again you use examples, statistics, and testimony);
- offer a counterposition that represents the opposing viewpoint fairly but which you effectively argue against;
- conclude with a restatement of the original proposition, perhaps saving one effective example as final clincher.

Remember to keep the structure of your argument clear and obvious so your readers will have no difficulty in following your line of reasoning. Carefully choose your examples, details, data, statistics, anecdotes, or testimony and make them easy to follow. If you do, your essay will demonstrate the careful logic required in argumentation. Even though persuasion relies more on emotion than logic, it too benefits from clear and orderly presentation.

For the rest of this introduction to argumentation and persuasion, we will focus on defining the terms and the situations where the precise meanings of these categories can help us to shape meaning in our writing. First of all, we require greater precision in determining exactly what *argumentation* and *persuasion* mean in relation to each other. Any writing we do is a rhetorical situation because we always have in mind an objective we want to achieve with our writing. Even when we have a grocery list, we have the purchase of bread, milk, and onions as our desired result; we communicate this knowledge to ourselves in direct terms so we can avoid misunderstanding and bringing home the wrong items. From this simple rhetorical situation, we move to such complex rhetorical situations as the debates over habitat preservation or endangered species. Argumentation refers to the logical side of a rhetorical situation, while persuasion indicates the emotional. This sort of division goes back to the earliest practitioners of rhetoric and indicates that old distinction between the head and the heart, between our logical faculty, which tells us is correct and supportable, and our emotional faculty, which tells us what we like or what we want to believe—whether or not that belief has any basis in logic.

For a quick test of this distinction, look at the ads in any glossy magazine. Or look at the automobile section of a Sunday newspaper. Do you find presentations that encourage a positive emotional response and others that encourage a response requiring some thinking to convince you? What sorts of devices do the ads use? What do these devices tell you about the assumptions of the advertiser and the intended audience?

Much of your writing will be like some of the ads you have seen: you will be mixing argument and persuasion, logic and emotion, in your presentations. In so doing, you will be facing three sorts of audiences: those who agree with your position; those who are neutral (who have not yet made up their minds or who have not thought much about the issue); and those who oppose your position. Like the people who generate the ads you have seen, you will need to get some sense of your audience. If your target audience is your classmates, you can question them directly about issues affecting you. Successful argumentation and persuasion require careful research so you are able to predict what your opponents will present, or so you will understand precisely the beliefs held by those who oppose your position and can attempt to sway them.

You might ask, "Can I cover all three audiences with a single presentation?" The answer is qualified: sometimes. You will have to determine exactly how strong your case is in relation to the arguments of your opponent. Your common sense and your past history with oral persuasion—getting your parents to let you spend the night at a friend's house, to take the car out alone, or to take an out-of-town trip—all of these at some time or other required you to present a logical argument, to make an emotional plea, or to combine the two. Do you remember how you had to adjust your plea to get your parents to respond? Was it enough just to say "I want!" or did you have to make concessions? Use such experiences to help you understand your reader's point of view. You have to anticipate responses to your arguments, and decide how best to respond to those arguments.

Remember, your tone and your manner of presentation must lead your reader to a point of accepting and perhaps even endorsing your presentation; simply telling your reader that you are correct or that any fool can see you're right is not likely to convince anyone to modify her or his position. You and your peer reviewers can help one another tremendously by playing devil's advocate and arguing the opposing viewpoint as you plan and draft your writing. This role playing is one of the best possible strategies for developing a strong presentation that addresses all the key ideas your opponent might use to counter your points.

The selections in this section reflect different modes of argument in relation to the earth and people's treatment of the planet. Much of the important knowledge to be gained in this chapter will come from your putting these arguments against one another and deciding which portions of each seem strongest to you. These areas of ecological and environmental concerns are far from settled today, as you can see if you look at a magazine such as *Wildlife Conservation,* where you can readily find up-to-date articles on pending legislation and political issues that have people sharply divided. After examining both sides in many of these issues, you can determine which side seems to have the stronger case.

Sometimes we feel that the claims or positions offered by science run counter to the evidence we intuit, or to what literature or religion tells us. In his essay, Lewis Thomas shows us how he uses his scientific training to interpret data in a way that accommodates his intuitive insights. Stegner, Walker, and Silko look at landscapes crucial to their worldviews and argue from these particular places how we all might see our worlds more clearly. Sahtouris begins her argument with the proposition that the earth is still a child as she considers our environmental crises from a truly global perspective. What sorts of differences in argument do you see between Foreman's specific recommendations for ecotage and Sahtouris' exhortation for us to adopt the Gaia viewpoint?

"The Music of *This* Sphere"
Lewis Thomas

An M.D. and President of the Memorial Sloan-Kettering Cancer Center in New York City, Lewis Thomas is also gifted as a writer. He can present accounts of complex facets of the world of nature in terms that enable considerable numbers of readers to understand scientific principles far better than they could before they read Thomas' essays. His collection of essays The Lives of a Cell *was published in 1974 and won the National Book Award for Arts and Letters in 1975; "The Music of* This *Sphere" appeared in that book.*

"Nature abhors a long silence," says Thomas, and he uses the ancient concept of unheard celestial music to suggest that only as we increase our scientific under-standing can we comprehend what our hearts tell us is true. Thomas uses varied evidence to argue his position as he alternates between human and natural worlds to explain how all the sounds we hear tell us about the nature of communication. Music, he argues, is vital to the human world.

EXPLORING JOURNAL

What sounds do you associate with the natural world? When do you most vividly hear sounds outside? What is the single most memorable sound you have heard in the woods, in your backyard, or on the street? Is it crickets, cicadas, a loon, coy-dogs, geese, or birds? In a focused freewriting entry, try to explain some of the reasons why this particular sound has made such an impression on you.

It is one of our problems that as we become crowded together, the sounds we make to each other, in our increasingly complex communication systems, become more random-sounding, accidental or incidental, and we have trouble selecting meaningful signals out of the noise. One reason is, of course, that we do not seem able to restrict our communication to information-bearing, relevant sig-nals. Given any new technology for transmitting information, we seem bound to use it for great quantities of small talk. We are only saved by music from being overwhelmed by nonsense.

It is marginal comfort to know that the relatively new science of bioacoustics must deal with similar problems in the sounds made by other animals to each other. No matter what sound-making device is placed at their disposal, creatures in general do a great deal of gabbling, and it requires long patience and observa-tion to edit out the parts lacking syntax and sense. Light social conversation, designed to keep the party going, prevails. Nature abhors a long silence.

Somewhere, underlying all the other signals, is a continual music. Termites make percussive sounds to each other by beating their heads against the floor in the dark, resonating corridors of their nests. The sound has been described as resembling, to the human ear, sand falling on paper, but spectrographic analysis of sound records has recently revealed a high degree of organization in the drum-ming; the beats occur in regular, rhythmic phrases, differing in duration, like notes for a tympani section.

From time to time, certain termites make a convulsive movement of their mandibles to produce a loud, high-pitched clicking sound, audible ten meters off. So much effort goes into this one note that it must have urgent meaning, at least to the sender. He cannot make it without such a wrench that he is flung one or two centimeters into the air by the recoil.

There is obvious hazard in trying to assign a particular meaning to this special kind of sound, and problems like this exist throughout the field of bioacoustics. One can imagine a woolly-minded Visitor from Outer Space, interested in human beings, discerning on his spectrograph the click of that golf ball on the surface of the moon, and trying to account for it as a call of warning (unlikely), a signal of mating (out of the question), or an announcement of territory (could be).

Bats are obliged to make sounds almost ceaselessly, to sense, by sonar, all the objects in their surroundings. They can spot with accuracy, on the wing, small insects, and they will home onto things they like with infallibility and speed. With such a system for the equivalent of glancing around, they must live in a world of ultrasonic bat-sound, most of it with an industrial, machinery sound. Still, they communicate with each other as well, by clicks and high-pitched greetings. Moreover, they have been heard to produce, while hanging at rest upside down in the depths of woods, strange, solitary, and lovely bell-like notes.

Almost anything that an animal can employ to make a sound is put to use. Drumming, created by beating the feet, is used by prairie hens, rabbits, and mice; the head is banged by woodpeckers and certain other birds; the males of death-watch beetles make a rapid ticking sound by percussion of a protuberance on the abdomen against the ground; a faint but audible ticking is made by the tiny beetle *Lepinotus inquilinus,* which is less than two millimeters in length. Fish make sounds by clicking their teeth, blowing air, and drumming with special muscles against tuned inflated air bladders. Solid structures are set to vibrating by toothed bows in crustaceans and insects. The proboscis of the death's-head hawk moth is used as a kind of reed instrument, blown through to make high-pitched, reedy notes.

Gorillas beat their chests for certain kinds of discourse. Animals with loose skeletons rattle them, or, like rattlesnakes, get sounds from externally placed structures. Turtles, alligators, crocodiles, and even snakes make various more or less vocal sounds. Leeches have been heard to tap rhythmically on leaves, engaging the attention of other leeches, which tap back, in synchrony. Even earthworms make sounds, faint staccato notes in regular clusters. Toads sing to each other, and their friends sing back in antiphony.

Birdsong has been so much analyzed for its content of business communication that there seems little time left for music, but it is there. Behind the glossaries of warning calls, alarms, mating messages, pronouncements of territory, calls for recruitment, and demands for dispersal, there is redundant, elegant sound that is unaccountable as part of the working day. The thrush in my backyard sings down his nose in meditative, liquid runs of melody, over and over again, and I have the strongest impression that he does this for his own pleasure. Some of the time he seems to be practicing, like a virtuoso in his apartment. He starts a run, reaches a midpoint in the second bar where there should be a set of complex harmonics, stops, and goes back to begin over, dissatisfied. Sometimes he changes his notation so conspicuously that he seems to be improvising sets of variations. It is a meditative, questioning kind of music, and I cannot believe that he is simply saying, "thrush here."

The robin sings flexible songs, containing a variety of motifs that he rearranges to his liking; the notes in each motif constitute the syntax, and the possibilities for variation produce a considerable repertoire. The meadow lark, with three hundred notes to work with, arranges these in phrases of three to six notes and elaborates fifty types of song. The nightingale has twenty-four basic songs, but gains wild variety by varying the internal arrangement of phrases and the length of pauses. The chaffinch listens to other chaffinches, and incorporates into his memory snatches of their songs.

The need to make music, and to listen to it, is universally expressed by human beings. I cannot imagine, even in our most primitive times, the emergence of talented painters to make cave paintings without there having been, near at hand, equally creative people making song. It is, like speech, a dominant aspect of human biology.

The individual parts played by other instrumentalists—crickets or earthworms, for instance—may not have the sound of music by themselves, but we hear them out of context. If we could listen to them all at once, fully orchestrated, in their immense ensemble, we might become aware of the counterpoint, the balance of tones and timbres and harmonics, the sonorities. The recorded songs of the humpback whale, filled with tensions and resolutions, ambiguities and allusions, incomplete, can be listened to as a *part* of music, like an isolated section of an orchestra. If we had better hearing, and could discern the descants of sea birds, the rhythmic tympani of schools of mollusks, or even the distant harmonics of midges hanging over meadows in the sun, the combined sound might lift us off our feet.

There are, of course, other ways to account for the songs of whales. They might be simple, down-to-earth statements about navigation, or sources of krill, or limits of territory. But the proof is not in, and until it is shown that these long, convoluted, insistent melodies, repeated by different singers with ornamentations of their own, are the means of sending through several hundred miles of undersea such ordinary information as "whale here," I shall believe otherwise. Now and again, in the intervals between songs, the whales have been seen to breach, leaping clear out of the sea and landing on their backs, awash in the turbulence of their beating flippers. Perhaps they are pleased by the way the piece went, or perhaps it is celebration at hearing one's own song returning after circumnavigation; whatever, it has the look of jubilation.

I suppose that my extraterrestrial Visitor might puzzle over my records in much the same way, on first listening. The 14th Quarter might, for him, be a communication announcing, "Beethoven here," answered, after passage through an undersea of time and submerged currents of human thought, by another long signal a century later, "Bartok here."

I, as I believe, the urge to make a kind of music is as much a characteristic of biology as our other fundamental functions, there ought to be an explanation for it. Having none at hand, I am free to make one up. The rhythmic sounds might be the recapitulation of something else—an earliest memory, a score for the transformation of inanimate, random matter in chaos into the improbable, ordered dance of living forms. Morowitz has presented the case, in thermodynamic terms, for the hypothesis that a steady flow of energy from the inexhaustible source of the sun to the unfillable sink of outer space, by way of the earth, is mathematically destined to cause the organization of matter into an increasingly ordered state. The resulting balancing act involves a ceaseless clustering of bonded atoms into molecules of higher and higher complexity, and the emergence of cycles for the storage and release of energy. In a nonequilibrium steady state, which is postulated, the solar energy would not just flow to the earth and radiate away; it is thermodynamically inevitable that it must rearrange matter into symmetry, away from probability, against entropy, lifting it, so to speak, into a constantly changing condition of rearrangement and molecular ornamentation. In such a system, the outcome is a chancy kind of order, always on the verge of descending into chaos, held taut against probability by the unremitting, constant surge of energy from the sun.

If there were to be sounds to represent this process, they would have the arrangement of the Brandenburg Concertos for my ear, but I am open to wonder whether the same events are recalled by the rhythms of insects, the long, pulsing runs of birdsong, the descants of whales, the modulated vibrations of a million locusts in migration, the tympani of gorilla breasts, termite heads, drumfish bladders. A "grand canonical ensemble" is, oddly enough, the proper term for a quantitative model system in thermodynamics, borrowed from music by way of mathematics. Borrowed back again, provided with notation, it would do for what I have in mind.

RESPONDING JOURNAL

Think of some ways in which animals in the wild might use sounds and sound systems to communicate certain emotions. Thomas suggests that they have certain signatures, much more than a simple statement of "_____'s here." Freewrite what your own sound would be like if you were being described by Thomas's extraterrestrial visitor. Give your own account of what your sound and mood would be like when that visitor hears "_____'s here."

QUESTIONS FOR CRITICAL THINKING

STRATEGY

1. If you are not familiar with the term "music of the spheres," look it up in a dictionary or encyclopedia. How does Thomas work against the meaning of this phrase in his essay? What effect does he achieve by doing so?
2. Look at the paragraph beginning "Almost anything that an animal can employ to make a sound is put to use." What function does this sentence

serve in relation to the rest of the paragraph? Does Thomas employ similar sentences in other paragraphs in the same way?

ISSUES

1. Thomas makes frequent comparisons between animal sounds and those made by humans. To what extent do you find his parallels valid? For example, how convincing is his case for animal Brandenburg Concertos?

2. Both Nabhan and Thomas, being scientists, want to explain the phenomena they observe. As they move toward their conclusions, they suggest at least tentative solutions to the dilemmas they present by classifying data in the bodies of their essays. Are the explanations equally convincing?

COLLABORATIVE WRITING ACTIVITY

Using this essay as a starting point, think of other examples from the natural world that suggest a pattern, a kind of organizing principle at work. Is there another kind of "music," a harmony or pattern or system of classification, that you think you might have detected? Explain your idea as best you can and bring it to your group, where you can try it out on the other members.

"The Dump Ground" from *Wolf Willow*
Wallace Stegner

Perhaps the most revered writer of the American west in the second half of the twentieth century, Wallace Stegner has written history and fiction about various aspects of the plains and mountain states. His study Beyond the Hundredth Meridian: John Wesley Powell and the Second Opening of the West *is one of the best works of American history; his fiction has won him both the Pulitzer Prize and the National Book Award. In* Joe Hill, *he has written a "biographical novel" of the famous labor leader of the west. He often brings the techniques of fiction to his historical accounts of the west in order to make them even more effective. Born in Iowa in 1909 and killed in an auto accident in 1993, he was a revered teacher of writing at Wisconsin, Harvard, and Stanford, where for many years he directed the writing program and where the Stegner Fellowships continue.*

Often his work shows a profitable mixing of different kinds of writing: the full title of Wolf Willow *is* A History, a Story, and a Memoir of the Last Plains

Frontier. *The first chapter, "History Is a Pontoon Bridge," allows him the opportunity to create a metaphor for the creation of knowledge out of his own experiences. The second chapter is reprinted here.*

EXPLORING JOURNAL

What comes to mind when you think of a dump or dumping ground? What sorts of information could you gain from one? On the other hand, what sorts of dangers might you face there? Freewrite for ten minutes on your ideas of dump grounds.

———————

One aspect of Whitemud's history, and only one, and a fragmentary one, we knew: the town dump. It lay in a draw at the southeast corner of town, just where the river left the Hills and where the old Mounted Police patrol trail (I did not know that that was what it was) made a long, easy, willow-fringed traverse across the bottoms. That stretch of the river was a favorite campsite for passing teamsters, gypsies, sometimes Indians. The very straw scattered around those camps, the ashes of those strangers' campfires, the manure of their teams and saddle horses, were hot with adventurous possibilities. The camps made an extension, a living suburb, of the dump ground itself, and it was for this that we valued them. We scoured them for artifacts of their migrant tenants as if they had been archaeological sites potent with the secrets of ancient civilizations. I remember toting around for weeks a broken harness strap a few inches long. Somehow or other its buckle looked as if it had been fashioned in a far place, a place where they were accustomed to flatten the tongues of buckles for reasons that could only be exciting, and where they had a habit of plating the metal with some valuable alloy, probably silver. In places where the silver was worn away, the buckle underneath shone dull yellow: probably gold.

Excitement liked that end of town better than our end. Old Mrs. Gustafson, deeply religious and a little raddled in the head, went over there once with a buckboard full of trash, and as she was driving home along the river she saw a spent catfish, washed in from the Swift Current or some other part of the watershed in the spring flood. He was two feet long, his whiskers hung down, his fins and tail were limp—a kind of fish no one had seen in the Whitemud in the three or four years of the town's life, and a kind that none of us children had ever seen anywhere. Mrs. Gustafson had never seen one like him, either. She perceived at once that he was the devil, and she whipped up the team and reported him, pretty loudly, at Hoffman's elevator.

We could still hear her screeching as we legged it for the river to see for ourselves. Sure enough, there he was, drifting slowly on the surface. He looked very tired, and he made no great effort to get away when we rushed to get an old rowboat, and rowed it frantically down to where our scouts eased along shore beckoning and ducking willows, and sank the boat under him and brought him ashore in it. When he died we fed him experimentally to two half-wild cats, who seemed to suffer no ill effects.

Upstream from the draw that held the dump, the irrigation flume crossed the river. It always seemed to me giddily high when I hung my chin over its plank edge and looked down, but it probably walked no more than twenty feet above the water on its spidery legs. Ordinarily in summer it carried six or eight inches of smooth water, and under the glassy surface of the little boxed stream the planks were coated with deep sun-warmed moss as slick as frogs' eggs. A boy could sit in the flume with the water walling up against his back, and grab a cross-brace above him, and pull, shooting himself sledlike ahead until he could reach the next cross-brace for another pull, and so on across the river in four scoots.

After ten minutes in the flume he would come out wearing a dozen or more limber black leeches, and could sit in the green shade where darning needles flashed blue, and dragonflies hummed and stopped in the air, and skaters dimpled slack and eddy with their delicate transitory footprints, and there pull the leeches off one by one, while their sucking ends clung and clung, until at last, stretched far out, they let go with a tiny wet *puk* and snapped together like rubber bands. The smell of the flume and the low bars of that part of the river was the smell of wolf willow.

But nothing else in the east end of town was as good as the dump ground. Through a historical process that went back to the roots of community sanitation, and that in law dated from the Unincorporated Towns Ordinance of the territorial government, passed in 1888, the dump was the very first community enterprise, the town's first institution.

More than that, it contained relics of every individual who had ever lived there. The bedsprings on which Whitemud's first child was begotten might be out there; the skeleton of a boy's pet colt; books soaked with water and chemicals in a house fire, and thrown out to flap their stained eloquence in the prairie wind. Broken dishes, rusty tinware, spoons that had been used to mix paint; once a box of percussion caps, sign and symbol of the carelessness that most of us had in matters of personal or public safety. My brother and I put some of them on the railroad tracks and were anonymously denounced in the *Leader* for nearly derailing the speeder of a section crew. There were also old iron, old brass, for which we hunted assiduously, by night conning junkmen's catalogs to find out how much wartime value there might be in the geared insides of clocks or in a pound of tea lead carefully wrapped in a ball whose weight astonished and delighted us.

Sometimes the unimaginable world reached out and laid a finger on us because of our activities on the dump. I recall that, aged about seven, I wrote a Toronto junk house asking if they preferred their tea lead and tinfoil wrapped in balls, or whether they would rather have it pressed flat in sheets, and I got back a typewritten letter in a window envelope advising me that they would be happy to have it in any way that was convenient to me. They added that they valued my business and were mine very truly. Dazed, I carried that windowed grandeur around in my pocket until I wore it out.

We hunted old bottles in the dump, bottles caked with filth, half buried, full of cobwebs, and we washed them out at the horse trough by the elevators, putting in a handful of shot along with the water to knock the dirt loose; and when we had shaken them until our arms were tired, we hauled them down in somebody's coaster wagon and turned them in at Bill Christenson's pool hall, where the smell of lemon pop was so sweet on the dark pool-hall air that it sometimes awakens me in the night even yet.

Smashed wheels of wagons and buggies, tangles of rusty barbed wire, the collapsed perambulator that the French wife of one of the town's doctors had once pushed proudly up the plank sidewalks and along the ditchbank paths. A welter of foul-smelling feathers and coyote-scattered carrion, that was all that remained of somebody's dream of a chicken ranch. The chickens had all got some mysterious pip at the same time, and died as one, and the dream lay out there with the rest of the town's short history to rustle to the empty sky on the border of the Hills.

There was melted glass in curious forms, and the half-melted office safe left from the burning of Joe Knight's hotel. On very lucky days we might find a piece of the lead casing that had enclosed the wires of the town's first telephone system. The casing was just the right size for rings, and so soft that it could be whittled with a jackknife. If we had been Indians of fifty years earlier, that bright soft metal could have enlisted our maximum patience and craft, and come out as ring and medal and amulet inscribed with the symbols of our observed world. Perhaps there were too many ready-made alternatives in the local drug, hardware, and general stores; in any case our artistic response was feeble, and resulted in nothing better than crude seal rings with initials or pierced hearts carved in them. They served a purpose in juvenile courtship, but they stopped a good way short of art.

The dump held very little wood, for in that country anything burnable got burned. But it had plenty of old metal, furniture, papers, mattresses that were the delight of field mice, and jugs and demijohns that were sometimes their bane, for they crawled into the necks and drowned in the rainwater or redeye that was inside.

If the history of Whitemud was not exactly written, it was at least hinted, in the dump. I think I had a pretty sound notion even at eight or nine of how significant was that first institution of our forming Canadian civilization. For rummaging through its foul purlieus I had several times been surprised and shocked to find relics of my own life tossed out there to blow away or rot.

Some of the books were volumes of the set of Shakespeare that my father had bought, or been sold, before I was born. They had been carried from Dakota to Seattle, and Seattle to Bellingham, and Bellingham to Redmond, and Redmond back to Iowa, and Iowa to Saskatchewan. One of the Cratchet girls had borrowed them, a hatchet-faced, thin, eager, transplanted Cockney girl with a frenzy for reading. Stained in a fire, they had somehow found the dump rather than come back to us. The lesson they preached was how much is lost, how much thrown aside, how much carelessly or of necessity given up, in the making of a new country. We had so few books that I knew them all; finding those thrown away was like finding my own name on a gravestone.

And yet not the blow that something else was, something that impressed me even more with how closely the dump reflected the town's intimate life. The colt whose picked skeleton lay out there was mine. He had been incurably crippled when dogs chased our mare Daisy the morning after she foaled. I had worked for months to make him well, had fed him by hand, curried him, talked my father into having iron braces made for his front legs. And I had not known that he would have to be destroyed. One weekend I turned him over to the foreman of one of the ranches, presumably so that he could be better cared for. A few days later I found his skinned body, with the braces still on his crippled front legs, lying on the dump. I think I might eventually have accepted the colt's death, and forgiven his killer, if it had not been for that dirty little two-dollar meanness that skinned him.

Not even finding his body cured me of going to the dump, though our parents all forbade us on pain of cholera or worse to do so. The place fascinated us, as it should have. For this was the kitchen midden of all the civilization we knew. It gave us the most tantalizing glimpses into our neighbors' lives and our own; it provided an aesthetic distance from which to know ourselves.

The town dump was our poetry and our history. We took it home with us by the wagonload, bringing back into town the things the town had used and thrown away. Some little part of what we gathered, mainly bottles, we managed to bring back to usefulness, but most of our gleanings we left lying around barn or attic or cellar until in some renewed fury of spring cleanup our families carted them off to the dump again, to be rescued and briefly treasured by some other boy. Occasionally something we really valued with a passion was snatched from us in horror and returned at once. That happened to the mounted head of a white mountain goat, somebody's trophy from old times and the far Rocky Mountains, that I brought home one day. My mother took one look and discovered that his beard was full of moths.

I remember that goat; I regret him yet. Poetry is seldom useful, but always memorable. If I were a sociologist anxious to study in detail the life of any community I would go very early to its refuse piles. For a community may be as well judged by what it throws away—what it has to throw away and what it chooses to—as by any other evidence. For whole civilizations we sometimes have no more of the poetry and little more of the history than this.

It is all *we* had for the civilization we grew up in. Nevertheless there was more, much more. If anyone had known that past, and told us about it, he might have told us something like this:

—————➤·◄—————

RESPONDING JOURNAL

Think about what you have thrown away in your lifetime. What have you given up that you would like to see again? Why would this now be so important to you? Explain yourself in a substantial journal entry of several paragraphs.

QUESTIONS FOR CRITICAL THINKING

STRATEGY

1. Does Stegner include enough concrete detail to make his writing effective? Which of his images seem to be the most effective? Explain your response.
2. Think about the order in which Stegner has presented his findings. What are the key elements in his decision to present his discoveries in the order he does?

ISSUES

1. "The town dump was our poetry and our history," says Stegner. Which of these two is more important? Why? How important do you think such a distinction is?
2. In his memoir of his childhood, *Wolf Willow,* Stegner ends his recollections of the dump ground with a transition to his next chapter. What do you think Stegner goes on to discuss in his next chapter? What has he prepared you for in the chapter you have just read?

COLLABORATIVE WRITING ACTIVITY

Think about things you have thrown away that would tell someone something about you. Look at how Stegner has handled bits of his own life in this selection to help you jog your memory and recall some of the details of the object.

Once all members of your group have read about their objects, work as a group to organize and describe what people have thrown away. What categories seem to come about from this activity? Provide a structure in which to present your group's work to the rest of the class. As you do so, think about the nature of your task: as you do with the final stages of editing, you are deciding on the most effective means possible of organizing and presenting your material for your readers.

"The Universe Responds"
Alice Walker

Alice Walker, author of The Color Purple, *also appears in this text in Chapter 8, where "Everything Is a Human Being" is reprinted. Both "The Universe Responds" and "Everything Is a Human Being" are included in her collection of essays entitled* Living by the Word.

EXPLORING JOURNAL

Have you ever felt the universe respond to a request or wish you made? Have you ever been sorely disappointed when the universe did not? Explore these possibilities in a ten-minute freewriting entry in your journal.

———◆———

To some people who read the following there will seem to be something special or perhaps strange about me. I have sometimes felt this way myself. To others, however, what I am about to write will appear obvious. I think our response to "strangeness" or "specialness" depends on where we are born, where we are raised, how much idle time we have had to watch trees (long enough at least to notice there is not an ugly one among them) swaying in the wind. Or to watch rivers, rainstorms, or the sea.

A few years ago, I wrote an essay called "Everything Is a Human Being," which explores to some extent the Native American view that all of creation is of one substance and therefore deserving of the same respect. I described the death of a snake that I caused and wrote of my remorse. I wrote the piece to celebrate the birth of Martin Luther King, Jr., and I read it first to a large group of college students in California. I also read it other places, so that by summer (I had written it in winter) it had been read three or four times, and because I cannot bear to repeat myself very much, I put it away.

That summer "my" land in the country crawled with snakes. There was always the large resident snake, whom my mother named "Susie," crawling about in the area that marks the entrance to my studio. But there were also lots of others wherever we looked. A black-and-white king snake appeared underneath the shower stall in the garden. A striped red-and-black one, very pretty, appeared near the pond. It now revealed the little hole in the ground in which it lived by lying half in and half out of it as it basked in the sun. Garden snakes crawled up and down the roads and paths. One day, leaving my house with a box of books in his arms, my companion literally tripped over one of these.

We spoke to all these snakes in friendly voices. They went their way. We went ours. After about a two-week bloom of snakes, we seemed to have our usual number: just Susie and a couple of her children.

A few years later, I wrote an essay about a horse called Blue. It was about how humans treat horses and other animals; how hard it is for us to see them as the suffering, fully conscious, enslaved beings they are. It also marked the beginning of my effort to become non-meat-eating (fairly successful). After reading this essay in public only once, this is what happened. A white horse came and settled herself on the land. (Her owner, a neighbor, soon came to move her.) The two horses on the ranch across the road began to run up to their fence whenever I passed, leaning over it and making what sounded to my ears like joyful noises. They had never done this before (I checked with the human beings I lived with to be sure of this), and after a few more times of greeting me as if I'd done something especially nice for them, they stopped. Now when I pass they look at me with the same reserve they did before. But there is still a spark of *recognition.*

What to make of this?

What I have noticed in my small world is that if I praise the wild flowers growing on the hill in front of my house, the following year they double in profusion and brilliance. If I admire the squirrel that swings from branch to branch outside my window, pretty soon I have three or four squirrels to admire. If I look into the eyes of a raccoon that has awakened me by noisily rummaging through the garbage at night, and acknowledge that it looks maddeningly like a mischievous person—paws on hips, masked eyes, a certain impudent stance, as it looks back at me—I soon have a family of raccoons living in a tree a few yards off my deck. (From this tree they easily forage in the orchard at night and eat, or at least take bites out of, all the apples. Which is not fun. But that is another story.)

And then, too, there are the deer, who know they need never, ever fear me.

In white-directed movies about the Indians of the Old West, you sometimes see the "Indians" doing a rain dance, a means of praying for rain. The message delivered by the moviemaker is that such dancing and praying is ridiculous, that either it will rain or it will not. All white men know this. The Indians are backward and stupid and wasting their time. But there is also that last page or so in the story of Black Elk, in which his anthropologist/friend John Neihardt goes with him on a last visit to the Badlands to pray atop Horney Peak, a place sacred to the Sioux. It is a cloudless day, but the ancient Black Elk hopes that the Great Spirit, as in the real "old" days, will acknowledge his prayer for the good of his people by sending at least a few drops of rain. As he prays, in his old, tired voice, mostly of his love of the Universe and his failure to be perfect, a small cloud indeed forms. It rains, just enough to say "Yes." Then the sky clears. Even today there is the belief among many indigenous holy people that when a person of goodness dies, the Universe acknowledges the spirit's departure by sending storms and rain.

The truth is, in the country, where I live much of the time, I am virtually overrun by birds and animals—raccoons, snakes, deer, horses (occasionally). During a recent court trial at which a neighbor and I both happened to find ourselves, her opening words of greeting included the information that two wild pigs she'd somehow captured had broken out and were, she feared, holed up somewhere on my land.

But at least, I thought, my house in the city is safe.

But no.

One night after dinner, as some friends were leaving my house, I opened my front door, only to have a large black dog walk gratefully inside. It had obviously been waiting quietly on the stoop. It came into the hallway, sniffed my hands, and prepared to make itself at home, exactly as if it had lived in my house all its life. There was no nervousness whatsoever about being an intruder. No, no, I said, out you go! It did not want to go, but my friends and I persuaded it. It settled itself at the door and there it stayed, barking reproachfully until I went to bed. Very late that night I heard its owners calling it. George! they called. George! Here, George! They were cursing and laughing. Drunk. George made no response.

I suddenly realized that George was not lost. He had run away. He had run away from these cursing, laughing drunks who were now trying to find him. This realization meant the end of sleep for me that night as I lay awake considering my responsibility to George. (I felt none toward his owners.) For George obviously "knew" which house was at least *supposed* to be a stop on the underground railroad, and had come to it; but I, in my city house, had refused to acknowledge my house as such. If I let it in, where would I put it? Then, too, I'm not particularly fond of the restlessness of dogs. The way they groan and fart in their sleep, chase rabbits in their dreams, and flop themselves over, rattling their chains (i.e., collars and dog tags). George had run away from these drunks who "owned" him, people no doubt unfit to own anything at all that breathed. Did they beat him? Did they tie him to trees and lampposts outside pubs (as I've so often seen done) while they went inside and had drink after drink? Were all the "lost" dogs one heard about really runaways? It hit me with great force that a dog I had once had, Myshkin, had undoubtedly run away from the small enclosed backyard in which he had been kept and in which he was probably going mad, whereas I had for years indulged in the fantasy that he'd been stolen! No dog in his right mind would voluntarily leave a cushy prison run by loving humans, right?

Or suppose George was a woman, beaten or psychologically abused by her spouse. What then? Would I let her in? I would, wouldn't I? But where to put George, anyway? If I put him in the cellar, he might bark. I hate the sound of barking. If I put him in the parlor, he might spread fleas. Who was this dog, anyway?

George stayed at my door the whole night. In the morning I heard him bark, but by the time I was up, he was gone.

I think I am telling you that the animals of the planet are in desperate peril, and that they are fully aware of this. No less than human beings are doing in all parts of the world, they are seeking sanctuary. But I am also telling you that we are connected to them at least as intimately as we are connected to trees. Without plant life human beings could not breathe. Plants produce oxygen. Without free animal life I believe we will lose the spiritual equivalent of oxygen. "Magic," intuition, sheer astonishment at the forms the Universe devises in which to express life—itself—will no longer be able to breathe in us. One day it occurred to me that if all the birds died, as they might well do, eventually, from the poisoning of their air, water, and food, it would be next to impossible to describe to our children the wonder of their flight. To most children, I think, the flight of a bird—if they'd never seen one fly—would be imagined as stiff and unplayful, like the flight of an airplane.

But what I'm also sharing with you is this thought: The Universe responds. What you ask of it, it gives. The military-industrial complex and its leaders and scientists have shown more faith in this reality than have those of us who do not believe in war and who want peace. They have asked the Earth for all its deadlier substances. They have been confident in their faith in hatred and war. The Universe, ever responsive, the Earth, ever giving, has opened itself fully to their desires. Ironically, Black Elk and nuclear scientists can be viewed in much the same way: as men who prayed to the Universe for what they believed they needed and who received from it a sign reflective of their own hearts.

I remember when I used to dismiss the bumper sticker "Pray for Peace." I realize now that I did not understand it, since I also did not understand prayer; which I know now to be the active affirmation in the physical world of our inseparableness from the divine; and everything, *especially* the physical world, is divine. War will stop when we no longer praise it, or give it any attention at all. Peace will come wherever it is sincerely invited. Love will overflow every sanctuary given it. Truth will grow where the fertilizer that nourishes it is also truth. Faith will be its own reward.

Believing this, which I learned from my experience with the animals and the wild flowers, I have found that my fear of nuclear destruction has been to a degree lessened. I know perfectly well that we may all die, and relatively soon, in a global holocaust, which was first imprinted, probably against their wishes, on the hearts of the scientist fathers of the atomic bomb, no doubt deeply wounded and frightened human beings; but I also know we have the power, as all the Earth's people, to conjure up the healing rain imprinted on Black Elk's heart. Our death is in our hands.

Knock and the door shall be opened. Ask and you shall receive.

Whatsoever you do the least of these, you do also unto me—and to yourself. For we are one.

"God" answers prayers. Which is another way of saying, "the Universe responds."

We are *indeed* the world. Only if we have reason to fear what is in our own hearts need we fear for the planet. Teach yourself peace.

Pass it on.

1987

RESPONDING JOURNAL

Have you had an experience where you see that further thinking causes you to modify an opinion you held earlier? Compose an entry of a few paragraphs on such an experience of revised thoughts in light of Walker's comments on her changed views.

QUESTIONS FOR CRITICAL THINKING

STRATEGY

1. How effective is Walker's first paragraph? Does it make you want to read on? Why or why not?

2. Think about the sorts of examples Walker uses to make her points. Do you see any difference between those from the animal world and those from the human world? Explain your response.

ISSUES

1. The praise Walker applies to the natural world is unusual. How comfortable are you with such pronouncements? Could you make those yourself? Do they contribute to the impression the essay makes on you? Explain your response.
2. Assess the parallel Walker makes between the military-industrial complex and Black Elk. Are they really so similar? Can you think of any parallels other than those she mentions?

COLLABORATIVE WRITING ACTIVITY

How can you "pass it on"? What do you feel you should do to teach the rest of the world the lessons necessary to make the world respond in a productive way? Share your response with your group; suggest the most workable points in one another's presentations. Draft as a group a plan for putting your ideas into action by letting the larger world know what you are doing. In that way, you will help them revise their thinking patterns.

from *Gaia: The Human Journey from Chaos to Cosmos*
Elisabet Sahtouris

Dr. Elisabet Sahtouris is a Greek scientist who offers here a reading of one of the most revolutionary concepts linking science with the physical planet: the Gaia theory, which sees the earth as a living entity perfectly capable of taking care of itself. Sahtouris follows the work of James E. Lovelock and Lynn Margulis, other adherents of the Gaia hypothesis, in building her argument that the ancient Greek account of mother earth, or Gaia, has much more to tell human society than a story of colossal creatures more closely related to giants in children's stories than to scientific validity.

Scientists in America have been of divided opinion on the Gaia hypothesis. Some feel it lacks verifiable data and therefore has little substantial basis, while others feel it carries the logical and emotional weight of a rational idea that we must all heed before we doom humanity by poisoning our mother the earth. Decide how far you are persuaded by the logic and the evidence of Sahtouris' claims. The text presented here is "A Twice-Told Tale," Chapter 1 of Gaia.

EXPLORING JOURNAL

What sorts of accounts of the earth and its creation or formation do you remember from myths or folklore you have read or heard recited? What connections do you think might exist between those accounts and important lessons about the scientific nature of the earth? Explore these ideas in a ten-minute freewrite.

———◦•◦———

Everyone knows that humanity is in crisis—politically, economically, spiritually, ecologically—any way you look at it. Many see humanity as close to suicide by way of our own technology; many others see humans as deserving God's or nature's wrath in retribution for our sins. However we see it, we are deeply afraid that we may not survive much longer. Yet our urge to survival is the strongest urge we have, and we do not cease our search for solutions in the midst of crisis.

The proposal made in this book is that we see ourselves in the context of our planet's biological evolution, as a still new, experimental species with development stages that parallel the stages of our individual development. From this perspective, humanity is now in adolescent crisis and, just because of that, stands on the brink of maturity—in a position to achieve true humanity in the full meaning of that word. Like an adolescent in trouble, we have tended to let our focus on the crisis itself or on our frantic search for particular political, economic, scientific, or spiritual solutions depress us and blind us to the larger picture, to avenues of real assistance. If we humbly seek help instead from the nature that spawned us, we will find biological clues to solving all our biggest problems at once. We will see how to make the healthy transition into maturity.

Some of these biological clues are with us daily, all our lives, in our own bodies; others can be found in various ages and stages of the larger living organism of which we are part—planet Earth. Once we see these clues, we will wonder how we could have failed to find them for so long.

The reason we *have* missed them is that we have not understood ourselves as living beings within a larger being, in the same sense that our cells are part of each of us. Our intellectual heritage for thousands of years, most strongly developed in the past few hundred years of science, has been to see ourselves as separate from the rest of nature, to convince ourselves we see *it* objectively—at a distance from ourselves—and to perceive or at least model it as a vast mechanism.

This objective mechanical worldview was founded in ancient Greece when philosophers divided into two schools of thought about the world—one that all nature, including humans, was alive and *self*-creative, ever making order from disorder; the other that the "real" world could be known only through pure reason, not through direct experience, and was God's geometric creation—

permanently mechanical and perfect behind our illusion of its disorder. This mechanical/religious worldview superseded the older one of living nature to become the foundation of the whole Western worldview up to the present.

Philosophers such as Pythagoras, Parmenides, and Plato were thus the founding fathers of our mechanical worldview, though Galileo, Descartes, and other men of the Renaissance translated it into the scientific and technological enterprise that has dominated human experience ever since.

What if things had gone the other way? What if Thales, Anaximander, and Heraclitus, the organic philosophers who saw all the cosmos as alive, had won the day back in that ancient Greek debate? What if Galileo, as he experimented with both telescope and microscope, had used the latter to seek evidence for Anaximander's theory of biological evolution here on earth, rather than looking to the skies for confirmation of Aristarchus's celestial mechanics? In other words, what if modern science and our view of human society had evolved from organic biology rather than from mechanical physics?

We will never know how the course of human events would have differed had they taken this path, had physics developed in the shadow of biology rather than the other way around. Yet it seems we were destined to find the biological path eventually, as the mechanical worldview we have lived with so long is now giving way to an organic view—in all fairness, an organic view made possible by the very technology born of our mechanical view.

The same technology that permits us to reach out into space has permitted us to begin seeing the real nature of our own planet—to discover that it is alive and that it is the only live planet circling our sun. The implications of this discovery are enormous, and we have hardly even begun to pursue them. We were awed by astronauts' reports that the earth looked from space like a living being, and were ourselves struck by its apparently live beauty when the visual images were before our eyes. But it has taken time to accumulate scientific evidence that the earth is a live planet rather than a planet with life upon it, and many scientists continue to resist the new conception because of its profound implications for change in all branches of science, not to mention all society.

The difference between a planet with life on it and a living planet is hard at first to understand. Take for example the word, the concept, the practice of "ecology," which has become familiar to us all within just the few short decades that we have been aware of our pollution and destruction of the environment on which our own lives depend. Our ecological understanding and practice has been a big, important step in understanding our relationship to our environment and to other species. Yet, even in our serious environmental concern, we still fall short of recognizing ourselves as part of a much larger living organism. It is one thing to be careful with our environment so it will last and remain benign; it is quite another to know deeply that our environment, like ourselves, is part of the body of an earth organism.

Despite the fact that our study of the earth as a live organism is still new, we already know something of its development as an "embryo" and of its present physiology. The earth transformed itself into a wealth of living species organized into environments in more or less the same way that fertile eggs transform themselves into a wealth of different cells organized into organs, except that in earth's case the organism matures without breaking its shell.

The earliest species into which the materials of the earth's crust transformed themselves created their own environments, and these environments in turn shaped the fate of species, much as cells create their surround and are created by it in our own embryological development. As for physiology, we already know that the earth regulates its temperature—as well as any of its warm-blooded creatures—such that it stays within bounds that are healthy for life despite the sun's steadily increasing heat. And just as our bodies continually renew and adjust the balance of chemicals in our skin and blood, our bones and other tissues, so does the earth continually renew and adjust the balance of chemicals in its atmosphere, seas, and soils. How these physiological systems work is now partly known, partly still to be discovered, as is also still the case with our bodies' physiological systems. Certainly it is ever more obvious that we are not studying the mechanical nature of Spaceship Earth but the self-creative, self-maintaining physiology of a live planet.

We still take the newer Live Earth concept—named Gaia after the earth goddess of early Greek myth—more as a poetic or spiritual metaphor than as a scientific reality. However, the name Gaia was never intended to suggest that the earth is a female being—the reincarnation of the Great Goddess or Mother Nature herself—or to start a new religion (though it would hardly hurt us to worship our planet as the greater Being whose existence we have intuited from time immemorial). It was intended simply to designate the concept of a live earth, in contrast to an earth with life upon it.

Actually, "Gaia," or the Roman form, "Gea," was an earlier name for our planet than "Earth." It was lost in the wandering of words from ancient Greek through other languages to English. In Greek, our planet has always been called Gaia in its alternate spelling "Ge," which did get into the English language words "geology," the formation of the earth; "geometry," the measurement of the earth; and "geography," the mapping of the earth. In accord with our own practice of calling planets by the names of Greek deities in their Roman versions, we really should call the earth Gea. Greek, like English, has always used the same word for earth-as-world and earth-as-ground—the ancient "Ge" that became the modern "Gi," pronounced "Yee." The English word "earth" came from an ancient Greek root meaning "working the ground," or earth—*ergaze*—which evolved into the name of the Nordic earth goddess, Erda, and then into the German Erde and the English Earth. Thus even the word "earth" implies a female deity.

With that digression intended to make the name Gaia more acceptable to those who still consider the name and image somehow inappropriate for a scientific concept, let us look also at the myth itself—the creation myth of Gaia's dance.

The story of Gaia's dance begins with an image of swirling mist in the black nothingness called Chaos by the ancient Greeks—an image reminding us of

modern photos of galaxies swirling in space. In the myth it is the dancing goddess Gaia, swathed in white veils as she whirls through the darkness. As she
becomes visible and her dance grows ever more lively, her body forms itself
into mountains and valleys; then sweat pours from her to pool into seas, and
finally her flying arms stir up a wind-sky she calls Ouranos—still the Greek
word for sky—which she wraps around herself as protector and mate.

Though she later banishes Ouranos—Uranus, in Latin—to her depths for
claiming credit for creation, their fertile union as Earth and Heaven brings forth
forests and creatures including the giant Titans in human form, who in turn
give rise to the gods and goddesses and finally to mortal humans.

From the start, says the myth—true to human psychology—people were
curious to know how all this had happened and what the future would bring. To
satisfy their curiosity, Gaia let her knowledge and wisdom leak from cracks in the
earth at places such as Delphi where her priestesses interpreted it for people.

Our curiosity is still with us thousands of years after this myth served as
explanation of the world's creation. And in a sense, Gaia's knowledge and wisdom are still leaking from her body—not just at Delphi, but everywhere we
care to look in a scientific study of our living planet.

The new scientific story of Gaian creation has other parallels to the ancient
myth. We now recognize the earth as a single self-creating being that came alive
in its whirling dance through space, its crust transforming itself into mountains
and valleys, the hot moisture pouring from its body to form seas. As its crust
became ever more lively with bacteria, it created its own atmosphere, and the
advent of sexual partnership finally did produce the larger life forms—the trees
and animals and people.

The tale of Gaia's dance is thus being retold as we piece together the scientific details of our planet's dance of life. And in its context, the evolution of
our own species takes on new meaning in relation to the whole. Once we truly
grasp the scientific reality of the Gaian organism and its physiology, our entire
worldview and practice are bound to change profoundly, revealing the way to
solving what now appear to be our greatest and most insoluble problems.

From a Gaian point of view, we humans are an experiment—a trial species
still at odds with ourselves and other species, still not having learned to balance
our own dance within that of our whole planet. Unlike most other species, we
are not biologically programmed to know what to do; rather, we are an experiment in free choice. This leaves us with enormous potential, powerful ego-
tism, and tremendous anxiety—a syndrome that is recognizably adolescent.

Human history may seem very long to us as we study all that has happened
in it, but we know only a few thousand years of it and have existed as humans
for only a few million years, while Gaian creation has been going on for billions of years. We have scarcely had time to come out of species childhood,
yet our social evolution has changed us so fast that we have leaped into our
adolescence.

Humans are not the first Gaian creatures to make problems for themselves
and for the whole Gaian system, as we will see. We are, however—unless

whales and dolphins beat us to it in past ages—the first Gaian creatures who can understand such problems, think about them, and solve them by free choice. In fact, the argument of this book is that our maturity as a species depends on our accepting the responsibility for our natural heritage of behavioral freedom by working consciously and cooperatively toward our own health along with that of our planet.

Our ability to be objective, to see ourselves as the "I" or "eye" of our cosmos, as beings independent of nature, has inflated our egos—"ego" being the Greek word for "I." We came to separate the "I" from the "it" and to believe that "it"—the world "out there"—was ours to do with as we pleased, telling ourselves we were either God's favored children or the smartest and most powerful naturally evolved creatures on earth. This egotistic attitude has been very much a factor in bringing us to adolescent crisis. And so an attitude of greater humility and willingness to accept some guidance from our parent planet will be an important factor in reaching our species maturity.

The tremendous problems confronting us now—inequality, hunger, the threat of nuclear annihilation and possibly irreversible damage to the natural world we depend on as much as any of our cells depend on the wholeness of our bodies for their life—are all of our own making. These problems have become so enormous that many of us believe we will never be able to solve them. Yet just at this time in our troubled world we stand on the brink of maturity, in a position to recognize that we are neither perfect nor omnipotent, but that we can learn a great deal from a parent planet that is also not perfect or omnipotent but has the experience of billions of years of overcoming an endless array of difficulties, small and great.

When we look anew at evolution, we see not only that other species have been as troublesome as ours, but that many a fiercely competitive situation resolved itself in a cooperative scheme. The kind of cells our bodies are made of, for example, began with the same kind of exploitation among bacteria that characterizes our historic human imperialism. And through the same technologies of transportation and communications first invented by those bacteria as they bound themselves into the cooperative venture that made our existence possible, we are uniting our selves into a single body of humanity that may make yet another new step in Gaian evolution possible. If we look to the lessons of evolution, we will gain hope that the newly forming worldwide body of humanity may also learn to adopt cooperation in favor of competition. The necessary systems have already been invented and developed; we lack only the understanding, motive, and will.

It may come as a surprise that nature has something to teach us about cooperative economics and politics. Sociobiologists who have told us much in recent decades about humanity's animal heritage have tended to paint us a bleak picture—calling on our evolutionary heritage as evidence that we will never cure ourselves of territorial lust and aggression toward one another, and that thus there will be no end to economic greed and political warfare. But it is the aim of this book to show that these sociobiologists have presented a misleading picture—as misleading as earlier scientists' one-sided view of all natural evolution as "red in tooth and claw," the hard and competitive struggle among individuals on which we have modeled our modern societies.

The new view of our Gaian earth in evolution shows, on the contrary, an intricate web of cooperative mutual dependency, the evolution of one scheme after another that harmonizes conflicting interests. The patterns of evolution show us the creative maintenance of life in all its complexity. Indeed nature is more suggestive of a mother juggling resources to ensure each family member's welfare as she works out differences of interest to make the whole family a cooperative venture, than of a rational engineer designing perfect machinery that obeys unchangeable laws. For scientists who shudder at such *anthropomorphism*—giving nature human form—let us not forget that *mechanomorphism*—giving nature mechanical form—is really no better than secondhand anthropomorphism, since mechanisms are human products. Is it not more likely that nature in essence resembles one of its own creatures than that it resembles in essence the nonliving product of one of its creatures?

The leading philosophers of our day recognize that the very foundations of our knowledge are quaking—that our understanding of nature as machinery can no longer be upheld. But those who cling to the old understanding seriously fear that all human life will break down without a firm foundation for our knowledge of nature in mathematical reference points and laws of physics. They fail to see what every child can see—that hummingbirds and flowers *work*, that nature does very well in utter ignorance of human conceptions of how it must work.

Machinery is in fact the very antithesis of life. One must always hope a machine, between its times of use, will not change, for only if it does not change will it continue to be of use. Left to its own devices, it will eventually be destroyed by its environment. Living organisms, on the other hand, cannot stay the same *without* changing constantly, and they use their environment to their advantage. To be sure, our machinery is getting better and better at imitating life; if this were not so, a mechanical science could not have advanced in understanding. But mechanical models of life continue to miss its essential self-creativity.

We are learning that there is more than one way to organize working systems, to produce order and balance; that the imperfect and flexible principles of nature lead to greater stability and resilience in natural systems than we have produced in ours—both technological and social—by following the mechanical laws we suppose to be natural.

We design our societies as though they were machinery; we make war on one another over who has the perfect social design. Our greatest conflict is over whether individuals should sacrifice their individual interest to the welfare of the whole or whether individual interest should reign supreme in the hope that the interests of the whole will thus take care of themselves.

No being in nature, outside our own species, is ever confronted with such a choice, and if we consult nature, the reason is obvious. The choice makes no sense, for neither alternative can work. No being in nature can ever be completely independent, although independence calls to every living being, whether it is a cell, a creature, a society, a species, or a whole ecosystem. Every being is part of some larger being, and as such its self-interest must be tempered by the interests of the larger being to which it belongs. Thus mutual consistency works itself out everywhere in nature, as we will see again and again in this book. For clues on organizing a workable economics and politics, we need not even look beyond our own bodies, with their cooperative diversity of cells and organs as a splendid example to us in working out our social future.

Diversity is crucial to nature, yet we humans seem desperately eager to eliminate it, in nature and in one another. This is one of the greatest mistakes we are making. We reduce complex ecosystems to one-crop "economies," and we do everything in our power to persuade or force others to adopt our languages, our customs, our social structures, instead of respecting theirs. Both practices impoverish and weaken us within the Gaian system.

We are right to worry about our survival, for we foolishly jeopardize it. We are wrong to devote our attention to saving or managing nature. Gaia will save herself—with or without us—and hardly needs advice or help in management. To look out for ourselves we would be wise to interfere as little as possible in her ways, and to learn as much as possible of them. Our technology has ravaged nature and continues to do so, but the ravages of technology are based on our unnatural greed, our profit motive. There is no intrinsic reason that we humans cannot develop a benign technology once we agree that our desire to maximize profits is completely at odds with nature's dynamic balance—that greed prevents health and welfare for all. No other creatures take more than they need, and this must be our first lesson. Our second is to learn and emulate nature's fine-tuned recycling system, largely powered by free solar energy.

The purpose of this book is to help pave the way to a happier and healthier future through an understanding of our relationship to the Gaian system that spawned us and of which we are part—a great being that, however it may annoy us, is not ours to dominate and control. We can damage it, but we cannot run it; we had better try to find out what it is all about and what we are doing, and may do, within it.

The aggressive and destructive motives of domination, conquest, control, and profit have been presented to us as human nature by historians as well as by sociologists. But mounting evidence from archaeology strongly suggests that human societies were for the greater part of civilized history based on cooperation and reverence for life and nature, not on competition and obsession with death and technology. It seems our human childhood—which lasted far longer than has our recent adolescence—was guided by religious images of a near and nurturing Mother Goddess before a cruel and distant Father God replaced her in influence. As we come out of adolescence we often recognize the value of what we were taught in childhood, and this new historical view of ourselves supports the general thesis of this book.

Like Gaian creation itself, human understanding or knowledge ever evolves. Parts of the story you are about to read will already have changed by the time you read it. Others will change in the years to come as new things about Earth-Gaia and about human history are discovered. Any of us is free to help find new pieces of the story, bring those we know up to date, and then reinterpret the evidence as a whole, for in the last analysis, every interpretation has its personal color and flavor.

The next chapter is concerned with cosmic beginnings as a living context for our living planet; succeeding chapters, up to half of this book, tell of Gaian evolution over billions of years before we humans become part of it. Those interested in the story of human society may be tempted to skip this part of the story, but the scientific account of evolution in this book is not separable from our human social history. The details of our biological heritage from ancient bacteria on are given because therein lie the clues to a better human future. It is only within this context that we can appreciate our newness and our differences from the rest of nature, to see at the same time how we can benefit from its vast experience to fit ourselves in more harmoniously.

It is on this that everything now depends; species suicide is our only alternative, and there is really no reason to make a dramatic adolescent exit instead of growing up, taking on adult responsibility, and reaping the pleasures of productive maturity. Let us then follow the evolution of Gaian creation and of our own history as social and technological creatures within this great dance of life. Let's see what meaning and guidance all this may give in our present crisis, to speed us on our way into full maturity—to a happier future in which we promote our own health and that of our planet within the greater cosmic dance.

RESPONDING JOURNAL

If you grant Sahtouris the possibility that her proposition is valid, what changes would we have to make as a society in America to adjust our habits to show our respect for Gaia more carefully? If America is too broad, think about what your own college community would have to change. Present your ideas in a journal entry of several paragraphs.

QUESTIONS FOR CRITICAL THINKING

STRATEGY

1. Sahtouris begins her essay, "Everyone knows that humanity is in crisis." How does such an opening affect you? Do you see any potential problems in such a statement? If you feel a problem exists, how might she correct it?

2. In her conclusion, Sahtouris makes an analogy between our present human behavior toward the earth and that of an adolescent in relation to a grown-up. How valid do you find such terms? Explain your response.

1. "The aggressive and destructive motives of domination, conquest, control and profit have been presented to us as human nature by historians as well as by sociologists." What are the implications of Sahtouris' statement? Do you agree totally with this concept? Are you more comfortable with an alternative thesis? Explain.

2. Sahtouris feels that we regard the "live earth" concept—that the planet is alive rather than having life on it—"more as a poetic or spiritual metaphor than as a scientific reality." What are the implications of these two extremes? How closely do they parallel persuasion and argument?

COLLABORATIVE WRITING ACTIVITY

Sahtouris claims that historically human societies were based on cooperation for a longer period than they have been on competition. How would you go about determining the validity of this statement? What would you do to research this idea further? What sorts of library activities would you have to perform? Create a log of your activities and what you found in your search so you can report back to your group; your recorder can then assemble a guide to library resources that will facilitate the entire class's checking of the likelihood of Sahtouris' claim.

from "Landscape, History, and the Pueblo Imagination"
Leslie Marmon Silko

Probably best known for her powerful 1977 novel Ceremony, *Leslie Marmon Silko is one of the most highly regarded contemporary female writers. Her Laguna/Pueblo ancestry perhaps predisposes her to think carefully about the nature of male-female relationships, especially as they relate to the foundation of tribal values, and she is characteristically focused on precisely how romantic relationships determine or are determined by gender roles in the tribal structure.*

In her short story "Storyteller," for example, Silko relates an account of intercultural relations; the central character is a Yupik who has to deal with Gussucks, a slang term for White people. The story depicts the Bureau of Indian Affairs as a federal agency that displays characteristic inability to comprehend that with which it has to deal. In such a situation, where the important players have to navigate their own destinies, an individual's common sense and survival skills are crucial. In the excerpt presented here, Silko explains Pueblo concepts of how one must relate to the land if one is to stay healthy. "Landscape, History, and the Pueblo Imagination" provides background on the concerns facing Native American women.

EXPLORING JOURNAL

What sorts of accounts would you expect to find in a "migration story"? What particular relevance might it have for a Native American? Explore this concept in a ten-minute freewrite.

THE MIGRATION STORY: AN INTERIOR JOURNEY

The Laguna Pueblo migration stories refer to specific places—mesas, springs, or cottonwood trees—not only locations which can be visited still, but also locations which lie directly on the state highway route linking Paguate village with Laguna village. In traveling this road as a child with older Laguna people I first heard a few of the stories from that much larger body of stories linked with the Emergence and Migration.[1] It may be coincidental that Laguna people continue to follow the same route which, according to the Migration story, the ancestors followed south from the Emergence Place. It may be that the route is merely the shortest and best route for car, horse, or foot traffic between Laguna and Paguate villages. But if the stories about boulders, springs, and hills are actually remnants from a ritual that retraces the creation and emergence of the Laguna Pueblo people as a culture, as the people they became, then continued use of that route creates a unique relationship between the ritual-mythic world and the actual everyday world. A journey from Paguate to Laguna down the long incline of Paguate Hill retraces the original journey from the Emergence Place, which is located slightly north of the Paguate village. Thus the landscape between Paguate and Laguna takes on a deeper significance: the landscape resonates the spiritual or mythic dimension of the Pueblo world even today.

Although each Pueblo culture designates a specific Emergence Place—usually a small natural spring edged with mossy sandstone and full of cattails and wild watercress—it is clear that they do not agree on any single location or

---

[1] The Emergence—All the human beings, animals, and life which had been created emerged from the four worlds below when the earth became habitable.

The Migration—The Pueblo people emerged into the Fifth World, but they had already been warned they would have to travel and search before they found the place they were meant to live.—Author's note.

natural spring as the one and only true Emergence Place. Each Pueblo group recounts its own stories about Creation, Emergence, and Migration, although they all believe that all human beings, with all the animals and plants, emerged at the same place and at the same time.[2]

Natural springs are crucial sources of water for all life in the high desert plateau country. So the small spring near Paguate village is literally the source and continuance of life for the people in the area. The spring also functions on a spiritual level, recalling the original Emergence Place and linking the people and the spring water to all other people and to that moment when the Pueblo people became aware of themselves as they are even now. The Emergence was an emergence into a precise cultural identity. Thus the Pueblo stories about the Emergence and Migration are not to be taken as literally as the anthropologists might wish. Prominent geographical features and landmarks which are mentioned in the narratives exist for ritual purposes, not because the Laguna people actually journeyed south for hundreds of years from Chaco Canyon or Mesa Verde, as the archaeologists say, or eight miles from the site of the natural springs at Paguate to the sandstone hilltop at Laguna.

The eight miles, marked with boulders, mesas, springs, and river crossings, are actually a ritual circuit or path which marks the interior journey the Laguna people made: a journey of awareness and imagination in which they emerged from being within the earth and from everything included in earth to the culture and people they became, differentiating themselves for the first time from all that had surrounded them, always aware that interior distances cannot be reckoned in physical miles or in calendar years.

The narratives linked with prominent features of the landscape between Paguate and Laguna delineate the complexities of the relationship which human beings must maintain with the surrounding natural world if they hope to survive in this place. Thus the journey was an interior process of the imagination, a growing awareness that being human is somehow different from all other life—animal, plant, and inanimate. Yet we are all from the same source: the awareness never deteriorated into Cartesian duality, cutting off the human from the natural world.

The people found the opening into the Fifth World too small to allow them or any of the animals to escape. They had sent a fly out through the small hole to tell them if it was the world which the Mother Creator had promised. It was, but there was the problem of getting out. The antelope tried to butt the opening to enlarge it, but the antelope enlarged it only a little. It was necessary for the badger with her long claws to assist the antelope, and at last the opening was enlarged enough so that all the people and animals were able to emerge up into the Fifth World. The human beings could not have emerged without the

---

[2] Creation—Tse'itsi'nako, Thought Woman, the Spider, thought about it, and everything she thought came into being. First she thought of three sisters for herself, and they helped her think of the rest of the Universe, including the Fifth World and the four worlds below. The Fifth World is the world we are living in today. There are four previous worlds below this world.—Author's note.

aid of antelope and badger. The human beings depended upon the aid and charity of the animals. Only through interdependence could the human beings survive. Families belonged to clans, and it was by clans that the human being joined with the animal and plant world. Life on the high arid plateau became viable when the human beings were able to imagine themselves as sisters and brothers to the badger, antelope, clay, yucca, and sun. Not until they could find a viable relationship to the terrain, the landscape they found themselves in, could they *emerge.* Only at the moment the requisite balance between human and *other* was realized could the Pueblo people become a culture, a distinct group whose population and survival remained stable despite the vicissitudes of climate and terrain.

Landscape thus has similarities with dreams. Both have the power to seize terrifying feelings and deep instincts and translate them into images—visual, aural, tactile—into the concrete where human beings may more readily confront and channel the terrifying instincts or powerful emotions into rituals and narratives which reassure the individual while reaffirming cherished values of the group. The identity of the individual as a part of the group and the greater Whole is strengthened, and the terror of facing the world alone is extinguished.

Even now, the people at Laguna Pueblo spend the greater portion of social occasions recounting recent incidents or events which have occurred in the Laguna area. Nearly always, the discussion will precipitate the retelling of older stories about similar incidents or other stories connected with a specific place. The stories often contain disturbing or provocative material, but are nonetheless told in the presence of children and women. The effect of these inter-family or inter-clan exchanges is the reassurance for each person that she or he will never be separated or apart from the clan, no matter what might happen. Neither the worst blunders or disasters nor the greatest financial prosperity and joy will ever be permitted to isolate anyone from the rest of the group. In the ancient times, cohesiveness was all that stood between extinction and survival, and, while the individual certainly was recognized, it was always as an individual simultaneously bonded to family and clan by a complex bundle of custom and ritual. You are never the first to suffer a grave loss or profound humiliation. You are never the first, and you understand that you will probably not be the last to commit or be victimized by a repugnant act. Your family and clan are able to go on at length about others now passed on, others older or more experienced than you who suffered similar losses.

The wide deep arroyo near the Kings Bar (located across the reservation borderline) has over the years claimed many vehicles. A few years ago, when a Viet Nam veteran's new red Volkswagen rolled backwards into the arroyo while he was inside buying a six-pack of beer, the story of his loss joined the lively and large collection of stories already connected with that big arroyo. I do not know whether the Viet Nam veteran was consoled when he was told the stories about the other cars claimed by the ravenous arroyo. All his savings of combat pay had gone for the red Volkswagen. But this man could not have felt any worse than the man who, some years before, had left his children and mother-in-law in his station wagon with the engine running. When he came out of the liquor store his station wagon was gone. He found it and its passengers upside down in the big arroyo. Broken bones, cuts and bruises, and a total wreck of the car. The big arroyo has a wide mouth. Its existence needs no explanation. People in the area regard the arroyo much as they might regard a living being, which has a certain character and personality. I seldom drive past that wide deep arroyo without feeling a familiarity with and even a strange affection for this arroyo. Because as treacherous as it may be, the arroyo maintains a strong connection between human beings and the earth. The arroyo demands from us the caution and attention that constitute respect. It is this sort of respect the old believers have in mind when they tell us we must respect and love the earth.

Hopi Pueblo elders have said that the austere and, to some eyes, barren plains and hills surrounding their mesa-top villages actually help to nurture the spirituality of the Hopi *way*. The Hopi elders say the Hopi people might have settled in locations far more lush where daily life would not have been so grueling. But there on the high silent sandstone mesas that overlook the sandy arid expanses stretching to all horizons, the Hopi elders say the Hopi people must "live by their prayers" if they are to survive. The Hopi way cherishes the intangible: the riches realized from interaction and interrelationships with all beings above all else. Great abundances of material things, even food, the Hopi elders believe, tend to lure human attention away from what is most valuable and important. The views of the Hopi elders are not much different from those elders in all the Pueblos.

The bare vastness of the Hopi landscape emphasizes the visual impact of every plant, every rock, every arroyo. Nothing is overlooked or taken for granted. Each ant, each lizard, each lark is imbued with great value simply because the creature is there, simply because the creature is alive in a place where any life at all is precious. Stand on the mesa edge at Walpai and look west over the bare distances toward the pale blue outlines of the San Francisco peaks where the ka'tsina spirits reside. So little lies between you and the sky. So little lies between you and the earth. One look and you know that simply to survive is a great triumph, that every possible resource is needed, every possible ally—even the most humble insect or reptile. You realize you will be speaking with all of them if you intend to last out the year. Thus it is that the Hopi elders are grateful to the landscape for aiding them in their quest as spiritual people.

———>•<———

RESPONDING JOURNAL

In a journal entry of several substantial paragraphs, explain your spiritual connections with the land. What sort of land, or what particular landscape, exerts a pull on you? What does this pull seem to be trying to persuade you to do?

QUESTIONS FOR CRITICAL THINKING

STRATEGY

1. How effective do you find the arroyo example Silko gives toward the end of her essay? Why might she use the example at this particular point?
2. Do you feel that Silko has sufficiently explained the idea of "emergence" narratives for her readers? What do you see as the most difficult part of this category to explain?

ISSUES

1. Silko writes, "Natural springs are crucial sources of water for all life in the high desert plateau country." What are the larger implications of this statement? What does this water represent? You might think of some other selections you have read that place a premium on such water sources.
2. Do you see any evidence of other people who, like the Hopi, do not wish to exchange a harsh life style in a difficult environment for one that is easier? When you think of people who might fit this description, decide whether they stay where they are because of a motive similar to the Hopis'.

COLLABORATIVE WRITING ACTIVITY

Explain a story you have heard to tell something important about a people or a cultural group. Washington Irving adds an important bit of Dutch folktale to his story of Rip Van Winkle when he tells of Henry Hudson's crew bowling in the Catskills; perhaps this is one way of accounting for the heavy thunderstorms that sometimes hit the area. Explain how the story of your choice might be a way of accounting for natural phenomena; once your group has finished, discuss how you would classify these groups according to their operation on either logical or emotional appeal.

SUGGESTIONS FOR FORMAL WRITING ASSIGNMENTS

1. Write an essay in which you examine evidence that has appeared after the publication of one of the essays in this book. Assess the present state of the issues you choose to discuss, and offer to your readers either an update of the original essay or a counterargument in which you present your findings and offer a different interpretation.

2. Write an essay in which you argue for or against the proposition that local governments should determine what happens to the natural resources to be

found within their area. Explore the proposition that whatever elements occur in a given area should be used primarily by the people living within that area, and that these people should have the right of deciding what ultimately becomes of those resources.

3. Is it possible to reconcile differing perspectives on land and resource use? For example, can you find a means of negotiation that will allow hunters and conservationists to agree on a statement regarding appropriate use of the environment? Explore this question in an essay; use as many different examples as you feel you need to make a strong case.

4. Characterize the Native American view of the land and human response to the land. Write an essay in which you argue for or against the Native American perspective as a logical basis for determining public policy with regard to land use. Be sure to anticipate potential disagreement with your points so you will be able to argue against them successfully.

5. Consider Sahtouris' position on our obligation to the planet and our concept of Gaia. What sort of argument can you develop for the earth as a body that requires the care and nurturing of all humans or as a self-regulating agency that possesses the power to fix itself? Write an essay in which you explain fully the position you take.

Chapter 12

USING RESEARCH TO CONSTRUCT ARGUMENTS: DIVERSE VIEWS ON LAND AND RESOURCE USE

1. What has your idea of Smokey the Bear been? To what extent has this bear affected the way you regard forested areas?

2. What parallel creatures seem to have been developed to present the views of other organizations or sell other products? Do they work as effectively as Smokey?

3. Explore how advertising uses images from the natural world—land and waterscapes, animals, plant life—to manipulate audience response.

Chapter 12

USING RESEARCH TO CONSTRUCT ARGUMENTS: DIVERSE VIEWS ON LAND AND RESOURCE USE

In Chapter 11, we saw the two means through which we convince our readers of our ideas in writing. Persuasion suggests the emotional aspect, while argumentation indicates the rational or logical side of case-making. While something of both is necessary in a convincing presentation, we have to be particularly careful in more complex or controversial cases to support our positions with sound logic. Argumentative writing requiring research in order to be convincing demands a greater degree of formality than essays not dependent on careful presentation for their effect. Therefore, when using research to construct arguments, you must observe more formal conventions of writing in order to appear competent and fair to your reader.

The twin issues of competency and fairness require care on your part as a writer. For instance, the question of animal vs. human rights has raised considerable emotion; on the one hand, animal rights activists argue that as living beings animals have rights equal to those of humans and to raise animals for laboratory testing or for their fur is criminal, while others claim that such testing insures safe drugs and other beneficial chemicals for humans. They argue as well that humans have always used animal skin for clothing. Anyone who has seen graphic depictions of baby seals being slaughtered for their hides knows the emotional power of those images to argue against such harvesting. At the same time, seeing a fine sealskin coat on a rack at a clothing store distances the activity of killing, particularly when such a beautiful garment is the result. The prospective shopper has the final product of the process—the coat—as the basis upon which to decide whether or not to buy. How does that person come to a decision?

The key to arguing successfully on a complicated topic is to present your case forcefully, yet in a way that does not alienate your reader. It may not be enough merely to present a compelling argument; you must control your tone so you do not seem either to look down at people who hold opposing viewpoints or to be just barely containing your contempt for the arguments some

people might advance against your position. In other words, you need to present your readers with an image of yourself as controlled, rational, and fair. You
have to avoid the "anyone who disagrees with me must be stupid" syndrome.
To be sure you are in fact avoiding this syndrome, you must be meticulous in
your presentation of the support you draw from your research.

To continue with this example, consider recent news coverage of the debate
between People for Ethical Treatment of Animals, an activist animal rights group,
and Putting People First, a group desiring to insure the rights of humans in relation to animals. Ingrid Newkirk, a member of People for Ethical Treatment of
Animals, has been quoted as equating broiler chickens with holocaust victims.
Kathleen Marquardt says that Putting People First "represents the average
American who drinks milk and eats meat, benefits from medical research, wears
leather, wool, and fur, hunts and fishes, owns a pet and goes to zoos."

In order to argue successfully for one side or the other, you would have to
determine first the precise terms on which you would base your argument.
Define what you mean by "animal rights." How far do such rights extend? Are the
two positions being treated fairly in the description above? Your research would
tell you how accurately these groups are being portrayed. You must consider the
research you offer in support of your argument; are you presenting data, examples, statistics, or expert testimony? You will most often use some combination
of these; the sources you consult will furnish numerous examples of arguments
constructed on solid evidence. Remember that any evidence you present must
pass the test of common sense as well as demonstrate expert substantiation.

To return to the three points of successful writing—thesis, audience,
role—you need to make certain that the concepts in your thesis are clear, that
you have treated your audience carefully and fairly by anticipating and answering their questions, and that your role as author includes conscientious
research as a means of substantiating the claims you make in your thesis.

The section in this text on documentation will show you how to present
correctly the results of your research. In order for you to begin that research,
however, you should know where to search efficiently. These are the first
sources to explore:

- *Reader's Guide to Periodical Literature* indexes popular periodicals
  (journals and magazines); a good first step for finding a range of the most
  current writing on a given topic.
- *New York Times Index* serves as the "newspaper of record," meaning
  the one you cite in formal research when you want to document an event
  or situation covered in newspaper accounts; balanced coverage of important events and issues on a daily basis.
- Card catalog provides guide to what is actually in your library; includes
  both periodicals and books.
- INFOTRAC accesses data on computer to give you not only citations of
  information but abstracts and reviews of sources as well; available in
  many public libraries, becoming more widely used in college libraries.

- Specialized indexes, such as the *Education Index* and the *Humanities and Social Sciences Index,* provide access to publications in specific fields.
- *Association for the Study of Literature and the Environment* (ASLE) *Bibliography 1990–1993* lists books, articles, and dissertations relating to study of literature and environment.

These sources should get you up and running. If you subscribe (or know someone who does) to computer networks such as Internet or the commercial services like America Online, CompuServe, or GEnie, you can access substantial databases that will allow you to do extensive research.

Once you have done your research, you will need to sort through it to see exactly what you need to write your essay. Always remember the cardinal rule of research: you have more information than you can use in your writing. In the course of your research, you may find that the thesis you originally had in mind has been altered. If so, do not be alarmed: this, too, is the normal course of research. You have modified your thesis in order to argue more effectively. In effect, you have learned to see your subject differently; you have modified your role as writer.

One useful way of thinking about such modifications of thesis and role in order to reach an audience more effectively is the sociological concept of *mental lenses.* In building their comprehension of the factors that influence our societies, sociologists require a set of principles to guide them. One such set of principles that often helps is a *paradigm;* according to Marvin E. Olsen, Dora G. Lodwick, and Riley E. Dunlap in their book *Viewing the World Ecologically,* a social paradigm is the "mental lens through which people view the world and that enables them to understand what they see." In other words, we unconsciously construct a paradigm whenever we look at the world; sociologists tell us that our environments determine to a large extent both what we see and what we make of what we see. Effective argumentation and persuasion depends on understanding not only our own mental lenses, but those of our audience as well. What do you think Kathleen Marquardt intends to convey about shared mental lenses in her comment about "the average American"?

The renewed ability to see created by looking through new mental lenses, or through the mental lenses of your opponent in an argument, carries over positively to your ability to argue and persuade effectively. Just as re-vision means a re-seeing, a fresh opportunity to see if your work is doing all that you require it to accomplish, so arguing and persuading require a new seeing, a test of mental lenses other than your own so that you will be able to anticipate your opponent's strongest points. You will probably want to return to this metaphor of lenses as you plan and draft your writing.

In persuasive or argumentative writing, you will have to make sure that any readers who do not already agree with your point or position will come over to your side by the time you finish your presentation. Most often, you will find convincing your readers easiest when you offer substantial evidence to support your claim. One sort of evidence that sways readers is causal; it explores the relation between cause and effect. Such writing explains how certain situations come about as a result of certain causes.

There are several methods of explaining the causal relationship. Since you want to find the strategy for development that works best with your thesis, you might find it helpful to think of this process as working in one of two ways: cause to effect, or effect to cause. For example, you might write about how people's careless disregard for what they throw away contributes to a general sense of disrespect for the land used up when we bury our trash in a landfill. Should you start your essay with the effect—the growing landfill problem—or with the cause—public carelessness about what they throw away?

Consider another example, this one drawn from an area that is still hotly debated. For the last twenty years, we have heard much about acid rain and its effects on land and water. If we wanted to make a strong statement about this phenomenon, should we start with a dramatic picture of former spots of natural beauty destroyed by the effects of acid rain and then explain what these places were like before the rain? Or do we first explain how sulphur emissions from manufacturing and power plants can damage land and water hundreds, even thousands, of miles away, and then show how this process has affected the Adirondacks or Swedish forests and lakes? Would the decision depend to some extent on the complexity of the thesis? That is, you might consider that a dramatic effect you can readily and clearly relate to a specific cause might be a good way to open your essay and present your thesis. Your ingenuity will lead you to other formulations of this basic principle, but the most basic possibilities for organizing cause and effect presentations look like these diagrams:

cause   →   effect   effect   →   cause

The diagrams are simple, but they often produce highly effective writing. Of course, a situation might have several causes or effects you would need to treat. Decide in such cases how best to organize your presentation to convey the force of your thesis. You might write an essay of causal analysis, or writing which explains how results come about through the operation of certain causes.

Another means of effective arguing based on research is the movement from specific points to general statements, or the opposite case of movement from general concepts to individual instances. Conventional rhetoric teaches that successful presentation of an argument involves inductive or deductive reasoning. Inductive reasoning is often associated with science and the scientific method, for induction is the process of observing facts or examples and then moving on to draw conclusions or form a hypothesis based on the observation of those details. Deductive reasoning, on the other hand, is the process of beginning with a general statement of recognized truth or validity and then moving on to specific applications of that idea. A claim is a statement the author offers as true and as the logical outcome of the evidence offered; a premise is the principle or generally held understanding upon which a claim is based. Induction and deduction, claim and premise look like this:

INDUCTION	DEDUCTION
evidence → claim	claim or premise → evidence
(discovery)	(application)

For example, if we see dead fish collecting in a river (evidence) we might state that someone is dumping pollutants into the river (claim). We might state that someone is dumping pollutants into the river (claim) or state that when people dump pollutants into a river, fish will die (premise) and then offer as proof for either of these deductive arguments the conclusion that the dead fish (evidence) supports the claim or premise of the argument. The result of the inductive argument is the discovery which comes from studying the evidence; the result of the deductive is the application of the general idea to the specific evidence. To see how clear understanding of this process can help us to become better and more forceful writers, we will return briefly to classical rhetoric.

Traditional study of the art of persuasion tells us that the results of induction are classified as reliable or unreliable. Reliable results are those that are reasonable and are true to the limited number of situations or amount of data presented. We cannot call these results true or untrue because we have not tested all possible eventualities; however, on the basis of our sampling, we see that they are consistent with our general premise. In many cases, this process of observing details in order to draw conclusions leads to the discovery of previously unrecognized information.

Deduction allows us to use our previous knowledge in new contexts. Since it moves toward application (in the opposite direction from induction), it shows us how to apply previously learned principles to new situations. The conclusion in a deductive argument is true or valid when that conclusion follows logically from the evidence presented; the conclusion is false or invalid when the conclusion does not follow logically from the evidence presented. Instances of illogical arguments include drawing conclusions based on incomplete evidence (presenting only part of the evidence) or based on inconsistent evidence (presenting evidence in a way that distorts or manipulates the evidence to create the desired effect). Remember that any lapse you can see in your argument will be evident to your reader as well.

The texts in this chapter indicate the difficulty of constructing arguments about the environment, the land, and our human interactions with our planet. Much of what writers discuss seems to have various, often diametrically opposed, viewpoints that could easily frustrate anyone attempting to make sense of the issues and find a comfortable position within the debate. Vice President Al Gore offers his sincere beliefs on what we must do to survive on our planet; Dixy Lee Ray offers her equally fervent beliefs, which conflict directly with Gore's. Both sides marshall impressive evidence to support their points. Animal rights issues suggest even more dramatic oppositions. Analysis of the factors that determine and shape how we regard our position in the universe helps us to see how we form our world views and opinions. Considering thesis, audience, and role helps us to see more clearly through both our own mental lenses and those of the authors of the sources we use to support our positions.

from *Earth in the Balance*
Al Gore

Vice President Al Gore developed an international reputation as a result of his crusading on behalf of the environment when he was a United States Senator from Tennessee. He began his political career in 1976, winning election to the House of Representatives, after seven years as a journalist. First winning his Senate seat in 1984, he became the most prominent environmentalist in that body; he continued his efforts through his successful campaign for the Vice Presidency in 1992.

His book Earth in the Balance *is the result of numerous concerns in his life. As a member of the House, Gore worked diligently at nuclear arms limitation. Growing up on a family farm had shown him much about respect for the land, and his work on the question of nuclear power gave him a global perspective for his views. Running for President in 1987, Gore quickly saw that his campaign platform's heavy orientation toward environmental issues was not winning him many voters, and so he dropped out of the race. His dramatic contact with environmental issues led him to write* Earth in the Balance, *which appeared in 1992; since assuming his new duties, he has had less to say about environmental issues.*

EXPLORING JOURNAL

Characterize the Vice President. What sort of a person does he seem to be? How much do you know about his environmental activism? What images of him do you think of first?

———◆———

[EDITOR'S NOTE: Gore uses the analogy of the original "Marshall Plan," a broad-ranging strategy that greatly aided European recovery from World War II, to outline his program for saving the physical planet from those forces threatening it. He has set five goals for his "Global Marshall Plan": stabilization of world population, efficient implementation of newly developed "environmentally appropriate technologies," thoroughly revised guidelines for determining economic impact on the environment, a fresh set of international accords and agreements relating to the environment, and "the establishment of a cooperative plan for educating the world's citizens about our global environment." The fifth goal is explained in the following section.]

V. A New Global Environmental Consensus

The fifth major goal of the Global Marshall Plan should be to seek fundamental changes in how we gather information about what is happening to the environment and to organize a worldwide education program to promote a more complete understanding of the crisis. In the process, we should actively search for ways to promote a new way of thinking about the current relationship between human civilization and the earth.

This is perhaps the most difficult and yet the most important challenge we face. If a new way of thinking about the natural world emerges, all of the other necessary actions will become instantly more feasible—just as the emergence of a new way of thinking about communism in Eastern Europe made feasible all of the steps toward democracy that had been "unthinkable" only a few months earlier. And indeed, the model of change we use in designing and implementing our strategy should be based on the assumptions that there is a threshold we must cross and that not very much change will be apparent and obvious until we reach that threshold, but when it finally is reached, the changes will be sudden and dramatic.

Central to any strategy for changing the way people think about the earth must be a concerted effort to convince them that the global environment is part of their "backyard"—as it really is. I have always been struck by the way a proposal for an incinerator or a landfill mobilizes a lot of people who do not want the offending entity near them. In the midst of such a controversy, no one seems to care much about the economy or unemployment rate; the only thing that matters is protecting their backyard. The famous "not in my backyard" syndrome, NIMBY, has been much maligned but is often on target and is an undeniably powerful political force. How might its energy be focused against threats to the environment? Is that possible? The key lies with the definition of "backyard," and in truth, our backyards are threatened by problems like global warming and ozone depletion.

An important step in the right direction would be to take a new approach to the collection of information about what precisely is happening to the global environment. As chairman of the Space Subcommittee in the Senate, I have strongly urged the establishment of the new program that NASA calls Mission to Planet Earth. Sally Ride, the first American woman in space, coined the phrase, and it is meant to be taken ironically. As she points out, we have under-taken highly sophisticated planetary studies by sending spacecraft into orbit around Mars and Venus, and we have used that unique perspective to study other, more distant planets. Yet we have not used the same techniques to improve our understanding of our own planet just when we desperately need to understand much more about the changes that are taking place.

Even more important than gathering new information, though, we must start to take action now—and the information collection system should enhance that goal. This conclusion carries with it two implications: first, the information should be collected as quickly as possible; and second, it should—wherever feasible—be collected in a manner that facilitates public education and fosters a greater understanding of what the new information means within the larger context of rapid global change.

In other words, the Mission to Planet Earth should be a Mission by the peo-ple of Planet Earth. Specifically, I propose a program involving as many coun-tries as possible that will use schoolteachers and their students to monitor the entire earth daily, or at least those portions of the land area that can be covered by the participating nations. Even relatively simple measurements—surface temperature, wind speed and direction, relative humidity, barometric pres-sures, and rainfall—could, if routinely available on a more nearly global basis, produce dramatic improvements in our understanding of climate patterns. Slightly more sophisticated measurements of such things as air and water pol-lutants and concentrations of CO_2 and methane would be even more valuable. But the first step is collecting the kind of rudimentary information necessary to monitor the environment closely, just as hospital emergency rooms monitor the vital signs of patients receiving intensive care.

The mass production of uniform instruments for this program could bring the unit costs down to trivial levels, and the instruments themselves could be designed to facilitate daily electronic polling or data collection. By deploying relatively cheap low orbit satellites capable of rapidly redistributing the infor-mation gathered from the many scattered monitoring stations, the data could be fed into regional, national, and global collation and analysis centers, where they could be studied and incorporated into computer models on a regular basis. As the schools gained experience and confidence, the range of activities in the program could be expanded to include, for example, soil sampling (to map soil types, monitor soil erosion rates, and measure residues of pesticides and salt) and an annual tree census, using sampling techniques that monitor deforestation and desertification.

If the program worked as planned, those involved might eventually be per-suaded to go even further and actually plant trees and establish nurseries for trees and crops indigenous to their individual areas. And a different sort of seed might be planted in the process: for example, the world's leading scientist on the problem of ozone depletion, Dr. Sherwood Rowland, first became interested in the atmospheric sciences as a youngster when he was asked to look after a backyard weather station by a neighbor who went on vacation for several weeks. The virtue of involving children from all over the world in a truly global Mission to Planet Earth is, then, threefold. First, the information is greatly needed (and the quality of the data could be assured by regular sampling). Second, the goals of environmental education could hardly be better served than by actually involving students in the process of collecting the data. And, third, the program might build a commitment to rescue the global environment among the young people involved.

There are now efforts to improve the Mission to Planet Earth, which NASA first organized along lines that resemble sprawling Defense Department weapons procurement programs; most of the money was budgeted for large pieces of hardware that will take ten to fifteen years to build and then deploy in space. We need the information faster and cheaper, if it is at all possible—and I am convinced it is. Toward that end, Senator Barbara Mikulski and I have been working together to force changes in the NASA program, with some success. Even as NASA is proposing new space platforms built by defense contractors to collect more data, the Bush administration is refusing to spend tiny sums of money to safeguard the valuable information already collected—by the Landsat system, for example, a series of satellites that have made a unique photographic record of the earth's surface for twenty years. The administration has allowed the data collected to go to waste and is now proposing to stop the launch of the next Landsat satellite, thus eliminating the chance to assemble new portraits of our planet and provide a rare and invaluable perspective on the changes we are causing to the earth's surface.

Another difficulty with the current design of the Mission to Planet Earth is that no one yet knows how to cope with the enormous volume of data that will be routinely beamed down from orbit. Nothing even approaching this amount of data has ever been dreamed about. In order to help organize it—and interpret it—I have proposed something called the Digital Earth program, which is designed to build a new global climate model capable of receiving data from several different sources that are not considered compatible by today's definitions; furthermore, Digital Earth would be designed to actually learn from its mistakes, when predictions based on information from the known climate record are run on the models of environmental change so that the results can be compared with what actually happened. Even though the global climate models all have serious limitations, they still give us the best information available about what is likely to happen to climate in the future, and I believe this new approach can substantially improve the quality and usefulness of the models.

Because of the unprecedented volume of data, it may also be necessary to disperse the means of storing and processing it much more widely. Most experts in the United States and Japan now believe in the inherent advantages of a computer architecture or system design known as massive parallelism, and massively parallel computers will undoubtedly play a key role in Mission to Planet Earth. These computers are valuable in another way too, for they provide a metaphor that I think is particularly useful in figuring out how to best cope with the task of collecting and processing the enormous quantity of data and how best, in the process, to change minds and hearts all over the world on the subject of the environment.

The power of massively parallel computers comes from their ability to process information, not in one central processing unit but in tiny, less powerful units throughout the computer's memory field in locations immediately next to the spot where the information itself is stored. For many applications, the inherent advantage of this design is crucial: the computer wastes less time and energy in retrieving raw data from the memory field, bringing it to the powerful central processor, waiting for processing, then taking the processed data back to the memory field to be stored again. By locating each small portion of the data with enough processing capacity to handle it, more data can be processed simultaneously, then transported only once, not twice, between the memory field and the center.

When you stop to think about this approach in generic terms, it seems obvious that both democracy, as a political system, and capitalism, as an economic system, work on the same principle and have the same inherent "design advantage" because of the way they process information. Under capitalism, for example, people free to buy and sell products or services according to their individual calculations of the costs and benefits of each choice are actually processing a relatively limited amount of information—but doing it quickly. And when millions process information simultaneously, the result is incredibly efficient decisions about supply and demand for the economy as a whole. Communism, in contrast, attempted to bring all of the information about supply and demand to a large and powerful central processor. Forced to deal with ever more complex information, the system's inherent inefficiencies led to its collapse and the collapse of the idea on which it was based.

Similarly, representative democracy operates on the still revolutionary assumption that the best way for a nation to make political decisions about its future is to empower all of its citizens to process the political information relevant to their lives and express their conclusions in free speech designed to persuade others and in votes—which are then combined with the votes of millions of others to produce aggregate guidance for the system as a whole. Other governments with centralized decision-making have failed in large part because they literally do not "know" what they or their citizens are doing.

Unfortunately, we are now on the verge of ignoring this powerful truth in designing the Mission to Planet Earth. The current plan is to bring all the data to a few large centers where they will be processed; somehow the results will then be translated into policy changes that are in turn shared around the world. The hope is that this mission will eventually help change thinking and behavior worldwide to the extent necessary to save the global environment.

The alternative approach—or architecture—that I am recommending here is to distribute the information collecting and processing capability in a "massively parallel" way throughout the world by involving students and teachers in every nation. This way, some of the essential work may well be accomplished much faster and much more efficiently—and we can then work to upgrade and improve the information handling capacity in each location. Furthermore, we ought to be establishing environmental training centers and technology assess-ment centers throughout those areas of the world (especially the Third World) where major environmental remediation efforts are needed and where major technology transfers from the West are expected.

In discussing information and its value, it is also worth remembering that some self-interested cynics are seeking to cloud the underlying issue of the environment with disinformation. The coal industry, for one, has been raising money in order to mount a nationwide television, radio, and magazine adver-tising campaign aimed at convincing Americans that global warming is not a problem. Documents leaked from the National Coal Association to my office reveal the depth of the cynicism involved in the campaign. For example, the strategy memorandum notes their "target groups" as follows: "People who respond most favorably to such statements are older, less-educated males from larger households, who are not typically active information-seekers . . . another possible target is younger, lower-income women [who are] likely to soften their support for federal legislation after hearing new information on global warming. These women are good targets for magazine advertisements."

In order to counter entrenched interests like this one, we will have to rely on the ability of an educated citizenry to recognize propaganda for what it is. And the economic and political stakes in this battle are so high, there will be a relentless onslaught of propaganda.

The key, again, will be a new public awareness of how serious is the threat to the global environment. Those who have a vested interest in the status quo will probably continue to be able to stifle any meaningful change until enough citizens who are concerned about the ecological system are willing to speak out and urge their leaders to bring the earth back into balance.

RESPONDING JOURNAL

Gore quotes astronaut Sally Ride's concept of a "Mission to Planet Earth" to help save our planet. What would your "Mission to Planet Earth" be? What mea-sures would you recommend? Explain the reasons for your recommendations in a substantial journal entry of two pages or more.

QUESTIONS FOR CRITICAL THINKING

STRATEGY

1. What does the "Mission to Planet Earth" assume about our relation to our planet? Do you agree with this estimate? Why or why not?
2. Think about Gore's metaphor of our planet being like a sick patient. On what basis does he make such a claim? How logical do you find it?

1. Review Gore's concept of a "Digital Earth program." How fully do you think Gore has thought through this concept? To what extent does the "Digital Earth" idea contribute to his argument?
2. Much of Gore's plan depends on world cooperation. What is your sense of the possibility of achieving such cooperation? Given your assessment of world cooperation, how successful are Gore's plans likely to be? What important variables are involved?

COLLABORATIVE WRITING ACTIVITY

Look at Gore's suggestion for students and teachers to monitor their planet daily. How might you have done this in high school? How might you do it now? What benefits could result from such activity? After listing your suggestions, bring them to your group for discussion. Together, plan an oral presentation to the rest of the class that will address what you might have done in the past and what you can do for the immediate future to help implement Gore's plan. You might offer at the end of the presentation a "minority report" if any members of the group have strong reservations about Gore's ideas.

"Who's Responsible for Overkill?" from *Environmental Overkill*
Dixy Lee Ray (with Lou Guzzo)

Dixy Lee Ray (1914–1994) was Governor of Washington and a former professor of zoology at the University of Washington in Seattle. She was a forthright and hard-working crusader for conservative causes; she served as Chair of the Atomic Energy Commission and won a United Nations Peace Prize.

Her first book on the environment, Trashing the Planet *(also written with Lou Guzzo), was a best-selling response to many environmentalists. In that first book she began her attacks on scientists who warned of the dangers of phenomena such as acid rain, the greenhouse effect, and the growing hole in the ozone layer.* Environmental Overkill *appeared in 1993.*

EXPLORING JOURNAL

Freewrite about the extent to which you think various aspects of the environmental movement have been overdone or pursued too zealously. Explore your sense of what the world needs to do about the environment and what you hear from the media about the state of the environment. Do you ever hear ideas that seem to run counter to your common sense?

———————

> *As the science editor at* Time, *I would freely admit that on this issue (the environment) we have crossed the boundary from news reporting to advocacy.*
>
> —CHARLES ALEXANDER, *TIME*[1]

> *I do have an axe to grind. . . . I want to be the little subversive person in television.*
>
> —BARBARA PYLE, CNN ENVIRONMENTAL DIRECTOR[2]

> *There is no such thing as objective reporting. . . . I've become even more crafty about finding the voices to say the things I think are true. That's my subversive mission.*
>
> —DIANNE DUMANOSKI, *BOSTON GLOBE* ENVIRONMENTAL REPORTER[3]

> *We in the press like to say we're honest brokers of information, and it's just not true. The press does have an agenda.*
>
> —BERNARD GOLDBERG, CBS, "48 HOURS"[4]

> *I'm not sure it is useful to include every single point of view.*
>
> —LINDA HARRAR, PBS PRODUCER[5]

> *It doesn't matter what is true; it only matters what people believe is true. . . . You are what the media define you to be. [Greenpeace] became a myth and a myth-generating machine.*
>
> —PAUL WATSON, CO-FOUNDER OF GREENPEACE[6]

> *It is journalistically irresponsible to present both sides [of the greenhouse, global warming theory] as though it were a question of balance. Given the distribution of views . . . it is irresponsible to give equal time to a few people standing out in left field.*
>
> —STEPHEN SCHNEIDER, NATIONAL CENTER FOR ATMOSPHERIC RESEARCH (IN *BOSTON GLOBE,* MAY 31, 1992)[7]

With attitudes like these shaping the news, is it any wonder that the public gets only the side that the media elite approve? It begs another crucial question:

Having stated our case against "environmental overkill," we are obliged to ask ourselves, "Who is responsible for it?"

Is it the educational process? At least a measure of the responsibility belongs to the schools, mainly because they have been targeted by a barrage of "learning materials" from the extreme environmental organizations and because they have been influenced by media reports. For the most part, teachers and reporters and editors have neither the time nor the patience to challenge the assertions put forth by those who *appear* to have credentials or have been presented as "experts." The truth about the environment belongs in the classroom, of course. But *the whole truth* is not always what students get.

Jonathan H. Adler, environmental policy analyst at the Competitive Enterprise Institute, put it this way:

> Most classroom environmental information, including most that is listed at the Environmental Protection Agency clearinghouse, comes from literature and teaching guides drafted and distributed by the major environmental groups. These materials include everything from the World Wildlife Fund's "Vanishing Rain Forests Education Kit" and the Chesapeake Bay Foundation's "What I Can Do to Save the Bay" to the Acid Rain Foundation's curriculum, "Air Pollutants and Trees," and the Sierra Club educational newsletter, "Sierraecology."[8]

We believe education's role in spreading the propaganda of the environmental extremists to the young and impressionable has been a large one. But we also believe that, without question, the primary culprits in misleading the public on environmental issues and producing the overkill—next to the environmental extremists, that is—are the electronic and print news media. Leading the way are the national television anchors and reporters and their local counterparts. They are the primary culprits because they have the largest audiences, by far, and thus the greatest influence.

Some examples:

1. From the beginning, the TV anchors and reporters adopted the line in news reports that the "Earth Summit" scheduled for Rio de Janeiro in June 1992, was a worldwide conference designed to "save the planet." Rarely, if ever, did a TV anchor or reporter question the nature of the United Nations Conference on the Environment and Development ("Earth Summit"), what the motives were behind it, who the planners were, and, most important of all, what the impact of the proposed treaties would be on the U.S. and other industrialized nations. Seldom has reportage on a single event been so one-sided.

As *Media Watch,* constant watchdog of all the American media, reported: "To reporters, the Earth Summit wasn't a forum for detailed reporting on the complexities of political and scientific debate on the environment. Instead, it was a laudable and idealistic gathering ruined by President Bush. The substance of the summit, the text of the treaties to be signed or rejected, took a back seat to style. Who was in favor of 'saving the planet'? And who was not?" Among the highlights of Rio bias:

Taking the position that the planet was in danger landed the media squarely in the liberal camp. In the days before the summit, anchors wallowed in the simplistic. On ABC's "World News Sunday" May 31 (1992), Forrest Sawyer stated: "The U.S. is under fire for standing in the way of efforts to protect the planet." CNN anchor Christiane Amanpour oozed like a U.N. press packet on June 3: "The Summit, with perhaps the loftiest goal ever, [seeks] to stop us from pushing our own planet toward environmental collapse."

The U.S. delegation was regularly described as "isolated" after it "watered down" a "global warming" treaty and refused to sign the biodiversity treaty, which was "designed to protect plants and animals." Almost every reporter, in print and broadcast, used this inaccurate shorthand. Few mentioned the actual text of the treaty which demanded that the U.S. hand foreign aid to Third World countries *with no conditions,* meaning they could not designate the money to protect plants and animals.[9]

In another vein, the Alar episode referred to in an earlier chapter is a textbook example of how a major TV network, CBS, and its progeny, "60 Minutes," can be manipulated to serve the selfish interests of a large and unscrupulous environmental organization, in this case the Natural Resources Defense Council. In an astounding story of press agentry and propaganda, the NRDC managed to talk "60 Minutes" into airing its outlandish and unsubstantiated charges against Alar, claiming that the growth-regulating compound used to prevent premature falling of apples was a serious cancer risk to children. All the media played the charges prominently in newscasts or in newspaper columns. Horrified parents across the country quit buying apples. And scores of apple growers were launched on the road to bankruptcy.[10]

In time it was discovered that the NRDC had based its Alar warning on a totally discredited animal test back in 1977 by a researcher named Bela Toth, who found that feeding extraordinary amounts of daminozide (Alar) and its derivative to mice produced tumors. A similar study 11 years earlier had revealed no cancers in other laboratory animals, even in rats! Other subsequent studies by reliable scientists brought similar "no cancer" results. But the militant fringe in the EPA refused to accept the legitimate studies, demanding that dosages of Alar be stepped up until researchers finally produced "one lonely mouse tumor at 22,000 times the maximum exposure that children would receive." In other words, it was physically impossible for any child to eat enough Alar-treated apples at one time even (for 22,000 times the exposure?) to incur the risk of growing a tumor.[11]

If the media had cared or had given truth and the other side a chance to be heard, TV, newspapers, and news magazines should have been demanding a thorough investigation of the NRDC and its influence on EPA practices. And, for that matter, on the national media.

2. When the issue of censorship arises, all facets of the news media explode in defensive anger. And rightly so. Where, then, were the "indignation" and the investigative arms of the national TV networks in particular and the American news media in general when the Public Broadcasting Service refused to run a highly praised British TV documentary, "The Greenhouse

Conspiracy"? The 55-minute production, which the *London Financial Times* called one of the best science documentaries made, was turned down because the educational network supported by U.S. tax dollars insisted it was "too one-sided."[12]

Utilizing the expertise of a large number of highly reputable scientists in the U.S. and Britain, "The Greenhouse Conspiracy" questions all the arguments for global warming and concludes that they are faulty and unsupported by reliable evidence.

While PBS was quick to reject "The Greenhouse Conspiracy," the broadcast network and its affiliated local stations had shown no qualms about running a large number of documentaries on the doom-and-gloom side of the environmental argument—among them a ten-part series, "The Race to Save the Planet"; "Crisis in the Atmosphere," and "After the Warming." All of them include many highly questionable conclusions that are fueled more by emotion and hysteria than scientific truth. One-sided?!

3. The Cable News Network matched the major TV networks' prejudice in favor of militant environmental action and frequently went them one better, probably because it hired an environmental director, Barbara Pyle, to manage its slanted, pro-environmental broadcast and documentaries. As CNN's vice president for environmental policy, she encouraged not only one-sided presentations on global warming, ozone "holes," and all the other scare stories in the environmental stable; she also targeted children through a special series of "eco-cartoon" programs under the label of "Captain Planet" and promoted "kiddie mail" to further CNN's doom-and-gloom agenda.

Here are a few among the many CNN pearls of "wisdom":

CNN anchor Bernard Shaw on global warming: "This is a story about human folly—mankind's attempt to engineer a better place to live, to improve upon nature with inventions such as refrigeration, foam packing, and electronics. But the man-made chemicals used in pursuit of the good life have all put life on earth in jeopardy. The chemicals have punched a hole in the sky. . . . Already there's a moral to the story, and that is: Nature may not always be able to recover from the abuses of modern civilization."[13]

Every one of these statements is wrong. And we were all under the impression that anchors, like other "news" reporters, were supposed to give us the news without opinion or prejudice!

CNN reporter Lucia Newman: "The problems are enormous and none of the solutions simple. But most are conscious that unless there's action, the planet may solve the problem—by simply making it impossible for people to live on it."[14]

CNN reporter Jill Dougherty: "Some environmentalists say that George Bush, the self-proclaimed 'environmental President,' should be called the extinction President."[15]

How else might one expect CNN newshands to behave with the likes of two militant Greens, Ted Turner and Jane Fonda, guiding the network's destiny?

4. But if you thought the TV reporters and anchors were alone in spooning out the one-sided opinions on environmental issues try these from the news magazines:

Time, in its usual fashion of mixing news and opinion in giving advice on the Rio conference, suggested: "Put an international tax on emissions of carbon dioxide and other greenhouse gases. . . . Find a way to put the brakes on the world's spiraling population, which will otherwise double by the year 2050. . . . Give the U.N. broad powers to create an environmental police force for the planet."[16]

Newsweek's Jerry Adler: "Almost alone among major nations, the United States retains a substantial constituency that is indifferent if not hostile toward environmental regulation—an attitude oddly shared by the GOP right wing and the leaders of the former Communist bloc. But this is increasingly a fringe position even among many of the business executives it is supposed to benefit."[17]

Newsweek reporter Sharon Begley: "When Bush shows up this week for a 40-hour appearance, even many of America's allies are going to greet him as the Grinch who stole the eco-summit. . . . America, in contrast, found itself in the role of cranky Uncle Scrooge."[18]

5. Another severe blot on the performance of America's news media came with the National Aeronautics and Space Administration's "ozone hole" reports late in 1991 and early in 1992. The first report—a press release proclaiming that the ozone "hole" had increased in size and would soon threaten the Northern Hemisphere in general and the U.S. in particular—became the No. 1 or No. 2 news item on TV news broadcasts across the U.S., from the networks to the local affiliates. It was front page news in virtually every newspaper in America and the main feature in the news magazines, as well. To say that Americans were frightened by the news-generated hysteria would be putting it mildly.

Then, a few weeks later, when a probing missile was sent up into the stratosphere to report back on what was really happening to the ozone layer, the next press release was issued quietly—and all but ignored by most of the media, except for *The Wall Street Journal* and a few columnists and commentators on the conservative side of things. Why? Because the new report was that the terrible "hole" hadn't materialized and the "danger" had passed! But the American public wasn't aware of NASA's duplicity, because it doesn't realize to this day that the news media had once again failed to do their job.[19]

6. The late Warren T. Brookes, courageous *Detroit News* columnist, whose investigations embarrassed many of the doomsaying environmentalists, pointed out several instances in which the bulk of the news media, electronic and print, failed to report the misdeeds or outlandish claims of the radical fringe.

He was one of those who chronicled the serious misjudgment by the federal Centers for Disease Control when it finally "admitted it had made a mistake when it shut down Times Beach, Missouri, in 1982 because of the dangers of dioxin." The CDC and the EPA both acknowledged after the damage was done that the danger was overstated and that "tens of billions of dollars in cleanup costs" were wasted across the nation on other projects whose cleanup actions were instigated by the example of Times Beach. The confession by the two agencies was reported almost casually by a few newspapers and TV stations— but it certainly didn't get the same play on news broadcasts and TV reports that the first scary Times Beach report had been given.[20]

Brookes also called attention to the fact that the news media all but ignored "the fraud of asbestos hysteria" that had been uncovered by another *Detroit News* reporter, Michael J. Bennett, and detailed both in the newspaper and later in a hard-hitting book everyone should read, *The Asbestos Racket.* This is a case of still another dereliction of duty by the media in turning their gaze from a situation that has cost American taxpayers billions already and will cost them billions more as more lawsuits are filed.

The element in Bennett's work that should have caught the attention of every American but didn't because the media ignored the investigation was this report: Through painstaking research Bennett discovered that the over-reaction of the EPA to the overblown "danger" of asbestos may have been the cause of the Challenger Space Shuttle disaster that brought death to seven American astronauts.

Because of an EPA ban on the use of asbestos, a non-asbestos containing putty was substituted which didn't have the insulating fire-retardant powers of asbestos. Bennett reported that the substitute putty used to seal the O-Rings was vulnerable to cracking in the extremely cold weather and led to the shuttle's crash on January 28, 1986.[21]

Why didn't the media follow up Bennett's investigation and question the EPA's panicky policy on asbestos? As Brookes pointed out, the finger should point to EPA Administrator William Reilly and his flip-flops on the asbestos issue. Although he re-affirmed the EPA's asbestos-ban policy in 1989, he acknowledged a year later that "the mere presence of asbestos poses no risks of harm to human health" and that removing it "may actually pose a greater health risk than simply leaving (it) alone."

Even more extraordinary was Reilly's admission that solid scientific documentation of the relative lack of risk in white chrysotile asbestos (95 percent of all that is in use) was available since the 1970s but was relentlessly dismissed by the EPA. Which William Reilly can you believe?[22]

How much more evidence do all the media need to convey the truth to viewers and readers?

Remember, this is the same Reilly who tried to talk his "boss," President Bush, into agreeing to become the banker to all the imagined environmental ills of the Third World—and much of the Second World, as well. Why the President didn't ask for Reilly's resignation remains one of the unanswered political questions of 1992.

7. One more sobering example should be offered to indicate how complacent the news media have become when it comes to covering "the other side" of environmental issues. Because it wanted the facts on acid rain and on the

environmentalist charges that acid rain was having a calamitous effect on lakes, rivers, forests, and even buildings and statues, Congress authorized a ten-year study, the National Acid Precipitation Assessment Program (NAPAP), by the most reputable scientists in America. More than half a billion dollars were spent on the massive scientific project.

The diligent NAPAP reported to Congress that it had found some acid-rain damage to some forests, lakes, and rivers—but that there was absolutely no "impending environmental disaster," as environmental groups had charged, and that "an expensive crash program to further accelerate the current rate of reduction of acid rain is not justified."[23]

Despite the well-researched report, Congress ignored it and passed the Clean Air Act of 1990, which will require the expenditure of $40 billion *needlessly* on projects designed to repair a crisis that doesn't exist! And yet the news media have chosen to ignore the effect of Congress' action, preferring instead to listen to the same political activists to whom the EPA listens.

Among the many reputable scientists who have tried in vain to break through the media's curtain of bias is Dr. Julian L. Simon, professor of business administration at the University of Maryland and the author of many books on demography and economics. Through years of painstaking research, he has provided convincing proof that the doomsday scenarios offered by government and the militant environmental groups—and dutifully reported by an obedient press—are without foundation.

Simon has methodically destroyed wild claims about disappearing agricultural land, soil erosion, diminishing natural resources, and the population "explosion." But, despite the compelling nature of his arguments, the news media have turned a deaf ear to his sound research.

"I come not in anger but in pain," Simon wrote. "Journalists take pride in their objectivity. But in reporting on population growth, natural resources, and the environment, objectivity goes out the window. The price in economic loss, misguided policies, and damage to national morale has yet to be calculated. But the costs may be fearfully high."[24]

Simon's compelling book dispelling environmental myths, *Population Matters,* should be required reading by all members of the media and, in fact, by every American who is sincerely interested in the environment and the nation's future.

Perhaps the most significant analysis of the media's love affair with the environmental extremists has been done by a doctoral threesome, S. Robert Lichter, Linda S. Lichter, and Stanley Rothman, in another book that should be on every media person's coffee table, *Watching America.*

Their profound, well-researched scientific study of the manner in which television has manipulated programs to suit the political and social beliefs of producers, editors, writers, and staff is an alarming revelation. The authors found, for example, that at least three-fourths of the creative leaders in TV acknowledge they are liberals and vote Democratic. Two-thirds of them believe the structure of American society is faulty and must be changed. Ninety-seven percent say women should have the right to decide whether they want to have an abortion, 80 percent believe there's nothing wrong with homosexual relations, and 51 percent see nothing wrong with adultery.[25]

Most significant, the TV people surveyed admit they work their ideas on social reform, mores, the environment, and other issues into the programs they create for their audiences. With that ultra-liberal background propelling TV programs, it's no surprise that so many television luminaries and staff people actively support the cause of the environmental extreme. Nor is it a surprise that Dr. Simon and all the other dedicated scientists who want to balance the environmental scales toward good science and logic are left out in the cold.

With the news media stacking the deck against common sense and scientific honesty in environmental affairs, it's little wonder, then, that polls indicate so many Americans believe "something must be done to protect the environment, no matter what the cost."

An August 1992 nationwide poll by the Wirthlin Group of McLean, Virginia, found that more Americans were "growing concerned about the environment and feel President Bush was not doing enough to protect it." Eighty percent said environmental improvements must be made regardless of the cost and that environmental standards cannot be too high. Sixty-seven percent said Bush was doing less than his share to reduce environmental problems. And 57 percent said environmental issues will be a "very important" factor in how they would vote in the 1992 presidential election.[26]

How can the electorate be expected to make truly intelligent decisions if the news media do not provide them with all the information on the environment, except what supports the most radical and outlandish environmentalist positions?

All of which brings us to some tips on what American TV watchers and newspaper and news-magazine readers should be alert to when they peruse the "news." To wit:

"Environmentalists say. . . ." Are the quoted "environmentalists" identified? If not, the article loses its validity and should not be taken seriously. If the environmentalists are identified, are they members of one of the militant environmental groups with a political agenda and a treasury that needs a continuing infusion of members and donations? Or are they independent scientists with no axe to grind and a reputation for being dedicated scientists, not mouthpieces for some political cause? Remember, each of us is an environmentalist at heart because we care about the planet to one degree or another. John Doe may be an "environmentalist" to Reporter Smith, but he doesn't necessarily speak for us. Reporters should be more careful and define the expertise of every person they have been accustomed to identifying as "an environmentalist" and letting it go at that. Even "scientist" is too broad a term in many cases. A laboratory technician, for example, can be called a scientist in, say, the field of pharmaceuticals, but may know comparatively little about another science, say, atmospherics.

"Critics say. . . ." Ditto above. When a lazy reporter needs a hook for an opinion, he often borrows this time-worn and totally unreliable ploy—without identifying the "critics" or indicating the nature of their expertise.

"Cancer-causing. . . ." Beware of this hyphenated tiger, which is the irresponsible reporter's (and radical environmentalist's) favorite scare phrase in his attempt to grab attention. This overused and usually false description precedes what the activist—or the reporter—is trying to depict as dangerous, poisonous, deadly, or whatever, even if it isn't, which is most often the case. In the lexicon of the reporter/activist, "cancer-causing" precedes such items as PCBs, pesticides, insecticides, herbicides, chemicals, gases, anything he chooses to infect with terror or apprehension.

"A commission or committee or group said. . . ." Many reporters have a habit of failing or refusing to identify the source of an important decision or study or research. Or, worse yet, having identified the commission, or committee, or group, fail to list its members so the viewer, listener, or reader can gauge for himself the validity of the report. "The Committee to Redress Grievances," the reporter will say or write, "declared today that it is dangerous and could possibly be fatal to eat oysters after 8 o'clock in the evening on a cold day." Really? Just who sit on the Committee to Redress Grievances? Are they marine scientists? Or three garage mechanics who go fishing every Saturday?

"Analysis. . . ." This catchall word has become very popular, particularly in the print media, as a device to inject a highly opinionated piece into the news columns—a piece that belongs on the editorial or op-ed page of a newspaper, not on page 1, 2, or 3. But in recent years, reporters have more and more laced their "straight news" reports with hidden or obvious expressions of opinion. That is especially true in TV reporting today. Communications schools at colleges and universities are primarily responsible, because many of them tell and teach students that activist or participatory journalism is the order of the day. In "activist" or "participatory" journalism, the reporter becomes an adjunct of the news itself and includes his opinions—read that "prejudices"—into the event he's reporting. The line between fact and opinion in news reporting has, as a consequence, become blurred and confusing to the viewer/listener/reader who wants information, not a sermon.

"May. . . ." It's a simple word, but a lazy or incompetent reporter will use it often to mask a lack of definitive information or his or her own inability or unwillingness to get more information or be more specific. "Magnets in the brain may link cancer and electricity," writes the hedging reporter. If he had done his interviewing and research properly, he would have been able to say or write it in this fashion: "Researchers conducting experiments with electricity and its effects on the human body are tracking down the possibility that excessive amounts of electrical waves could affect a person's health. Some scientists in the study are suggesting the possibility that tumors might result, but no proof of it has been found. The research might, in fact, determine that electricity has no effect at all on the human body, several scientists have indicated." Just tell

it like it is, as the saying goes; don't manufacture scare words for the sake of jarring the viewer or reader.

"Conventional wisdom. . . ." This phrase, rapidly growing in popularity with reporters, anchors, and editors, is acceptable—until it is used as an excuse for the reporter's not having done his homework. If his dispatch is based on official, statistical, or legal information, he should spell it out for the viewer or reader, not cover his faulty workmanship with such a generalized reference as "conventional wisdom says this or that is so."

All these—and many, many more not listed here—are common devices that contribute heavily to environmental overkill in the news. Good reporters will avoid them. In fact, good reporters will refuse to be manipulated by anyone, let alone militant environmentalists.

In addition to such telltale words that should warn us of danger ahead in news reports, there are certain subtle techniques reporters use that are tip-offs to trouble ahead in their reports. For example:

The Second-Day Lead: In broadcast and print media, a "first-day lead" is a straightforward report of an event that has just happened. No frills. No extras. No interpretations. It might read: "Senator James Smith today said he would mount a campaign in the Senate to overturn the acid-rain provisions of the Clean Air Act because they will result in billions of dollars in unnecessary expense to taxpayers."

In the past, that report would ride out the day in the news just as is. The next day a lead devoted to reaction or rebuttal would be customary and proper. That is what's known in media as the second-day lead.

What some reporters frequently do—unintentionally or with subtle prejudice at work—is water down the impact of the first-day lead by introducing a counter or rebuttal immediately. For example: "Senator James Smith said today he intends to knock the acid-rain provisions out of the Clean Air Act, but environmentalists declared he's just grandstanding because he's thinking of entering the race for Governor and needs the publicity."

What has happened is that the reporter has punctured Senator Smith's balloon before it gets off the ground by twinning it immediately with a reaction that should have gone farther down in the story or used as a new lead the next day. In theater, it would be called upstaging. In journalism, it's sneaky and unfair. Watch for it; you'll hear and read the mechanism used frequently.

The Planted Paragraph: It's one of the most common devices in communications. An Associated Press reporter entered the following information after the first few paragraphs of a Washington report on a dispute between the Agriculture Department and the General Accounting Office on testing pesticides in fruits and vegetables:

> "There are more than 60 pesticides that are known or suspected carcinogens used in foods," said Lawrie Mott, a senior scientist at the Natural Resources Defense Council specializing in pesticides and children's health. Ms. Mott said there is a special danger to children because they eat or drink some pesticide-laden foods—such as apple juice—in disproportionate amounts compared with adults.[27]

The unsubstantiated Mott statement was not only out of place in the report. It demonstrated the reporter's prejudice in favor of the NRDC position and his penchant for quoting a "scientist." He failed to mention her relevance to the report or to explain that the NRDC is a highly politicized environmental group with its own bias on pesticides, Alar, and similar subjects. Worst of all, Ms. Mott's statement stands alone, without presentation of the other side of the argument. Typical bad reporting.

The News that Isn't: Obviously, a great deal of news is not reported, which is a problem in itself. But what of the news that is unnecessary and "over-covered"—a report that suits a reporter's or editor's formula or prejudice about "what the public oughta know"? Here's a good example, known not only to many reporters and editors but to political environmentalists who include nuclear power on their hit lists, despite the fact that nuclear power is the friendliest energy to the environment:

> PORTSMOUTH, N.H., Reuters—A fire broke out on the grounds of the Seabrook nuclear power plant Friday, officials said. The small blaze occurred about 50 feet from the plant's cooling tower and less than 500 feet from the core reactor building, said Rob Williams, spokesman for the plant's manager, North Atlantic Energy Service Core.
>
> No radiation was released and the plant's operation was unaffected, Williams said. . . . The fire broke out in a portable diesel-driven air compressor being used for exterior sandblasting of the cooling tower building. Williams said the fire was a minor incident, which caused no injuries. The blaze was extinguished within 11 minutes by an on-site emergency team.[28]

The Reuter report was distributed nationwide. Why? Because anything that happens at a nuclear plant, no matter how minor or uninteresting, is reported, thanks to the media's "overkill" attitude about things nuclear. Let another fire happen in a coal mine and kill a dozen people and it won't get the national distribution of an insignificant incident at a nuclear plant.

The point here is that too many persons in the media treat anything nuclear with the same disdain and suspicion as they do other targets of the environmental lobby. Their personal prejudices dictate their reaction to the news.

Will the biased news coverage never end? Certainly no time soon. We can be sure the growing ranks of political environmentalists and pressure groups will think up new horror stories to replace those now in vogue. What can we expect in the future in the realm of environmental overkill ? Here are a few issues that have begun to capture the fancy of the doom-and-gloom organizations:

Electromagnetic Fields (EMFs): We have detailed this issue in an earlier chapter, but it's important to stress the point that news media reports have frequently not told the whole story. The best science acknowledges that further research is needed but cautions that no serious danger is evident at this time. Still, the media can't resist sending out alarms and frightening people unnecessarily.[29]

The Chlorine "Peril": Despite the fact that chlorine in America's drinking water has undoubtedly saved millions of lives in the 20th century from a variety of once-waterborne diseases—like typhoid, cholera, gastroenteritis, and dysentery—we're now told that people who drink chlorinated water *may* be facing a 21 percent greater risk of developing bladder cancer and a 38 percent higher risk of rectal cancer. There's that word, "may," again, this time applied by a Dr. Robert Morris of the Medical College of Wisconsin to findings he and a research team came up with. Water officials everywhere remain skeptical and want to see more evidence before they propose any action on chlorine, which has been credited with saving more lives than perhaps any other single chemical. At least one water-quality manager has said that even if the risk of cancer should be supported by more research, that risk is far outweighed by the benefits provided by chlorine.[30] Is there a strong hint of environmental overkill here? Time will tell.

Lead Poisoning: One of the EPA's newest bogeymen is lead poisoning, which is reputed to be—by those who are ringing the alarm over it, that is—a growing danger to children particularly. Several scientists were at each other's legal throats in a developing donnybrook over research on lead poisoning. One faction says it's terribly dangerous if not controlled, while the other insists the problem has been seriously overblown and should be put back into proper perspective. At last reports, the EPA apparently wasn't waiting to find out who would win the lead poisoning battle. It was already using the work of the "terribly dangerous" advocates in writing regulations on the use of lead.[31] What could one expect, given the shoot-from-the-hip, environmental-overkill nature of the EPA?

In the meantime, it's appropriate to end this discussion of the role of education and media in overkill with this warning from Dr. Bernard Cohen, University of Pittsburgh physicist and one of America's most distinguished nuclear scientists:

> Our government's science and technology policy is now guided by uninformed and emotion-driven public opinion, rather than by sound scientific advice. Unless solutions can be found to this problem, the U.S. will enter the 21st century declining in wealth, power, and influence. . . . The coming debacle is not due to the problems the environmentalists describe, but to the policies they advocate.[32]

That should spell out the peril inherent in continued "environmental overkill."

NOTES

1. "Journalists and Others for Saving the Planet," David Brooks quoting Charles Alexander of *Time* at summer 1989 environmental conference, *Wall Street Journal,* October 5, 1989.

2. *American Spectator,* quoted by Micah Morrison, July 1991.
3. Ibid.
4. "The Media's Middle Name Is Not Objectivity," Harry Stein quoting Bernard Goldberg, *TV Guide,* June 13–19, 1992.
5. "Editor's Comment," by Chuck Diaz quoting Linda Harrar, *Speak Up America,* September 1, 1992.
6. Ibid.
7. *Boston Globe,* May 31, 1992.
8. "Little Green Lies: The Environmental Miseducation of America's Children," Jonathan H. Adler, *Policy Review,* Summer 1992.
9. "Rio Reductionism," *Media Watch,* July 1992.
10. "How a PR Firm Executed the Alar Scare," *Wall Street Journal,* October 3, 1989.
11. "How the Media Launched the Hysteria About Alar," Warren T. Brookes, *Detroit News,* February 25, 1990.
12. *The Christian Science Monitor,* Richard Miniter, February 17, 1991.
13. *Media Watch,* June 8, 1992, referring to CNN News Telecast, Bernard Shaw, May 26.
14. Ibid, June 22, 1992, CNN "Agenda Earth" Telecast, Lucia Newman, June 3, 1992.
15. Ibid, CNN World News, Jill Dougherty, May 30.
16. Ibid, June 8, 1992, quoting *Time,* June 1, 1992.
17. Op. cit., quoting *Newsweek,* Jerry Adler, June 1, 1992.
18. Op. cit., quoting *Newsweek,* Sharon Begley, June 15 and June 22, 1992.
19. "Hole in Ozone Didn't Develop," *Wall Street Journal,* May 1, 1992. (Associated Press, Knight-Ridder News Service, Reuters, and other news services noted the development almost as a non-news event, most of them twinning the brief NASA press release with the disclaimer by NASA scientists that "the threat would return.")
20. Brookes, Warren T., *Detroit News,* December 23, 1991.
21. *The Asbestos Racket,* Michael J. Bennett, Free Enterprise Press, 1991.
22. *Access to Energy,* Dr. Petr Beckmann, December 1990.
23. "Save the Planet, Sacrifice the People: The Environmental Party's Bid for Power," Edward C. Krug, soil scientist, *Imprimis,* July 1991.
24. *Population Matters,* by Dr. Julian Simon, Transaction Publishers, 1990.
25. *Watching America: What Television Tells Us About Our Lives,* S. Robert Lichter, Linda S. Lichter, Stanley Rothman, 1992.
26. "Bush Not Doing Enough to Protect Environment, Poll Says," *Reuters,* August 6, 1992.
27. "GAO Recommends Suspending Pesticide Testing Program," Associated Press, December 19, 1991.
28. "Fire at Seabrook Nuclear Plant," Reuter, July 6, 1992.
29. *Access to Energy,* Dr. Petr Beckmann, July 1990 and May 1992.
30. "Chlorinated Water Holds Risk for Cancer, Researchers Find," *Seattle Times,* Junly 2, 1992.
31. "A Lack of Integrity Also Poisons the Air," Warren T. Brookes, *Detroit News,* November 11, 1991.
32. "The Global Warming Panic," Warren T. Brookes quoting Dr. Bernard Cohen, *Forbes,* December 25, 1989.

RESPONDING JOURNAL

How much information or knowledge did you gain from reading this essay? How does this new information make you feel? Did it change your estimation of the news media? Explain your response in a journal entry of at least a page.

QUESTIONS FOR CRITICAL THINKING

STRATEGY

1. Does the series of quotations opening the essay convince you of "environmental overkill"? How effectively does this technique serve to introduce the chapter? Explain your response.
2. Look at the quote from Jerry Adler on page 511. How comfortable are you with the author's characterization of such efforts as "propaganda of the environmental extremists"?

ISSUES

1. The outcry over the chemical Alar being used to make apples look more appealing seems now to have been an unfounded scare, at least according to Ray's account. Look up her sources and check other reactions to the Alar question. What is your considered opinion? What do you think caused the great public outcry over Alar?
2. The media certainly does determine to a large extent the factual information we receive about our planet. To what extent do you think media advocate certain positions? Do you see a difference between print media such as newspapers or magazines and electronic media such as network news? Is the environment an issue that seems to encourage certain responses? Provide examples to support your responses.

COLLABORATIVE WRITING ACTIVITY

Draft a letter to the author in response to the arguments Ray presents here. Assume that this chapter has come to you as an editor and that you are to check her presentation for the soundness with which it argues from cause to effect. Tell her how well she has demonstrated "environmental overkill" and suggest the revisions she should make to strengthen her argument, if you feel she needs to do so.

Pair up with another member of your group. This person will be your associate editor. She or he will look over your suggestions and offer responses: what else could you suggest, what should you elaborate, what seems too harsh? You will then do the same for your partner. After you two have discussed this exchange, revise your letter to your author accordingly. Circulate the letters around the class; what are some of the most interesting ideas to come out of this activity?

"Letting Go: The Virtue of Vacant Ground"
Janet Kauffman

Janet Kauffman was born in the farmlands of eastern Pennsylvania and moved to Michigan as a young woman. In Michigan, she began farming and teaching in addition to her writing. She combines her interests effectively, as the following essay shows. Her poetry is collected in her books Writing Home, The Weather Book, *and* Where the World Is. *Her fiction appears in her books* Places in the World a Woman Could Walk *and* Collaborators. *In 1985 the Academy-Institute of Arts and Letters awarded her its Rosenthal Award.*

In her writing, Kauffman confronts some of the crucial issues concerning land use today. As a country, America seems conditioned to want to produce more and more, to measure well-being in terms of how much more we do today than we did yesterday. Kauffman questions not only this attitude but the deeper beliefs that produce such thoughts.

EXPLORING JOURNAL

Try a ten-minute freewrite on the idea of vacant ground around you. What vacant lots are around you? What used to stand on that lot? If you think of recent construction or activity on what used to be vacant ground, explain what is on the ground now, and what had previously been there. What is gained and what is lost by the conversions to or from vacant ground?

⟫◦◦⟪

Drive around southern Michigan these days, and you'll see the new look of farmland. It's wild. Unkempt. Downright gorgeous.

Where, in the past, you'd have seen a regimen of rows, now you see weedy sprawl. Where there were singular crops, there's a riot of undergrowth, wind-seeded. There is raggedness, mess, variety, mix. On its own, out of human control, farmland demonstrates an abundance—flourishing, bizarre, rank, twisted, vital. And the feel, in a way, is more urban than rural: boom and surprise and decay, all in one.

With the farm economy aggravated by grain surpluses, with corn and bean prices in steep decline, it's not unusual in southern Michigan to see a countryside visibly changed by the abandonment of farms, or the abandonment of traditional field practices. A lot of places around here, everything's gone to seed. Government set-aside programs have opened up huge spaces to cover crops; and many farmers, on their own, have begun to let marginal fields lie fallow— just let them go. The gravely hills, glacial dumps, the undrained low-lying clays—they're easy to give up now. It's cheaper, most of the time, not to till and plant them.

If you examine any square foot of unfarmed ground—a square foot which a few years ago would have contained some dirt and a couple of cornstalks— you will find this: quack grass, of course, like a mat, and milkweed, ragweed, or pigweed, lamb's-quarters, maybe some bergamot. Weeds. All weeds. Nothing planted, nothing for profit. Nobody cultivates weeds. But in this economy—here is the virtue, to start—they cost the farmer nothing.

The virtues that follow, and accumulate, are more crucial, more enticing, in the long run; but the fact that weeds are cost-efficient—no production costs, no harvest costs—is good enough to begin with. Farmers here, a few, have learned how to tolerate the velvetleaf or goldenrod, rampant in a vacant field, and have figured out how to see something in the scene besides trouble.

To somebody who hasn't farmed, the scenery is unambiguous, spectacular: weeds bloom in field after field of flower. This time of year, late July, the commonplace explosion of Queen Anne's lace—hand-size white bursts at the roadsides—proliferates outward, into acres of old fields. It's a painterly, splotchy wash. Although ragweed and goldenrod haven't flowered yet, they've shot up to chest height, green and feathery, so there is an expansion and bunching of foliage and vegetation, on to the horizon. In the most meager soils, stalks, like stubble, appear—the flower bits of blue; and sweet clover takes over gravel slopes, a serious tangle of tough stem and small, indeterminate leaf.

Any parcel of marginal, uncultivated land breaks out in these stupendous varieties of weed, diverse but not as random as one might think. Without the inhibition of herbicides or the row arrangement of cultivation, plants emerge according to preferences for soil type, composition, elevation, moisture. By the looks of things—the colors, the textures—you can spot even slight gradations in slope, and all the shifts from sand to loam to gravel to muck.

Looking across these fields is an education in the designs of ease. In natural selection. Adaptation. Variety. In wildness. In one of the plain pleasures of human vision: seeing something not human.

Gone wild, even the first year, a field becomes a thigh-high jungle. Anybody trying to walk through these waste places understands there is no vacancy, none, in "vacant ground."

By vacant ground, we mean, of course, unfarmed, uncultivated land. Uncontrolled by us. *Not* farming marginal farmlands, letting some land go wild or lie fallow, was common farm practice, a regenerative practice, before the use of chemical fertilizers. Now, with some choice in the matter, in a redefined economy, farmers again can acknowledge the value of vacant ground, the virtue—why not call it that?—of letting go.

My concern here, the more I look around, is not agricultural but moral: a concern for how human beings use land.

Farming, like mining or house-building or any construction, is one way humans use land; it is no more "natural" a land use than oil-drilling. So I am not talking about farming as pastoral, or profitable, or good for the soul. Not at all. Farming, here, is *provocateur*—it's made me think. And what I am thinking about is how human beings inhabit the earth.

I have lived in the hummocky glacial dumping ground of southern Michigan for more than sixteen years, almost as long as I lived in the limestone-bedded and rolling farmlands of Lancaster County, Pennsylvania, where I grew up. And while Michigan isn't a wilderness, it certainly is wild, unruly, as scenery, compared to the farm landscapes I knew as a child. This part of Michigan—a good bit of it swampland when white settlers first arrived, a place notorious for mosquitoes and bad air—remains, because of the problems with drainage, and the more extreme ranges of seasons and rainfall, much less domesticated, much less rich than eastern Pennsylvania. Not many farms here have the neat arrangement of barns, with good paint, the encompassing fields and pastures, that show up on the postcards of Lancaster County. From a Michigan point of view, though—and that's what I have now—the pastoral aura, the precision, of postcard farms looks more like tyranny than bliss.

A good Lancaster County farm had few hedgerows, very small woodlots, lawn to the barns and lawn around the barns. No weeds. Tangle was akin to sin. Land was tillable and therefore the land was tilled. Whatever grew was planted. And always for human use. (When I moved to Michigan and first saw the huge leaves of wild burdock, I thought, good lord, *rhubarb!* The mad Germanic notion that if it was there, it had been planted, and could probably be cooked.) In Lancaster County it would have been unseemly, a serious lapse in responsibility, to let a field, even a rocky slope, grow up in weeds. The same obsession with order and control of the landscape continues there now, even though most of the farms, because they were "scenic," are gone—sold to developers and subdivided and recut with streets of houses. Each development wanted a view of the next farm. No postcards now. But the order persists, and it follows from the same tyrannical, familiar assumption: humans have the right, the obligation, to work and rearrange, and "order" the natural world. The wild is a place to be tamed. It is an arrogant designation of priority—make the world over for humans. Americans, seeing landscape from the beginning as real estate, are scrupulous about dealing in it, and in prosperous areas all over the country, a fierce moral judgment falls on "waste" places, scrubland, even the vacant lots in developments, where the ragweed's got a good hold. For God's sake, do something with it! everybody says.

In Michigan, too, you hear remnants of a wish for control. Shreds of it. When I called my neighbor to borrow a cultipacker—the wide iron cylinder that would press down the seeds of a new prairie patch—her question about the coneflowers and partridge pea, predictably, was, "So, will the stuff spread?" She knew it would. And her question carried the comic knowingness—the willful resignation—that comes with living where human control isn't absolute. Where weeds win.

There's a kind of what-the-hell feel to farming here. You do what you can, but the uneven ground, the clay pockets, the swamps—they're a powerful opposition. You compromise. You give in. Sometimes, you just let go. You can't delude yourself into thinking that farming is God's work, or a good man or woman's work. You lose faith in the idea of beauty as a mowed yard. You don't look at an aerial picture of your farm and think, ah, peace. Harmony.

Instead, you haul rocks. You pull out mulberry and locust trees that sprout in fields. You spray quack grass. You say, "So, will the stuff spread?" And some-times—when it finally makes sense—you let cracked drainage tiles stay cracked. You let the swamp be a swamp. You let the locusts sprout and take over a stony field. You stop baling hay in the wet low places where the bales came in damp and molded anyway.

I've done every one of these things, and every one is a mindful struggle. Things look worse, for a while. In a way. When I stopped farming some fields, I minded the scrappiness as much as any dead-end farmer. I thought about putting up signs—Wildlife Preserve—as an excuse, an explanation. And one of these days I might get around to doing it. Not as an excuse anymore, but because it's the truth.

Taking the care (and it *is* care, not lassitude) to hang on to productive land and let go of the hills, the holes, the margins, some farmers have found it's even possible to make money. One farmer near here, who reduced his acreage to the good acreage, makes more money farming 140 acres than another farmer makes working 1,000 acres of unselected fields, farmed straight through, no matter what the lay of the land or the condition of the soil. But there is the "clean" farm, spread out. People say, oh, take a picture. We like these pictures, the photogenic. And in farming, that usually means as complete a control of natural conditions as possible.

A number of untamed, unadulterated beauties have survived in this coun-try, singled out and preserved, protected. When the dream of wilderness con-fronts the dream of the tamed farm (Americans know how to split the soul), we've often done a good job saving our "natural wonders." But to be wonders, wild places must meet substantial requirements as dramatic (Yosemite), spec-tacularly bizarre (Yellowstone), grand (the Grand Canyon). Even if we've done fairly well preserving the dramatic and starkly beautiful, we'd also do well to attend to ordinary places, the nondramatic, waste and wayside places—all habi-tats and sceneries.

To a farmer, it sounds simple. We must finally recognize the rights of the earth. Civil rights, human rights, women's rights, animal rights—all these move outward, expand, in implication, if we keep at it, toward the planet's rights. The human compulsion—centuries old—to use land rather than inhabit it must ultimately appear barbaric, an extreme form of domination and exploitation.

We have grown rich on tyrannical ideas: the idea that the world is a resource, full of goods, for human consumption; the idea that tillable ground should be tilled—an idea that cleared the pines, utterly, from the state of Michigan in the last century; an idea that is clearing jungles and "bringing life" to some deserts today and causing desertification elsewhere. We have an arse-nal of ideas about land use possibly as dangerous to human life on the planet as the use of nuclear arms.

And worse, we have no global plan, no serious national debate concerning the preservation of land and landscape. In fact, in the development and use of coastlines, for instance (and lakeshores, and water rights), recent court decisions persist in supporting private landowners' rights rather than the broader "public interest"—although even that interest, too often one state's or one nation's interest, can be extremely narrow. Americans who regard as sacred the right to hold private property must also ask at some point, what are the rights of the property itself? This is not simply an environmental question, although environmentalists are the ones we most often hear from (the ones who, like Earth First protesters in the Northwest lumbering districts, are labeled "environmental guerrillas"). What are the rights of the earth itself? And another way of asking it is: how should humans inhabit a world not wholly human?

After all, this is a world of rock and water and air. It is elemental. It is not ours.

Our rights cannot be exclusive. Human habitation of the planet must be based on mutuality, not domination. Feminists know this territory. Some farmers do. If we care about the land, it will be necessary to redefine whole economies, not just the farm economy. A complex, solid economy could certainly grow around a policy of cooperation with natural environments. Why haven't we proposed such politics—on as grand a scale as national defense—when our own species is at stake?

We are primitives in our thinking about economies. We are babes. We believe the world is ours, like a heated house. Landscape is sold for holiday viewing. This culture and this economy promote the idea of all-terrain vehicles and disregard the idea of terrain. From an ATV, the landscape is backdrop, nothing more. It's scenery on an ad. You roar through it. That's it.

Almost every April, guys on three-wheelers rip through the beech woods at the back of this farm and cut down the stream bank. With the trillium and rue anemone, it's a pretty place to ride through, with enough rocks in the water for a scare. Once when my son and I caught up with the machines—there were three of them, stalled in some swamp willows, as far into scrub as they could drive—it was clear these were men prepared for combat: they had the camouflage waterproof gear, the helmets and goggles, buckled gloves.

We smugly dismiss the last century, its imperialisms and arrogances and abuses, as unenlightened. They were blind. We can see the arrogance and abuse. But in our own world, we see again, once again, progress. *Development* is one of our favorite, most blessed, words. Real estate development. Third World development. Arctic development. Development for tourism.

Instead of new housing starts (more developments, more subdivisions) as one of the measures of an economy's health, why not new reclamation starts? Why not?

After all, this is a world of rock and water and air. It is elemental. It is not ours.

What do we want to *do* with it? Because we are conscious creatures, the entire planet—the universe—has become a place for the pleasure of the human mind. And being human, we *must* range and speculate. We must terrify ourselves with our thinking. This is our art.

But in the dailiness of human life, in the physical world of carbon and hydrogen, oxygen and uranium,, we may not range thoughtlessly or speculate endlessly to our gain. The world is not ours to use up, or blanket with our debris, or despoil.

In this century, we have made of the world an elaborate, grotesque, noxious cake—it's layered and layered with richness and artificial decoration: a global and decadent art, the mind can say. Who can eat it? Who wants to? What is the appetite that cooked it up, or could be satisfied by it? Set the globe in a great gallery, and it would be something to see.

But life is not art. Or rather, it *dare not* be. We dare not let it be, or the world will be lost. Lost to us, and lost to all artless things—the matter—of the earth. Some knowledge—death (which is inevitable)—cannot be lived (except through art), and therefore it must be known and accepted. Other knowledges—violence, destruction, tyranny (which is not inevitable)—cannot be lived with for long (except through art), and therefore these must be known, and rejected.

It is possible to reconceive the world. It's been done. In the past, humans changed their minds; they went from heresy to new belief. It became apparent: the Earth, the Ptolemaic center of things, was not the Galilean center. It became apparent: slavery, economically feasible, was not morally tolerable.

It must also become apparent: the physical destruction of the planet—right here, an easy mark—is a crime.

In the terror of ancient times, humans could live intimately with the natural world. Without technologies, they no doubt lived in awe. And in peril.

With our technologies—ones of incalculable power: earth-shattering, planet-altering; and ones of incredible potential: earth-restoring, planet-preserving—we can rediscover an intimacy, a mutuality with the natural world, that is not primitive (though based in part on fear), but *knowing*. It might even be possible to relearn a life of awe. And inhabit landscape without violation. With the least violation.

There will be nothing simple about living generously, coherently, and intimately with the natural world.

If I'm happy to see a few Midwest fields go wild, it's a small thing. I know. But re-viewing, re-conceiving the land we inhabit, is not a small thing. It is not nostalgia to sing the praises of vacant ground. It is not longing for the past, but an immoderate and profound desire for the future, that leads a person to say—about wasteland and wetland and any steep slope and any undeveloped shoreline—let it go. Let it be.

RESPONDING JOURNAL

In a journal entry, answer Kauffman's question about the earth: "What do we want to *do* with it? " Begin your thinking about this question by reviewing what she says, and then use this opportunity to agree or disagree with her comments. Be as concrete in your response as you can.

QUESTIONS FOR CRITICAL THINKING

STRATEGY

1. Look at Kauffman's repeated sentences, "After all, this is a world of rock and water and air. It is elemental. It is not ours." How does she use these repetitions in her writing? Do they serve a structural purpose? What advantages do the repetitions provide her?
2. Look at Kauffman's first paragraph. Does the final sentence fragment change the mood of the rest of the paragraph? How does this shift work? Explain how effective you find this device. Does she use similar devices in other paragraphs?

ISSUES

1. "Our rights cannot be exclusive," Kauffman says. What does she mean? How effectively has she demonstrated her point to you? What other instances can you think of to support her argument or to argue the opposite side?
2. Kauffman says that with our great technologies, we can inaugurate an age of *knowing*. Explain what she means, and explain how far you are willing to agree with her argument.

COLLABORATIVE WRITING ACTIVITY

Brainstorm a list of possibilities for constructing an ad similar to the one Kauffman describes for ATVs, only make this a pro-earth, pro-saving the environment ad. How would it differ from the sort of ad she describes for an ATV? Bring your list of suggestions to your group, and then discuss the possibilities together. Decide on an effective image, and work together to create the accompanying text to impress your ideas upon your audience, which will be the larger class. After all groups have made their presentations, discuss how your various ads might work in a larger context. How would the local community respond to them? In what national or local magazines or papers could you imagine them appearing? You might want to follow up this activity with a journal entry focused on what you learned from the experience.

"A Realtor Runs Through It"
Jessica Maxwell

As the author of the "True Nature" column which appears monthly in Audubon *magazine, Jessica Maxwell has achieved popularity as a columnist for the variety of subjects she tackles. She has written on the problems of relocating beavers from one ecosystem to another; she has covered the regulation of salmon fishing as opposed to reestablishing the runs of the fish; she has explained the monitoring of moose equipped with radio collars to study the relationship between moose life and old growth forests; she has described the activities of the Tyee Club, a salmon fishing club that is also one of the earliest fish conservation agencies in America.*

Such wide and varied interests indicate the array of information she must master in order to write effectively. Not content with reporting on interesting aspects of life in the natural world, she also writes of her own experiences in the natural world. Her recounting of her explorations on the California coast with a sister and indulgent father works as effectively as a personal essay as it does a nature essay. In "A Realtor Runs Through It," Maxwell strikes an effective tone as she presents evidence of a potentially disturbing situation occurring not only in Montana but in much of the rural United States.

EXPLORING JOURNAL

Think about the title of Maxwell's article; what text does it make you recall? Judging from the addition of "Realtor," explain in an entry what you think the essay will cover.

Montana has been cut, quartered, and sold to wanna-be ranchers.

The best way to understand Montana is to get lost in it. Luckily, this is not hard to do. I myself had been driving for miles and still hadn't seen a trace of the turnoff my hand-drawn map swore was there.

"You say you got a map?" the old rancher hollered across the dust rising between his beat-up pickup and my prissy rented Explorer. He had a face Ralph Lauren would kill for—brown as the earth, with lines running through it like riverbeds and eyes that matched the sky. His name, he said, was Reid.

"Well, let's see it," said Reid, climbing out of the truck. His running buddy, Bud, jumped out behind him.

We were standing in the middle of a gravel road on the back of some serious ranch land about 50 miles south of Great Falls. It had rained all summer, and the place was laid out north, south, east, and west with the greenest of grasses bracketed by great silver bolls of sage. The country looked exactly as you'd expect Montana to look—big biscuit clouds rising in a blue oven of a sky above a prairie that seemed to tuck and roll forever. Already I had crossed the Missouri River once, the Smith twice, and half a dozen little feeder creeks in between. I'd seen a whole herd of pronghorn, one young golden eagle, and a road-kill porcupine. The elegant dihedral of a northern harrier had stopped me cold, and two red foxes had stared unblinking while I took their picture. Reid's truck was the only other vehicle on the road.

"Well, let's see here now," he said as he and Bud scanned my map. They studied it right-side up, then sideways, then right-side up again. Finally they shook their heads as if it just wasn't possible to fit a realistic replica of any part of Montana onto an 8 1/2- by 11-inch piece of paper.

"Where we are now isn't even on this thing. *Where* is it that you're tryin' to go?"

"The Rahrs'," I said. But neither Bud nor Reid had ever heard of them.

"Oh, for cryin' out loud," Reid said. "Look at this, Bud. There's Gaddis Hill—so *that's* where you're wantin' to go. Didn't the Fronhoffers used to own a place over there? Then they sold it—who was it they sold it to?"

Bud raked his left hand across his head. Three of his fingertips were missing. "Let's see . . . it was . . . it was the Gruels!" he announced. "That was it—the Gruels."

"But how do I *get* there?" I interrupted.

They tried to give me directions for five minutes before they gave up and Reid said: "We're retired. We're just screwin' around today, so we'll take you there. Let's go." I followed their palomino dust geyser for an hour down miles of gravel roads that cut through that argentine country with few signs of homes or cattle until we finally arrived at the gate I'd been looking for. Bud and Reid stayed until we figured out the stubborn lock and spent another 20 minutes talking fishing.

This has been the true nature of Montana for a long, long time, a patient, quiet place ruled by its terrain and proud of having a people-to-land ratio lower than the Milwaukee Brewers' current batting average. Its human culture evolved in the void, away from the cities and toughened by the real work of ranch life. Seasons mattered, weather mattered, land mattered, and people mattered a lot because there were so few.

"Implicated as we westerners are in this sperm, blood, and guts business of ranching," writes Gretel Ehrlich in *The Solace of Open Spaces*, "and propelled forward by steady gusts of blizzards, cold fronts, droughts, heat, and wind, there's a ceremonial feel to life on a ranch. It's raw and impulsive but the narrative thread of birth, death, chores, and seasons keeps tugging at us until we find ourselves braided inextricably into the strand."

Only now is that strand unraveling.

Given the current premium on clean air and open space, ranchers can make a lot more money selling their land these days than they can raising cattle. Buyers of means—Ted Turner, Liz Claiborne, Tom Brokaw—have purchased whole ranches, which they themselves visit only occasionally. Their presence—or lack thereof—might not nourish the local culture, but it has become part of the new glue that's holding Montana's wildlands together, especially if they have the good sense to grant conservation easements to land trusts. Ted Turner donated one to the Nature Conservation. Tom Brokaw and his partners granted one to the Montana Land Reliance, one of the most successful land trusts in the country, as did my friends the Rahrs. This means that none of their ranches can be developed, not now, not ever.

Meanwhile, the converse is occurring at a frightening speed: All over Montana, developers of lesser renown are also buying up big ranches, hacking them into 20-acre "ranchettes." and selling the dream of the American West to the urban disenfranchised, piece by eco-busting piece, without so much as a whisper of environmental review.

For two decades, under the Montana Subdivision and Platting Act of 1973, any land parcel of more than 20 acres was exempt from both local and state review. This oversight amounted to a real estate free-for-all during which an estimated 92 percent of the state's subdivided land escaped monitoring of any sort while realtors hawked it at prices often 10 times the going rate of rangeland, which is about $80 an acre. Thus began the salvaging of Montana's incomparable open spaces. Its watersheds have been assaulted, its critical riparian zones fractured, and its wildlife migration routes irrevocably violated so that newcomers seeking the good life could build a cabin by a creek where the deer and the antelope play—all for about $1,000 an acre, 5 percent down, $300 a month, and $10 a year in taxes.

It's a wonderful dream, a vision as American as the Old West itself, and it surely makes a certain class of realtor wake up each morning humming the Hallelujah Chorus. But you multiply that dream by thousands upon thousands of city-weary citizens who would just love to own a little piece of Montana, and you've got a land-use nightmare on your hands that could turn the whole state into suburbia in nothing flat.

"Suburbia with really bad roads," said Janet Ellis as the Explorer lurched along the rutted gravel mess that passed for a road in the Skelly Gulch subdivision, just 11 miles outside Helena. A biologist by training and an environmental lobbyist since 1983, Janet is the program director for the Montana Audubon Council, which opened its first staffed office in 1989. She is, by all accounts, responsible for seeing to it that in April, after a 20-year fight, the Montana legislature finally passed amendments rendering that old Subdivision and Platting Act essentially ineffective. Their most important achievement was closing the 20-acre loophole by redefining a subdivision as 160 acres or less.

"We still have to live with the thousands of twenty-acre plots already filed at Clerk and Recorder Offices throughout the state," Janet says. "But from now on, any new subdivisions under a hundred and sixty acres will have to be reviewed by local government."

Under Janet's guidance, the Montana Audubon Council's next project is to launch an education program to teach county commissioners the importance of environmental review.

"The problem with twentying is that it breaks up open space really quickly," Janet explained as she scanned a map of the Skelly Gultch subdivision. "And it causes a lot of roading and erosion." She studied the map right-side up, then sideways, then, squinting, right-side up again.

"Well, according to this, *that* is a building site," she finally said, pointing in the direction of the cliff to our right. I gasped. It was 90 degrees of crumbling limestone, and the narrow so-called road we were on was its only access.

"I'd like to see a fire truck try and put out a fire up there," Janet said as we walked over to investigate. "I don't think one could even make it down this road." The Skelly Gulch developers did not return telephone calls or respond to written questions about the subdivision.

Fire protection is just one of the services no one seems to consider when they sell or buy a 20-acre ranchette. "Before the new bill, anything over twenty acres was taxed as agricultural land," Janet explained. "With such a low tax rate there's no money for basic services. People move here expecting the urban amenities they're used to—well-maintained roads, police and fire protection, schools, school busses, libraries. In a lot of cases, Montana tax-payers end up paying for these things."

Then there's the question of installing water, power, sewage, and phone service in undeveloped rural areas. According to a special report called "Behind the Dream" that ran in *The Billings Gazette* on August 16, 1992, landowners at the Wild Horse Ranch, south of the town of Forsyth, pay a contractor between $2,000 and $2,500 a year just to grade a 25- to 30-mile dirt road. Wells cost from a few thousand dollars to $20,000 to drill. Septic systems cost between $5,000 and $6,000. The Mid-Yellowstone Electric Cooperative would need $550,000 collectively from property owners to bring power to the development—and a 12.5 kilowatt generator costs about $6,800. The one couple who inquired about having a phone installed was told by the Range Telephone Cooperative in Forsyth that it would cost them $10,000.

All of which Montana subdivision realtors often fail to tell their customers, many of whom sell everything they own to buy their $20,000 piece of Montana. When they can't find jobs and are unable to live off the land, they default on their contracts and lose their property. "Most go back to where they came from," according to *The Billings Gazette*.

It is particularly vexing that many ranchettes are sold sight unseen. Janet recalled one woman from the East Coast who wanted to buy some property where she could fly-fish. Over the phone, a realtor sold her acreage on Silver Bow Creek, in the Clark Fork River basin. Then she received a notice from Montana's Hazardous Waste Department about a meeting with other creekside property-owners. It turned out that the Clark Fork River basin is one of the biggest Superfund sites in the country. It was contaminated by the Anaconda mine and smelting complex, which was shut down in 1983.

"There hadn't been any fish in that creek for a hell of a long time," Janet declared.

"Here's a brochure by the company that's marketing Skelly Gulch," she offered. A six-point bull elk stared back at me from beneath the headline "Aspen Gold, Ranches of Montana." Inside, I was encouraged to "reach for the sky" on my own Montana ranch, guaranteed to have "well-maintained roads with year-round access," as well as building sites with "numerous springs and creeks" that "attract abundant wildlife including deer, elk, and wild turkeys" to which I would have "exclusive hunting privileges" not to mention "excellent fishing in Montana's blue ribbon trout streams."

"Isn't it a little presumptuous," I asked her, "to call twenty acres a ranch in the first place?"

"Especially considering that in Montana you could graze maybe one horse on it," Janet replied. "Or probably not, without supplemental feed."

Real ranchers know that horses in Montana—and cows—need at least 40 acres each, which is why real ranches in Montana consist of thousands and thousands of acres. And why a law defining a subdivision as 160 acres or less is definitely a step in the right direction. And why selling entire ranches to people who can afford them is even better, especially if those people happen to believe in the virtues of conservation easements.

But you can hardly fault less well-off urbanites for coming to their senses and feeling, maybe for the first time, bone-deep longing to live in a place with a sky like a blue lake and a horizon filled with elk and lodgepole pine. What, oh what, do we say to the newly converted whom we, as conservationists, have worked so hard to convert, especially if they don't have $200,000 to spend on a piece of land?

Lane Coulston, owner of American Conservation Real Estate (ACRE), a four-year-old Helena-based company, has an idea. "We work with both buyers and sellers who share the same good values," explains Coulston, a native Montanan with a background in biological sciences. "Those include sound ecosystem management and protection of open space and wildlife habitat. We encourage people who own land in areas that have already been subdivided to form their own community land trust and donate conservation easements to a shared pool to benefit each other and the land."

"Instead of buying a twenty," echoes Rock Ringling, lands director for the Montana Land Reliance, "you can get four people together to buy a hundred and sixty acres, do a concentrated development in one corner, and put an easement on the remaining property. You really can do development in a sensible, holistic manner, so that your footprint on the land is at a minimum. Even if you go ahead and buy forty acres, you can still walk down to the creek to fish rather than build your casting platform right on the banks of the Madison River like so many folks are doing."

But a house every 20 or 40 acres—no matter where you put it—still kicks the teeth out of open space. It seems clear that everyone's dream of a cabin in the woods by a creek somewhere in Montana simply needs to be reconsidered.

Bob Kiesling agrees. Founder of Montana and Wyoming's Nature Conservancy and one of the state's most ardent land-use activists, Kiesling has long been concerned about what he calls "the irrationality of land ownership in the West."

"It's simply not our mission here to fulfill those kinds of dreams for everyone," he says, "and I don't think sacrificing the landscape to do it is an appropriate course of action—it is, in fact, the worst kind of land use. That's why we have a lot of public access. That's why the Forest Service has campgrounds. Ownership and occupancy of rural land is just not in the cards for everybody, and it shouldn't be. If someone's dream is compelling enough, they'll find a way to make it happen, but it simply is not society's obligation."

⇒•◦•⇐

RESPONDING JOURNAL

Evaluate the lead idea in Maxwell's article: "Montana has been cut, quartered, and sold to wanna-be ranchers." Does she prove the dramatic point she suggests in this sentence? On what sorts of evidence does she base her claim?

QUESTIONS FOR CRITICAL THINKING

STRATEGY

1. After her dramatic opening, Maxwell states, "The best way to understand Montana is to get lost in it." Explain how well you feel her strategy works in this essay.
2. What sort of evidence does Maxwell seem to feel is most compelling? How do you know? Evaluate her logic in assessing the effectiveness of the presentation she makes.

ISSUES

1. If Montana has actually been as much "a patient, quiet place" as Maxwell suggests, why has it changed so suddenly? How convincingly does Maxwell explain the new developments?
2. Draw up a chart that indicates the competing arguments for the use of Montana land. Decide the most logical basis on which to establish the conflicting viewpoints.

COLLABORATIVE WRITING ACTIVITY

Draft a letter of advice to people who want to move to Montana. What should they know about the situation with regard to land ownership? What sorts of evidence would you present them? After you draft your response, bring it to your group so you can combine your work into a group effort that will offer a complete view of the conflicting issues.

"The Ethics of Respect for Nature"
Paul W. Taylor

Paul W. Taylor, Professor Emeritus of Philosophy at Brooklyn College, presents his evaluation of the considerations involved in a careful and responsible examination of human ethics in response to nature. Since its first publication in 1981, this essay has become a standard work in the articulation of the most basic relationship between humans and the natural world.

Despite the close reasoning of Taylor's essay and the careful way in which he presents the rationale and the causes for each of his statements about environmental ethics, he presents his case clearly and avoids philosophical jargon. For example, when he distinguishes his position (arguing for a life-centered theory of environmental ethics) from that of William Frankena (who holds that a creature's being assigned moral value depends on its sentience, or ability to feel or perceive through the senses), he spells out his reasons—the causes for the final effect of his theory—in detail.

EXPLORING JOURNAL

What do you know about ethics as a general topic? As a course of study? How do you believe an ethics of response to nature should be constructed? Explain what you think your "ethics of response to nature" is right now; present your ideas in a journal entry of at least 1-1/2 to 2 pages.

I. HUMAN-CENTERED AND LIFE-CENTERED SYSTEMS OF ENVIRONMENTAL ETHICS

In this paper I show how the taking of a certain ultimate moral attitude toward nature, which I call "respect for nature," has a central place in the foundations of a life-centered system of environmental ethics. I hold that a set of moral norms (both standards of character and rules of conduct) governing human treatment of the natural world is a rationally grounded set if and only if, first, commitment to those norms is a practical entailment of adopting the attitude of respect for nature as an ultimate moral attitude, and second, the adopting of that attitude on the part of all rational agents can itself be justified. When the basic characteristics of the attitude of respect for nature are made clear, it will be seen that a life-centered system of environmental ethics need not be holistic or organicist in its conception of the kinds of entities that are deemed the appropriate objects of moral concern and consideration. Nor does such a system require that the concepts of ecological homeostasis, equilibrium, and integrity provide us with normative principles from which could be derived (with the addition of factual knowledge) our obligations with regard to natural ecosystems. The "balance of nature" is not itself a moral norm, however important may be the role it plays in our general outlook on the natural world that underlies the attitude of respect for nature. I argue that finally it is the good (well-being, welfare) of individual organisms, considered as entities having inherent worth, that determines our moral relations with the Earth's wild communities of life.

In designating the theory to be set forth as life-centered, I intend to contrast it with all anthropocentric views. According to the latter, human actions affecting the natural environment and its nonhuman inhabitants are right (or wrong) by either of two criteria: they have consequences which are favorable (or unfavorable) to human well-being, or they are consistent (or inconsistent) with the system of norms that protect and implement human rights. From this human-centered standpoint it is to humans and only to humans that all duties are ultimately owed. We may have responsibilities *with regard to* the natural ecosystems and biotic communities of our planet, but these responsibilities are in every case based on the contingent fact that our treatment of those ecosystems and communities of life can further the realization of human values and/or human rights. We have no obligation to promote or protect the good of nonhuman living things, independently of this contingent fact.

A life-centered system of environmental ethics is opposed to human-centered ones precisely on this point. From the perspective of a life-centered theory, we have prima facie moral obligations that are owed to wild plants and animals themselves as members of the Earth's biotic community. We are morally bound (other things being equal) to protect or promote their good for *their* sake. Our duties to respect the integrity of natural ecosystems, to preserve endangered species, and to avoid environmental pollution stem from the fact that these are ways in which we can help make it possible for wild species populations to achieve and maintain a healthy existence in a natural state. Such obligations are due those living things out of recognition of their inherent worth. They are entirely additional to and independent of the obligations we owe to our fellow humans. Although many of the actions that fulfill one set of obligations will also fulfill the other, two different grounds of obligation are involved. Their well-being, as well as human well-being, is something to be realized *as an end in itself.*

If we were to accept a life-centered theory of environmental ethics, a profound reordering of our moral universe would take place. We would begin to look at the whole of the Earth's biosphere in a new light. Our duties with respect to the "world" of nature would be seen as making prima facie claims upon us to be balanced against our duties with respect to the "world" of human civilization. We could no longer simply take the human point of view and consider the effects of our actions exclusively from the perspective of our own good.

II. THE GOOD OF A BEING AND THE CONCEPT OF INHERENT WORTH

What would justify acceptance of a life-centered system of ethical principles? In order to answer this it is first necessary to make clear the fundamental moral attitude that underlies and makes intelligible the commitment to live by such a system. It is then necessary to examine the considerations that would justify any rational agent's adopting that moral attitude.

Two concepts are essential to the taking of a moral attitude of the sort in question. A being which does not "have" these concepts, that is, which is unable to grasp their meaning and conditions of applicability, cannot be said to have the attitude as part of its moral outlook. These concepts are, first, that of the good (well-being, welfare) of a living thing, and second, the idea of an entity possessing inherent worth. I examine each concept in turn.

1. Every organism, species population, and community of life has a good of its own which moral agents can intentionally further or damage by their actions. To say that an entity has a good of its own is simply to say that, without reference to any *other* entity, it can be benefitted or harmed. One can act in its overall interest or contrary to its overall interest, and environmental conditions can be good for it (advantageous to it) or bad for it (disadvantageous to it). What is good for an entity is what "does it good" in the sense of enhancing or preserving its life and well-being. What is bad for an entity is something that is detrimental to its life and well-being.[1]

We can think of the good of an individual nonhuman organism as consisting in the full development of its biological powers. Its good is realized to the extent that it is strong and healthy. It possesses whatever capacities it needs for successfully coping with its environment and so preserving its existence throughout the various stages of the normal life cycle of its species. The good of a population or community of such individuals consists in the population or community maintaining itself from generation to generation as a coherent system of genetically and ecologically related organisms whose average good is at an optimum level for the given environment. (Here *average good* means that the degree of realization of the good of *individual organisms* in the population or community is, on average, greater than would be the case under any other ecologically functioning order of interrelations among those species populations in the given ecosystem.)

The idea of a being having a good of its own, as I understand it, does not entail that the being must have interests or take an interest in what affects its life for better or for worse. We can act in a being's interest or contrary to its interest without its being interested in what we are doing to it in the sense of wanting or not wanting us to do it. It may, indeed, be wholly unaware that favorable and unfavorable events are taking place in its life. I take it that trees, for example, have no knowledge or desires or feelings. Yet it is undoubtedly the case that trees can be harmed or benefitted by our actions. We can crush their roots by running a bulldozer too close to them. We can see to it that they get adequate nourishment and moisture by fertilizing and watering the soil around them. Thus we can help or hinder them in the realization of their good. It is the good of trees themselves that is thereby affected. We can similarly act so as to further the good of an entire tree population of a certain species (say, all the redwood trees in a California valley) or the good of a whole community of plant life in a given wilderness area, just as we can do harm to such a population or community.

When construed in this way, the concept of a being's good is not coextensive with sentience or the capacity for feeling pain. William Frankena has argued for a general theory of environmental ethics in which the ground of a creature's being worthy of moral consideration is its sentience. I have offered some criticisms of this view elsewhere, but the full refutation of such a position, it seems to me, finally depends on the positive reasons for accepting a life-centered theory of the kind I am defending in this essay.[2]

It should be noted further that I am leaving open the question of whether machines—in particular, those which are not only goal-directed, but also self-regulating—can properly be said to have a good of their own.[3] Since I am concerned only with human treatment of wild organisms, species populations, and communities of life as they occur in our planet's natural ecosystems, it is to those entities alone that the concept "having a good of its own" will here be applied. I am not denying that other living things, whose genetic origin and environmental conditions have been produced, controlled, and manipulated by humans for human ends, do have a good of their own in the same sense as do wild plants and animals. It is not my purpose in this essay, however, to set out or defend the principles that should guide our conduct with regard to their good. It is only insofar as their production and use by humans have good or ill effects upon natural ecosystems and their wild inhabitants that the ethics of respect for nature comes into play.

2. The second concept essential to the moral attitude of respect for nature is the idea of inherent worth. We take that attitude toward wild living things (individuals, species populations, or whole biotic communities) when and only when we regard them as entities possessing inherent worth. Indeed, it is only because they are conceived in this way that moral agents can think of themselves as having validly binding duties, obligations, and responsibilities that are *owed* to them as their *due*. I am not at this juncture arguing why they *should* be so regarded; I consider it at length below. But so regarding them is a presupposition of our taking the attitude of respect toward them and accordingly understanding ourselves as bearing certain moral relations to them. This can be shown as follows:

What does it mean to regard an entity that has a good of its own as possessing inherent worth? Two general principles are involved: the principle of moral consideration and the principle of intrinsic value.

According to the principle of moral consideration, wild living things are deserving of the concern and consideration of all moral agents simply in virtue of their being members of the Earth's community of life. From the moral point of view their good must be taken into account whenever it is affected for better or worse by the conduct of rational agents. This holds no matter what species the creature belongs to. The good of each is to be accorded some value and so acknowledged as having some weight in the deliberations of all rational agents. Of course, it may be necessary for such agents to act in ways contrary to the good of this or that particular organism or group of organisms in order to further the good of others, including the good of humans. But the principle of moral consideration prescribes that, with respect to each being an entity having its own good, every individual is deserving of consideration.

The principle of intrinsic value states that, regardless of what kind of entity it is in other respects, if it is a member of the Earth's community of life, the realization of its good is something *intrinsically* valuable. This means that its good is prima facie worthy of being preserved or promoted as an end in itself and for the sake of the entity whose good it is. Insofar as we regard any organism, species population, or life community as an entity having inherent worth, we believe that it must never be treated as if it were a mere object or thing whose entire value lies in being instrumental to the good of some other entity. The well-being of each is judged to have value in and of itself.

Combining these two principles, we can now define what it means for a living thing or group of living things to possess inherent worth. To say that it possesses inherent worth is to say that its good is deserving of the concern and consideration of all moral agents, and that the realization of its good has intrinsic value, to be pursued as an end in itself and for the sake of the entity whose good it is.

The duties owed to wild organisms, species populations, and communities of life in the Earth's natural ecosystems are grounded on their inherent worth. When rational, autonomous agents regard such entities as possessing inherent worth, they place intrinsic value on the realization of their good and so hold themselves responsible for performing actions that will have this effect and for refraining from actions having the contrary effect.

III. THE ATTITUDE OF RESPECT FOR NATURE

Why should moral agents regard wild living things in the natural world as possessing inherent worth? To answer this question we must first take into account the fact that, when rational, autonomous agents subscribe to the principles of moral consideration and intrinsic value and so conceive of wild living things as having that kind of worth, such agents are *adopting a certain ultimate moral attitude toward the natural world.* This is the attitude I call "respect for nature." It parallels the attitude of respect for persons in human ethics. When we adopt the attitude of respect for persons as the proper (fitting, appropriate) attitude to take toward all persons as persons, we consider the fulfillment of the basic interests of each individual to have intrinsic value. We thereby make a moral commitment to live a certain kind of life in relation to other persons. We place ourselves under the direction of a system of standards and rules that we consider validly binding on all moral agents as such.[4]

Similarly, when we adopt the attitude of respect for nature as an ultimate moral attitude we make a commitment to live by certain normative principles. These principles constitute the rules of conduct and standards of character that are to govern our treatment of the natural world. This is, first, an *ultimate* commitment because it is not derived from any higher norm. The attitude of respect for nature is not grounded on some other, more general, or more fundamental attitude. It sets the total framework for our responsibilities toward the natural world. It can be justified, as I show below, but its justification cannot consist in referring to a more general attitude or a more basic normative principle.

Second, the commitment is a *moral* one because it is understood to be a disinterested matter of principle. It is this feature that distinguishes the attitude of respect for nature from the set of feelings and dispositions that comprise the love of nature. The latter stems from one's personal interest in and response to the natural world. Like the affectionate feelings we have toward certain individual human beings, one's love of nature is nothing more than the particular way one feels about the natural environment and its wild inhabitants. And just as our love for an individual person differs from our respect for all persons as such (whether we happen to love them or not), so love of nature differs from respect for nature. Respect for nature is an attitude we believe all moral agents ought to have simply as moral agents, regardless of whether or not they also love nature. Indeed, we have not truly taken the attitude of respect for nature ourselves unless we believe this. To put it in a Kantian way, to adopt the attitude of respect for nature is to take a stance that one wills it to be a universal law for all rational beings. It is to hold that stance categorically, as being validly applicable to every moral agent without exception, irrespective of whatever personal feelings toward nature such an agent might have or might lack.

Although the attitude of respect for nature is in this sense a disinterested and universalizable attitude, anyone who does adopt it has certain steady, more or less permanent dispositions. These dispositions, which are themselves to be considered disinterested and universalizable, comprise three interlocking sets: dispositions to seek certain ends, dispositions to carry on one's practical reasoning and deliberation in a certain way, and dispositions to have certain feelings. We may accordingly analyze the attitude of respect for nature into the following components. (a) The disposition to aim at, and to take steps to bring about, as final and disinterested ends, the promoting and protecting of the good of organisms, species populations, and life communities in natural ecosystems. (These ends are "final" in not being pursued as means to further ends. They are "disinterested" in being independent of the self-interest of the agent.) (b) The disposition to consider actions that tend to realize those ends to be prima facie obligatory *because* they have that tendency. (c) The disposition to experience positive and negative feelings toward states of affairs in the world *because* they are favorable or unfavorable to the good of organisms, species populations, and life communities in natural ecosystems.

The logical connection between the attitude of respect for nature and the duties of a life-centered system of environmental ethics can now be made clear. Insofar as one sincerely takes that attitude and so has the three sets of dispositions, one will at the same time be disposed to comply with certain rules of duty (such as nonmaleficence and noninterference) and with standards of character (such as fairness and benevolence) that determine the obligations and virtues of moral agents with regard to the Earth's wild living things. We can say that the actions one performs and the character traits one develops in fulfilling these moral requirements are the way one *expresses* or *embodies* the attitude in one's conduct and character. In his famous essay, "Justice as Fairness," John Rawls describes the rules of the duties of human morality (such as fidelity, gratitude, honesty, and justice) as "forms of conduct in which recognition of others as persons is manifested."[5] I hold that the rules of duty governing our treatment of the natural world and its inhabitants are forms of conduct in which the attitude of respect for nature is manifested.

IV. THE JUSTIFIABILITY OF THE ATTITUDE OF RESPECT FOR NATURE

I return to the question posed earlier, which has not yet been answered: why *should* moral agents regard wild living things as possessing inherent worth? I now argue that the only way we can answer this question is by showing how adopting the attitude of respect for nature is justified for all moral agents. Let us suppose that we were able to establish that there are good reasons for adopting the attitude, reasons which are intersubjectively valid for every rational agent. If there are such reasons, they would justify anyone's having the three sets of dispositions mentioned above as constituting what it means to have the attitude. Since these include the disposition to promote or protect the good of wild living things as a disinterested and ultimate end, as well as the disposition to perform actions for the reason that they tend to realize that end, we see that such dispositions commit a person to the principles of moral consideration and intrinsic value. To be disposed to further, as an end in itself, the good of any entity in nature just because it is that kind of entity, is to be disposed to give consideration to *every* such entity and to place intrinsic value on the realization of its good. Insofar as we subscribe to these two principles we regard living things as possessing inherent worth. Subscribing to the principle is what it *means* to so regard them. To justify the attitude of respect for nature, then, is to justify commitment to these principles and thereby to justify regarding wild creatures as possessing inherent worth.

We must keep in mind that inherent worth is not some mysterious sort of objective property belonging to living things that can be discovered by empirical observation or scientific investigation. To ascribe inherent worth to an entity is not to describe it by citing some feature discernible by sense perception or inferable by inductive reasoning. Nor is there a logically necessary connection between the concept of a being having a good of its own and the concept of inherent worth. We do not contradict ourselves by asserting that an entity that has a good of its own lacks inherent worth. In order to show that such an entity "has" inherent worth we must give good reasons for ascribing that kind of value to it (placing that kind of value upon it, conceiving of it to be valuable in that way). Although it is humans (persons, valuers) who must do the valuing, for the ethics of respect for nature, the value so ascribed is not a human value. That is to say, it is not a value derived from considerations regarding human well-being or human rights. It is a value that is ascribed to nonhuman animals and plants themselves, independently of their relationship to what humans judge to be conducive to their own good.

Whatever reasons, then, justify our taking the attitude of respect for nature as defined above are also reasons that show why we *should* regard the living things of the natural world as possessing inherent worth. We saw earlier that since the attitude is an ultimate one, it cannot be derived from a more fundamental attitude nor shown to be a special case of a more general one. On what sort of grounds, then, can it be established?

The attitude we take toward living things in the natural world depends on the way we look at them, on what kind of beings we conceive them to be, and on how we understand the relations we bear to them. Underlying and supporting our attitude is a certain *belief system* that constitutes a particular world view or outlook on nature and the place of human life in it. To give good reasons for adopting the attitude of respect for nature, then, we must first articulate the belief system which underlies and supports that attitude. If it appears that the belief system is internally coherent and well-ordered, and if, as far as we can now tell, it is consistent with all known scientific truths relevant to our knowledge of the object of the attitude (which in this case includes the whole set of the Earth's natural ecosystems and their communities of life), then there remains the task of indicating why scientifically informed and rational thinkers with a developed capacity of reality awareness can find it acceptable as a way of conceiving of the natural world and our place in it. To the extent we can do this we provide at least a reasonable argument for accepting the belief system and the ultimate moral attitude it supports.

I do not hold that such a belief system can be *proven* to be true, either inductively or deductively. As we shall see, not all of its components can be stated in the form of empirically verifiable propositions. Nor is its internal order governed by purely logical relationships. But the system as a whole, I contend, constitutes a coherent, unified, and rationally acceptable "picture" or "map" of a total world. By examining each of its main components and seeing how they fit together, we obtain a scientifically informed and well-ordered conception of nature and the place of humans in it.

This belief system underlying the attitude of respect for nature I call (for want of a better name) "the biocentric outlook on nature." Since it is not wholly analyzable into empirically confirmable assertions, it should not be thought of as simply a compendium of the biological sciences concerning our planet's ecosystems. It might best be described as a philosophical world view, to distinguish it from a scientific theory or explanatory system. However, one of its major tenets is the great lesson we have learned from the science of ecology: the interdependence of all living things in an organically unified order whose balance and stability are necessary conditions for the realization of the good of its constituent biotic communities.

Before turning to an account of the main components of the biocentric outlook, it is convenient here to set forth the overall structure of my theory of environmental ethics as it has now emerged. The ethics of respect for nature is made up of three basic elements: a belief system, an ultimate moral attitude, and a set of rules of duty and standards of character. These elements are connected with each other in the following manner. The belief system provides a certain outlook on nature which supports and makes intelligible an autonomous agent's adopting, as an ultimate moral attitude, the attitude of respect for nature. It supports and makes intelligible the attitude in the sense that when an autonomous agent understands its moral relations to the natural world in terms of this outlook, it recognizes the attitude of respect to be the only *suitable* or *fitting* attitude to take toward all wild forms of life in the Earth's biosphere. Living things are now viewed as *the appropriate objects of the attitude of respect* and are accordingly regarded as entities possessing inherent worth. One then places intrinsic value on the promotion and protection of their good. As a consequence of this, one makes a moral commitment to abide by a set of rules of duty and to fulfill (as far as one can by one's own efforts) certain standards of good character. Given one's adoption of the attitude of respect, one makes that moral commitment because one considers those rules and standards to be valid and binding on all moral agents. They are seen as embodying forms of conduct and character structures in which the attitude of respect for nature is manifested.

This three-part complex which internally renders the ethics of respect for nature is symmetrical with a theory of human ethics grounded on respect for persons. Such a theory includes, first, a conception of oneself and others as persons, that is, as centers of autonomous choice. Second, there is the attitude of respect for persons as persons. When this is adopted as an ultimate moral attitude it involves the disposition to treat every person as having inherent worth or "human dignity." Every human being, just in virtue of her or his humanity, is understood to be worthy of moral consideration, and intrinsic value is placed on the autonomy and well-being of each. This is what Kant meant by conceiving of persons as minds in themselves. Third, there is an ethical system of duties which are acknowledged to be owed by everyone to everyone. These duties are forms of conduct in which public recognition is given to each individual's inherent worth as a person.

This structural framework for a theory of human ethics is meant to leave open the issue of consequentialism (utilitarianism) versus non-consequentialism (deontology). That issue concerns the particular kind of system of rules defining the duties of moral agents toward persons. Similarly, I am leaving open in this paper the question of what particular kind of system of rules defines our duties with respect to the natural world.

V. THE BIOCENTRIC OUTLOOK ON NATURE

The biocentric outlook on nature has four main components. (1) Humans are thought of as members of the Earth's community of life, holding that membership on the same terms as apply to all the nonhuman members. (2) The Earth's natural ecosystems as a totality are seen as a complex web of interconnected elements, with the sound biological functioning of each being dependent on the sound biological functioning of the others. (This is the component referred to above as the great lesson that the science of ecology has taught us.) (3) Each individual organism is conceived of as a teleological center of life, pursuing its own good in its own way. (4) Whether we are concerned with standards of merit or with the concept of inherent worth, the claim that humans by their very nature are superior to other species is a groundless claim and, in the light of elements (1), (2), and (3) above, must be rejected as nothing more than an irrational bias in our own favor.

The conjunction of these four ideas constitutes the biocentric outlook on nature. In the remainder of this paper I give a brief account of the first three components, followed by a more detailed analysis of the fourth. I then conclude by indicating how this outlook provides a way of justifying the attitude of respect for nature.

VI. HUMANS AS MEMBERS OF THE EARTH'S COMMUNITY OF LIFE

We share with other species a common relationship to the Earth. In accepting the biocentric outlook we take the fact of our being an animal species to be a fundamental feature of our existence. We consider it an essential aspect of "the human condition." We do not deny the differences between ourselves and other species, but we keep in the forefront of our consciousness the fact that in relation to our planet's natural ecosystems we are but one species population among many. Thus we acknowledge our origin in the very same evolutionary process that gave rise to all other species and we recognize ourselves to be confronted with similar environmental challenges to those that confront them. The laws of genetics, of natural selection, and of adaptation apply equally to all of us as biological creatures. In this light we consider ourselves as one with them, not set apart from them. We, as well as they, must face certain basic conditions of existence that impose requirements on us for our survival and well-being. Each animal and plant is like us in having a good of its own. Although our human good (what is of true value in human life, including the exercise of individual autonomy in choosing our own particular value systems) is not like the good of a nonhuman animal or plant, it can no more be realized than their good can without the biological necessities for survival and physical health.

When we look at ourselves from the evolutionary point of view, we see that not only are we very recent arrivals on Earth, but that our emergence as a new species on the planet was originally an event of no particular importance to the entire scheme of things. The Earth was teeming with life long before we appeared. Putting the point metaphorically, we are relative newcomers, entering a home that has been the residence of others for hundreds of millions of years, a home that must now be shared by all of us together.

The comparative brevity of human life on Earth may be vividly depicted by imagining the geological time scale in spatial terms. Suppose we start with algae, which have been around for at least 600 million years. (The earliest protozoa actually predated this by several *billion* years.) If the time that algae have been here were represented by the length of a football field (300 feet), then the period during which sharks have been swimming in the world's oceans and spiders have been spinning their webs would occupy three quarters of the length of the field; reptiles would show up at about the center of the field, mammals would cover the last third of the field; hominids (mammals of the family *Hominidae*) the last two feet; and the species *Homo sapiens* the last six inches.

Whether this newcomer is able to survive as long as other species remains to be seen. But there is surely something presumptuous about the way humans look down on the "lower" animals, especially those that have become extinct. We consider the dinosaurs, for example, to be biological failures, though they existed on our planet for 65 million years. One writer has made the point with beautiful simplicity:

> We sometimes speak of the dinosaurs as failures; there will be time enough for
> that judgment when we have lasted even for one tenth as long. . . .[6]

The possibility of the extinction of the human species, a possibility which starkly confronts us in the contemporary world, makes us aware of another respect in which we should not consider ourselves privileged beings in relation to other species. This is the fact that the well-being of humans is dependent upon the ecological soundness and health of many plant and animal communities, while their soundness and health does not in the least depend upon human well-being. Indeed, from their standpoint the very existence of humans is quite unnecessary. Every last man, woman, and child could disappear from the face of the Earth without any significant detrimental consequence for the good of wild animals and plants. On the contrary, many of them would be greatly benefitted. The destruction of their habitats by human "developments" would cease. The poisoning and polluting of their environment would come to an end. The Earth's land, air, and water would no longer be subject to the degradation they are now undergoing as the result of large-scale technology and uncontrolled population growth. Life communities in natural ecosystems would gradually return to their former healthy state. Tropical forests, for example, would again be able to make their full contribution to a life-sustaining atmosphere for the whole planet. The rivers, lakes, and oceans of the world would (perhaps) eventually become clean again. Spilled oil, plastic trash, and even radioactive waste might finally, after many centuries, cease doing their terrible work. Ecosystems would return to their proper balance, suffering only the disruptions of natural events such as volcanic eruptions and glaciation. From these the community of life could recover, as it has so often done in the past. But the ecological disasters now perpetrated on it by humans—disasters from which it might never recover—these it would no longer have to endure.

If, then, the total, final, absolute extermination of our species (by our own hands?) should take place and if we should not carry all the others with us into oblivion, not only would the Earth's community of life continue to exist, but in all probability its well-being would be enhanced. Our presence, in short, is not needed. If we were to take the standpoint of the community and give voice to its true interest, the ending of our six-inch epoch would most likely be greeted with a hearty "Good riddance!"

VII. THE NATURAL WORLD AS AN ORGANIC SYSTEM

To accept the biocentric outlook and regard ourselves and our place in the world from its perspective is to see the whole natural order of the Earth's biosphere as a complex but unified web of interconnected organisms, objects, and events. The ecological relationships between any community of living things and their environment form an organic whole of functionally interdependent parts. Each ecosystem is a small universe itself in which the interactions of its various species populations comprise an intricately woven network of cause-effect relations. Such dynamic but at the same time relatively stable structures as food chains, predator-prey relations, and plant succession in a forest are self-regulating, energy-recycling mechanisms that preserve the equilibrium of the whole.

As far as the well-being of wild animals and plants is concerned, this ecological equilibrium must not be destroyed. The same holds true of the well-being of humans. When one views the realm of nature from the perspective of the biocentric outlook, one never forgets that in the long run the integrity of the entire biosphere of our planet is essential to the realization of the good of its constituent communities of life, both human and nonhuman.

Although the importance of this idea cannot be overemphasized, it is by now so familiar and so widely acknowledged that I shall not further elaborate on it here. However, I do wish to point out that this "holistic" view of the Earth's ecological systems does not itself constitute a moral norm. It is a factual aspect of biological reality, to be understood as a set of causal connections in ordinary empirical terms. Its significance for humans is the same as its significance for nonhumans, namely, in setting basic conditions for the realization of the good of living things. Its ethical implications for our treatment of the natural environment lie entirely in the fact that our *knowledge* of these causal connections is an essential *means* to fulfilling the aims we set for ourselves in adopting the attitude of respect for nature. In addition, its theoretical implications for the ethics of respect for nature lie in the fact that it (along with the other elements of the biocentric outlook) makes the adopting of that attitude a rational and intelligible thing to do.

VIII. INDIVIDUAL ORGANISMS AS TELEOLOGICAL CENTERS OF LIFE

As our knowledge of living things increases, as we come to a deeper understanding of their life cycles, their interactions with other organisms, and the manifold ways in which they adjust to the environment, have become more fully aware of how each of them is carrying out its biological functions according to the laws of its species-specific nature. But besides this, our increasing knowledge and understanding also develop in us a sharpened awareness of the uniqueness of each individual organism. Scientists who have made careful studies of particular plants and animals, whether in the field or in laboratories, have often acquired a knowledge of their subjects as identifiable individuals. Close observation over extended periods of time has led them to an appreciation of the unique "personalities" of their subjects. Sometimes a scientist may come to take a special interest in a particular animal or plant, all the while remaining strictly objective in the gathering and recording of data. Nonscientists may likewise experience this development of interest when, as amateur naturalists, they make accurate observations over sustained periods of close acquaintance with an individual organism. As one becomes more and more familiar with the organism and its behavior, one becomes fully sensitive to the particular way it is living out its life cycle. One may become fascinated by it and even experience some involvement with its good and bad fortunes (that is, with the occurrence of environmental conditions favorable or unfavorable to the realization of its good). The organism comes to mean something to one as a unique, irreplaceable individual. The final culmination of this process is the achievement of a genuine understanding of its point of view and, with that understanding, an ability to "take" that point of view. *Conceiving of it as a center of life, one is able to look at the world from its perspective.*

   This development from objective knowledge to the recognition of individuality, and from the recognition of individuality to full awareness of an organism's standpoint, is a process of heightening our consciousness of what it means to be an individual living thing. We grasp the particularity of the organism as a teleological center of life, striving to preserve itself and to realize its own good in its own unique way.

It is to be noted that we need not be falsely anthropomorphizing when we conceive of individual plants and animals in this manner. Understanding them as teleological centers of life does not necessitate "reading into" them human characteristics. We need not, for example, consider them to have consciousness. Some of them may be aware of the world around them and others may not. Nor need we deny that different kinds and levels of awareness are exemplified when consciousness in some form is present. But conscious or not, all are equally teleological centers of life in the sense that each is a unified system of goal-oriented activities directed toward their preservation and well-being.

When considered from an ethical point of view, a teleological center of life is an entity whose "world" can be viewed from the perspective of *its* life. In looking at the world from that perspective we recognize objects and events occurring in its life as being beneficent, maleficent, or indifferent. The first are occurrences which increase its powers to preserve its existence and realize its good. The second decrease or destroy those powers. The third have neither of these effects on the entity. With regard to our human role as moral agents, we can conceive of a teleological center of life as a being whose standpoint we can take in making judgments about what events in the world are good or evil, desirable or undesirable. In making those judgments it is what promotes or protects the being's own good, not what benefits moral agents themselves, that sets the standard of evaluation. Such judgments can be made about anything that happens to the entity which is favorable or unfavorable in relation to its good. As was pointed out earlier, the entity itself need not have any (conscious) *interest* in what is happening to it for such judgments to be meaningful and true.

It is precisely judgments of this sort that we are disposed to make when we take the attitude of respect for nature. In adopting that attitude those judgments are given weight as reasons for action in our practical deliberation. They become morally relevant facts in the guidance of our conduct.

IX. The Denial of Human Superiority

This fourth component of the biocentric outlook on nature is the single most important idea in establishing the justifiability of the attitude of respect for nature. Its central role is due to the special relationship it bears to the first three components of the outlook. This relationship will be brought out after the concept of human superiority is examined and analyzed.[7]

In what sense are humans alleged to be superior to other animals? We are different from them in having certain capacities that they lack. But why should these capacities be a mark of superiority? From what point of view are they judged to be signs of superiority and what sense of superiority is meant? After all, various nonhuman species have capacities that humans lack. There is the speed of a cheetah, the vision of an eagle, the agility of a monkey. Why should not these be taken as signs of *their* superiority over humans?

One answer that comes immediately to mind is that these capacities are not as *valuable* as the human capacities that are claimed to make us superior. Such uniquely human characteristics as rational thought, aesthetic creativity, autonomy and self-determination, and moral freedom, it might be held, have a higher value than the capacities found in other species. Yet we must ask: valuable to whom, and on what grounds?

The human characteristics mentioned are all valuable to humans. They are essential to the preservation and enrichment of our civilization and culture. Clearly it is from the human standpoint that they are being judged to be desirable and good. It is not difficult here to recognize a begging of the question. Humans are claiming human superiority from a strictly human point of view, that is, from a point of view in which the good of humans is taken as the standard of judgment. All we need to do is to look at the capacities of nonhuman animals (or plants, for that matter) from the standpoint of *their* good to find a contrary judgment of superiority. The speed of the cheetah, for example, is a sign of its superiority to humans when considered from the standpoint of the good of its species. If it were as slow a runner as a human, it would not be able to survive. And so for all the other abilities of nonhumans which further their good but which are lacking in humans. In each case the claim to human superiority would be rejected from a nonhuman standpoint.

When superiority assertions are interpreted in this way, they are based on judgments of *merit.* To judge the merits of a person or an organism one must apply grading or ranking standards to it. (As I show below, this distinguishes judgments of merit from judgments of inherent worth.) Empirical investigation then determines whether it has the "good-making properties" (merits) in virtue of which it fulfills the standards being applied. In the case of humans, merits may be either moral or nonmoral. We can judge one person to be better than (superior to) another from the moral point of view by applying certain standards to their character and conduct. Similarly, we can appeal to nonmoral criteria in judging someone to be an excellent piano player, a fair cook, a poor tennis player, and so on. Different social purposes and roles are implicit in the making of such judgments, providing the frame of reference for the choice of standards by which the nonmoral merits of people are determined. Ultimately such purposes and roles stem from a society's way of life as a whole. Now a society's way of life may be thought of as the cultural form given to the realization of human values. Whether moral or nonmoral standards are being applied, then, all judgments of people's merits finally depend on human values. All are made from an exclusively human standpoint.

The question that naturally arises at this juncture is: why should standards that are based on human values be assumed to be the only valid criteria of merit and hence the only true signs of superiority? This question is especially pressing when humans are being judged superior in merit to nonhumans. It is true that a human being may be a better mathematician than a monkey, but the monkey may be a better tree climber than a human being. If we humans value mathematics more than tree climbing, that is because our conception of civilized life makes the development of mathematical ability more desirable than the ability to climb trees. But is it not unreasonable to judge nonhumans by the values of human civilization, rather than by values connected with what it is for a member of *that* species to live a good life? If all living things have a good of their own, it at least makes sense to judge the merits of nonhumans by standards derived from *their* good. To use only standards based on human values is already to commit oneself to holding that humans are superior to nonhumans, which is the point in question.

A further logical flaw arises in connection with the widely held conviction that humans are *morally* superior beings because they possess, while others lack, the capacities of a moral agent (free will, accountability, deliberation, judgment, practical reason). This view rests on a conceptual confusion. As far as moral standards are concerned, only beings that have the capacities of a moral agent can properly be judged to be *either* moral (morally good) *or* immoral (morally deficient). Moral standards are simply not applicable to beings that lack such capacities. Animals and plants cannot therefore be said to be morally inferior in merit to humans. Since the only beings that can have moral merits *or be deficient in such merits* are moral agents, it is conceptually incoherent to judge humans as superior to nonhumans on the ground that humans have moral capacities while nonhumans don't.

Up to this point I have been interpreting the claim that humans are superior to other living things as a grading or ranking judgment regarding their comparative merits. There is, however, another way of understanding the idea of human superiority. According to this interpretation, humans are superior to nonhumans not as regards their merits but as regards their inherent worth. Thus the claim of human superiority is to be understood as asserting that all humans, simply in virtue of their humanity, have *a greater inherent worth* than other living things.

The inherent worth of an entity does not depend on its merits.[8] To consider something as possessing inherent worth, we have seen, is to place intrinsic value on the realization of its good. This is done regardless of whatever particular merits it might have or might lack, as judged by a set of grading or ranking standards. In human affairs, we are all familiar with the principle that one's worth as a person does not vary with one's merits or lack of merits. The same can hold true of animals and plants. To regard such entities as possessing inherent worth entails disregarding their merits and deficiencies, whether they are being judged from a human standpoint or from the standpoint of their own species.

The idea of one entity having more merit than another, and so being superior to it in merit, makes perfectly good sense. Merit is a grading or ranking concept, and judgments of comparative merit are based on the different degrees to which things satisfy a given standard. But what can it mean to talk about one thing being superior to another in inherent worth? In order to get at what is being asserted in such a claim it is helpful first to look at the social origin of the concept of degrees of inherent worth.

The idea that humans can possess different degrees of inherent worth originated in societies having rigid class structures. Before the rise of modern democracies with their egalitarian outlook, one's membership in a hereditary class determined one's social status. People in the upper classes were looked up to, while those in the lower classes were looked down upon. In such a society one's social superiors and social inferiors were clearly defined and easily recognized.

Two aspects of these class-structured societies are especially relevant to the idea of degrees of inherent worth. First, those born into the upper classes were deemed more worthy of respect than those born into the lower orders. Second, the superior worth of upper class people had nothing to do with their merits nor did the inferior worth of those in the lower classes rest on their lack of merits. One's superiority or inferiority entirely derived from a social position one was born into. The modern concept of a meritocracy simply did not apply. One could not advance into a higher class by any sort of moral or nonmoral achievement. Similarly, an aristocrat held his title and all the privileges that went with it just because he was the eldest son of a titled nobleman. Unlike the bestowing of knighthood in contemporary Great Britain, one did not earn membership in the nobility by meritorious conduct.

We who live in modern democracies no longer believe in such hereditary social distinctions. Indeed, we would wholeheartedly condemn them on moral grounds as being fundamentally unjust. We have come to think of class systems as a paradigm of social injustice, it being a central principle of the democratic way of life that among humans there are no superiors and no inferiors. Thus we have rejected the whole conceptual framework in which people are judged to have different degrees of inherent worth. That idea is incompatible with our notion of human equality based on the doctrine that all humans, simply in virtue of their humanity, have the same inherent worth. (The belief in universal human rights is one form that this egalitarianism takes.)

The vast majority of people in modern democracies, however, do not maintain an egalitarian outlook when it comes to comparing human beings with other living things. Most people consider our own species to be superior to all other species and this superiority is understood to be a matter of inherent worth, not merit. There may exist thoroughly vicious and depraved humans who lack all merit. Yet because they are human they are thought to belong to a higher class of entities than any plant or animal. That one is born into the species *Homo sapiens* entitles one to have lordship over those who are one's inferiors, namely, those born into other species. The parallel with hereditary social classes is very close. Implicit in this view is a hierarchical conception of nature according to which an organism has a position of superiority or inferiority in the Earth's community of life simply on the basis of its genetic background. The "lower" orders of life are looked down upon and it is considered perfectly proper that they serve the interests of those belonging to the highest order, namely humans. The intrinsic value we place on the well-being of our fellow humans reflects our recognition of their rightful position as our equals. No such intrinsic value is to be placed on the good of other animals, unless we choose to do so out of fondness or affection for them. But their well-being imposes no moral requirement on us. In this respect there is an absolute difference in moral status between ourselves and them.

This is the structure of concepts and beliefs that people are committed to insofar as they regard humans to be superior in inherent worth to all other species. I now wish to argue that this structure of concepts and beliefs is completely groundless. If we accept the first three components of the biocentric outlook and from that perspective look at the major philosophical traditions which have supported that structure, we find it to be at bottom nothing more than the expression of an irrational bias in our own favor. The philosophical traditions themselves rest on very questionable assumptions or else simply beg the question. I briefly consider three of the main traditions to substantiate the point. These are classical Greek humanism, Cartesian dualism, and the Judeo-Christian concept of the Great Chain of Being.

The inherent superiority of humans over other species was implicit in the Greek definition of man as a rational animal. Our animal nature was identified with "brute" desires that need the order and restraint of reason to rule them (just as reason is the special virtue of those who rule in the ideal state). Rationality was then seen to be the key to our superiority over animals. It enables us to live on a higher plane and endows us with a nobility and worth that other creatures lack. This familiar way of comparing humans with other species is deeply ingrained in our Western philosophical out-look. The point to consider here is that this view does not actually provide an argument *for* human superiority but rather makes explicit the framework of thought that is implicitly used by those who think of humans as inherently superior to nonhumans. The Greeks who held that humans, in virtue of their rational capacities have a kind of worth greater than that of any nonrational being, never looked at rationality as but one capacity of living things among many others. But when we consider rationality from the standpoint of the first three elements of the ecological outlook, we see that its value lies in its importance for *human* life. Other creatures achieve their species-specific good without the need of rationality, although they often make use of capacities that humans lack. So the humanistic outlook of classical Greek thought does not give us a neutral (non-question-begging) ground on which to construct a scale of degrees of inherent worth possessed by different species of living things.

The second tradition, centering on the Cartesian dualism of soul and body, also fails to justify the claim to human superiority. That superiority is supposed to derive from the fact that we have souls while animals do not. Animals are mere automata and lack the divine element that makes us spiritual beings. I won't go into the now familiar criticisms of this two-substance view. I only add the point that, even if humans are composed of an immaterial, unextended soul and a material, extended body, this in itself is not a reason to deem them of greater worth than entities that are only bodies. Why is a soul substance a thing that adds value to its possessor? Unless some theological reasoning is offered here (which many, including myself, would find unacceptable on epistemological grounds), no logical connection is evident. An immaterial something which thinks is better than a material something which does not think only if thinking itself has value, either intrinsically or instrumentally. Now it is intrinsically valuable to humans alone, who value it as an end in itself, and it is instrumentally valuable to those who benefit from it, namely humans.

For animals that neither enjoy thinking for its own sake nor need it for living the kind of life for which they are best adapted, it has no value. Even if "thinking" is broadened to include all forms of consciousness, there are still many living things that can do without it and yet live what is for their species a good life. The anthropocentricity underlying the claim to human superiority runs throughout Cartesian dualism.

A third major source of the idea of human superiority is the Judeo-Christian concept of the Great Chain of Being. Humans are superior to animals and plants because their Creator has given them a higher place on the chain. It begins with God at the top, and then moves to the angels, who are lower than God but higher than humans, then to humans, positioned between the angels and the beasts (partaking of the nature of both), and then on down to the lower levels occupied by nonhuman animals, plants, and finally inanimate objects. Humans, being "made in God's image," are inherently superior to animals and plants by virtue of their being closer (in their essential nature) to God.

The metaphysical and epistemological difficulties with this conception of a hierarchy of entities are, in my mind, insuperable. Without entering into this matter here, I only point out that if we are unwilling to accept the metaphysics of traditional Judaism and Christianity, we are again left without good reasons for holding to the claim of inherent human superiority.

The foregoing considerations (and others like them) leave us with but one ground for the assertion that a human being, regardless of merit, is a higher kind of entity than any other living thing. This is the mere fact of the genetic makeup of the species *Homo sapiens.* But this is surely irrational and arbitrary. Why should the arrangement of genes of a certain type be a mark of superior value, especially when this fact about an organism is taken by itself, unrelated to any other aspect of its life? We might just as well refer to any other genetic makeup as a ground of superior value. Clearly we are confronted here with a wholly arbitrary claim that can only be explained as an irrational bias in our own favor.

That the claim is nothing more than a deep-seated prejudice is brought home to us when we look at our relation to other species in the light of the first three elements of the biocentric outlook. Those elements taken conjointly give us a certain overall view of the natural world and of the place of humans in it. When we take this view we come to understand other living things, their environmental conditions, and their ecological relationships in such a way as to awake in us a deep sense of our kinship with them as fellow members of the Earth's community of life. Humans and nonhumans alike are viewed together as integral parts of one unified whole in which all living things are functionally interrelated. Finally, when our awareness focuses on the individual lives of plants and animals, each is seen to share with us the characteristic of being a teleological center of life striving to realize its own good in its own unique way.

As this entire belief system becomes part of the conceptual framework through which we understand and perceive the world, we come to see ourselves as bearing a certain moral relation to nonhuman forms of life. Our ethical role in nature takes on a new significance. We begin to look at other species as we look at ourselves, seeing them as beings which have a good they are striving to realize just as we have a good we are striving to realize. We accordingly develop the disposition to view the world from the standpoint of their good as well as from the standpoint of our own good. Now if the groundlessness of the claim that humans are inherently superior to other species were brought clearly before our minds, we would not remain intellectually neutral toward that claim but would reject it as being fundamentally at variance with our total world outlook. In the absence of any good reasons for holding it, the assertion of human superiority would then appear simply as the expression of an irrational and self-serving prejudice that favors one particular species over several million others.

Rejecting the notion of human superiority entails its positive counterpart: the doctrine of species impartiality. One who accepts that doctrine regards all living things as possessing inherent worth—the *same* inherent worth, since no one species has been shown to be either "higher" or "lower" than any other. Now we saw earlier that, insofar as one thinks of a living thing as possessing inherent worth, one considers it to be the appropriate object of the attitude of respect and believes that attitude to be the only fitting or suitable one for all moral agents to take toward it.

Here, then, is the key to understanding how the attitude of respect is rooted in the biocentric outlook on nature. The basic connection is made through the denial of human superiority. Once we reject the claim that humans are superior either in merit or in worth to other living things, we are ready to adopt the attitude of respect. The denial of human superiority is itself the result of taking the perspective on nature built into the first three elements of the biocentric outlook.

Now the first three elements of the biocentric outlook, it seems clear, would be found acceptable to any rational and scientifically informed thinker who is fully "open" to the reality of the lives of nonhuman organisms. Without denying our distinctively human characteristics, such a thinker can acknowledge the fundamental respects in which we are members of the Earth's community of life and in which the biological conditions necessary for the realization of our human values are inextricably linked with the whole system of nature. In addition, the conception of individual living things as teleological centers of life simply articulates how a scientifically informed thinker comes to understand them as the result of increasingly careful and detailed observations. Thus, the bio-centric outlook recommends itself as an acceptable system of concepts and beliefs to anyone who is clear-minded, unbiased, and factually enlightened, and who has a developed capacity of reality awareness with regard to the lives of individual organisms. This, I submit, is as good a reason for making the moral commitment involved in adopting the attitude of respect for nature as any theory of environmental ethics could possibly have.

X. MORAL RIGHTS AND THE MATTER OF COMPETING CLAIMS

I have not asserted anywhere in the foregoing account that animals or plants have moral rights. This omission was deliberate. I do not think that the reference class of the concept, bearer of moral rights, should be extended to include nonhuman living things. My reasons for taking this position, however, go beyond the scope of this paper. I believe I have been able to accomplish many of the same ends which those who ascribe rights to animals or plants wish to accomplish. There is no reason, moreover, why plants and animals, including whole species populations and life communities, cannot be accorded *legal* rights under my theory. To grant them legal protection could be interpreted as giving them legal entitlement to be protected, and this, in fact, would be a means by which a society that subscribed to the ethics of respect for nature could give public recognition to their inherent worth.

There remains the problem of competing claims, even when wild plants and animals are not thought of as bearers of moral rights. If we accept the bio-centric outlook and accordingly adopt the attitude of respect for nature as our ultimate moral attitude, how do we resolve conflicts that arise from our respect for persons in the domain of human ethics and our respect for nature in the domain of environmental ethics? This is a question that cannot adequately be dealt with here. My main purpose in this paper has been to try to establish a base point from which we can start working toward a solution to the problem. I have shown why we cannot just begin with an initial presumption in favor of the interests of our own species. It is after all within our power as moral beings to place limits on human population and technology with the deliberate intention of sharing the Earth's bounty with other species. That such sharing is an ideal difficult to realize even in an approximate way does not take away its claim to our deepest moral commitment.

NOTES

1. The conceptual links between an entity *having* a good, something being good *for* it, and events doing good *to* it are examined by G. H. Von Wright in *The Varieties of Goodness* (New York: Humanities Press, 1963), chaps. 3 and 5.

2. See W. K. Frankena, "Ethics and the Environment," in K. E. Goodpaster and K. M. Sayre, eds., *Ethics and Problems of the 21st Century* (Notre Dame: University of Notre Dame Press, 1979), pp. 3–20. I critically examine Frankena's views in "Frankena on Environmental Ethics," *Monist,* vol. 64, no. 3 (July 1981), 313–324.

3. In the light of considerations set forth in Daniel Dennett's *Brainstorms: Philosophical Essays on Mind and Psychology* (Montgomery, Vermont: Bradford Books, 1978), it is advisable to leave this question unsettled at this time. When machines are developed that function in the way our brains do, we may well come to deem them proper subjects of moral consideration.

4. I have analyzed the nature of this commitment of human ethics in "On Taking the Moral Point of View," *Midwest Studies in Philosophy,* vol. 3, *Studies in Ethical Theory* (1978), pp. 35–61.

5. John Rawls, "Justice As Fairness," *Philosophical Review* 67 (1958): 183.

6. Stephen R. L. Clark, *The Moral Status of Animals* (Oxford: Clarendon Press, 1977), p. 112.

7. My criticisms of the dogma of human superiority gain independent support from a carefully reasoned essay by R. and V. Routley showing the many logical weaknesses in arguments for human-centered theories of environmental ethics. R. and V. Routley, "Against the Inevitability of Human Chauvinism," in K. E. Goodpaster and K. M. Sayre, eds., *Ethics and Problems of the 21st Century* (Notre Dame: University of Notre Dame Press, 1979), pp. 36–59.

8. For this way of distinguishing between merit and inherent worth, I am indebted to Gregory Vlastos, "Justice and Equality," in R. Brandt, ed., *Social Justice* (Englewood Cliffs, N.J.: Prentice-Hall, 1962), pp. 31–72.

RESPONDING JOURNAL

What has Taylor's essay caused you to think about? What adjustments in your thinking might you attribute to Taylor's ideas? What might be some of the benefits for writing about the natural world that could come from some study of ethics? Explain your answer in a journal entry of at least a page.

QUESTIONS FOR CRITICAL THINKING

STRATEGY

1. Much of the structure of Taylor's essay derives from its nature as an explanation of a philosophical position. Explain the strengths and weaknesses you find in Taylor's essay in terms of its structure. Given the nature of his subject, explore your reasons for evaluating his work either positively or negatively.

2. On page 547, Taylor explains the concept of human superiority, the fourth component of the biocentric outlook, as "the single most important idea in establishing the justifiability of the attitude of respect for nature." What benefits does he derive from positioning his

most important support at this point in his essay? Do you feel that another organizational pattern might have helped you to read and understand the essay better? Explain your response.

ISSUES

1. Have you ever thought of the human point of view in terms of the ethics of the natural world? Does Taylor's essay suggest that we might not be quite so important in the overall scheme of things? Do any of his points cause you to look at the natural world and human behavior any differently? If so, which ones?
2. How valid do you find the parallel between class distinctions in American society and the distinctions between different forms of life? How do we get the idea that some have more worth or value than others? Where might you see such distinctions being displayed on your campus?

COLLABORATIVE WRITING ACTIVITY

Now that you have read Taylor's presentation, present an application of your own ethical principles to the natural world. You must decide on fundamental issues—for example, whether or not you agree with Taylor's attempt to discredit the idea of human superiority—in order to be certain the causes you offer will satisfy your reader as sufficient to justify your conclusion. Offer one example with one attendant chain of reasons or causes for your position; you may imitate Taylor's means of proceeding if it makes your task easier.

Your group function is to compile your individual responses into a coherent statement of your group's beliefs regarding the ethics of respect for nature. You will end up with a position paper that will run at least two to three pages, if not considerably longer. Allow your individual examples to help you with organizing your presentation, but you will have to decide in group discussion how best to represent the thinking of your individual group members as you complete the assignment. Many group members may well modify their positions as a result of the discussion created in the fulfilling of this assignment.

from *The Case for Animal Rights*
Tom Regan

Tom Regan is one of the leading academic crusaders for the rights of animals. In The Case for Animal Rights *(1983) and his earlier book* All That Dwell Therein *(1982), he explores the traditions governing our conceptions*

and treatments of animals. An eloquent spokesperson for the rights of ani-
mals, Regan is often quoted on the subject.

He believes firmly that no animal should be used for any sort of scientific
experimentation, no animal should be raised in a commercial growing ven-
ture, and no animal should be subject to hunting or trapping. While other
writers might modify these positions in the interests of a "greater human
good," Regan believes such distinctions are ethically unjustified. He states his
case forcefully and unapologetically.

EXPLORING JOURNAL

What specific rights do you think animals have? Does it matter if the animal is
an insect, a bird, a domestic cat or dog, or a bear? How about a fish? Does the
situation determine the rights of an individual animal? Explore these possibilities in a freewriting activity for twelve minutes or longer if you find yourself on
a roll.

THE RIGHTS VIEW

. . . The rights view is not opposed to efforts to save endangered species. It
only insists that we be clear about the reasons for doing so. On the rights view,
the reason we ought to save the members of endangered species of animals is
not because the species is endangered but because the individual animals have
valid claims and thus rights against those who destroy their natural habitat, for
example, or who would make a living off their dead carcasses through poaching and traffic in exotic animals, practices that unjustifiably override the rights
of these animals. But though the rights view must look with favor on any
attempt to protect the rights of any animal, and so supports efforts to protect
the members of endangered species, these very efforts, aimed specifically at
protecting the members of species that are endangered, can foster a mentality
that is antagonistic to the implications of the rights view. If people are encouraged to believe that the harm done to animals matters morally *only when* these
animals belong to endangered species, then these same people will be encouraged to regard the harm done to *other* animals as morally acceptable. In this
way people may be encouraged to believe that, for example, the trapping of
plentiful animals raises no serious moral question, whereas the trapping of rare
animals does. This is not what the rights view implies. The mere size of the relative population of the species to which a given animal belongs makes no difference to the grounds for attributing rights to that individual animal or to the
basis for determining when that animal's rights may be justifiably overridden or
protected.

Though said before, it bears repeating: *the rights view is not indifferent to*
efforts to save endangered species. It supports these efforts. It supports them,

however, not because these animals are few in number; primarily it supports them because they are equal in value to all who have inherent value, ourselves included, sharing with us the fundamental right to be treated with respect. Since they are not mere receptacles or renewable resources placed here for our use, the harm done to them as individuals cannot be justified merely by aggregating the disparate benefits derived by commercial developers, poachers, and other interested third parties. That is what makes the commercial exploitation of endangered species wrong, not that the species are endangered. On the rights view, the same principles apply to the moral assessment of rare or endangered animals as apply to those that are plentiful, and the same principles apply whether the animals in question are wild or domesticated.

The rights view does not deny, nor is it antagonistic to recognizing, the importance of human aesthetic, scientific, sacramental, and other interests in rare and endangered species or in wild animals generally. What it denies is that (1) the value of these animals is reducible to, or is interchangeable with, the aggregate satisfaction of these human interests, and that (2) the determination of how these animals should be treated, including whether they should be saved in preference to more plentiful animals, is to be fixed by the yardstick of such human interests, either taken individually or aggregatively. Both points cut both ways, concerning, as they do, both how animals may and how they may not be treated. In particular, any and all harm done to rare or endangered animals, done in the name of aggregated human interests, is wrong, according to the rights view, because it violates the individual animal's right to respectful treatment. With regard to wild animals, the general policy recommended by the rights view is: *let them be!* Since this will require increased human intervention in *human* practices that threaten rare or endangered species (e.g., halting the destruction of natural habitat and closer surveillance of poaching, with much stiffer fines and longer prison sentences), the rights view sanctions this intervention, assuming that those humans involved are treated with the respect they are due. Too little is not enough.

. . . The difficulties and implications of developing a rights-based environmental ethic . . . should be abundantly clear by now and deserve brief comment before moving on. The difficulties include reconciling the *individualistic* nature of moral rights with the more *holistic* view of nature emphasized by many of the leading environmental thinkers. Aldo Leopold is illustrative of this latter tendency. "A thing is right," he states, "when it tends to preserve the integrity, stability, and beauty of the biotic community. It is wrong when it tends otherwise."[1] The implications of this view include the clear prospect that the individual may be sacrificed for the greater biotic good, in the name of "the integrity, stability, and beauty of the biotic community." It is difficult to see how the notion of the rights of the individual could find a home within a view that, emotive connotations to one side, might be fairly dubbed "environmental fascism." To use Leopold's telling phrase, man is "*only* a member of the biotic team,"[2] and as such has the same moral standing as any other "member" of "the team." If, to take an extreme, fanciful but, it is hoped, not unfair example, the situation we faced was either to kill a rare wildflower or a (plentiful) human being and if the wildflower, as a "team member," would contribute more to "the integrity, stability, and beauty of the biotic community" than the human, then presumably we would not be doing wrong if we killed the human and saved the wildflower. The rights view cannot abide this position, not because the rights view categorically denies that inanimate objects can have rights (more on this momentarily) but because it denies the propriety of deciding what should be done to individuals who have rights by appeal to aggregative considerations, including, therefore, computations about what will or will not maximally "contribute to the integrity, stability, and beauty of the biotic community." Individual rights are not to be outweighed by such considerations (which is not to say that they are never to be outweighed). Environmental fascism and the rights view are like oil and water: they don't mix.

The rights view does not deny the possibility that collections or systems of natural objects might have inherent value—that is, might have a kind of value that is not the same as, is not reducible to, and is incommensurate with any one individual's pleasures, preference-satisfactions, and the like, or with the sum of such goods for any number of individuals. The beauty of an undisturbed, ecologically balanced forest, for example, might be conceived to have value of this kind. The point is certainly arguable. What is far from certain is how moral rights could be meaningfully attributed to the *collection* of trees or the ecosystem. Since neither is an individual, it is unclear how the notion of moral rights can be meaningfully applied. Perhaps this difficulty can be surmounted. It is fair to say, however, that no one writing in this important area of ethics has yet done so.[3]

Because paradigmatic right-holders are individuals, and because the dominant thrust of contemporary environmental efforts (e.g., wilderness preservation) is to focus on the whole rather than on the part (i.e., the individual), there is an understandable reluctance on the part of environmentalists to "take rights seriously," or at least a reluctance to take them as seriously as the rights view contends we should. But this may be a case of environmentalists not seeing the forest for the trees—or, more accurately, of not seeing the trees for the forest. The implications of the successful development of a rights-based environmental ethic, one that made the case that individual inanimate natural objects (e.g., *this* redwood) have inherent value and a basic moral right to treatment respectful of that value, should be welcomed by environmentalists. If individual trees have inherent value, they have a kind of value that is not the same as, is not reducible to, and is incommensurate with the intrinsic values of the pleasures, preference-satisfactions, and the like, of others, and since the rights of the individual never are to be overridden merely on the grounds of aggregating such values for all those affected by the outcome, a rights-based environmental ethic would bar the door to those who would uproot wilderness in the name of "human progress," whether this progress be aggregated economic, educational, recreational, or other human interests. On the rights view, assuming this could be successfully extended to inanimate natural objects, our general policy

regarding wilderness would be precisely what the preservationists want—
namely, let it be! Before those who favor such preservation dismiss the rights
view in favor of the holistic view more commonly voiced in environmental circles, they might think twice about the implications of the two. There is the
danger that the baby will be thrown out with the bath water. A rights-based
environmental ethic remains a live option, one that, though far from being
established, merits continued exploration. It ought not to be dismissed out of
hand by environmentalists as being in principle antagonistic to the goals for
which they work. It isn't. Were we to show proper respect for the rights of the
individuals who make up the biotic community, would not the *community* be
preserved? And is not that what the more holistic, systems-minded environmentalists want? . . .

NOTES

1. Aldo Leopold, *A Sand County Almanac* (New York: Oxford University Press, 1949),
   p. 217.
2. Aldo Leopold, *A Sand County Almanac,* p. 209, emphasis added.
3. For further remarks on these matters, see my "What Sorts of Beings Can Have Rights?"
   and "The Nature and Possibility of an Environmental Ethic," both in Regan, *All That
   Dwell Therein* (Berkeley: University of California Press, 1982).

RESPONDING JOURNAL

Regan points out that we often think of animals as *"our resources."* To what
extent does this idea determine your concept of animals? Have you rethought
your position as a result of reading Regan's comments? Do you find lapses in his
logic, or does your common sense tell you there are flaws in his reasoning?
Explore your reactions in a journal entry of at least a page.

QUESTIONS FOR CRITICAL THINKING

STRATEGY

1. This excerpt from Regan's book focuses on the rights of endangered
   species as opposed to other, presently unendangered, species. Do you
   feel that his organization and explanation sufficiently present his view
   that much current endangered species legislation is at odds with ethics?
   What leads you to your conclusion?
2. Are you clear on the application of the quote from Aldo Leopold on page
   558 to the concept of individual vs. community rights? Try providing a
   statement of this idea you would offer to Regan as proof that you understand his reasoning.

ISSUES

1. Regan feels that if we observe the proper rights of each individual, then the community of which that individual is a part would benefit. He even uses the terms "individualistic" and "holistic" to suggest this distinction. How do you see these ideas worked out in your community?

2. Regan argues that the animal rights movement is analogous to the human rights movement. Explain whether you accept such an argument categorically, or if you would modify it in certain ways. If you choose to modify his position, explain how you would give Reagan reasons or causes for your ethical position.

COLLABORATIVE WRITING ACTIVITY

As a group, decide on some species of animals you might offer as examples of your positions on the ethical treatment of animals. Each member should take a different species. You will probably want to include various categories of species, such as endangered, wild but not endangered, domestic, and other classifications you will identify. After you have articulated your statement of the rights of that species, present it to the group so you can agree on an introduction to your individual presentations; this introduction will serve as the preface to your collection of views. If possible, print this work on a laser printer so that you can devise a cover and publish it as widely as possible as a statement of your group's position on animal rights.

SUGGESTIONS FOR FORMAL WRITING ASSIGNMENTS

1. Gore and Ray look at the world and the question of how we should relate to our environment in very different ways. Assume that you have been asked to moderate at a meeting between these two very impassioned writers. Write an essay in which you tell them how they might work together for the benefit of the country and the planet by combining their ideas. In order to accomplish this assignment, you will have to negotiate carefully between reverence and overkill.

It might help to think about the implications of these two highly successful writers. Both want to identify the causes of the present state of affairs in the country and the world environment. They suggest how we might massage these causes to bring about more helpful effects. Their methods are very different, but they claim the same objective. Explain to them how they might best achieve their ends.

2. Imagine the effects that Homer Simpson, or another television sitcom figure, would like to see in the world. What is important to him? What does he seem

to value on the earth? What causes would have to be operational for Homer to have his wish? How does his daughter Lisa serve as a potential guide? If you are not familiar with Homer, think of a self-centered male whose comforts such as naps, food, and television are often more important to him than his family. What are the implications for the natural world of such an attitude? Explore these ideas in a substantial essay; you might even present your argument in the form of a sitcom episode, complete with appropriate commercials.

3. In the case of animal rights, think of a world of animals without human presence. The animals live by themselves. What would the animals do? Would they survive as well? What does our presence in their world do to them?

Now that you have imagined this situation, think of our present world with humans deciding the fate of animals. Use the ideas you have derived from your reading and your thinking about a human-less world to compose an essay of at least 500 words in which you explain your argument of how animals should be accorded rights. You might find a cause-and-effect relationship helpful in organizing your ideas, particularly in terms of ethical positions.

4. Write an essay in which you explore the qualities of a worldview necessary to provide a sustainable environmental policy for your community. What problems currently face your community? How must you resolve those problems? Whose aid must you enlist? How will you shape the dominant worldview to achieve agreement and commitment? Try to offer a plan of action that will show how the modified worldview will implement the effects necessary for the well-being of the environment. In so doing, you will be using a cause-and-effect pattern to argue for the land.

Appendix

RESEARCH AND DOCUMENTATION GUIDE

Research means the work you do to find information about a subject in which you are interested or to supplement or confirm knowledge you already have. Most often, research means going to a facility such as a library where you can find sources of information. Ten years ago books and periodicals (magazines, journals, and newspapers—those texts which are published on a regular and recurring basis) were the main staple of a library and thus of a research project. Today, however, we have computer-assisted data banks and texts in CD-ROM format that have altered significantly the patterns of carrying out research. This section covers the basic concepts and practices of research, provides examples of sample research activities and related writing, and demonstrates two forms of documentation: those of the Modern Language Association (MLA) and the American Psychological Association (APA). This information will provide the foundation for the research you will need to do not only in this class but in your other courses as well.

Even though you may be writing essays of less than 1,000 words, almost all of the selections in this text are several times that length, and some are much longer. In order to write an extended informative essay or argument, you will have to have either extensive personal experience or material gained through research, or both. When you use information other than what you knew before you began your work, you must document that information when you present it in your writing. You must acknowledge your source for any direct quotations, ideas, or specific phrases that are not your own.

The two most widely used formats for documentation are explained and illustrated in the *MLA Handbook for Writers of Research Papers*, 4th ed. (New York: Modern Language Association of America, 1995) for the MLA style and in the *Publication Manual of the American Psychological Association,* 4th ed. (Washington, DC: APA, 1994) for the APA format. The MLA format is widely used in the humanities for essays in the fields of languages and literature, philosophy, history, music, and American studies. The APA format serves social science fields such as anthropology, psychology, sociology, political science, and education.

Documentation is the primary means of demonstrating to your reader that you have considered your subject thoroughly and that you are qualified to offer an opinion. Documentation is the proof that you have considered and weighed the ideas of others. This activity of showing your substantiation appears either within the body of your text, as in the case of parenthetical citations, or at the end of your text in footnotes or endnotes. You will document direct quotation, paraphrase, and summary; in short, you must provide documentation whenever you are indebted for an idea to the work of another writer. In direct quotation you indicate that you are reproducing precisely the words of the author by placing the text being quoted within quotation marks, or in the case of longer quotations, by indenting and setting it off from the rest of your text.

In paraphrase, you are putting the writing of the source into your own words. When you do so, you are using approximately the same amount of space that the original author took to explain the idea being paraphrased. It is crucial that you keep a careful record of the material you take from the source so you will be able easily to tell whether you are taking words directly from the source (in which case you will use quotation marks) or you are paraphrasing the source in your notes. It is often a good idea to copy source material carefully onto your notecards or computer as directly quoted material because you can always paraphrase later, and your later research may send you back to the earlier source for clarification.

Summarizing is the condensation of the source into your own words. Summarizing always involves a reduction of material: you may summarize a paragraph of the source into a single sentence of your own, you may summarize a chapter of the original into a single paragraph, or you may summarize the contents of a book in a page or two. In each case, you document to allow your reader to find readily the source for the ideas you are discussing.

Remember, you can never get yourself in trouble or diminish the effect of your writing by documenting your indebtedness to the works of others. But you can get into serious difficulty by failing to document fully what you have used from the work of others. Such a failure to document indebtedness is plagiarism.

Plagiarism takes several forms. While it is always the offering as one's own the work of another, it can be obvious or subtle. Clear instances of plagiarism include deliberate presentation of someone else's work as your own, as in the case of submitting of a paper written by someone else. Another case would be the presentation of source material without any documentation. Such a presentation offers the work as the plagiarizing writer's, when it is actually the work of the original writer who is now being plagiarized. The third level of plagiarism is either unwitting borrowing or simply sloppy work. Such plagiarism occurs when you have an idea that came from another writer, but which you did not recognize as the work of another when writing your essay. This form of stealing also occurs when you use a quote from the source and forget to indicate either that it is a direct quotation or that it comes from another writer. Sloppiness occurs when you fail to document precisely any words borrowed from the original author. In this case, plagiarizing can occur even in a footnote: for example, if I offer a footnote and suggest that I am paraphrasing by using no quotation marks when in fact I am quoting material directly from my source—two or more words together—then I am plagiarizing. Brief phrases of two or more words taken directly from the source must be placed in quotation marks; even a single word, if it is an unusual word or connotation or clearly owes something to the author's invention, must be placed within quotation marks.

In order to help you avoid inadvertent plagiarism, here are examples of paraphrase and summary done both correctly and incorrectly. Notice how the second version—the plagiarized one—does not acknowledge the distinctive phrasing of the original.

> As more and more people enter the lists of the affluent, pressure on better liv-ing spaces and recreation spots free of technological taint increases. And as environmental degradation becomes more and more pervasive—involving such global concerns as acid rain, the greenhouse effect, and the pollution of the oceans—the general public is mobilized to the cause of environmentalism. At this point, the public must face the dilemma: Does the system that feathers the nest create conditions that ultimately make life unpleasant or unbearable? Can the system be adjusted to compensate for the side effects of progress, or must the system be thoroughly overhauled, if not replaced? What are the lim-its to the liberal faith in technology? (Killingsworth and Palmer 6)

The essay in which the quote appears would have a "Works Cited" page at the end with the following entry:

Killingsworth, M. Jimmie, and Jacqueline S. Palmer. *Ecospeak: Rhetoric and Environmental Politics in America*. Carbondale: Southern Illinois UP, 1992.

CORRECT PARAPHRASE

Increases in the American standard of living have provided various benefits for many citizens, but now some of these people are finding that they have to compete for desirable places to live and play. Increasing pollution in various forms like acid rain, the greenhouse effect, and fouled oceans is beginning to wake people up to the problems facing them; the question is, though, do peo-ple care enough about these problems to do something about them? Will Americans adjust their lifestyles to correct these problems, or will their "liberal faith in technology" cause them to overlook the danger signals? (Killingsworth and Palmer 6)

INCORRECT PARAPHRASE

Increases in the American standard of living have provided various benefits for many citizens, but now some of these people have to compete for better liv-ing spaces and places to play free of technological taint. The general public is mobilized to the cause of environmentalism. Increasing pollution in various forms like acid rain, the greenhouse effect, and fouled oceans is beginning to wake people up to the problems facing them; the question is, though, do people care enough about these problems to do something about them? Will Americans adjust their lifestyles to correct these problems, or will their liberal faith in technology cause them to overlook the danger signals? (Killingsworth and Palmer 6)

In the correct paraphrase, the writer uses his or her own words to restate accurately the ideas in the original paragraph, but the writer of the incorrect paraphrase makes two serious mistakes. First, that writer uses the exact words from the original without quotation marks to show that these are not his or her own words. This is plagiarism. To avoid plagiarism, the phrases "better living spaces" and "free of technological taint" must be enclosed in quotation marks because these words appear exactly in this manner in the original text. Similarly, the clause "the general public is mobilized to the cause of environmentalism" cannot be used as a sentence without putting it within quotation marks.

Second, the writer makes the serious mistake of misrepresenting the meaning of the original text. By quoting the clause apart from the original context— a context suggesting a developing situation rather than a status already reached—the writer has falsified the sense of the text.

Finally, common phrases such as "acid rain" and "greenhouse effect" are so widely used that they would not have to be placed within quotation marks. Because these phrases are in our everyday vocabularies, they are not the contribution of the original authors in the quoted passage.

CORRECT SUMMARY

Killingsworth and Palmer argue that increasing affluence will place more environmental pressure on the most desirable places to live and play. They question, however, whether Americans will negotiate successfully between comfort and a healthy environment (6).

INCORRECT SUMMARY

Killingsworth and Palmer argue that people will put more pressure on better living spaces and recreation spots free of technological taint. They question, however, whether Americans will come to terms with progress and realize the dangers of not recognizing the limits to the liberal faith in technology (6).

The second summary is a plagiarism because the phrases "pressure on better living spaces and recreation spots free of technological taint" and "limits to the liberal faith in technology" are quoted directly from the source and are not placed within quotation marks.

Notice that a set of parentheses follows both the correct summary and the incorrect summary. Within these parentheses you see the proper documentation of the source; this practice of parenthetical documentation within the text is now used almost universally. The number "6" indicates the page number; the names of the authors do not appear in the parenthetical documentation because they have already been mentioned in the paragraph. In the examples of correct and incorrect paraphrase, the parenthetical documentation includes the authors' names because they were not provided in the paraphrases.

In the past, footnotes were routinely used for documentation; however, they are used today only for special explanations and not for routine documentation of sources. Footnotes often make it difficult for a reader to follow an argument because the need to keep looking at the bottom of a page or the end of a chapter to find source documentation can be distracting and annoying. With parenthetical documentation next to the quoted or paraphrased material, the reader can follow the writer's argument more readily. Such a practice eliminates the need to repeat material as was formerly done when a researcher would provide full citations for sources in footnotes and in a bibliography at the end of the paper. With in-text documentation, the sources are listed together at the end of the essay in a "works cited" or "bibliography" section.

Documentation requires careful observation of the specifics required by the style sheet—the precise requirements established by the authors of that guide. The two most frequently required style sheets or manuals are published by the Modern Language Association (MLA) and the American Psychological Association (APA); during your college career you will likely be required to submit work in both formats. In the following sections, you will see the important differences in documentation practices between the two.

DOCUMENTATION WITHIN THE TEXT: MLA

When using the MLA format, you will cite in the body of your text the author and page number of each source to which you are indebted. Most often, this information in parentheses will come at the end of a sentence or, in the case of extensive summary or paraphrase, at the end of a paragraph if that paragraph is substantially indebted to a single source. You will also provide a "Works Cited" page or pages at the end of your essay and possibly an "Endnotes" page if you used any explanatory footnotes. The following examples illustrate different kinds of citations you may need to use in the MLA format.

Standard practice dictates that you indicate the source you are citing with the author's last name and page number:

He begins his study with a chapter entitled "A Semidesert with a Desert Heart" (Reisner 1).

Note that the end punctuation does not appear until after the parenthetical cita-tion. If you had provided the author's name in the text of the essay, you would omit it in the parenthetical citation.

> Reisner begins his study with a chapter entitled "A Semidesert with a
>
> Desert Heart" (1).

If more than one work by Reisner had been used in your paper, you would indi-cate which one you were citing by mentioning the short title of the work and the page number. A comma separates the author and short title. Such a citation would read:

> He begins his study with a chapter entitled "A Semidesert with a Desert
>
> Heart" (Reisner, *Cadillac* 1).

The in-text citation form is the same for both books and periodicals.

DOCUMENTATION WITHIN THE TEXT: APA

When using the APA format, you will cite in the body of your paper the last name of your source. You will include page numbers in the case of direct quo-tations, but if you are summarizing or paraphrasing, you will not include a page number reference when using this documentation format. Only in the case of a long source work where your reader might have difficulty in locating the mate-rial you have paraphrased would you consider including a page number in the text. References to your sources in parentheses will come at the point in your sentence closest to where the actual information is used. Thus, you will often have your parenthetical citations in the middle of sentences. You will provide a "References" section at the end of your essay to cite full documentation of the sources used in writing your essay. The following examples provide illustra-tions of the different forms of citation you will need to use with the APA format.

The APA format differs from the MLA format in two main ways: APA requires date of publication of a work and places a comma between author's last name and date of publication. MLA requires author's last name but not the date of publication. APA format requires a page or chapter reference only when a specific reference is being made, while the MLA format usually indicates a page reference. Thus, the APA format takes these forms:

> Teddy Roosevelt and Gifford Pinchot failed to prevent Los Angeles
>
> water interests from diverting the Owens River for personal gain (Reisner,
>
> 1987, pp. 84–87).

> or

> Reisner argues that Teddy Roosevelt and Gifford Pinchot failed to pre-
>
> vent Los Angeles water interests from diverting the Owens River for per-
>
> sonal gain (1987, pp. 84–87).

or

The story of battles over water rights in the American West is fascinating, but it does not inspire confidence that policy-making bodies always act with the public good in mind (Reisner, 1987).

Reisner explains the Owens Valley water rights fight so effectively that even the most intricate details of the opposing arguments make sense to a reader unfamiliar with the water wars (1987, chap. 2).

With five or fewer authors, list all of their names in the first parenthetical citation. You can refer to them with the name of the first author and "et al." in subsequent citations. For example, use (Andrews, Jones, McNeil, Fisch, and Healy, 1990) for the first reference, and (Andrews et al., 1990) for a subsequent reference.

When you wish to refer to more than one source in a footnote, arrange your sources in alphabetical order and separate them with semicolons: (Duncan, 1987; Stegner, 1990; Woods, 1988). If you are referring to more than one work by the same author, list them in the order of their dates of publication and separate them with commas: (Stegner, 1990, 1992).

MLA FORMAT FOR BIBLIOGRAPHY

According to MLA guidelines, each entry should be double-spaced and listed alphabetically according to the author's last name. The first line of each entry should begin at the left-hand margin with subsequent lines indented five spaces or one-half inch. There are three primary parts that comprise each entry—the author's full name, title, and publication information. Capitalize the first, last, and all major words of the title; do not capitalize articles, prepositions, or conjunctions.

A Book by a Single Author

Johnson, Cathy. *A Naturalist's Cabin: Constructing a Dream*. New York:

Plume, 1991.

Note that in most cases a book publisher's name can be abbreviated to a single word.

A Book by a Single Author Published before 1900

Higginson, Thomas Wentworth. *A Book of American Explorers*. New

York, 1877.

Note that for books published before 1900, you should omit the publisher's name and use a comma after the name of the city.

A Book Edited by a Single Editor

Anderson, Lorraine, ed. *Sisters of the Earth: Women's Prose and Poetry*

about Nature. New York: Vintage, 1991.

Note that the only difference between author and editor citations is the abbreviation "ed." after a single editor or "eds." after two or more editors.

A Book by Two Authors

Davidson, James West, and John Rugge. *Great Heart: The History of a*

Labrador Adventure. New York: Penguin, 1988.

A Book by Multiple Authors

Belenky, Mary Field, Blythe McVicker Clinchy, Nancy Rule Goldberger,

and Jill Mattruck Tarule. *Women's Ways of Knowing: The*

Development of Self, Voice, and Mind. New York: Basic, 1986.

or

Belenky, Mary Field, et al. *Women's Ways of Knowing: The Development*

of Self, Voice, and Mind. New York: Basic, 1986.

Note that with three or more authors or editors, you have the option of listing them all or listing the first and using "et al."

A Book by a Corporate Author

National Research Council. *Prudent Practices for Disposal of Chemicals*

from Laboratories. Washington Nat. Acad., 1983.

Note that corporate titles are indexed according to the first word—excluding articles—of the corporate title and not, as is the case with authors, by last name.

A Work in a Collection

LaChapelle, Dolores. "Mountains Constantly Walking." *Talking on the*

Water: Conversations about Nature and Creativity. Ed. Jonathan

White. San Francisco: Sierra Club, 1994. 157–179.

A Preface, Introduction, Foreword, or Afterword

Flader, Susan L., and J. Baird Callicott. Introduction. *The River of the*

Mother of God and Other Essays. By Aldo Leopold. Madison: U of

Wisconsin P, 1991. 3–31.

Note that "University" and "Press" are abbreviated "U" and "P."

A Work Reprinted from Another Source

Snyder, Gary. "The Woman Who Married a Bear." *The Practice of the*
Wild. New York: North Point P, 1990. 155-174. Rpt. in *On Nature's*
Terms: Contemporary Voices. Eds. Thomas J. Lyon and Peter Stine.
College Station: Texas A & M UP, 1992. 123-138.

A Work in Multiple Volumes

Durant, Will, and Ariel Durant. *The Age of Napoleon*. New York: Simon,
1975. Vol. 11 of *The Story of Civilization*. 11 vols. 1935-75.

or

Durant, Will, and Ariel Durant. *The Story of Civilization*. 11 vols. New
York: Simon, 1935-75.

A Translation

Leprohon, Pierre. *The Italian Cinema*. Trans. Roger Greaves and Oliver
Stallybrass. New York: Praeger, 1972.

A Pamphlet

United States Military Academy. Department of Geography and Computer
Science. *An Orientation Tour of the Mid-Hudson Valley*. West
Point: USMA, 1985.

Note that a pamphlet is treated the same way as a book.

An Article in an Encyclopedia

Craker, Lyle E. "Basil." *The World Book Encyclopedia*. 1993 ed.

An Article in a Periodical with Continuous Pagination through a Volume

Addison, Catherine. "Once Upon a Time: A Reader-Response Approach to
Prosody." *College English* 56 (1994): 655-678.

Note that there is no punctuation between the periodical title and the volume
number.

An Article in a Periodical Paginated Separately in Each Issue

Hamill, Pete. "When the Air Was Clean." *Audubon* 95.1 (1993): 38-49.

Note that there is no punctuation between the periodical title and the volume
number. "95.1" means "volume 95, issue number 1."

An Article in a Monthly Periodical

Hiss, Tony. "How Now, Drugged Cow?" *Harper's Magazine* Oct. 1994:
80+.

Note that when an article's pages are not consecutive, indicate the first page followed by "+."

An Article in a Weekly Periodical

White, Jack E. "The Beauty of Black Art." *Time* 10 Oct. 1994: 66-73.

An Article in a Weekly Newspaper

Dial, Harold G. "Native American Day of Prayer Planned at Washington
Monument." *Carolina Indian Voice* 3 Nov. 1994: 5.

An Article in a Daily Newspaper

Roberts, Chalmers M. "Averell Harriman at 90." *Washington Post*
8 Nov. 1981: C5.

A Letter to the Editor

Okun, Daniel A. Letter. *New York Times* 28 May 1994: 18.

A Review

O'Conner, Patricia T. Rev. of *Politically Correct Bedtime Stories*, by
James Finn Garner. *New York Times Book Review* 15 May 1994: 3.

An Interview

Abbey, Edward. Interview. *Headed Upstream: Interviews with
Iconoclasts*. By Jack Loeffler. Tucson: Harbinger House, 1989.
3-19.

Dunlavy, Francine. Telephone interview. 20 May 1990.

Material in Other Reference Works

"Critical Technologies." *The World Almanac and Book of Facts*. 1992 ed.

"Stratification." *The Random House Dictionary of the English Language*.
1987 ed.

Note that for well-known and frequently used reference works, you follow the same procedure as you would for an encyclopedia.

Stevenson, L. Harold, and Bruce C. Wyman. "Food Chain." *The Facts on*

   File Dictionary of Environmental Science. New York: Facts on File,

   1991.

Vane-Wright, R. I. "Vulnerable Butterflies." *The Atlas of Endangered*

   Species. Ed. John A. Burton. New York: Macmillan, 1991.

Note that with more unusual reference works you should provide full citations to assist your reader.

APA FORMAT FOR BIBLIOGRAPHY

As with MLA, the APA bibliography format organizes entries according to the last name of the author; in cases where no author is indicated, alphabetize entries according to title. APA bibliography format differs from MLA in that APA emphasizes the date of publication by placing it right after the author's name. APA style capitalizes only the first word of the title and the first word of the sub-title when a work has a subtitle. Proper names are capitalized wherever they appear. Also, note that the author's initial rather than full first name is used.

A Book by a Single Author

Johnson, C. (1991). *A naturalist's cabin: Constructing a dream*. New

   York: Plume.

A Book with Two Authors

Davidson, J. W., & Rugge, J. (1988). *Great heart: The history of a*

   Labrador adventure. New York: Viking Penguin.

Note that in APA format, an ampersand "&" is used instead of the "and" found in the MLA format.

An Edited Work

Anderson, L. (Ed.). (1991). *Sisters of the earth: Women's prose and*

   poetry about nature. New York: Vintage Books.

Note that in APA format, publishers' names are presented in full.

A Book by a Corporate Author

National Research Council. (1983). *Prudent practices for disposal of chem-*

   icals from laboratories. Washington, DC: National Academy Press.

A Work in a Collection

LaChapelle, D. (1994). Mountains constantly walking. In J. White (Ed.),
Talking on the water: Conversations about nature and creativity
(pp. 157–179). San Francisco: Sierra Club Books.

A Preface, Introduction, Foreword, or Afterward

Flader, S. L., & Callicott, J. B. (1991). Introduction. In *The river of the
mother of god and other essays* by A. Leopold. Madison: University
of Wisconsin Press.

A Work in Multiple Volumes

Durant, W., & Durant, A. (1935–1975). *The story of civilization* (Vols.
1–11). New York: Simon & Schuster.

A Translation

Leprohon, P. (1972). *The Italian cinema* (R. Greaves & O. Stallybrass,
Trans.). New York: Praeger.

A Pamphlet

United States Military Academy, Dept. of Geography and Computer
Science. (1985). *An orientation tour of the mid-Hudson valley*.
West Point, NY: USMA.

Article in Encyclopedia

Craker, L. E. (1993). Basil. In *The world book encyclopedia* (Vol. 2,
p. 501). New York: World Book Encyclopedia.

An Article in a Periodical with Continuous Pagination through a Volume

Addison, C. (1994). Once upon a time: A reader-response approach to
prosody: *College English, 56,* 655–678.

Note that in APA format, you italicize the volume number of a periodical.

An Article in a Periodical Paginated Separately in Each Issue

Hamill, P. (1993). When the air was clean. *Audubon, 95* (1), 38–49.

An Article in a Monthly Periodical

Hiss, T. (1994, October). How now, drugged cow? *Harper's Magazine*,

pp. 80+.

An Article in a Newspaper

Dial, H. G. (1994, November 3). Native American day of prayer planned

at Washington Monument. *Carolina Indian Voice*, p. 5.

Chalmers, M. R. (1981, November 8). Averell Harriman at 90.

Washington Post, p. C5.

A Review

O'Conner, P. T. (1994, May 15). [Review of the book *Politically correct*

bedtime stories]. *New York Times Book Review*, 3.

An Interview

The APA style sheet suggests omitting personal interviews from your list of
works cited. You should, of course, provide in the body of your text the information that you would normally provide in a citation: name of the person interviewed, date, and telephone or personal interview.

INDEX

LITERARY CREDITS

Excerpt from *The Journey Home* by Edward Abbey. Used by permission of Dutton Signet, a division of Penguin Books USA Inc.

Gloria Anzaldúa. "Preface" and "The Homeland, Aztlan/*El otro Mexico*" © 1987 Gloria Anzaldúa. Reprinted with permission from Aunt Lute Books.

Jimmy Santiago Baca, *Martin & Meditations on the South Valley*. Copyright © 1987 by Jimmy Santiago Baca. Reprinted by permission of New Directions Publishing Corp.

Doug Bandow, "Ecoterrorism: The Dangerous Fringe of the Environmental Movement," from Heritage *Backgrounder* #764. Reprinted by permission of The Heritage Foundation.

Dave Barry, "Know Your Nature Well Before Hiking," reprinted by permission: Tribune Media Services.

Stephen Vincent Benét. "The Ballad of William Sycamore" by Stephen Vincent Benét. Copyright 1922 by Stephen Vincent Benét. Copyright renewed © 1950 by Stephen Vincent Benét. Reprinted by permission of Brandt & Brandt Literary Agents, Inc.

Marcia Bonta, from "January Journal." Reprinted by permission of Marcia Bonta.

Rachel Carson. From *The Sea Around Us*, revised edition, by Rachel L. Carson. Copyright © 1950, 1951, 1961 by Rachel L. Carson; renewed 1979, 1989 by Roger Christie. Reprinted by permission of Oxford University Press, Inc.

Tim Cahill. From *A Wolverine Is Eating My Leg* by Tim Cahill. Copyright © 1989 by Tim Cahill. Reprinted by permission of Vintage Books, a division of Random House, Inc.

Ella Cara Deloria. Reprinted from *Waterlily* by Ella Cara Deloria, by permission of the University of Nebraska Press. Copyright © 1988 by the University of Nebraska Press.

Marie De Santis, "Last of the Wild Salmon," reprinted by permission of the author.

Bill Devall, "Ecotourism," from *Living Richly in an Age of Limits* by Bill Devall, published by Gibbs N. Smith, Inc., 1993. Used by Permission.

Annie Dillard. "Heaven and Earth in Jest" from *Pilgrim at Tinker Creek* by Annie Dillard. Copyright © 1974 by Annie Dillard. Reprinted by permission of HarperCollins Publishers, Inc.

Dayton Duncan. From *Out West* by Dayton Duncan. Copyright © 1987 by Dayton Duncan. Used by permission of Viking Penguin, a division of Penguin Books USA Inc.

Gregg Easterbrook, from "Everything You Know about the Environment Is Wrong," reprinted by permission of the author.

Gretel Ehrlich. "From a Sheepherder's Notebook: Three Days," from *The Solace of Open Spaces* by Gretel Ehrlich. Copyright © 1985 by Gretel Ehrlich. Used by permission of Viking Penguin, a division of Penguin Books USA Inc.

Clarissa Pinkola Estes. From *Women Who Run with the Wolves* by Clarissa Pinkola Estes. Copyright © 1991 by Clarissa Pinkola Estes. Reprinted by permission of Ballantine Books, a division of Random House Inc.

John Fire/Lame Deer. From *Lame Deer: Seeker of Visions*. Copyright © 1972 by John Fire/Lame Deer and Richard Erdoes. Reprinted by permission of Simon & Schuster, Inc.

Linda Flowers, from *Throwed Away*. Reprinted by permission of The University of Tennessee Press. From *Throwed Away: Failures of Progress in*

ILLUSTRATION CREDITS

(in order of appearance)

Thomas Gainsborough, *Robert Andrews and His Wife,* 1760. Reproduced by courtesy of the Trustees, The National Gallery, London.

Marion Post Wolcott, *Pursglove, West Virginia,* 1938. Reproduced from the Collections of the Library of Congress.

Dorothea Lange, *Westward to the Pacific Coast on U.S. 80,* 1938. Reproduced from the Collections of the Library of Congress.

Solomon D. Butcher, *Harvey Andrews family at the grave of their child Willie,* 1887. Solomon D. Butcher Collection, Nebraska State Historical Society.

Walker Evans, *Graveyard and Steel Mill, Bethlehem, Pennsylvania,* 1936. Reproduced from the Collections of the Library of Congress.

Randy Lee White, *Custer's Last Stand—Revised,* 1980. Collection of the Museum of Fine Arts, Museum of New Mexico, Santa Fe (gift of the artist.)

Dorothea Lange, *Childress County, Texas, June 1938.* Reproduced from the Collections of the Library of Congress.

Asher B. Durand, *Kindred Spirits,* 1849. Collection of The New York Public Library (Astor, Lenox and Tilden Foundations).

Ernest Blumenschein, *Sangre de Christo Mountains,* 1924. Courtesy of The Anschutz Collection, Denver, Photography James O. Milmoe.

Georgia O'Keeffe, *Red Hills, Lake George,* 1927. The Phillips Collection, Washington, D.C.

George Inness, *The Lackawanna Valley,* 1855. © 1995 Board of Trustees, National Gallery of Art, Washington, D.C. (gift of Mrs. Huttleston Rogers).

Jerry Bywaters, *Oil Field Girls,* 1940. Archer M. Huntington Art Gallery, The University of Texas at Austin (Michener Collection Acquisition Fund, 1984). Photography by George Holmes.

Grandma Moses, *Hoosick Falls in Winter,* 1944. Copyright 1990, Grandma Moses Properties Co., New York. Reproduction copyright The Phillips Collection, Washington, D.C.

Karl Bodmer, *Mih-Tutta-Hang-Kusch, Mandan Village,* 1833–34. Joslyn Art Museum, Omaha, Nebraska (gift of the Enron Art Foundation). Photography by Bernard O. Milmoe.

Maynard Dixon, *Open Range,* 1942. Museum of Western Art, Denver.

Arthur Rothstein, *Displaced Sharecropper,* 1939. Reproduced from the Collections of the Library of Congress.

Alexandre Hogue, *Erosion #2 Mother Earth Laid Bare,* 1938. The Philbrook Museum of Art, Tulsa, Oklahoma.

John K. Hilliers, *Hopi Pueblo of Walpi,* 1876. National Anthropological Archives, Smithsonian Institution.

Samuel Colman, *Storm King on the Hudson,* 1866. National Museum of American Art, Washington, D.C. Photography courtesy of Art Resource, New York.

Thomas Worthington Whittredge, *Crossing the Ford, Platte River, Colorado,* 1868–70. The Century Association, New York.

Jack Delano, *Puxnawatsney, PA,* 1940. Reproduced from the Collections of the Library of Congress.

Arthur Rothstein, *Copper Mine and Miners' Homes, Meaderville, Montana,* 1939. Reproduced from the Collections of the Library of Congress.

Smokey Bear poster, 1956. Courtesy of USDA Forest Service.